STUDIES IN DEVELOPMENTAL NEUROBIOLOGY

Essays in Honor of Viktor Hamburger

VIKTOR HAMBURGER.
Photograph by W. M. Cowan

STUDIES IN DEVELOPMENTAL NEUROBIOLOGY

Essays in Honor of Viktor Hamburger

edited by

W. MAXWELL COWAN
Director, Weingart Laboratory
for Developmental Neurobiology
The Salk Institute for Biological Studies
San Diego

New York Oxford
OXFORD UNIVERSITY PRESS
1981

Library of Congress Cataloging in Publication Data
Main entry under title:

Studies in developmental neurobiology.

Bibliography: p.
Includes index.
1. Developmental neurology—Addresses, essays, lectures. 2. Neurobiology—Addresses, essays, lectures. 3. Hamburger, Viktor, 1900– . I. Hamburger, Viktor, 1900– . II. Cowan, William Maxwell, 1931– . [DNLM: 1. Nervous system—Growth and development—Essays. 2. Nervous system—Embryology—Essays. WL 101 S932]
QP363.5.S88 599.03′33 80–28232
ISBN 0–19–502927–5

Printing (last digit): 9 8 7 6 5 4 3 2 1

Printed in the United States of America

For
Doris, Carola, and Margaret

Preface

This collections of essays had its origin in a telephone conversation between the editor and Mr. Jeffrey House of the Oxford University Press. Professor Hamburger, or "Viktor" as he is affectionately known to all his friends, had hoped for some years to bring his academic career appropriately to an end by writing a monograph that would provide a synoptic view of the principal conceptual issues in developmental neurobiology. A few years ago he graciously invited me to join him in this undertaking, and we spent three summers together (in La Jolla, Boulder, and Tübingen) discussing the issues and reviewing the field. Shortly after the second of these "working vacations," Jeffrey House indicated that the Oxford University Press might be willing to publish such a monograph. Later, when it became evident that the pressure of other commitments would make it impossible for us to complete such a volume, Jeffrey not only released us from the obligation of producing a manuscript, but readily acquiesced in the suggestion that an alternative format might be a *Festschrift* on the occasion of Viktor's 80th birthday. The idea behind this revised proposal was not to attempt to cover the entire field of neuroembryology (such an undertaking would probably require two or three volumes) but rather to present a collection of essays by some of Viktor's closest friends in the field. At one and the same time, it was felt that such a collection would be a fitting tribute to his remarkable contributions to the subject and would provide a broad overview of many of the subjects that are of current interest in developmental neurobiology. The suggestion met with a warm response from all those who were invited to participate; as one author wrote: "Considering how much I have benefited from my association with Viktor, it will be a pleasure to do something for him in return."

To the general reader this collection of essays may seem to be little more than a smorgasbord of unrelated papers, linked to one another only by the respect and affection their authors share for a distinguished scholar. To a significant degree this is true, and since few restraints were placed upon the contributors, a good deal of heterogeneity is to be expected. Yet, at the same time it is our hope that collectively the papers will provide a broad synopsis of some of the work currently being done in developmental neurobiology. And if any apology were needed for the diversity of the material covered, it is to be found in Viktor Hamburger's own range of interests.

Something of their breadth—and of the far-reaching significance of his achievements—is to be found in the first essay. There I have attempted to provide a brief survey of Viktor's contributions to neuroembryology from his earliest studies as a graduate student in Spemann's laboratory to his most recent work, some 60 years later. Rita Levi-Montalcini, Viktor's longtime friend and collaborator, has contributed a more personal reminiscence of her work and association with Viktor. Viktor himself has recently covered much of the same ground in an essay he wrote for the *Annual Review of Neuroscience* (1980; 3:269–78). Together these two articles provide an intriguing insight into one of the most successful scientific collaborations, and an interesting series of personal reminiscences about one of the more important discoveries made in the field during the last thirty years. The third article, by Salome Gluecksohn-Waelsch, again strikes a personal note and serves to remind us that throughout his career Viktor has had a profound and lasting influence on each of his students. It also reminds us that one of his less well-known research interests has been in the area of developmental genetics, which is now experiencing a renaissance, almost as dramatic as that in neurobiology.

These partly biographical essays are followed by a series of articles that review at varying length some of the research topics that have engaged Viktor's attention for so many years. Chief among them has been his interest in the development of the spinal cord in general and of the natural history of the spinal motoneuron in particular. Lynn Landmesser has addressed the important problem of the specificity of neuromuscular innervation and Ronald Oppenheim the phenomenon of naturally occurring cell death which, as Viktor showed some years ago, serves to play such a critical role in determining the final number of motoneurons. Anne Bekoff's essay reviews what is now known of the ontogeny of motor coordination, and what has been contributed to this field since Viktor's work in the 1960s on the onset of behavior in chick embryos. There follows a short review of the earliest stages of muscle development in the chick by Guido Filogamo, one of Viktor's many Italian friends and admirers.

Peggy Hollyday and Paul Grobstein provide an interesting comparison between the work on myoneural specificity and the patterning of connections in the retinotectal system, a subject on which Viktor himself never worked, but one in which he always showed the greatest interest, and which he encouraged many of us to examine. Sansar Sharma follows this with a short review of some of the more recent work on the establishment of retinal polarity in the frog, which calls in question certain widely accepted views that relate the acquisition of positional information in the retina to the cessation of DNA synthesis by the first retinal ganglion cells. This paper also provides an interesting alternative explanation for experiments that are based on "compound eyes." This group of papers concludes with an interesting discussion of the whole issue of synaptic specificity by Dale Purves, whose work on the superior cervical ganglion has done so much to clarify the issues involved in this problem and has provided some of the firmest experimental evidence for neuronal specificity in the peripheral nervous system.

The past decade has witnessed a revival of interest in the phenomenon of selective cell aggregation generally, and more specifically on the capacity of neurons to aggregate together to form discrete neural entities such as peripheral ganglia, or central nuclear groups and cortical layers. This important topic is reviewed by David Gottlieb and Luis Glaser. A quite different, but equally important, recent advance has been the analysis of the development of membrane properties in neurons. This field is covered in a splendid review by two of Viktor's friends, Nick Spitzer and Janet Lamborghini, at La Jolla. Many workers have felt that the obvious way to study the development of the nervous system (and even behavior) is to use the same types of genetic approach that have proved so successful in molecular biology. Gunther Stent's essay provides a salutary reminder that this is not necessarily the case: there is a large and at present unbridgeable gap between the analysis of genes (that code for particular proteins) and an understanding of the complex sequence of events that lead to the formation of a nervous system.

The next group of three papers includes: a broad overview of the role of the Schwann cell in the economy of the peripheral nervous system by Mary and Richard Bunge; an examination of certain developmental aspects of the neocortex by Ted Jones; and some observations on the development of the dentate gyrus by my colleagues Brent Stanfield and David Amaral, and myself. Collectively, these essays show some of the ways in which the general principles of developmental neurobiology are now being applied to the analysis of the development of specific regions of the central nervous system. And, while it remains true that we are far from being able to give a complete account of the development of any component of either the central or the peripheral nervous system—even at a descriptive level—it is also true that we have made considerable progress toward this goal in the past few years.

Finally, the volume ends with a short essay by Viktor's longtime friend and colleague, Tom Hall, who has shared for many years Viktor's interest in the history of biology in general and of neuroscience in particular.

On behalf of all the contributors I should like, with respect and affection, to present these essays to Viktor, whom we know and admire both as a distinguished scholar, and as a generous friend.

Had it not been for Jeffrey House's gentle persistence and encouragement this volume would probably not have been undertaken; it certainly could not have been brought to fruition without his continued help; I am grateful to him for his encouragement and forebearance at every stage in this endeavor. I should also like to thank my secretary, Ruth Hanna-Amaral for her careful typing (and retyping) of some of the manuscripts and for the gracious way in which she seems to make everything I undertake both possible and worthwhile.

W. M. C.

La Jolla
September 1980

Contents

Contributors

D. G. AMARAL

Salk Institute for Biological Studies
San Diego, California

ANNE BEKOFF

Department of Environmental, Population and Organismic Biology
University of Colorado
Boulder, Colorado

MARY B. BUNGE

Department of Anatomy and Neurobiology
Washington University School of Medicine
St. Louis, Missouri

RICHARD P. BUNGE

Department of Anatomy and Neurobiology
Washington University School of Medicine
St. Louis, Missouri

W. MAXWELL COWAN

Salk Institute for Biological Studies
San Diego, California

GUIDO FILOGAMO

Instituto di Anatomia Umana Normale
Università di Torino
Torino, Italy

LUIS GLASER

Department of Biological Chemistry
Division of Biology and Biomedical Sciences
Washington University School of Medicine
St. Louis, Missouri

SALOME GLUECKSOHN-WAELSCH

Department of Genetics
Albert Einstein College of Medicine
Bronx, New York

DAVID I. GOTTLIEB

Departments of Biological Chemistry and Neurobiology
Division of Biology and Biomedical Sciences
Washington University School of Medicine
St. Louis, Missouri

PAUL GROBSTEIN

Department of Pharmacological and Physiological Sciences
University of Chicago
Chicago, Illinois

THOMAS S. HALL

Department of Biology
Washington University
St. Louis, Missouri

MARGARET HOLLYDAY

Department of Pharmacological and Physiological Sciences
University of Chicago
Chicago, Illinois

E. G. JONES

Department of Anatomy and Neurobiology
Washington University School of Medicine
St. Louis, Missouri

Lynn Landmesser

Biology Department
Yale University
New Haven, Connecticut

Janet E. Lamborghini

Biology Department
University of California at San Diego
La Jolla, California

Rita Levi-Montalcini

Laboratory of Cell Biology
Rome, Italy

Ronald W. Oppenheim

Neuroembryology Laboratory
Dorothea Dix Hospital
Raleigh, North Carolina

Dale Purves

Department of Physiology and Biophysics
Washington University School of Medicine
St. Louis, Missouri

S. C. Sharma

Department of Ophthalmology
New York Medical College
Valhalla, New York

Nicholas C. Spitzer

Biology Department
University of California at San Diego
La Jolla, California

B. B. Stanfield

Salk Institute for Biological Studies
San Diego, California

Gunther S. Stent

Department of Molecular Biology
University of California at Berkeley
Berkeley, California

STUDIES IN DEVELOPMENTAL NEUROBIOLOGY
Essays in Honor of Viktor Hamburger

1 / Viktor Hamburger's Contribution to Developmental Neurobiology: An Appreciation

W. MAXWELL COWAN

FOR MORE THAN FIFTY YEARS Viktor Hambürger has had a significant impact on the field of developmental neurobiology. From the time of his first publication, an experimental study of the effects of the nervous system upon limb development in amphibians, which appeared in 1925, to his most recent contribution on the transport of nerve growth factor from the limb buds of chicks to the related sensory ganglion cells, which was published in 1981, he has either been directly associated with, or has indirectly influenced, almost every major development in neuroembryology. Since it is given to relatively few to be so totally identified with any scientific endeavor, it may be of interest to consider the problems that have engaged his attention for almost half a century and to review the many discoveries that will always be associated with his name.

Viktor's[1] interest in developmental biology can be traced back to his youth, when as a teenager he explored the local fauna in the ponds surrounding his home town of Landeshut in Silesia. He was intrigued by the many varieties of freshwater clams and sponges, but especially by the developing frogs and salamanders, many of which he reared in his own aquarium. Out of this early interest in natural history there grew an increasing fascination with the problems of development which later led him to spend two semesters at Heidelberg University studying developmental genetics. However, when the time came to choose a university in which to do his Ph.D. he chose not to remain at Heidelberg but rather to go to Freiburg, largely on the grounds that the hiking and skiing in the surrounding mountains were said to be a good deal better! He had, of course, inquired about the possibilities for research in biology at the University of Freiburg, and was fortunate in that his aunt, Dr. Clara Hamburger, a senior Research Assistant in the Zoological Institute in Heidelberg, was aware of the exciting things that were going on in Hans Spemann's laboratory. At her urging, he applied to work in Spemann's department at Freiburg and was immediately exposed to the most exciting work in developmental biology. In retrospect, only his later move to the United States was to have a more lasting effect on his career than this almost fortuitous decision to go to Freiburg. For, it brought him not only into close association with the greatest of European experimental embryologists (whose work was soon to culminate in the discovery of the organizer), but also with the greatest of American embryologists, Ross G. Harrison, who at that time spent most summers in Spemann's department.

Like most of the graduate students in Spemann's laboratory, Viktor was anxious to work on the organizer, but he was actively discouraged by Spemann's remark that there "are already too many people hanging from the dorsal lip of the blastopore." At Spemann's suggestion he turned his attention from the problems of embryonic induction, to what seemed, at the time, the more prosaic and certainly less-exciting problems of neural development. The specific problem he was assigned by Spemann was to check a recently reported observation that eye removal in amphibian embryos could result in a variety of abnormalities in the growth of the limbs. Although the notion that the removal of an eye can affect limb development now seems highly improbable, at the time it probably seemed no less improbable than that an entire second embryo could be formed by transplanting a small piece of blastoporal tissue! As it happened, it did not take Viktor long to show that if care were taken to raise the embryos appropriately, the limbs could develop perfectly normally in the absence of one or both eyes, and that the anomalous limb growth previously reported was due to some minor aberration in the medium in which the embryos had been raised and was quite unrelated to the earlier surgical intervention (Hamburger, 1925). Again, although in retrospect this hardly seems a promising starting point for a career in neuroembryology, it was to mark the beginning of Viktor's lifelong interest in two related issues: in what way is the development of the limbs dependent upon their innervation? and to what extent is the development of the spinal cord and of the related spinal ganglia affected by the presence or absence of the limbs?

Viktor's first important contribution to neuroembryology was made in the late 1920s when he was working as a research associate in Otto Mangold's laboratory at the Kaiser Wilhelm Institute for Biology in Berlin-Dahlem. This was the demonstration that innervation is not essential for limb development: he showed this by removing the limb-innervating segments of the developing spinal cord together with the adjoining portion of the neural crest from which the sensory innervation of the limbs is derived. Using the simple, but remarkably effective, tools introduced by Spemann (glass needles, hair loops, etc.), he was able to produce a large number of embryos with aneurogenic limbs that were essentially normal from the point of view of their skeletal and muscular development. Of course, in time, the muscles in the nerveless limbs became atrophic, but these early experiments established once and for all the developmental independence of the various tissues that comprise the limb, from the nervous system (Hamburger, 1927, 1928). An interesting ancillary observation that emerged from some of the experiments in which the denervation of the limb was incomplete, was the finding that the surviving nerve fibers often formed surprisingly normal patterns (including characteristic limb plexuses) in the proximal part of the limb even though they often failed to reach their peripheral targets (Hamburger, 1928). The nature of the factors that are responsible for directed axonal outgrowth and the formation of nerve patterns was to command his attention from time to time over the next three decades.

In 1932 Viktor received a Rockefeller Travelling Fellowship to work for a year in F. R. Lillie's laboratory at the University of Chicago. At that time Lillie's laboratory was one of the few in which chick embryos were being used for developmental studies, but the only analytic technique commonly used there involved the transplantation of tissues or organs onto the chorio-allantoic membrane. Although this was useful for short-term studies of the inherent capacities of the organs or tissue for self-determination, in time the transplants became grossly distorted and difficult to analyze. Viktor was quick to realize that this approach was unlikely to be of great use in neuroembryological studies so he explored the possibility of doing intraembryonic surgery in chicks using the tools he had learned to make in Freiburg.

From a personal point of view, the coming to power of the National Socialist Party in Germany under Adolf Hitler in 1933, and with it, those uncontrolled outbursts of anti-Semitism that were to drive so many of Germany's intellectuals into exile, were to have a momentous effect on his life and work. He had come to the United States in 1932 with no thought of leaving Germany permanently, but in April 1933 he received a letter from the Dean of the Faculty at Freiburg informing him that he had been dismissed from his position as an Assistant in the Zoology Department, and as *Privatdozent,* in accordance with the newly promulgated law "for the cleansing of the professions." At the same time Spemann wrote to say that since the universities were state controlled, there was absolutely nothing he could do to circumvent this decision, and that Viktor should endeavor to find a position in the United States. Fortunately, his support from the Rockefeller Foundation was continued for a further year, which made it possible for him to continue his work at Chicago until the summer of 1935 when he moved to Washington University as an Assistant Professor of Zoology. He was to remain at Washington University for the rest of his career, serving as chairman of the Department from 1941 to 1966, when he was appointed Mallinckrodt Distinguished Service Professor.

Viktor's experience at Chicago had convinced him of the advantage of the chick for neuroembryological studies and although he was later to publish *A Manual of Experimental Embryology* (1st ed., 1942; 2nd, rev. ed., 1960), which included descriptions of many of the classic experiments on amphibian development, his own work from the mid-1930s onward was, with one brief interlude, confined to avian embryos. The reasons for this were partly technical—chick embryos are easily accessible and survive experimental manipulation quite well, and partly neuroanatomical—the structural organization of the avian nervous system is not unlike that of mammals and in its response to most experimental procedures a good deal easier to interpret than that of most lower vertebrates. It would be fair to say that, with the possible exception of Frank Lillie, Viktor more than any other person, put chick embryology on the map, and it is only appropriate that in 1977 his staged series of chick embryos (prepared with the help of H. L. Hamilton, 1951) was recognized by the Institute of Scientific Information as a "citation classic," being one of the 500 most-cited scientific papers in the 15-year period from 1961 through 1975.

The first problem Viktor studied in the chick concerned the effects of early limb extirpation on the development of the lateral motor columns of the spinal cord and on the related sensory ganglia (Hamburger, 1934). The problem had been studied some years earlier by one of Lillie's students, Miss M. L. Shorey,[2] using electrocoagulation of the developing limbs, but it seemed worthwhile repeating the experiments with the more refined methods that Viktor had introduced. The results he obtained were clear-cut: the lateral motor columns, which contain the motoneurons that innervate the limb musculature, were virtually wiped out in those cases in which the limbs had been completely removed; and the sensory ganglia at the same segmental levels were substantially reduced in size. Two general conclusions were drawn from these experiments. First, the development of the lateral motor columns is critically dependent upon the integrity of the limb musculature; and second, the development of the sensory neurons in the spinal ganglia is totally independent of the growth of the motoneurons. However, the interpretation of the findings was less clear. In keeping with contemporary notions about the control of neurogenesis, Viktor was inclined to attribute the observed reduction in the size of the motor columns and the sensory ganglia either to some remote effect on cell proliferation or to a failure in the recruitment of undifferentiated (or uncommitted) neurons in the spinal cord.

Some years before Detwiler[3] (1920) had shown that when a supernumerary limb is transplanted close to the normal limb in *Ambystoma,* there is an appreciable increase in the size of the associated sensory ganglia. Viktor repeated this experiment in the chick and was able to show a similar (although less-marked) increase in the size of the motor columns in the spinal cord and an enlargement of the adjoining sensory ganglia at the level of the supernumerary limb (Hamburger, 1939). Again, Viktor interpreted these observations as being due either to an increase in cell proliferation or to increased cell recruitment. As we shall see, in the light of later evidence, Viktor had reason to revise his interpretation of these experimental findings, but even at that time he was perceptive enough to note that the observations clearly implied that there must be some form of retrogradely transported signal from the "periphery" to the developing neural centers.

The innervation of the supernumerary limbs provided Viktor with another opportunity to examine the factors involved in establishing nerve patterns (Hamburger, 1939). Interestingly, he found that although the supernumerary limbs were innervated in part from "non-limb segments," the general arrangement of the nerves in the limb and their branching patterns were essentially normal. The clear conclusion from this is that the patterning of axons in the limb is not due to some intrinsic factor within the axons but is largely shaped by extrinsic clues provided by the major limb structures. This finding prompted Viktor to examine the development of aneurogenic limbs in the chick (Hamburger and Waugh, 1940). He found that it was relatively easy to produce such limbs by simply implanting limb buds into the celomic cavity. As he had earlier found in amphibians, the overall appearances of the skeletal and muscular components of the nerveless limbs were relatively normal, although ankylosis of the joints was common.

In the mid-1940s, while he was demonstrating a series of sections of a 10-mm pig embryo to a group of undergraduate students, Viktor noticed for the first time that there was a dense accumulation of mitotic figures in the dorsal (or alar) part of the spinal cord, whereas in the ventral (or basal) half of the ventricular zone they were relatively sparse. This chance observation set him thinking about the regulation of cell proliferation in the nervous system and about the factors that might determine its patterning in space and time. The actual process of cell division in the ventricular zone had been carefully described (and correctly interpreted in terms of the interkinetic migration of the cell nuclei) by Sauer some thirteen years earlier (Sauer, 1935),[4] but apart from Coghill's (1933)[5] finding that in *Ambystoma* proliferation proceeds in a general rostral to caudal sequence, little attention seems to have been paid to the temporospatial patterning of mitosis in the nervous system. Viktor accordingly set out to systematically examine the distribution of mitotic figures in the chick spinal cord over the entire period of cell proliferation (Hamburger, 1948).

This resulted in the important finding that although in the chick there are no very striking rostrocaudal differences in mitotic activity—and, in particular, there is no obvious enhancement of proliferation in the limb innervating segments—there are major temporal differences along the ventrodorsal dimension. With the help of two research associates Viktor counted literally thousands of mitotic figures and computed the "mitotic densities" (i.e., the numbers of mitoses/1000 μm^2 of ventricular surface) at each of several stages in development. This analysis showed clearly that cell proliferation in the region of the basal plate (from which the principal motor components of the cord are developed) reaches its peak some three to four days earlier than in the alar plate (which gives rise to most of the "sensory components" of the cord). Indeed, at a stage when the development of the motor system of the cord is essentially complete and the axons of the motoneurons have already entered the base of the limb, proliferation in the alar region is still far from reaching its peak rate. The full implications of this finding did not emerge for some years, but the primacy of the basal plate (and of motor regions in general) was firmly established and has subsequently been found to hold true in all vertebrates.

The next major event in Viktor's career was again the result of one of those lucky accidents that seem to favor the prepared, persistent, and deserving mind. Quite by chance he stumbled upon a group of papers by Rita Levi-Montalcini and her mentor, Guiseppe Levi, which had been published during the war in the Belgian *Archives de Biologie*[6,7] and in the proceedings of the Pontifical Academy of Science,[8] on the effects of limb ablation upon the developing spinal ganglia. The findings in these studies (which, as Rita has pointed out,[9] had been stimulated by Viktor's 1934 paper) confirmed that limb ablation in the chick results in a marked atrophy of the related ganglia; but, whereas Viktor had interpreted his findings in terms of a failure of recruitment of "undifferentiated" cells, the Italian workers had found that it was due to a secondary degeneration of previously differentiated ganglion cells. A lesser man might well have been chagrined over such a difference in interpretation, or perhaps inclined to dismiss it because of its relative obscurity; but not Viktor. With characteristic forthrightness he contacted Dr. Levi-Montalcini, persuaded her to come to St. Louis to reexamine the problem in his laboratory, and obtained the necessary funds from the Rockefeller Foundation to make it possible. So began one of the longest and most fruitful collaborations in neuroembryology; for like Viktor's "one-year" visit to the United States 15 years earlier, Rita remained at Washington University until her retirement some 30 years later.

Their first collaborative study on the effects of the limb-bud extirpation and supernumerary limb transplantations on the sensory ganglia resulted in a paper that is now widely recognized as one of the classics in the field of neuroembryology (Hamburger and Levi-Montalcini, 1949). Among other things it established first, that limb extirpation does not act primarily upon cell proliferation or neuronal differentiation in the ganglia but on the maintenance of differentiated cells, and second that during normal development there is a substantial degree of cell death in the sensory ganglia. The finding of degenerating neurons in the developing nervous system was not new—Collin[10] had reported it more than 40 years earlier—but this was the first study to draw attention to the magnitude and widespread nature of the phenomenon. Viktor was to return to the problem of the trophic interaction of the periphery upon the nervous system some years later, but before that was to occur, there was an exciting interlude that led to the discovery of the celebrated nerve growth factor, or NGF as it is commonly known.

The fact that peripheral tissues could have such a profound effect on the developing nervous system led Viktor to suggest to one of his graduate students, E. Bueker, that it might be interesting to see if other rapidly growing tissues could exert a similar influence. A rapidly growing mouse sarcoma was chosen as the test tissue, and in one of those bold and unpredictable experiments that often change the course of a field, Bueker implanted a portion of the tumor into the celomic cavity of a series of early chick embryos.[11] The grafted tumors grew rapidly and within a few days became densely innervated by fibers from the nearby sensory ganglia. The ganglia themselves were considerably enlarged; indeed, they were somewhat larger than the ganglia Viktor and Rita had observed after supernumerary limb transplants.

With Bueker's permission, they repeated the experiments (Levi-Montalcini and Hamburger, 1951) with a number of refinements. While analyzing the experimental material, Rita made the seminal observation that the tumor masses were filled with nerve fibers derived not only from sensory ganglia but even more strikingly from the sympathetic ganglia, some of which were more than seven times their normal volume. And, even more importantly, she noted that many of the prevertebral sympathetic ganglia that did not enter into the neurotization of the implanted tumor were also enlarged, apparently in response to a diffusible agent released by the tumor. That this was so was convincingly demonstrated in a second series of experiments in which the tumor was implanted in an *extraembryonic* site, on the chorio-allantoic membrane, and resulted in a similar hyperplastic response in the sympathetic ganglia (Levi-Montalcini and Hamburger, 1953).

The further story of the development of an *in vitro* bioassay system for the analysis of the growth-promoting factor, its successful isolation and purification by the biochemist Stanley Cohen, and the finding that it is present in large amounts in snake venom and in the submandibular salivary glands of male mice, has frequently been told,[12] and need not be repeated here. Partly for personal reasons, and partly because he felt he had little to contribute to the biochemical phase of the work, Viktor graciously stood aside and turned from the NGF to the problem of cell-strain specificity in the nervous system (Hamburger, 1962) and to his longtime interest in the factors that determine peripheral nerve patterns.

While working on the sensory ganglia, Rita had drawn his attention to the fact that the trigeminal ganglion, like the dorsal root ganglia, consists of two distinct populations of cells: an outer or ventrolateral group of large, early differentiating neurons, and a group of dorsomedially placed cells that are appreciably smaller and appear later in development. In the spinal ganglia both groups of cells are of neural crest origin, but the trigeminal ganglion had been known for some time to have a dual origin—partly from the neural crest and partly from the trigeminal placode. The possibility that the two cell populations were derived from different sources was intriguing and Viktor set out to study this experimentally by appropriate extirpation experiments (Hamburger, 1961). Removal of the rostral medulla with the associated neural crest resulted in glanglia composed only of large early-formed neurons; removal of the placode gave ganglia with only small cells. And interestingly, the former (but not the latter) had peripheral nerve distributions that were essentially like those seen in normal animals (although, of course, much reduced in size). Why one class of fiber should be able to follow an apparently predetermined set of pathways, while another arising in the same general region (but from a different class of cell) is unable to identify the appropriate "preneural pathway"—but can attach itself to fibers that have grown out earlier—is not known; it remains one of the many unresolved problems about axonal path finding. And as Viktor has frequently pointed out, both in his research reports and in several reviews of the problems of neuroembryology (Hamburger, 1956, 1957, 1962), we know as little about how nerve fibers find their way to their target loci as we know about how they identify the appropriate targets and innervate them.

The late 1950s were a singularly difficult period for Viktor, and after completing the work on the trigeminal ganglion, he was temporarily at a loss to know what problems he should next investigate. Several decades before, as he had contemplated his future career in science, he had outlined a program of research on the development of the nervous system, which was to culminate in an attack on the complex problems associated with the ontogeny of behavior. His thinking along these lines had been greatly influenced by the work of the early ethologists and especially by Konrad Lorenz, whose concept of "innate releasing mechanisms" seemed to fit so well what little was known about the earliest phases in embryonic behavior. After coming to the United States, he had the good fortune to meet G. E. Coghill and soon became an admirer of his analytic studies of larval salamander behavior, which had led to the general view that the structural differentiation of the nervous system precedes the onset of functional behavior but is itself little influenced by function. In retrospect, Viktor was perhaps equally fortunate in being unaware of the views of many developmental psychologists who held that all behavior was either reflex or learned; if he had, he might well have been discouraged from entering this controversial field—it was not in his nature to seek out controversy or to promote it.

As it was, he approached the onset of behavior in the chick with an open, but not an unprepared, mind. His first objective was simply to observe what types of behavior normal chicks display at each developmental stage, and then to proceed, using his well-tested experimental procedures, to analyze the structural bases for the observed behavioral repertoire. Chick embryos are particularly well suited for studies of this kind. They can be readily observed, for hours at a time, through openings in the shell, and, as Viktor had well shown, they are amenable to a variety of experimental manipulations. So with a postdoctoral fellow, Martin Balaban,[13] he set out to see what chick embryos do *in ovo* (Hamburger and Balaban, 1963). The first thing that struck them was that the embryos appeared to move spontaneously or, at least, in the absence of any sign of external stimulation. Next, they noted that up to the 13th day of incubation the movements occurred, not continuously but periodically, with phases of activity alternating with periods of inactivity that became progressively shorter as development proceeded. After the 13th day, activity appeared to be more or less continuous until a day or two before hatching, when motility declined sharply. Perhaps what was most interesting was the observation that these spontaneous movements seemed to be quite uncoordinated; the legs, wings, head, beak, and eyelids appeared to move randomly, and there was no indication of patterned sequences of temporally-related movements.

It was only after making these key observations that Viktor read (or, more probably, reread with new insight) W. Preyer's generally forgotten work on chick behavior published as early as 1885. In his monograph *Specielle Physiologie des Embryo,* Preyer states (with none of the false modesty that attenuates most modern writing): "I consider this [i.e., the discovery of a 'pre-reflexogenic period'] one of the most important in the entire field of the physiology of the embryo." In one of those unique flashes of insight that so often characterize great scientists, everything seemed to fall in place. Long (on the scale of embryonic time) before the dorsal part of the spinal cord is developed, cell proliferation and migration in the basal plate have ended, the motor columns of the cord have become established, and the axons of the motoneurons have grown out to innervate the peripheral musculature. Motility in the chick (and interestingly this seemed likely to be true of all amniotes as his experience with the 10-mm pig embryo had first suggested) must be autonomously generated and totally independent of sensory input.

The notion of the primacy of motor activity over sensory input ran counter to the ideas of most developmental psychologists and, not surprisingly, met with considerable opposition. Where others might have attempted to gain support for their views by polemical writing, or trumpet blowing at scientific meetings, Viktor returned to his laboratory to subject his (and Preyer's) hypothesis to the test of experiment. The first step in the experimental analysis was to determine to what extent the observed motility was influenced from suprasegmental levels. This was done by making a large number of spinal embryos in which several segments of the cervical cord were removed (Hamburger et al., 1965). The early, apparently spontaneous movements (which has come to be termed Type I motility) were quite unaffected by spinal transection; and it was not until about the tenth day of incubation that the first indication of a suprasegmental influence became apparent. When this first appears, it manifests itself as a reduction in the *amount* of time the spinal embryos spend in activity. Oppenheim was later to show that the *number* of movements made by a spinal animal per minute is actually unchanged, which suggests that the primary effect of this suprasegmental influence is to modulate the temporal pattern of activity rather than to introduce new components into the behavioral repertoire.

The second step in the analysis was to establish that Type I motility does indeed occur in the absence of sensory input (Hamburger, Wenger, and Oppenheim, 1966). To do this, it was necessary to remove the entire dorsal half of the spinal cord at lumbosacral levels, together with the neural crest from which the sensory ganglia are derived. To eliminate descending inputs from higher levels, several thoracic segments were also extirpated. Although later motility in these chronic preparations was diminished (in large part because of progressive ankylosis of many of the joints), the jerky, uncoordinated movements typical of Type I motility occurred in both limbs, in an essentially normal manner. Later experiments, which involved bilateral extirpation of the primordia of the trigeminal and of the acousticovestibular ganglia (Hamburger and Narayanan, 1969), answered the criticism that head-stroking movements by the limbs provided the embryos with a form of self-stimulation.

With a succession of postdoctoral fellows, and a group of colleagues who were familiar with electrophysiological techniques, Viktor proceeded to demonstrate several correlations: first, that the motoneurons in the ventral horn of the spinal cord discharge spontaneously and at random intervals (Provine et al., 1970; Sharma et al., 1970); second, that the spontaneous bursts of activity that can be recorded extracellularly from the anterior horn coincide, in time, with periods of Type I activity (Sharma et al, 1970); third, that bursts of activity that begin at one level of the cord can be propagated to other levels; and four, that such periods of activity occur even during curarization.

The biological significance of this type of spontaneous activity is still unknown[14] (Viktor believes it may serve to prevent joint fusion), but it clearly can play no part in such adaptive processes as hatching or early posthatching behavior. Viktor accordingly turned his attention to the analysis of hatching, a phenomenon that had been observed for centuries but, surprisingly, had never been carefully described, let alone formally analyzed. Hatching behavior turns out to involve a number of complex and highly coordinated movements that serve to bring the chick into the "hatching position" with the beak pressed against the shell. This is followed by a series of coordinated prehatching movements that involve most parts of the embryo from the 17th day of incubation onwards. Together, these lead to the hatching act itself on the 20th or 21st day. This climax is characterized by a succession of sharp thrusts of the beak against the shell, accompanied by coordinated rotations of the whole body; the thrusting movements lead to a cracking of the shell around two-thirds of its circumference and, in time, to the escape of the head and limbs as the shell cap breaks loose (Hamburger and Oppenheim, 1967).

It had generally been assumed that such complex and evidently integrated behaviors arise out of, and are shaped by, the earlier uncoordinated movements. That it actually arises more or less *de novo* and is not simply a coordinated elaboration of Type I activity was a wholly unexpected finding. But its implication seems inescapable: the entire structural framework for this (and possibly for every other) behavior is laid down prior to the first appearance of the behavior and is only minimally influenced, if at all, by the behavior in question (Hamburger, 1968).

After a brief attempt at conventional reflexology (involving tactile stimulation of the head, trunk, and limbs at different stages in development), which yielded little, Viktor undertook an electromyographic analysis of intralimb coordination (Bekoff, Stein, and Hamburger, 1975). This revealed that as early as the seventh day of incubation the limb flexors and extensors are reciprocally linked, and from this it follows that the necessary motoneurons and interneuron pools are synaptically connected at a very early stage in spinal cord development. To document this, Viktor made one brief incursion into electron microscopy. With the help of Bob Skoff, he was able to show that some synapses are present as early as the fourth day of incubation, almost immediately after the first movements are observed (Hamburger and Skoff, 1974). It is, of course, difficult to determine the origin of presynaptic processes in random, ultrathin sections, or in short sequences of serial sections, and it is evident that the analysis of the synaptic organization of the developing spinal cord and brain has barely begun. Wisely, Viktor decided not to pursue this problem, but rather to return to the more manageable issue of the dependency of the motor cell columns upon the limb musculature.

His experience with the sensory ganglia had convinced him that the periphery exerts no significant effect upon the early phases of neurogenesis—proliferation, migration, and selective cell aggregation—but that it is essential for the maintenance of the differentiated neurons. This had led him in the late 1950s, to reexamine the development of the motor system. His approach was essentially the same as before, but in the light of later experience he paid particular attention to the occurrence of cell degeneration, both during normal development and after limb extirpations (Hamburger, 1958). The results of this new study were clear. The lateral motor column was found to be numerically complete by the sixth day of incubation and, up to this stage, unaffected by limb extirpation. However, over the course of the next three days there was a massive neuronal degeneration on the side of the limb ablation, which effectively wiped out the entire lateral motor column. The occurrence of this neuronal degeneration in his experimental animals, coincided with the appearance of an appreciable number of degenerating cells in the motor columns of *normal* embryos. This suggested that there are two causally related phenomena at work: one that is responsible for the naturally occurring cell death, and one that accounts for the experimentally induced neuronal degeneration. The finding of a programmed period of naturally occurring cell death suggested an alternative explanation for the apparent hyperplasia seen in cases with supernumerary limbs: it could well be due, not to the recruitment of additional cells (either by proliferation or differentiation) but to the regulation of cell death.

To confirm this possibility, it was obviously necessary to determine the magnitude of the naturally occurring cell loss. If naturally occurring cell death involved only a relatively small number of neurons, it clearly could not account for the observed hyperplasia. Undeterred by the enormous amount of work that this would necessitate, in 1975 Viktor counted the numbers of motoneurons present in the lateral motor columns of the hindlimb innervating segments in a series of no fewer than 60 embryos. But the effort was amply repaid. For the first time there was clear evidence that between the sixth and tenth days of incubation, about 40 percent of the motoneurons in the lumbosacral cord degenerate.

The next step in the analysis was to determine whether all or only some of the cells that might be expected to die can be "rescued" by an appropriately placed supernumerary limb. With Peggy Hollyday (whose graduate work had involved transplanting limbs in amphibian larvae) he was able to show that in every case with a successful limb transplant, the lateral motor column on the side of the transplanted limb was enlarged and that the number of surviving motoneurons was increased by between 11 and 27 percent (Hollyday and Hamburger, 1976). Since the term hyperplasia (which implies an increased *production* of cells) was obviously inappropriate in the light of these findings, Viktor suggested that the decreased cell death should be referred to as *hypothanasia* and the phenomenon of naturally-occurring cell death as *neurothanasia.* To his regret (but with a resigned acceptance of the general decline in classical education) neither term seems to have caught on.

An interesting secondary finding in the study of both supernumerary limbs was the observation that the additional motoneurons are not confined to one region of the motor column but seem to be distributed throughout its extent. As this had important implications for the question of motoneuronal specificity, it clearly called for further study. Accordingly a further series of embryos with supernumerary limbs was prepared, and small amounts of the enzyme-marker horseradish peroxidase were injected into the selected muscles of both the normal and the adjacent supernumerary limbs. When the distribution of the resulting retrogradely labeled motoneurons was plotted, the findings were quite unexpected: the neurons innervating the muscles in the supernumerary limb were neither confined to the same motoneuron pools as those that innervate the homologous muscles in the control limb nor were they segregated within a "foreign" motoneuron pool in the region of the cord normally responsible for the innervation of the proximal (thigh) muscles (Hollyday, Hamburger, and Farris, 1977). The consistency of this finding, from experiment to experiment, implied that despite the apparent "mismatch" in the innervation, the organization of the motoneurons that innervated the supernumerary limb musculature was not random but rather was determined by some form of "hierarchical" specification.

Having clarified this problem, which had engaged his mind for more than 40 years, Viktor might well have decided to turn off the incubator and to put away his microscope. But, no: he was now anxious to return to the spinal ganglia, to see if he could identify *in vivo* a clear developmental role for NGF. The availability of I^{125}-NGF, and the fact that it was known to be retrogradely transported in other systems, prompted his next set of experiments. The labeled NGF was to be injected at an early stage into limb buds in order to see if it would significantly reduce the massive cell death that normally occurs in the ganglia. With Judy Brunso-Bechtold he has recently succeeded in showing that labeled NGF can be selectively transported back to the sensory ganglia from the limbs (Brunso-Bechtold and Hamburger, 1979). But to demonstrate that the exogenously supplied NGF can maintain at least some proportion of the cells that might otherwise be expected to die will require more cell counts. And as this essay is being written, this is exactly what Viktor is doing![15]

In presenting this synoptic view of Viktor's contribution to developmental neurobiology, I have intentionally omitted all reference to his other contributions. But it would be misleading not to mention that at different times in his career he has worked on such diverse topics as color vision in teleosts, on viral-induced birth defects, and on several different aspects of developmental genetics. And throughout most of his active career, he has carried a heavy teaching load and, for many years, the substantial administrative responsibilities of a department chairman.

By nature quiet and retiring, he never sought the limelight and consistently eschewed the accolades that so many seek. Yet, appropriately, he has not been without honor or recognition. In 1953 he was elected to the National Academy of Sciences; in 1959 to the American Academy of Arts and Sciences; in 1979 he was the recipient of the Wakeman Award; and in 1978 he was awarded an honorary degree of Doctor of Science by Washington University. In 1950, and again in 1951, he served as President of the Society for Growth and Development (now the Society for Developmental Biology); in 1955 he was elected to the presidency of the American Society of Zoologists; and in 1960 he was installed as Vice President and Chairman of the Zoology section of the American Association for the Advancement of Science. But perhaps the recognition he appreciates most is the respect and admiration of his students and peers. His career has been an inspiration to successive generations of students and to a host of colleagues. While, arguably some few of his contemporaries have accomplished more, none is more highly regarded or more genuinely loved. For, as Tom Hall, his close colleague of more than 30 years, has put it, "With Viktor one cannot simply have an 'affair of the mind,' inevitably it becomes an 'affair of the heart.' "

NOTES

1. In preparing this essay I have drawn heavily both on a personal memoir that Viktor wrote three or four years ago at the urging of some of his colleagues and on his F. O. Schmitt Lecture published in 1977. I have taken the liberty of rearranging the material to suit the chronological pattern that seemed appropriate for this review and, with Viktor's permission, have referred to some incidents that he had recounted from time to time, but had not included in his memoir. I have also chosen to refer to him throughout this essay by his first name since that is how he is known to almost everyone. The references to his own work that are cited are listed in the accompanying bibliography.
2. M. L. Shorey (1907) The effect of the destruction of peripheral areas on the differentiation of the neuroblasts. *J. Exp. Zool.*, 7:25–86.
3. S. R. Detwiler (1920) On the hyperplasia of nerve centers resulting from excessive peripheral overloading. *Proc. Nat. Acad. Sci. U.S.A.* 6:96–101.
4. F. C. Sauer (1935) Mitosis in the neural tube. *J. Comp. Neurol.*, 62:377–406.
5. G. E. Coghill (1933) Correlated anatomical and physiological studies of the growth of the nervous system of amphibia. XI. The proliferation of cells in

the spinal cord as a factor in the individuation of reflexes of the hindleg of *Ambystoma punctatum* Cope. *J. Comp. Neurol.*, 57:327–47.

6. R. Levi-Montalcini and G. Levi (1942) Les conséquences de la destruction d'un territoire d'innervation periphérique sur le développments des centres nerveux correspondents dans l'embryon de poulet. *Arch. de Biol.* (Liège), 53:537–45.
7. R. Levi-Montalcini and G. Levi (1943) Recherches quantitatives sur la marche des processes de différentiation des neurons dans les ganglions spinaux de l'embryon de poulet. *Arch. de Biol.* (Liège), 54:189–206.
8. R. Levi-Montalcini and G. Levi (1944) Correlazioni nello svilluppo tra varie parti del sistema nervoso. I. Consequenze della demolizione dell' abozzo di un arto sui centri nervosi nell' embrione di pollo. *Commentat. Pontif. Acad. Sci.*, 8:527–68.
9. R. Levi-Montalcini (1975) NGF: An uncharted route. In *The Neurosciences: Paths of Discovery*, ed. F. G. Worden, J. P. Swazey, and G. Adelman, pp. 244–65. Cambridge: MIT Press.
10. R. Collin (1906) Récherches cytologiques sur le development de la cellule nerveuse. *Névraxe*, 8:185–309.
11. E. Bueker (1948) Implantation of tumors in the hind limb field of the embryonic chick and the developmental response of the lumbosacral nervous system. *Anat. Rec.*, 102:369–89.
12. See R. Levi-Montalcini (1966) The Nerve Growth Factor: Its mode of action on sensory and sympathetic nerve cells. Harvey Lect., 60:217–59; R. Levi-Montalcini and P. U. Angeletti (1968) Nerve Growth Factor. *Physiol. Rev.*, 48:534–69; R. Levi-Montalcini (1975). (See note 9 above.)
13. After withdrawing from the "NGF-saga" (as he has termed it), Viktor collaborated with a large number of colleagues, most of whom came to his laboratory as postdoctoral fellows. For convenience, in what follows I have referred to many of these collaborative studies as though he, alone, had been responsible for the work. Since most of his collaborators would be quick to acknowledge that the ideas—if not all the technical approaches—were Viktor's, I am sure they will forgive me this convenient form of shorthand. The actual participants in each study are identified in the accompanying list of Viktor's collected scientific papers.
14. However, it is clearly important, as Viktor was later to show when he observed comparable spontaneous, random, and uncoordinated movements in rat fetuses both *in utero* and after exteriorization (Narayanan, Fox, and Hamburger, 1971).
15. While this volume was in preparation a paper describing the results of these experiments appeared in the first issue of the *Journal of Neuroscience* (Hamburger, Brunso-Bechtold, and Yip, 1981).

VIKTOR HAMBURGER
BIBLIOGRAPHY*

Hamburger, V. 1925. Über den Einfluss des Nervensystems auf die Entwicklung der Extremitäten von Rana fusca. *Roux Archiv.* 105:149–201

Hamburger, V. 1926. Versuche über Komplementär-Farben bei Ellritzen *(Phoxinus laevis). Z. Vergl. Physiol.* 4:286–304

Hamburger, V. 1927. Entwicklungsphysiologische Beziehungen zwischen den Extremitäten der Amphibien und ihrer Innervation. *Naturwissenschaften,* 15:657–81

Hamburger, V. 1928. Die Entwicklung experimentell erzeugter nervenloser und schwach innervierter Extremitäten von Anuren. *Roux Archiv.* 114:272–362

Hamburger, V. 1929. Experimentelle Beiträge zur Entwicklungsphysiologie der Nervenbahnen in der Froschextremität. *Roux Archiv.* 119:47–99

Hamburger, V. 1930. Der Farbensinn der Fische. *Naturforscher* 7:128–33; 172–80

Hamburger, V. 1934. The effects of wing bud extirpation in chick embryos on the development of the central nervous system. *J. Exp. Zool.* 68:449–94

Hamburger, V. 1935. Malformations of hind limbs in species hybrids of *Triton taeniatus x Triton cristatus. J. Exp. Zool.* 70:43–54

Hamburger, V. 1936. The larval development of reciprocal species hybrids of *Triton taeniatus* (and *Tr. palmatus*) *x Triton cristatus. J. Exp. Zool.* 73:319–73

Hamburger, V. 1938. Morphogenetic and axial self-differentiation of transplanted limb primordia of two-day chick embryos. *J. Exp. Zool.* 77:379–97

Hamburger, V. 1939. The development and innervation of transplanted limb primordia of chick embryos. *J. Exp. Zool.* 80:347–89

Hamburger, V. 1939. Motor and sensory hyperplasia following limb bud transplantations in chick embryos. *Physiol. Zool.* 12:268–84

Hamburger, V. 1939. A study of hereditary Chondrodystrophia in the chick (Creeper Fowl) by means of embryonic transplantation. *Proc. Soc. Exp. Biol. Med.* 41:13–14

Silber, R., V. Hamburger 1939. The production of duplicitas cruciata and multiple heads by regeneration in *Euplanaria tigrina. Physiol. Zool.* 12:285–300

Hamburger, V., Waugh, M. 1940. The primary development of the skeleton in nerveless and poorly innervated limb transplants of chick embryos. *Physiol. Zool.* 13:367–80

Rudnick, D., Hamburger, V. 1940. On the identification of segregated phenotypes in progeny from Creeper fowl matings. *Genetics* 25:215–24

Hamburger, V. 1941. Transplantation of limb primordia of homozygous and heterozygous chondrodystrophic ("Creeper") chick embryos. *Physiol. Zool.* 14:355–64

Brown, M. G., Hamburger, V., Schmitt, F. O. 1941. Density studies on amphibian embryos with special reference to the mechanism of organizer action. *J. Exp. Zool.* 88:353–72

Hamburger, V. 1942. The developmental mechanics of hereditary abnormalities in the chick. *Biol. Symp.* 6:311–34

* Excluding abstracts, book reviews, and occasional nonscientific studies.

Hamburger, V. 1942. *A Manual of Experimental Embryology.* Chicago: Univ. of Chicago Press. 213 pp.

Gayer, K., Hamburger, V. 1943. The developmental potencies of eye primordia of homozygous Creeper chick embryos tested by orthotopic transplantation. *J. Exp. Zool.* 93:147–83

Hamburger, V. 1944. Developmental physiology. *Ann. Rev. Physiol.* 6:1–24

Hamburger, V., Keefe, E. L. 1944. The effects of peripheral factors on the proliferation and differentiation in the spinal cord of chick embryos. *J. Exp. Zool.* 96:223–42

Hamburger, V. 1946. Isolation of the brachial segments of the spinal cord of the chick embryo by means of Tantalum foil blocks. *J. Exp. Zool.* 103:113–42

Hamburger, V. 1947. Experimental embryology. In *Encyclopaedia Britannica,* pp. 974–80 (revised 1967)

Hamburger, V., Habel, K. 1947. Teratogenetic and lethal effects of influenza-A and mumps viruses on early chick embryos. *Proc. Soc. Exp. Biol. Med.* 66:608–17

Hamburger, V. 1948. The mitotic patterns in the spinal cord of the chick embryo and their relation to histogenetic processes. *J. Comp. Neurol.* 88:221–84

Hamburger, V., Levi-Montalcini, R. 1949. Proliferation, differentiation and degeneration in the spinal ganglia of the chick embryo under normal and experimental conditions. *J. Exp. Zool.* 111:457–502

Hamburger, V., Levi-Montalcini, R. 1950. Some aspects of neuroembryology. In *Genetic Neurology,* ed. P. Weiss, pp. 128–60. Chicago: Univ. of Chicago Press

Hamburger, V., Hamilton, H. 1951. A series of normal stages in the development of the chick embryo. *J. Morphol.* 88:49–92

Levi-Montalcini, R., Hamburger, V. 1951. Selective growth stimulating effects of mouse sarcoma on the sensory and sympathetic nervous system of the chick embryo. *J. Exp. Zool.* 116:321–62

Hamburger, V. 1952. Development of the nervous system. *Ann. N.Y. Acad. Sci.* 55:117–32

Hamburger, V. 1953. Growth correlations between the nervous system and peripheral structures. *Scientia* 47:1–6

Levi-Montalcini, R., Hamburger, V. 1953. A diffusible agent of mouse sarcoma, producing hyperplasia of sympathetic ganglia and hyperneurotization of viscera in the chick embryo. *J. Exp. Zool.* 123:233–88

Hamburger, V. 1954. Trends in experimental neuroembryology. In *Biochemistry of the Developing Nervous System,* ed. H. Waelsch, pp. 52–73. Proc. 1st Int. Neurochem. Symp.

Cohen, S., Levi-Montalcini, R., Hamburger, V. 1954. A nerve growth-stimulating factor isolated from sarcomas 37 and 180. *Proc. Natl. Acad. Sci.* U.S.A. 40:1014–18

Levi-Montalcini, R., Meyer, H., Hamburger, V. 1954. *In vitro* experiments on the effects of mouse sarcomas 180 and 37 on the spinal and sympathetic ganglia of the chick embryo. *Cancer Res.* 14:49–57

Hamburger, V. 1955. *Analysis of Development,* Co-ed B. Willier, P. Weiss. Philadelphia: W. B. Saunders, 735 pp.

Hamburger, V. 1955. Regeneration in the central nervous system of reptiles and of birds. In *Regeneration in the Central Nervous System,* ed. W. F. Windle, pp. 47–53. Springfield, Ill.: Thomas

Holtfreter, J., Hamburger, V. 1955. Amphibians. In: *Analysis of Development,* ed. B. Willier, P. Weiss and V. Hamburger, pp. 230–96

Hamburger, V. 1956. Developmental correlations in neurogenesis. In *Cellular Mechanisms in Differentiation and Growth,* ed. D. Rudnick, pp. 191–212. 14th Growth Symp., Princeton U. Press

Hamburger, V. 1957. The life history of a nerve cell. *Am. Sci.* 45:263–77

Hamburger, V. 1957. The concept of "development" in biology. In *The Concept of Development,* ed. D. B. Harris, pp. 49–58. Minneapolis: Univ. of Minnesota Press

Hamburger, V. 1958. Regression versus peripheral control of differentiation in motor hypoplasia. *Am. J. Anat.* 102:365–410

Hamburger, V. 1960. *Manual of Experimental Embryology,* rev. ed. Chicago: Univ. of Chicago Press. 221 pp.

Hamburger, V. 1961. Experimental analysis of the dual origin of the trigeminal ganglion in the chick embryo. *J. Exp. Zool.* 148:91–124

Hamburger, V. 1962. Specificity in neurogenesis. *J. Cell. Comp. Physiol.,* 60:81–92

Hamburger, V. 1963. Some aspect of the embryology of behavior. *Q. Rev. Biol.* 38:342–65

Hamburger, V., Balaban, M. 1963. Observations and experiments on spontaneous rhythmical behavior in the chick embryo. *Dev. Biol.* 7:533–45

Hamburger, V. 1964. Ontogeny of behaviour and its structural basis. In *Comparative Neurochemistry,* ed. Richter. pp. 21–34. Proc. 5th Int. Neurochem. Symp. Oxford: Pergamon.

Hamburger, V., Balaban, M., Oppenheim, R., Wenger, E. 1965. Periodic motility of normal and spinal chick embryos between 8 and 17 days of incubation. *J. Exp. Zool.* 159:1–13

Hamburger, V., Wenger, E., Oppenheim, R. 1966. Motility in the chick embryo in the absence of sensory input. *J. Exp. Zool.* 162:133–60

Hamburger, V., Oppenheim, R. 1967. Prehatching motility and hatching behavior in the chick. *J. Exp. Zool.* 166:171–204

Hamburger, V. 1968. Emergence of nervous coordination. Origins of integrated behavior. *Dev. Biol. Suppl.* 2:251–71

Hamburger, V. 1969. Hans Spemann and the organizer concept. *Experientia* 25:1121–25

Hamburger, V., Narayanan, C. H. 1969. Effects of the deafferentation of the trigeminal area on the motility of the chick embryo. *J. Exp. Zool.* 170:411–26

Hamburger, V. 1970. Embryonic motility in vertebrates. In *The Neurosciences: Second Study Program,* ed. F. O. Schmitt, pp. 141–51. New York: Rockefeller Univ. Press

Provine, R. R., Sharma, S. C., Sandel, T., Hamburger, V. 1970. Electrical activity in the spinal cord of the chick embryo, *in situ. Proc. Natl. Acad. Sci.* U.S.A. 65:508–15

Sharma, S. C., Provine, R. R., Hamburger, V., Sandel, T. T. 1970. Unit activity in the isolated spinal cord of the chick embryo, *in situ. Proc. Natl. Acad. Sci.* U.S.A. 66:40–47

Hamburger, V. 1971. Development of embryonic motility. In *The Biopsychology of Development,* eds. E. Tobach, L. R. Aronson and E. Shaw, pp. 45–66. New York: Academic

Narayanan, C. H., Fox, M. W., Hamburger, V. 1971. Prenatal development of spontaneous and evoked activity in the rat *(Rattus norwegicus albinus). Behaviour* 40:100–134

Narayanan, C. H., Hamburger, V. 1971. Motility in chick embryos with substitution of lumbosacral by brachial and brachial by lumbosacral cord segments. *J. Exp. Zool.* 178:415–32

Hamburger, V. 1973. Anatomical and physiological basis of embryonic motility in birds and mammals. In *Studies on the Development of Behavior and the Nervous System,* ed. G. Gottlieb, 1:51–76. New York: Academic

Hamburger, V., Skoff, R. 1974. Fine structure of dendritic and axonal growth cones in embryonic chick spinal cord. *J. Comp. Neurol.* 153:107–48

Hamburger, V. 1975. Cell death in the development of the lateral motor column of the chick embryo. *J. Comp. Neurol.* 160:535–46

Hamburger, V. 1975. Changing concepts in developmental biology. *Perspect. Biol. Med.* 18:162–78

Hamburger, V. 1975. Fetal behavior. In *The Mammalian Fetus,* ed. E. S. E. Hafez, pp. 68–81. Springfield, Ill.: Thomas

Bekoff, A., Stein, P. S. G., Hamburger, V. 1975. Coordinated motor output in the hindlimb of the 7-day chick embryo. *Proc. Natl. Acad. Sci.* U.S.A., 72:1245–48

Hollyday, M., Hamburger, V. 1976. Reduction of the normally occurring motor neuron loss by enlargement of the periphery. *J. Comp. Neurol.* 170:311–20

Hamburger, V. 1977. The developmental history of the motor neuron. (The F. O. Schmitt Lecture in Neuroscience) *Neurosci. Res. Program Bull. Suppl.* 15:1–37

Hollyday, M., Hamburger, V. 1977. An autoradiographic study of the formation of the lateral motor column in the chick embryo. *Brain Res.* 132:197–208

Hollyday, M., Hamburger, V., Farris, J. 1977. Localization of motor neuron pools supplying identified muscles in normal and supernumerary legs of chick embryos. *Proc. Natl. Acad. Sci.* U.S.A. 74:3582–86

Brunso-Bechtold, J. K., Hamburger, V. 1979. Retrograde transport of nerve growth factor in chicken embryo. *Proc. Natl. Acad. Sci.* U.S.A. 76:1494–96

Hamburger, V. 1980a. Trophic interactions in neurogenesis: a personal historical account. *Ann. Rev. Neurosci.* 3:269–78

Hamburger, V. 1980b. Prespecification and plasticity in neurogenesis. In *Nerve Cells, Neurotransmitters and Behavior,* ed. R. Levi-Montalcini, pp. 433–50. Amsterdam: Elsevier

Hamburger, V., Bruno-Bechtold, J. K., Yip, J. 1981. Neuronal death in the spinal ganglia of the chick embryo and its reduction by nerve growth factor. *J. Neurosci.* 1:60–71

2 / *One of Hans Spemann's Pupils*

RITA LEVI-MONTALCINI

A SUMMER DAY IN 1940

I first met Viktor in a cattle car in northern Italy. It was on a day in that fateful June of 1940 when Mussolini declared war against France, which had already been stabbed to death by Hitler's army. Sharing a train with cows was only a minor discomfort compared with those about to afflict the daily life of the Italian nation throughout a catastrophic five-year war. But, in midsummer of 1940, Il Duce was triumphantly predicting "instant victory" over loudspeakers that blared from the Palazzo Venezia.

I was sitting on the floor of one of those railway cars, which had neither seats nor side walls with conventional doors (niceties unknown to cows), and my legs were dangling out in the open air. The train was running at a slow speed across the country between Turin and the small village I was heading for. I was young and I enjoyed the fresh air and this rather unusual and somewhat dangerous way of traveling. While contemplating the yellowing corn and the bright red poppies, I was reading a reprint lent to me by Giuseppe Levi on the effects of wing bud extirpation on the development of the central nervous system of chick embryos. The article was dated 1934, and as Levi had informed me, it had been written by a pupil of Hans Spemann (Hamburger, 1934). In the eyes of Levi, who was a great admirer of Spemann, this was the main, if not the only, merit of the article. I confess that I started reading it with only lukewarm interest, being more fascinated by the beauty of the passing scenery than by the account of the changes called forth by intervention in the developing nervous system of the embryo; but the crystal clarity of the writing slowly prevailed over my admittedly never-too-strong feelings for rural life. I could certainly not have anticipated on that day the far-reaching consequences of this first encounter with the chick embryo—and with Viktor. Though I was to meet the former in the flesh on subsequent days, seven years were to elapse first—and many extremely traumatic experiences were to be endured—before I actually met Viktor. By that time, I had learned a great deal about the chick embryo though nothing about Viktor, who remained for me, until the day of our first personal encounter, "one of Spemann's pupils," according to Levi's definition which spelled out in these three words his admiration for "Spemann the Master"—and his concept that a pupil is simply a master's hand with no brain of its own.

OCTOBER 10, 1947

Was "The Spirit of St. Louis," the train that took me from New York to St. Louis on that mild October day in 1947, as comfortable and luxurious as it has remained forever imprinted in my memory? Or did I remodel it according to my joyful anticipation of this long-dreamed of experience in the new world?

It was about 4:00 P.M. when I arrived at the Union Station and asked a kindly taxi driver to take me to Washington University. On the way to the university the proud St. Louis born driver pointed out to me the beauty of the city and first of all the three-million-dollar cathedral, which I admired as we drove up Lindell Avenue. According to my driver, it was even larger and more beautiful than St. Peter's in Rome! At the entrance of Rebstock Hall, which housed the Zoology Department, a slender, tall young woman in a lab coat met me; her blond hair was combed in tresses around her austerely erect head. She muttered a few words, which I did not understand and she motioned me to the upper floor. There, standing in the library, was Viktor, consulting a book. "Dr. Levi-Montalcini has arrived," Florence Moog announced briefly and went back to her laboratory. Viktor's gaze took in my slight Mediterranean frame and rested with some amusement on my hands, which were moving in an effort to convey what I could not express in my broken English. He understood that I was thirsty and showed me how to press a button on a water-cooler fountain, which was placed unobtrusively in a corner of the corridor. The jet of ice-cold water springing into my dry mouth anticipated many similar and far more valuable experiences that I was to enjoy in the days and years to come.

Viktor then invited me to his office. Sitting in front of him, with the disadvantage of being confronted by his six-foot two-inch, gentle and imposing figure partly neutralized by the chairs, I could now submit him to the same objective examination that he had performed on me following Florence's introduction. I never found out if I had passed his test; he certainly had mine. I realized, with no uncertainty, that I had landed in the right place, a place I had dreamed of ever since the turbulent days of the war when, in the small confines of my bedroom-laboratory, I had gone over again and again the pages of the article which I had first read in that cattle car, comparing my own findings with those described there. While I was lost in these thoughts and savoring in advance the pleasure of working in this novel and attractive environment that I loved at first sight, Viktor was glancing at me with a friendly smile, which further encouraged my hopeful expectations.

I gladly accepted his invitation to walk with him to his home, to meet his wife and two daughters. The leaves on the trees of the Washington University campus were a melange of gold-red colors shining in a glorious midwestern sunset that I shall never forget. Many similar sunsets were to follow, but none in my memory is so brilliant as that one, the first that I saw while we were walking across the campus. Martha Hamburger received me in a kind, informal way; I was attracted by her open and sympathetic approach to the frail foreign woman I was. At the same time, I sensed immediately the extreme tension that was soon to disrupt, like a hurricane, the apparently peaceful course of her life. Doris, a pretty 17-year-old-blond teenager, did not pay too much attention to me and cheerfully took leave of her parents to spend the evening with a boyfriend who was waiting for her at the door. The spirited and beautiful nine-year-old Carola, who had just washed her dark hair and combed it like a turban on her head, threw a rather disapproving glance at me and did not care to sit at the dining table with a person so ignorant as to be unable to understand and speak English. After dinner, Viktor drove me to Delmar Boulevard where he had found me a room in a boardinghouse run by a middle-class, midwestern family. There I would come in contact, he told me, with the genuine all-American natives, with all their virtues and their shortcomings. The husband, a member of the Lions' Club, as his wife proudly informed me, was paralyzed and aphasic and uttered only a few unintelligible sounds from his armchair. The wife was a stocky, gray-haired woman who was happy to exchange words with a foreigner. She was, as I understood, a conscientious churchgoer and a fervent Republican. She also hated Jews, the pests of the world. But I was too happy and too tired to be disturbed by her beliefs, which differed from mine. I told myself that if half of Viktor's predictions had materialized in such a surprisingly short time, the others would also come true. They did, and in fact they greatly outnumbered the "first half."

ONE MONTH LATER

Barely a month elapsed from that memorable tenth of October 1947 when I first became aware of one of the finest traits in Viktor's character, one I was to appreciate time and time again in the future. This was his whole-hearted joy over fortunate turns in a colleague's or pupil's research, and his instant grasp of possible new avenues in their work.

It was a November afternoon. I had taken possession of my workbench and a microscope in an office close to Viktor's and I was inspecting, in a somewhat cursory way, my latest serially sectioned and silver-stained four-day to six-day chick embryos. The silver reaction was perfect. The motoneurons in the spinal cord stood out against the pale yellow background of the dense matrix of the neural tube, as if etched with India ink by a very gifted craftsman. In the cervical segment of the spinal cord of four-and-a-half-day embryos, dead cells exhibiting a pitch-black color and a round shape and cell debris were scattered in large numbers among a few healthy nerve cells. To my mind, still imbued with the memory of the war, it was a battlefield strewn with corpses. But in five-day and six-day embryos, the dead bodies had been removed by a process that I discovered in the following days. The residual healthy nerve cells had gathered together and built a slender motor column much thinner than the corresponding column at the other segments of the spinal cord. In the thoracic segment of the four-and-a-half-day embryo, I contemplated an entirely different scene. The ventral motor column consisted of two distinct nerve cell populations: one compact, small group was positioned in the ventrolateral aspect of the tube while the other consisted of loosely assembled nerve cells at an early stage of differentiation. Their very intense affinity for silver revealed their elongated fishlike silhouette; the long axis of their bodies was directed toward the midline. The microscopic landscape was teeming with life in sharp contrast to the panorama offered by the cervical segments. In five-day and six-day embryos the elongated nerve cells had moved farther away from their early lateral location and now appeared either as isolated units or assembled in rows oriented between their earlier position and the mediodorsal edge of the still-elongated central canal. Each cell exhibited a thin, apical, antennalike thread and trailed behind it a long filament that I identified as the axon.

I saw in these apparently random and chaotic cell movements a highly organized process remarkably similar in its overall pattern to the displacement of hundreds of moving units along a preestablished route under the control of an invisible headquarters. Like soldiers in a war game or like armies of ants or termites, the neuroblasts moved mostly in compact columns, each slender spindle-shaped cell following the other. From my previous training with Giuseppe Levi in Turin, I recognized the migratory cells as belonging to the preganglionic visceral nucleus first identified by Terni in older embryos where they form two small cell aggregates located near the central canal. An inspection of the brachial and lumbar spinal cord segments showed that at these levels nerve cells of the motor column underwent neither massive death as in the cervical segment nor migratory activity as in the thoracic segment. They aggregated in two groups, a larger and a smaller one that went through a steplike uneventful differentiative process *in situ*. The lateral group gave origin to robust fibers that assembled in large nerves and innervated the limb primordia, while the medial group produced thinner nerves that innervated the axial musculature. At the sacral level a similar migratory process, even if less dramatic and extensive than at the thoracic level, was also apparent.

I sketched on a piece of cardboard what my imagination—more than my eyes—had seen in the developing spinal cord and in a state of great excitement I knocked at Viktor's door. Viktor leaned over my sketch and then over the microscope and became instantly infected by my enthusiasm. "*This,*" he said, "is a breakthrough in the understanding of neurogenetic processes."

To explain the massive cell death in the cervical spinal cord segment, I submitted the hypothesis that the cells doomed to death belonged to an abortive visceromotor column in the cervical region. (Later, this hypothesis received support from the experimental studies of one of our students who transplanted the cervical segment in place of the thoracic segment of the neural tube in two-day embryos [Shieh, 1951].) It had taken only half an hour to visualize the elaborate development pattern of the spinal cord, which proved to be similar if not identical to that which I later found in the mammalian spinal cord, but it took almost a year before I had collected all pieces of evidence through laborious and time-consuming staining and reconstructions performed on the spinal cord fixed at short time intervals between the third and seventh day of incubation and sectioned in transversal, frontal, and sagittal planes.

UNDER THE PORCH: 6060 WASHINGTON AVENUE

During that humid and hot summer of 1948, three of us used to sit in the shade on the porch of the Hamburger's pleasant residence at 6060 Washington Avenue: an old gentleman, whom Viktor spoke to most of the time in German and called "Vater" even when he switched to English; Viktor; and I. No one could have doubted the identity of the gentle old man. In spite of his rather short stature—he was barely taller than I—the penetrating and friendly gaze of his eyes, the fine features of his nose and mouth, which betrayed his Jewish descent, and, most of all, the dry witty humor that flavored his always benevolent remarks, were all too similar to Viktor's look and manner not to reveal at first sight the bond of kinship. Martha had left with Carola to spend the summer in Germany; Doris was also away from St. Louis, and Viktor was in charge of the house. He took care of it in the same relaxed and competent way he did the Department of Zoology, of which he was chairman.

Most evenings I shared the frugal dinner with Viktor and Vater, and then, while old Mr. Hamburger read one of his beloved classic German books and looked fondly at his son, Viktor and I spent the evening sitting around a table on the porch, fully absorbed in the process of writing two papers: one on proliferation, differentiation, and degeneration in the spinal cord of the chick embryo under normal and experimental conditions (Hamburger and Levi-Montalcini, 1949); and the other on the origin and development of the visceral system in the spinal cord of the same embryos (Levi-Montalcini, 1950). The latter was the result of my laborious reconstruction studies mentioned above. In those evening hours I realized that writing a scientific paper can be even more time consuming, though no less rewarding, than working at a laboratory bench. And I experienced again the same pleasure that had decided my future career as when I was struggling to keep my balance while holding tightly onto an iron bar in that cattle car and at the same time reading the article lent to me by Giuseppe Levi. If Viktor had contented himself in that paper with a description of the bare facts, if he had merely stated that the ablation of a limb bud in the embryo entails the depletion of nerve cells in the spinal cord and the dorsal root ganglia innervating the limb, I would have probably paid only passing attention to his description of this neurogenetic process and pursued further the analysis of the developmental behavior of the chick embryo, which had already absorbed two years of my attention. It was, however, his sharp, talmudiclike approach to the problem, that awakened my curiosity and spurred my desire to repeat the experience. The crystal clarity that had impressed me in Viktor's writing was even more sparkling and far more enjoyable in the endless exchange of views that took place as we discussed the results of experiments I had performed along the same line in my secluded one-room laboratory, which had been built Robinson Crusoe fashion in a village house near Turin where we had taken refuge during the war. The evening hours that I enjoyed on the porch at 6060 Washington Avenue were the first of many dividends from the time spent in those simple surroundings.

THE OPENING OF THE NGF SAGA

One morning in the fall of 1949 Viktor called me to his office and showed me an article written a year earlier by one of his former students, Elmer

Bueker. Bueker (1948) had performed the now well-known and celebrated experiment of transplanting fragments of three mouse tumors into the body wall of three-day chick embryos. One of the tumors, a Rous sarcoma, produced extensive hemorrhage and the death of the host, another was reabsorbed, and the third not only became established but underwent vigorous growth. Five days later, embryos bearing transplants of this last-mentioned tumor, a mouse sarcoma known as sarcoma 180, were sacrificed, sectioned serially, stained with hematoxylin, and inspected under the light microscope. At the level of the transplant, Bueker saw that sensory fibers from adjacent dorsal root ganglia had gained access to the tumor and that the ganglia of origin of these fibers appeared larger than the contralateral ganglia innervating the limb. Bueker submitted the hypothesis that chemical properties of the sarcoma favored the entrance of sensory nerve fibers and their branching among neoplastic cells. This peripheral effect in turn was responsible for the overgrowth of the ganglia confronted with a larger than normal area for innervation.

This finding struck me as being of extraordinary interest, and Viktor and I decided, after having received Bueker's consent (he was at that time no longer investigating this effect), to repeat his experiments. Shortly afterwards, Viktor left St. Louis for a semester in Cambridge, where he had been invited by F. O. Schmitt of the Massachusetts Institute of Technology to help develop a new curriculum in biology. During this time, the first mice to bear transplants of mouse sarcoma 180 were shipped from the Jackson Memorial Laboratory and reached Rebstock Hall. Through their small, inquisitive red eyes they were busy cheerfully inspecting their new surroundings.

The Tumor's Effect

A cursory examination of chick embryos bearing grafts of mouse sarcoma 180, sectioned serially and stained with the specific silver technique, persuaded me of the correctness of Bueker's report. Sensory nerve fibers had branched in the tumor, and the sensory ganglia that gave origin to these fibers were larger than the contralateral ganglia innervating the limb tissues. A careful inspection of an extensive series of experimental embryos sacrificed between the end of the 5th and 15th day of incubation called my attention to some additional and most unusual features of the tumor's effects: (1) of the two neuronal cell populations present in the sensory ganglia, only the mediodorsal population appeared enormously enlarged; (2) sympathetic ganglia underwent an even more dramatic increase in volume than sensory ganglia as proven by the fact that in older embryos they reached a volume about sixfold larger than the contralateral ganglia, an overgrowth that had no parallel in any previously observed effect produced by transplantation of embryonic tissues; (3) motor nerve cells in the lateral horn of the spinal cord at the level of the tumor transplant not only did not participate in its innervation but also underwent a striking atrophy as if the tumor had prevented their further differentiation and growth; (4) sympathetic and sensory fibers closely intermingled with each other and branched profusely among neoplastic tissues without, however, establishing synaptic connections with the cells.

I kept Viktor informed weekly of the progress of my studies and of my growing interest in this extraordinary effect. Upon his return to St. Louis in the spring of 1950, Viktor shared my enthusiasm and my belief that the growth response elicited by the tumor differed in many respects from those called forth by supernumerary limbs. And yet, even conceding that the effects of mouse sarcoma 180 were much more extensive than those elicited by transplantation of additional limbs and differed in many respects from the former, we were still not prepared to see in this response any flagrant deviation from normality, so difficult is it to evaluate with an unprejudiced mind results that do not fit into preconceived schemes. These findings are reported in our joint article published in 1951 (Levi-Montalcini and Hamburger, 1951).

In the spring of 1951 we received from the Jackson Memorial Laboratory a fortuitous and fortunate shipment of mice that bore transplants of either sarcoma 180 or sarcoma 37. (The two tumors exhibit almost identical morphological features, and both have been carried for many years in the laboratory after originating from two mouse mammary carcinomas.) When transplanted into chick embryos, sarcomas 180 and 37 exhibited more invasive and more chaotic growth patterns than those used previously. So new were the features of the growth response elicited by these tumors that they called for a revision of the hypothesis submitted earlier. In embryos bearing transplants of one or the other of these tumors, I saw, to my astonishment, that the viscera were filled with robust nerve fiber bundles emerging from the greatly enlarged prevertebral sympathetic ganglia. Some of these nerves had even gained entrance into the veins and obliterated the lumen of these vessels with large neuromas. Since this aberrant and highly atypical peripheral distribution of sympathetic nerve fibers was also apparent in organs and in parts of the circulatory system remote from the tumor, such as the head region, I submitted the hypothesis that this effect was due to the release of a humoral growth factor by neoplastic cells into the embryonic circulation.

This hypothesis of a humoral growth factor was fully confirmed by transplants of fragments of mouse sarcoma onto the chorioallantoic membrane of four-day chick embryos. In such a position the embryonic and neoplastic tissues shared the circulation but there was no direct contact between them. In these embryos the same extraordinary effect materialized: sympathetic nerve fibers produced by the hyperplastic and hypertrophic sympathetic ganglia massively invaded the embryonic viscera and entered the vessels. In describing this phenomenon some months later, while I was still under the fascination of this stupendous, newly discovered effect, I stated "all barriers seem to have broken down and the organs surrender to the invading fibers" (Levi-Montalcini, 1952). The effect was described in greater detail in a joint paper with Viktor (Levi-Montalcini and Hamburger, 1953).

The next, and most difficult, problem was to identify the hypothetical "tumor factor." I realized that this was a prohibitive task with little or no chance of success if studies in the developing embryo, and I conceived the idea of exploring it *in vitro.* In Turin I had become familiar with the tissue culture technique while preparing my M.D. thesis under Prof. Giuseppe Levi. In that laboratory I had met, and become a close friend of, Hertha Meyer, a former valiant assistant of the tissue culture expert, Emil Fischer in Berlin. Hertha had joined Professor Levi in Turin after Hitler came to power and had moved to Rio de Janeiro shortly before the beginning of the war. In Rio she was in charge of the Tissue Culture Section of the Institute of Biophysics directed by Professor Carlos Chagas. After a short exchange of letters with Hertha and Professor Chagas and with a travel grant by the Rockefeller Foundation, I boarded a plane for Brazil in the late fall of 1952.

Letters from Brazil

Not long ago, Viktor generously returned to me, although with regret as he avowed, the 11 letters that I had sent him from Rio de Janeiro, together with the collection of my India ink drawings depicting the *in vitro* effect of mouse sarcoma on sensory ganglia dissected out from eight-day chick embryos and cultured for 24, 48, and 72 hours in proximity of the neoplastic tissue. In control cultures I had combined the same ganglia with normal embryonic chick or mouse tissues (Fig. 2-1). The neatly handwritten letters, each four to five pages long, which spanned the entire period that I spent in Brazil (October 9–December 29), had been kept safely since 1952 in a drawer of Viktor's desk. They exude either enthusiasm or desperation, depending on the results of the day that favored or discouraged my hypothesis that the "halo effect" produced by ganglia facing fragments of mouse—I first used this term to describe the tumor's effect in my letter of December 15—was due to the same factor that caused *in vivo* the overgrowth of sensory and sympathetic ganglia and the massive invasion of viscera by fibers produced by the sympathetic nerve cells. I had in fact discovered, to my great disappointment, that mouse tissues, unlike chick tissues, also enhanced to a moderate extent nerve fiber outgrowth from the same ganglia *in vitro.* This finding seemed to me to cast some doubt on the specificity of the "halo effect" elicited by the tumor. Six years would elapse before the explanation of

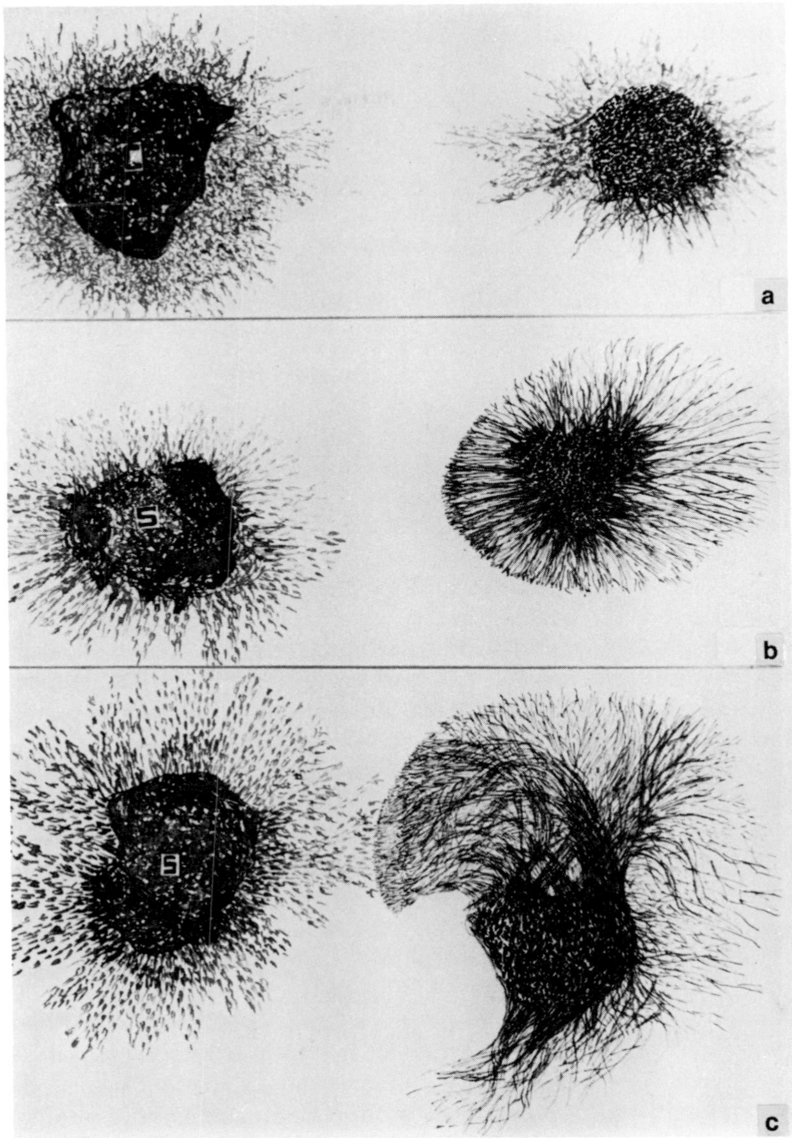

Figure 2.1 India ink drawings sent to Viktor Hamburger from Rio de Janeiro in December 1952, illustrating the "fibrillar halo effect" elicited by fragments of mouse sarcoma 180 on sensory ganglia from eight-day chick embryos. (a) indicates control culture of a fragment of a chick embryo's heart (left) and sensory ganglion, after 24 hours *in vitro.* Fibroblasts and very few nerve fibers are produced by the ganglion on the first day of culture.

(b) and (c) show 24- and 48-hour cultures respectively of fragments of mouse sarcoma 180 (left) and sensory ganglia: (b) typical fibrillar halo produced by the ganglion facing an explant of the mouse sarcoma after 24 hours of culture; (c) the neurotrophic effect elicited by the mouse tumor on nerve fibers produced by the sensory ganglion.

this mysterious and perturbing "mouse effect" was forthcoming. However, after examining more than 600 cultures and comparing the effects of the mouse tumor with those called forth by normal mouse tissues, I became convinced that the fibrillar halo produced by ganglia facing the tumor or cultured in a medium enriched with a tumor extract was due to the still mysterious growth promoting factor released by the neoplastic cells (Levi-Montalcini et al., 1954). My last letter from Rio de Janeiro to Viktor dated shortly before leaving Brazil for Peru and Ecuador, for what I considered a well-deserved reward, proclaims my joy in less than subdued tones.

Back in St. Louis

On December 15 I had written to Viktor: "I am very glad that you hired a biochemist by the name of Stanley Cohen. From what you write, I believe that he is the right person to help us to solve this difficult problem [the identification of the tumor growth promoting factor]."

Few statements could have been more prophetic than this. In contacting Stanley, Viktor's flair for finding the right person at the right moment showed itself at its best. In the case of Stanley, or Stan as we called him, the choice proved to be perfect almost beyond expectation. Stan, then a young biochemist who had earned his Ph.D. in biochemistry at Columbia University and had recently completed an investigation with Martin Kamen, who was at that time in St. Louis, was available for adoption. Viktor had informed him about my discovery of the "halo effect" elicited by mouse sarcoma *in vitro,* and he was willing to accept the challenge of identifying the mysterious tumor factor. This happened soon after my return to St. Louis in January 1953. One year later an article by Stanley, myself, and Viktor brought the news that the nerve growth-stimulating factor (later to be shortened to nerve growth factor or NGF) was a nucleoprotein fraction isolated from mouse sarcoma 180 (Cohen, Levi-Montalcini, Hamburger, 1954). With the subsequent discoveries in 1956 and in 1958 that much larger quantities of a protein molecule endowed with identical properties as the mouse factor can be extracted from snake venom and mouse salivary glands, it was realized that NGF was a protein and not a nucleoprotein (Cohen and Levi-Montalcini, 1956; Cohen, 1958; Levi-Montalcini, 1958).

Before these two memorable events, which opened the way to the chemical characterization of NGF, Viktor, without making any mention of his intention, slowly withdrew from the scene. The reason was not that the problem had lost its fascination for him. On the contrary, I believe that he had never been more thrilled than by the new results that came each day and by the expectation of those that were to come. But, he felt that his pupil needed to come out from beneath the shadow cast by his personality and his widely established reputation.

This decision, which I admired and regretted at the same time, did not bring to an end our association—indeed, it remained unaltered until I left Washington University in July 1977, but it did set Viktor on a different trail.

THE LONG JOURNEY INSIDE THE DEVELOPING NERVOUS SYSTEM

While NGF's uncharted route was slowly unfolding and alternate waves of euphoria and depression swept the first and second floors of Rebstock Hall, Viktor on the third floor became engaged in a grandiose project that was to absorb all his time and thoughts from the early 1960s to the present day. What in previous years had been a somewhat restricted approach to only a few of the hundred problems raised by the developing nervous system became a massive frontal attack carried on for two decades with unrelented vigor at all levels from nerve cells to neuronal circuits and behavior.

It is fortunate for the historian (for I see myself in this role) that Viktor himself in his most lucid style has reviewed many times the most relevant findings of these investigations, which touch upon practically all aspects of developmental neurobiology (Hamburger 1962, 1963, 1968, 1970, 1975a, 1975b, 1977, 1980). But although the present account owes much to his own elaboration of the significance of these studies, I shall attempt to convey to the reader the overall view of this scientific adventure from my vantage point as historian and as an eye-witness of it. Valiant participants in its different moments were: M. Balaban, A. Beckoff, F. J. Brunso-Bechtold, J. Decker, M. Hollyday, C. H. Narayanan, R. Oppenheim, R. Provine, S. Sharma, R. Skoff, P. Stein, and E. Wenger.

Beginning of the Journey

It was with nostalgic feelings, and the revival of emotions of a long-gone past, that I saw Viktor set himself to a task that I had considered as my life goal about 20 years before falling in love with the humoral nerve growth factor.

This early phase of my scientific activity had taken place in the basement of the Clinic of Neurology and Psychiatry of the University of Turin. At that time I had joined a brilliant young neurophysiologist and clinician, Fabio Visintini, in the project of exploring the development of motility of the chick embryo at the structural, physiological, and behavioral levels. We had proven to our great satisfaction that the motility of three-and-a-half-days to four-day chick embryos is neurogenic rather than myogenic in nature as claimed by other investigators, and we had also shown that the motility of four-day to seven-day embryos results from the spontaneous activity of spinal motor neurons and is not reflexogenic in nature. This investigation, which I deeply enjoyed and which was the object of a long article published in a Swiss journal (Visintini and Levi-Montalcini, 1939), came to an abrupt end when Mussolini launched his racial campaign barring non-Aryans from access to academic institutions. It was at that time that Viktor came to my rescue with his 1934 article, which not only spurred my curiosity but also suggested exploration of a problem that I could investigate in my bedroom-laboratory without having to resort to pieces of equipment, such as the oscilloscope, which were no longer available to me. Fortunately, our Swiss article came to the attention of Viktor, and I like to believe that it impressed him almost as much as his 1934 article impressed me in 1940. Leaving this point unsettled, it remains certain that as my interest switched from problems of behavior to the study of exogenous factors influencing nerve cell growth and differentiation, Viktor's interest two decades later shifted in the opposite direction. The results of his remarkable investigations, reported in several articles that appeared between 1963 and 1967 (Hamburger, 1963, 1964; Hamburger and Balahan, 1963; Hamburger et al., 1965; Hamburger, E. Wenger, and R. Oppenheim, 1966; Hamburger and Oppenheim, 1967; Decker and Hamburger, 1967; Bekoff, P. S. G. Stein, and V. Hamburger, 1975), will remain landmarks in the field of behavioral studies, combining as they do the rigorous tools of analysis, sharp thinking, and a master's skill in extracting general principles from seemingly simple and, to other eyes, unrevealing performances of the chick embryo, the object of Viktor's love. In rereading these articles I became equally impressed by the punctilious description of every minute detail of the embryo's early motility, which I remembered all too well from my past experience, and of the subsequent sequence of events, which climax in the prehatching and hatching behavior. His superb use of the English language conveys to the reader the exact meaning of every performance, or the "lack of performance" in the cyclic repertory of activity and inactivity phases. I would not hesitate to define these studies as classical examples of scientific thinking and writing, were I not discouraged by the witty definition that Viktor gave to the word "classical." "Classical," writes Viktor, "refers to a period of major achievement in a branch of science or of other human endeavor; a period in which a field has reached maturity and gained considerable impact on ways of thinking. Classical is nowadays also used occasionally in a derogatory way, meaning old-fashioned, kind of dusty, too far removed from present-day frontiers to be of much interest" (Hamburger, 1972). Later on, I shall make further comments on the significance of these studies in the more general field of neurobiology.

Unveiling the Secret of the Neuron-to-Muscle Relationship

If the analysis of the early motility of the chick embryo and of the prehatching and hatching performances impresses me for its precision, insight, and ability to unravel the intricate maneuvers of the embryo that enable it to escape from the shell, I confess my bias in favor of the more recent frontal attack on one of the fundamental problems in neurogenesis; the building of peripheral and central neuronal circuits. This problem, which is unique to the nervous system, has become in recent years the object of innumerable investigations carried out simultaneously in different laboratories and on different levels from the cellular to the behavioral.

In the case of Viktor and of his young associate, Margaret (Peggy) Hollyday, the approach was that of the experimental neuroembryologist, that is, the analysis of normal embryos and of embryos bearing supernumerary leg transplants. They were submitted to injections of horseradish peroxidase (HRP) into individual muscles, allowed to survive for a few hours, and then processed for identification of the nerve cells, which had become labeled through retrograde axonal transport of HRP. The problem that Viktor and Margaret set themselves to solve was to identify the source of origin of motor nerves that innervate individual muscles in the supernumerary transplanted leg. The test object was of course the chick embryo. First, the researchers ascertained that a leg grafted rostrally to a host leg is endowed with spontaneous motor activity and does not differ in its morphological appearance from its host leg. Then, they proceeded to identify in the fully functional transplants the origin of the nerves innervating the gastrocnemius, an ankle extensor that is normally innervated by a motor pool located in the caudal segments of the lateral motor column. As for the same muscle in the transplant, its innervation might have been provided by any one of three possible sources: from the same motor pool as the homologous muscle of the host leg; from neurons distributed at random in the lateral motor column; or from a nongastrocnemius pool of neurons that would either be taken over entirely by the supernumerary muscle or would innervate both the transplant and its own host leg muscle. The motor pool that gives rise to the innervation of the transplanted gastrocnemius was identified by using the method of retrograde axonal transport of HRP. The results unequivocally favored the third hypothesis. In all six cases studied by Peggy and Viktor the nerves to the transplanted gastrocnemius originated in the same rostral motor pool, located medially in the lateral motor column and separated from the more caudally located pool of neurons innervating the host gastrocnemius. Since the rostral motoneurons innervating the gastrocnemius in the supernumerary leg would have normally supplied a thigh muscle (as ascertained from a preliminary HRP mapping of the motor pools in the lumbar motor columns) the results gave unequivocal evidence for a change for the destination of nerve fibers when they are faced with supernumerary muscles. "If prespecification exists," states Viktor in his Schmitt's Lecture, "it is modifiable under experimental conditions" (p. 18). These findings are in agreement with Sperry's chemoaffinity theory (1963), which was designed originally to explain selective synapse formation, and was later extended to the patterning of nerve pathways (Sperry, 1965).

AN OVERVIEW OF VIKTOR'S MAIN CONTRIBUTIONS TO DEVELOPMENTAL NEUROEMBRYOLOGY

A comparison of Viktor's frequently quoted article of 1934 with his article of 1977 on the identification of motoneurons innervating individual muscles in the host and in the supernumerary leg of chick embryos offers a panoramic view of the refinement in techniques and of the markedly increased sophistication that took place in these four decades in approaching problems of developmental neurobiology. The comparison affords an appraisal not only of the progress that occurred during this period but also of the changes in concepts and goals of these two scholars who are similar, yet different. Conceding that the low analytical resolution power of the ablation experiments of 1934 did set limits, and rather severe ones, to the field of inquiry, we should at the same time acknowledge that the delineation of these boundaries was also one of the major merits of this investigation, which prepared the ground for the more sophisticated studies that followed.

A major drawback in the approach to neurobiological problems during the 19th and early 20th century on the part of researchers dealing with the problem of neurogenesis or with the immensely more complex brain–mind problem was their naive attempt to find "the answer" by resorting to philosophical speculations or to the use of crude and fully inadequate experimental procedures. When Viktor undertook the study of the effects of limb ablation on the sensory and motor nerve centers providing its innervation in the chick embryo and, at the same time, tackled the more ambitious goal of uncovering the mechanism of action of peripheral areas of innervation on nerve centers, he had chosen a problem that had already been the object of rather extensive explorations in amphibian larvae. Besides its already-mentioned rigorous analytical approach—far superior to that of preceding studies, the major merit of Viktor's article was to have analyzed this effect in detail and in depth without attempting to force the results into preconceived schemes. Another merit was the choice of the object of investigation: the chick embryo in preference to the amphibian larvae. Although the former is more difficult to manipulate than the latter, its nervous system is far more amenable to a detailed analysis in view of its more elaborate organization and segregation into well-defined nerve cell pools. An additional advantage that Viktor did not exploit at the time, but one that was fully appreciated in subsequent years, is the high affinity of the avian nerve cell populations to silver, which permits the visualization of these cells and of their axons to an extent unattainable in the amphibian nervous system.

Subsequent to 1934, Viktor studied the problem of the effects of exogenous and endogenous factors in the differentiation of sensory and motor spinal systems over and over again with progressively more-refined techniques between 1939 and 1958 (Hamburger, 1939; Hamburger and Keefe, 1944; Hamburger, 1948; Hamburger, 1958). In these papers he focused his attention mainly on the analysis of the occurrence of normal cell death and its possible cause in developing spinal motoneurons.

In 1962 Viktor took a leave of absence from problems of nerve cell growth and differentiation to concentrate all his efforts on behavioral studies, which will be commented on below. It was not until 13 years later, in 1975, that he found the time had come to reinvestigate, with more subtle tools and with a refreshingly new and more sophisticated approach, the problem of neuron-to-target cell interaction during neurogenesis. His renewed interest was spurred by two major developments that had taken place in the meantime in neurobiological sciences. The first was the tremendous impact of Sperry's chemoaffinity theory in the field of experimental neurogenesis; the second was the availability of new histochemical techniques with a much higher resolution power than those available only a decade earlier. Last but not least was Viktor's fortunate association with a brilliant young postdoctoral scientist, Peggy Hollyday, to whom goes part of the credit for the merit of these investigations.

At first Viktor and Peggy prepared the ground for a frontal attack on the above-cited problem by undertaking a systematic study of the "rescue effects" called forth by an enlarged peripheral field of innervation on nerve cells otherwise doomed to death owing to failure in the competition for synaptic sites or other trophic factors released by postsynaptic target cells (Hollyday and Hamburger, 1976). The second preparatory study consisted of an autoradiographic determination of the time of origin of the lateral motor column of the chick embryo spinal cord (Hollyday and Hamburger, 1977). Perhaps even more important than their own investigations for launching this new project were the most accurate and valuable electrophysiological mapping studies of the segmental source of innervation of many of the major muscles in the chick embryo's leg by Landmesser and Morris (1975).

The results of these studies, already summarized briefly in a previous section, brought new support to the chemoaffinity theory, which had been developed first on entirely theoretical grounds in the rather remote field of behavioral studies but had resulted in a new insight into the respective roles of prespecification and plasticity in neurogenesis. Peggy and Viktor coined the fortunate expression "selective mismatching" to designate the connection of particular motor clusters in the lateral motor column of the spinal cord with given muscles of supernumerary legs. In his outstanding article (1980b), Viktor conceives of the chemoaffinity hypothesis in terms of a hierarchical order of specifications. At successive points of their peripheral branching on their way to individual muscles, the nerve fibers are provided with increasingly more specific cues.

If the article of 1977 opens an entirely new chapter in the field of developmental neurobiology rather than brings to completion an investigation started 43 years earlier, the studies Viktor pursued between his first and his more recent period of research activity, as already mentioned, were devoted to the study of the development of behavior of the chick embryo and to a lesser extent to that of the mammalian fetus (Hamburger, 1975). "What has impressed me most in all phases of our investigations," wrote Viktor in his lecture in honor of F. O. Schmitt," is the primacy of activity over reactivity or response. This, to me, has become symbolic of animal life, in general. The elemental force which embryos and fetuses express freely in their spontaneous activity, sheltered as they are in the egg or uterus, has perhaps remained throughout evolution, the biological mainspring of creative activity in animals and man, and autonomy of action is also the mainspring of freedom" (Hamburger, 1975). Translated into structuralist terms, this captivating and imaginative suggestion agrees with one of the fundamental tenets of this school of thought, namely, the existence of innate ideas and of knowledge without learning, or of mental activity independent of, and neither determined by nor ancillary to, sensory information.

In an overall view of Viktor's contributions it is impossible not to mention his scholarly essays that reveal no less than his professional achievements the breadth of his humanistic as well as scientific interests. The essays on Goethe, on Cajal, and on Malpighi, and the very recent one on his friend Roger Sperry, all written in Viktor's inimitable style, belong to the field of art as well as of science. Very few scientists have either his magic touch or his ability to find the time to indulge in the luxury of contemplating and of understanding the past and present accomplishments of other scientists.

FORTY YEARS LATER—NOT A FAREWELL

It was a summer day when we first met in that railway cattle car. The insolent scarlet red of the poppies in the corn fields was competing for my attention with your crystal-clear account of the effects of limb ablation on the sensory ganglia and the motor spinal system of the chick embryo. Not too far from the train, the war still raged on the French front and the fate of Western civilization was dangerously threatened by the triumphant advance of the new German barbarians, as it has been so often in the past by other hordes of barbarians on the battered lands of the Old Continent.

But this dark hour also passed away, and the poppies are blooming again in the fields of Europe and elsewhere in an atmosphere that is apparently calm but in reality deeply perturbed, in this summer of 1980. At present, you are pursuing the same train of thoughts as you have for all these past years, there in the pleasant surroundings of your office on the Washington University campus or on the porch of your residence at 740 Trinity Avenue, one that differs only in minor details from the porch at 6060 Washington Avenue, less than a mile to the east.

Your hair has grown whiter than when we first met 33 years ago (Figs. 2-2 and 2-3), but your thinking (as well as the way of conveying it) has gained ever greater precision and depth—a living challenge to the concept that equates aging with loss of brain power. Your curiosity in regard to the endless mysteries that surround us, from those amenable to the analytical approach that you love, such as the hatching behavior of the chick embryo, to those that are not, such as the mind–brain problem, is if anything sharper than it was when we first met. At the same time, your tolerance of human weakness and your loyalty to your numerous friends are even stronger than they used to be. Likewise, the flavor of your dry, benevolent humor, which I appreciated from the day of my first encounter with you and with "Vater" is even more appealing to me now than when I first tasted it. Your approach toward mankind at large and toward the scientific community in particular is colored with compassion, an approach that I love all the more since it is seldom, if ever, found in fellow scientists, absorbed as they often are in self-assertion and in the race for success, which leave neither much time nor room for living a decent and worthwhile life.

The feared years of old age came for both of us. But in your case the fall has the splendid red and gold colors of a midwestern sunset, which is perhaps more glorious, though less celebrated, than the green parade of spring.

And it does not matter that the probability of my coming back to St. Louis, like the pattern of my commuting—which I so much cherished—between our two continents is now decreasing exponentially and will eventually reach zero. St. Louis and our friendship are parts of myself as long as I live and there is, between ourselves, no farewell. Living in an Einsteinian world—and, even more importantly, being aware of it—

Figure 2.2 Photograph of some participants in the Conference on Genetic Neurology, organized by P. Weiss at the University of Chicago in March 1949. *Top row from left:* V. Hamburger, J. Piatt, R. Sperry, J. Z. Young. *Bottom row from left:* R. Gerard, H. Hyden, R. Levi-Montalcini, P. Weiss, F. O. Schmitt, J. Boeke.

Figure 2.3 V. Hamburger and R. Levi-Montalcini. Monsanto Laboratory, Washington University, St. Louis. May 1977. (Reproduced from the *Science Year,* [1978 Volume]. Field Enterprise Educational Corporation, Chicago.)

we also know that this world is devoid of significance. Space and time, present and future, have merged together in a unifying concept of this splendid and awesome mystery we live in, and if our emotionally loaded archaic brain still refuses to acknowledge it, our rational brain accepts it and willingly submits to these newly discovered, more flexible and yet even more impenetrable, rules. Viktor, we can therefore look at this world, at our friendship, at the past we so much enjoyed, and at the future, which may or may not materialize, in a *sub-specie-aeternitatis* frame of mind, in a crystal-clear atmosphere uncorrupted by the turbulence of human passions and sorrows.

REFERENCES

Bekoff, A., Stein, P. S. G., Hamburger, V. 1975. Coordinated motor output in the hindlimb of the 7-day chick embryo. *Proc. Natl. Acad. Sci. U.S.A.* 72:1245–48

Bueker, E. D. 1948. Implantation of tumors in the hind limb field of the embryonic chick end developmental response of the lumbosacral nervous system. *Anat. Rec.* 102:369–90

Cohen, S. 1958. A nerve growth-promoting protein. In *Chemical Basis of Development,* ed. W. D. McElroy, B. Glass, pp. 665–67. Baltimore: Johns Hopkins Press

Cohen, S., Levi-Montalcini, R. 1956. A nerve growth-stimulating factor isolated from snake venom. *Proc. Natl. Acad. Sci. U.S.A.* 42:571–74

Cohen, S., Levi-Montalcini, R., Hamburger, V. 1954. A nerve growth-stimulating factor isolated from sarcomas 37 and 180. *Proc. Natl. Acad. Sci. U.S.A.* 40:1014–18

Decker, J. D., Hamburger, V. 1967. The influence of different brain regions on periodic motility of the chick embryo. *J. Exp. Zool.* 165:371–84

Hamburger, V. 1934. The effects of wing bud extirpation on the development of the central nervous system in chick embryos. *J. Exp. Zool.* 68:449–94

Hamburger, V. 1939. Motor and sensory hyperplasia following limb bud transplantation in chick embryos. *Physiol. Zool.* 12:268–84

Hamburger, V. 1948. The mitotic patterns in the spinal cord of the chick embryo and their relation to histogenetic processes. *J. Comp. Neur.* 88:221–84

Hamburger, V. 1958. Regression versus peripheral control of differentiation in motor hypoplasia. *Am. J. Anat.* 102:365–410

Hamburger, V. 1960. Goethe and the natural sciences (unpublished)

Hamburger, V. 1963. Some aspects of the embryology of behavior. *Q. Rev. Biol.* 38:342–65

Hamburger, V. 1968. The beginning of co-ordinated movements in the chick embryo. In *Growth of the Nervous System, A Ciba Foundation Symposium,* ed. G. E. W. Wolstenholme, M. O'Connor, pp. 99–109. London: J. & A. Churchill

Hamburger, V. 1968. Malpighi the master. *Q. Rev. Biol.* 43:175–78

Hamburger, V. 1970. Embryonic motility in vertebrates. In *The Neurosciences: Second Study Program,* ed. F. O. Schmitt, pp. 141–51. New York: Rockefeller Univ. Press

Hamburger, V. 1972. Reflections on "classical" experimental embryology and on neurogenesis. Personal Philosophy of Science Series, Washington Univ., School of Medicine (unpublished)

Hamburger, V. 1975. Fetal Behavior. In *The Mammalian Fetus,* ed. E. S. E. Hafez, pp. 68–91. Springfield, Ill.: Thomas

Hamburger, V. 1975. Changing concepts in developmental neurobiology. *Perspec. Biol. Med.* 18:162–78

Hamburger, V. 1977. The developmental history of the motor neuron. The F. O. Schmitt Lecture in Neuroscience, 1976. *Neurosci. Res. Program Bull.* 15:1–37

Hamburger, V. 1980a. Trophic interactions in neurogenesis: A personal historical account. *Ann. Rev. Neurosci.* 3:269–78

Hamburger, V. 1980b. Prespecification and plasticity in neurogenesis. In *Nerve Cells, Neurotransmitters and Behavior,* ed. R. Levi-Montalcini, pp. 433–50 Amsterdam: Elsevier

Hamburger, V. 1980c. S. Ramón y Cajal, R. G. Harrison and the Beginnings of Neuroembryology. *Perspec. Biol. Med.* 23

Hamburger, V., Balaban, M. 1963. Observations and experiments on spontaneous rhythmical behavior in the chick embryo. *Dev. Biol.* 7:533–45

Hamburger, V., Balaban, M., Oppenheim, R., Wenger, E. 1965. Periodic motility of normal and spinal chick embryos between 8 and 17 days of incubation. *J. Exp. Zool.* 159:1–14

Hamburger, V., Levi-Montalcini, R. 1949. Proliferation, differentiation and degeneration in the spinal ganglia on the chick embryo under normal and experimental conditions. *J. Exp. Zool.* 111:457–502

Hamburger, V., Levi-Montalcini, R. 1950. Some aspects of neuroembryology. In *Genetic Neurology,* ed. P. Weiss, pp. 128–60. Chicago: Univ. of Chicago Press

Hamburger, V., Sperry, R. 1979. *Neurosci. Newsl.* 10:5–6

Hamburger, V., Wenger, E., Oppenheim, R. 1966. Motility in the chick embryo in the absence of sensory input. *J. Exp. Zool.* 162:133–60

Hollyday, M., Hamburger, V. 1976. Reduction of the naturally occurring motor neuron loss by enlargement of the periphery. *J. Comp. Neurol.* 170:311–20

Hollyday, M., Hamburger, V. 1977. An autoradiographic study of the formation of the lateral motor column in the chick embryo. *Brain Res.* 132:197–208

Hollyday, M., Hamburger, V., Fabris, J. M. G. 1977. Localization of motor pools supplying identified muscles in normal and supernumerary legs of chick embryos. *Proc. Natl. Acad. Sci. U.S.A.* 74:3582–86

Landmesser, L., Morris, D. 1975. The development of functional innervation in the hind-limb of the chick embryo. *J. Physiol.* (London) 249:301–26

Levi-Montalcini, R. 1950. The origin and development of the visceral system in the spinal cord of the chick embryo. *J. Morphol.* 86:253–84

Levi-Montalcini, R., 1952. Effects of mouse tumor transplantation on the nervous system. *Ann. N.Y. Acad. Sci.* 55:330–43

Levi-Montalcini, R. 1958. Chemical stimulation of nerve growth. In *Chemical Basis of Development,* ed. W. D. McElroy, B. Glass, pp. 646–54. Baltimore: Johns Hopkins Press

Levi-Montalcini, R., Hamburger, V. 1951. Selective growth-stimulating effects of sarcoma on the sensory and sympathetic system of the chick embryo. *J. Exp. Zool.* 116:321–62

Levi-Montalcini, R., Hamburger, V. 1953. A diffusible agent of mouse sarcoma producing hyperplasia of sympathetic ganglia and hyperneurotization of the chick embryo. *J. Exp. Zool.* 123:233–88

Levi-Montalcini, R., Meyer, H., Hamburger, V. 1954. In-vitro experiments on the effects of mouse sarcoma 180 and 37 in the spinal and sympathetic ganglia of the chick embryo. *Cancer Res.* 14:49–57

Shieh, P. 1951. The neoformation of cells of preganglionic type in the cervical spinal cord of the chick embryo following the transplantation to the thoracic level. *J. Exp. Zool.* 117:359–95

Sperry, R., 1963. Chemoaffinity in the orderly growth of nerve fibers and connections. *Proc. Natl. Acad. Sci.* U.S.A. 50:703–10

Sperry, R. 1965. Embryogenesis of Behavioral Nerve Nets. In *Organogenesis,* ed. R. L. De Haan, H. Ursprung, pp. 161–86. New York: Holt, Rinehart and Winston

Visintini, F., Levi-Montalcini, R. 1939. Relazione tra differenziazione strutturale e funzionale dei centri e delle vie nervose nell'embrione di pollo. *Archiv. Suisses Neurol. Psychiatr.* 43:1–44

3 / *Viktor Hamburger and Dynamic Concepts of Developmental Genetics*

SALOME GLUECKSOHN-WAELSCH

THE FIELD OF DEVELOPMENTAL GENETICS has undergone significant changes since the days when Boveri (1902), Wilson (1925), and Morgan (1934) first addressed the question of the relationship between embryology and genetics and stressed their interdependence in the analysis of numerous problems of biology. Prominent among these were problems concerning the role of chromosomes and genes in the control of normal and abnormal development, mechanisms of gene action during development, and the puzzling paradox between genetic equality of the mitotically derived embryonic cells, on the one hand, and the phenotypic diversity of the embryo's differentiated cell types, on the other.

The dynamic approach to the analysis of development which characterized experimental embryology in the first part of this century led to the formulation of concepts and ideas that strongly influenced studies of developmental genetics. By analogy with other areas of genetics, especially biochemical genetics, where mutations affecting biochemical processes provided tools for the analysis of normal biochemical pathways, the existence of developmental mutants in a variety of organisms was exploited towards the identification of normal mechanisms of development. This exploitation of developmental mutants was particularly useful when researchers dealt with organisms not easily accessible to experimental manipulation. In the interpretation of results obtained in studies of developmental genetics, the analytical insight gained by experimental embryologists played a guiding and decisive role. My own approach to problems of mammalian developmental genetics was strongly determined by my fortunate exposure to the environment of Spemann's Zoologisches

Institut in Freiburg im Breisgau where Viktor Hamburger was Privatdozent and where I spent my years as a graduate student.

I met Viktor Hamburger in 1928 when I first came to Freiburg in the hope of being able to study under Hans Spemann. Viktor was at that time Privatdozent in Spemann's Institute and among his responsibilities was that for the Grosses Praktikum, a six-day-a-week, all-day (and all-night), year-long laboratory course in zoology for graduate students. It was he who had to interview each student applicant and decide about admission to the course for which I had applied. Since the Professor himself was safely secluded behind closed doors, it was up to the Privatdozent to guide and instruct the graduate students, of whom there were only a handful. Spemann was available to the students only when he made rounds once a day and stopped at each student's desk for a brief exchange of questions and comments.

Therefore my introduction to experimental embryology (Entwicklungsmechanik), the new, fast-growing, and exciting biological science of that time, and to its concepts, problems, and methodology, came from Viktor Hamburger and our frequent, extended discussions throughout those years. The intellectual and scientific environment created by Viktor Hamburger succeeded in acquainting graduate students with the significance of concepts of cell and tissue interactions, and instilling in them curiosity concerning the role of such interactions in developmental mechanisms. Thus, embryology became for all of us a truly dynamic science, quite different from the static view of development as presented in the descriptive studies of sequential stages that formed the basis of most courses and textbooks of embryology at that time. In addition, Viktor Hamburger was the only embryologist in Spemann's Institute fully aware of the relevance and the importance of principles of the young science of genetics for theories of development. I have always felt that this awareness reflected the experience during his postdoctoral years at the Kaiser Wilhelm Institute in Dahlem where he had come to know Richard Goldschmidt and his theories of physiological genetics. The role of genes in developmental mechanisms came under particularly lively discussion in connection with the exciting results obtained in the Freiburg Institute by Spemann and Schotté (1932) in their xenoplastic transplantation experiments between frogs (i.e., anurans) and newts (i.e., urodeles). They discovered that the genetic constitution of the inducing host tissue determined the regional specificity of the organs (e.g., head organs), while the genetic constitution of the transplanted tissue (belly ectoderm) determined the species-specific type of the induced structures. Thus head structures typical of anurans (i.e., horny "teeth" and suckers) were induced in the belly ectoderm of anurans transplanted to the head region of urodeles, which themselves normally form so-called "balancers." The reciprocal experiment gave equivalent results. Certainly, the interpretative discussions of these experiments, in which my memory ascribes a leading intellectual role to Viktor Hamburger, provided a good introduction to the concepts of developmental genetics. My own later work in the field of mammalian developmental genetics, carries the imprint of the effects that Viktor Hamburger's conceptual contributions and interpretations in various areas of experimental embryology had on the ideas governing the analytical approaches of his disciples.

The strong influence of experimental embryology on progress in mammalian developmental genetics is apparent from the vast literature in the field. Studies of some of the most important and most productive experimental systems in mammalian developmental genetics bear testimony to the impact of the principles of experimental embryology on developmental genetics. Among the most striking examples in support of this is the analysis of the *T*-locus and its effects on embryonic development and differentiation in the mouse. Analytical studies of the relevant mutations relied heavily on results obtained with the primary system of analysis in experimental embryology, the developing nervous and axial systems in amphibians (Gluecksohn-Waelsch and Erickson, 1970).

Inductive interaction of mesoderm and ectoderm resulting in the latter's differentiation into neural plate, neural tube, and finally nervous system was revealed experimentally in Amphibian embryos (for a review, see Holtfreter and Hamburger, 1955) and later in teleosts and birds (for a review, see Rudnick, 1955). Even though it could be expected that the mammalian nervous system would be subject to the action of similar developmental mechanisms, proof of this could not be obtained easily. Experimental barriers inherent in the mammalian embryo prevented the easy manipulation carried out with the Amphibian or chick embryo. Consequently, knowledge of mammalian developmental mechanisms remained limited. However, the causal analysis of development in mammals was helped greatly by the use as experimental tools of mutations with effects on developmental processes. The first analytical study of this kind was that of a dominant mutation *(T)* at the *T*-locus in the mouse, which Chesley carried out in 1935 in the laboratory of L. C. Dunn at Columbia University. Chesley (1935) described morphological and histological investigations of *T/T* homozygous mouse embryos obtained in intercrosses of heterozygotes *(T/+)* and reported the practically complete absence of notochord in these homozygotes. Under the influence, no doubt, of the ideas generated by the studies in experimental embryology (Entwicklungsmechanik), Chesley ascribed the severe abnormalities of nervous system and somites of *T/T* mouse embryos to the failure of inductive interactions to take place between the ectoderm and the notochord as required for normal differentiation. The logical conclusion of this was the recognition that normal development of the mammalian embryo required the existence of an inductive relationship between notochord and the primordia of other axial structures, and that therefore the notochord played a role in processes of organization of the mammalian embryo similar to that identified with techniques of experimental embryology in amphibians. Chesley's analytical study laid the groundwork for subsequent investigations over many years, which dealt with the effects of various other mutations at and near the *T*-locus in the mouse (Gluecksohn-Schoenheimer, 1949). Abnormalities of several recessive and two dominant mutations in the *T*-locus region were interpreted as affecting various aspects of the developmental mechanics of the early embryo. Failure of organization and differentiation characterize one such homozygous type *(t^0/t^0)* at a stage when in the normal embryo separation and organization into specific areas occur; abnormalities of entodermal cells were tentatively interpreted as indicating a specific role of the entoderm in the differentiation of the notochord-mesoderm. The effects of yet another dominant mutation *(Fu^{Ki})* near the *T*-locus were interpreted in terms of the concept of embryonic regulation. In this case, striking abnormalities, such as duplications of different parts of the embryo, characterize the homozygotes Fu^{Ki}/Fu^{Ki}) and include embryonic structures such as the nervous system, somites, vertebral axis, and heart. Their striking resemblance to the double monsters obtained in constriction experiments of amphibian embryos in the two-cell stage led to the suggestion that an "organizer" region analogous to that identified experimentally in amphibians existed in mammalian embryos and that its normal functioning was severely affected in Fu^{Ki}/Fu^{Ki} embryos. Furthermore, the very existence of embryonic duplications seemed to indicate the expression of strong regulative properties in the mammalian embryo since obviously the formation of double structures required a considerable amount of regulation, that is, reprogramming, a characteristic property also of other vertebrate embryos. There is no doubt that all of these interpretations of mutational effects on the developmental mechanism were strongly influenced by the orientation of the particular investigators and their view of development as depending on a series of inductive interactions.

Whereas the primary goal of studying mutations at the *T*-locus in the mouse was the identification of causal mechanisms of mammalian development with the mutations serving as tools, it was hoped also that further analytical studies of their abnormal effects might provide knowledge in a more general way concerning the biochemical and molecular basis of normal embryonic inductive interactions, which had remained obscure. This applies particularly to the phenomenon of primary induction, that is, the induction of the ectoderm by the chordamesoderm to form neural tissue. The notochord-mesoderm appears to be the developmental system affected particularly by certain mutations at and near the *T*-locus. It was therefore hoped that the identification of the mode of action of *T*-locus genes—and the nature of their gene products—might provide leads towards the molecular analysis of normal inductive mechanisms and that such analysis might produce significant contributions to the general problem of embryonic induction.

However, in spite of the fact that various laboratories have devoted much effort for many years to the possible identification of the mechanisms of action of the *T*-locus, not much progress has been registered at this time. The suggestion that *T*-locus products might be located at the cell surface could not be substantiated by supporting evidence and has remained controversial. Some reports have claimed to have identified cell surface antigens coded for by *T*-locus genes (Yanagisawa et al., 1974), but they have not provided any evidence that these might actually be responsible for cellular interactions in the course of inductive events during embryonic differentiation. Furthermore, these reports have not been confirmed in other laboratories (Goodfellow, 1979). Obviously the experimental confirmation of the existence of antigens as *T*-locus products would have contributed greatly to the molecular analysis of normal embryonic induction. Thus, developmental genetics, which owes a debt of gratitude to experimental embryology for having provided many ideas and concepts that have promoted causal analytical studies of mutant effects, has not yet been able to "pay back" this debt by contributing its share of further analytical advances concerning differentiation on the molecular level.

Various "minor" systems of inductive interactions have been objects of investigations in developmental genetics and have also leaned heavily on concepts of experimental embryology. Among others, these systems include the development of the lens, the ear, and the red blood cell system. Another example is provided by an analysis of the lethal effects of a radiation-induced chromosomal deletion in the mouse that in homozygotes causes early embryonic death. Histological studies of such mutant embryos indicate interference with inductive interactions between inner cell mass and trophectoderm, which is thought to be essential for the differentiation of the mouse embryo following uterine implantation (Lewis et al., 1976). All these studies provide additional examples of mutual interaction between the fields of developmental genetics and experimental embryology.

An additional system will be mentioned here in some detail. It illustrates particularly well how the strong and lasting influence of the intellectual and scientific training received under Viktor Hamburger's tutelage was expressed in the analytical approaches to problems of developmental genetics developed by one of his disciples. This system is that of kidney development in the mouse as studied with the help of a lethal mutation causing absence of kidneys in homozygous newborn. The urogenital system has received much attention as an object of studies of cell differentiation. It has been shown that the two primordia of the kidney, the ureteric bud and the metanephrogenic mesenchyme, which arise independently, are able to differentiate only if they are permitted to undergo mutual inductive interactions (Grobstein, 1955). This developmental system therefore represents an excellent model for the further analysis of mechanisms of induction. Once more it was hoped that the study of a mutation that seemed to interfere with the essential inductive interaction might help to identify the specific mechanisms instrumental in the corresponding normal inductive processes, since histological studies excluded degenerative processes as possibly being responsible for the failure of kidney differentiation. The question whether the potential for inductive interaction existed between the two mutant kidney primordia was subjected to an experimental test. In a series of experiments, mutant and normal kidney rudiments were exposed to each other in all possible combinations in an organ culture system (Gluecksohn-Waelsch and Rota, 1963). The results showed clearly that both types of mutant rudiments were perfectly capable of differentiating normally if the experimental conditions provided the necessary and essential prerequisites for inductive interaction with the other type of rudiment. It seems therefore that the mutation does not interfere with the actual mechanisms of inductive interactions but with a temporal factor that normally assures the essential synchronization and the actual coming together of the inductive and the reacting systems during differentiation. This temporal factor can be manipulated and thus corrected in the *in vitro* system. The particular mutant system serves to emphasize the significance of the dimension of time in inductive interactions, which in turn is under genetic control (Gluecksohn-Waelsch, 1964).

At this time, studies in developmental genetics have failed to provide any evidence of the existence of a gene or a gene complex that directly controls embryonic induction on any level. The highly complex nature of inductive processes as stressed in many of Viktor Hamburger's writings would hardly lead one to expect such simple relationships. Our ignorance of the genetic control of development, however, extends beyond the areas of embryonic induction. Equally unknown are any general principles that govern the genetic control of morphogenesis in higher organisms, a situation expressed eloquently in a posthumously published lecture by Boris Ephrussi (1980).

Viktor Hamburger himself has made significant contributions to developmental genetics by his experimental studies of the Creeper mutation in the chick. In the title of one of his papers (Hamburger, 1942) he refers to the "developmental mechanics" of the genetic defects as the object of his investigations. In the text, he talks of the "complex interplay of actions and reactions between different parts of the developing embryo," a concept that guides his analytical studies. Hamburger's own approach to a specific problem in developmental genetics is testimony to the lasting influence of the concepts of Entwicklungsmechanik. His finishing sentence is as true today as it was at that time: "The complete story of the mode of gene action must be written jointly by geneticists, embryologists and physiologists."

My current investigations concern a system that has provided certain additional information concerning mechanisms of cellular differentiation and their genetic control in mammals. It is that of a series of lethal deletion alleles in the mouse. These interfere with the differentiation of subcellular membranous organelles as well as with biochemical differentiation of certain enzymes and proteins in liver and kidney cells only. The same subcellular organelles remain completely unaffected in other cell types, and, with the exception of the affected enzymes, the majority of liver specific enzymes also remain normal. The coordinate genetic control of tissue-specific biochemical as well as ultrastructural membrane differentiation raises questions concerning the causal sequence of gene effects and the possible causal relationship of biochemical differentiation and morphogenesis (Gluecksohn-Waelsch, 1979). Recently unpublished cell hybridization experiments of mutant mouse cells with rat hepatoma cells have shown that interaction of the two genomes succeeds in restoring the expression of at least one of the mouse enzymes deficient in the mutants. These results and others indicate that the deleted mouse gene sequences include regulatory genes involved in the differentiation of specific enzymes. The experimental system provided by the lethal deletion alleles offers a model for the study of cell interactions and the relative roles of structural and regulatory genes in processes of differentiation. Its study may also contribute to the analysis of structure and organization of gene sequences in eukaryotic chromosomes. Continued investigations of this promising system are expected to yield further information about developmental mechanisms and their genetic control in high eukaryotes.

If ever a history of ideas in developmental genetics were to be written, it would have to begin with an account of Boveri's beautiful work on sea urchins (1902). Furthermore it would no doubt include as one of its most important chapters an account of the intellectual role that "inductive interaction" between the fields of genetics and embryology has played in the analysis of developmental mechanisms and their genetic control in higher organisms. A few illustrations of this were presented in the examples described earlier. At the time of the flowering of experimental embryology, with Viktor Hamburger as one of its chief proponents, interest focused primarily on inductive mechanisms on the level of tissues and organs. Once again the interaction between geneticists and developmental biologists is particularly strong and lively as emphasis is placed on problems of cellular differentiation and its genetic control. Concepts derived from both fields interplay and guide methodology and interpretation of experimental approaches on the level of the cell, particularly the study of somatic cell genetics. The trend in contemporary developmental genetics resembles that in other areas of biology and aims at the identification of the molecular nature of developmental mechanisms and interactions and their genetic control, and the subsequent description of cell differentiation in molecular terms. As yet, the goal is far removed, but the systems briefly described here are part of a particular experimental approach, which together with many others, may help to unravel the puzzle of development, its mechanisms, and genetic control.

The early days of developmental genetics of higher organisms owe a great debt of gratitude to those who were imaginative and farsighted enough to extend their interest in developmental mechanisms to the level of their genetic control. Among these, Viktor Hamburger ranks at the top. If my own search for further knowledge of developmental mechanisms and their genetic control in mammals has been at all successful and made any significant contributions, it is due to Viktor whose teaching provided the high standards, excitement, stimulation, and conceptual understanding necessary to sustain the effort.

ACKNOWLEDGMENTS

The work of the author discussed in this paper was supported generously by grants from the National Institutes of Health and the American Cancer Society. Thanks are due to colleagues and friends who read the manuscript and made useful comments: Paul Blanc, Leslie Pick, and Lee Silver.

REFERENCES

Boveri, Th. 1902. Über mehrpolige Mitosen als Mittel zur Analyse des Zellkerns. *Verh. Phys. Med. Ges. Würzburg Neue Folge* 35:67–90

Chesley, P. 1935. Development of the short-tailed mutant in the house mouse. *J. Expt. Zool.* 70:429–59

Ephrussi, B. 1980. Mendelism and the new genetics. *Somat. Cell Genet.* 5:681–95

Gluecksohn-Schoenheimer, S. 1949. Causal analysis of mouse development by the study of mutational effects. *Growth Sympos.* 9:163–76

Gluecksohn-Waelsch, S. 1964. Genetic control of mammalian differentiation. *Genet. Today, Proc. 11th Int. Congr. Genet.* 209–19

Gluecksohn-Waelsch, S. 1979. Genetic control of morphogenetic and biochemical differentiation: Lethal albino deletions in the mouse. *Cell* 16:225–37

Gluecksohn-Waelsch, S., Erickson, R. P. 1970. The T-locus of the mouse: Implications for mechanisms of development. *Curr. Top. Dev. Biol.* 5:281–316

Gluecksohn-Waelsch, S., Rota, T. R. 1963. Development in organ tissue culture of kidney rudiments from mutant mouse embryos. *Dev. Biol.* 7:432–44

Goodfellow, P. N., Levinson, J. R., Gable, R. J., McDevitt, H. O. 1979. Analysis of anti-sperm sera for T/t locus-specific antibody. *J. Reprod. Immunol.* 1:11–21

Grobstein, C. 1955. Inductive interaction in the development of the mouse metanephros. *J. Exp. Zool.* 130:319–40

Hamburger, V. 1942. The developmental mechanics of hereditary abnormalities in the chick. *Biol. Symp.* 6:311–36

Holtfreter, J., Hamburger, V. 1955. Embryogenesis: Progressive differentiation. Amphibians. In *Analysis of Development,* ed. B. H. Willier, P. A. Weiss, V. Hamburger, pp. 230–96. Philadelphia: W. B. Saunders.

Lewis, S. E., Turchin, H. A., Gluecksohn-Waelsch, S. 1976. The developmental analysis of an embryonic lethal (c^{6H}) in the mouse. *J. Embryol. Exp. Morphol.* 36:363–71

Morgan, T. H. 1934. *Embryology and Genetics.* New York: Columbia Univ. Press

Rudnick, D. 1955. Embryogenesis: Progressive differentiation. Teleosts and Birds. In *Analysis of Development,* ed. B. H. Willier, P. A. Weiss, V. Hamburger, pp. 297–314. Philadelphia: W. B. Saunders

Spemann, H., and Schotté, O. 1932. Über xenoplastische Transplantation als Mittel zur Analyse der embryonalen Induktion. *Naturwiss enschaften* 20:463–67

Wilson, E. B. 1925. *The Cell in Development and Heredity.* 3rd ed. New York: Macmillan.

Yanagisawa, K., Bennett, D., Boyse, E. A., Dunn, L. C., DiMeo, A. 1974. Serological identification of sperm antigens specified by lethal t-alleles in the mouse. *Immunogenetics* 1:57–67

4 / *Pathway Selection by Embryonic Neurons*

LYNN LANDMESSER

INTRODUCTION

The tip of a growing axon, at least in tissue culture, is a curiously active structure that continuously extends and retracts filopodia as it traces out its pathway. These exploratory like processes occurring at the growth cone both *in vivo* (Speidel, 1942) and in tissue culture (Harrison, 1910; Luduena and Wessels, 1973; Letourneau, 1975; see Johnson and Wessells, 1980, for a review) have caused many workers, beginning with Ramón y Cajal (1905, 1909), to confer upon the growth cone an important role: that of actively seeking and responding to environmental cues that enable it to reach and to finally synapse with its proper target. Yet when one carefully considers the data on regenerating, and especially on developing systems, there is scant evidence to support this role.

This lack of relevant data is at least partly because most studies have focused on defining connectivity rather than the pathway taken by the process of one cell to reach another. For example, although there are numerous analyses of connectivity in the vertebrate retinotectal system following a variety of experimental perturbations (see Willshaw and Malsberg, 1978, for a review), only recently have workers begun to study the pathways taken by retinal axons in normal development (Bunt and Horder, 1978; Horton, Greenwood, and Hubel, 1978; Scholes, 1979; Rusoff and Easter, 1979) or following experimental alterations (Attardi and Sperry, 1963; Straznicky, Gaze and Horden, 1979; Meyer, 1980). However, there is no doubt that in many systems mature pathways are as precise and reproducible as the ultimate connectivity patterns.

Two main questions need to be answered for developing systems:

first, what mechanisms are used by axons to get from their point of origin to their appropriate target, which may be several to many millimeters away; and second, what role, if any, does precise pathway selection play in generating specific connectivity patterns? With respect to mechanism, one extreme view would be to consider that axons reach their appropriate targets simply because they are generated at the right time and place in the developing embryo. Any axon at point *A* would invariably reach point *B* either by growing along nonspecific mechanical guides (i.e., oriented collagen bundles or blood vessels) or by other events associated with tissue morphogenesis (Horder, 1978). At the other extreme is the idea that growth cones are able to recognize and respond to specific chemical cues in the environment through which they grow (Ramón y Cajal, 1905, 1909). Interactions between fibers may also be of importance, especially in systems where arrays of fibers grow out together (Hope, Hammond, and Gaze, 1976; Cook, 1980).

Some years ago Viktor Hamburger made several relevant observations while studying the dual origin of the trigeminal ganglion in the chick (Hamburger, 1961). This sensory ganglion, which is of both placodal and neural crest origin, supplies three main nerve trunks: ophthalmic, maxillary, and mandibular. He found that the depleted ganglia that formed following either neural crest or placode ablation were often severely displaced, yet the axons that emerged from them formed the typical stereotyped nerve pattern once they reached their normal territory. He proposed that the axons grew somewhat randomly until they encountered tracks that had been laid down in the tissue primordia during morphogenesis. Further, these tracks did not appear to be totally nonspecific, because he found that following placode removal, small cells of neural crest origin only extended their axons down the mandibular pathway and never down the ophthalmic or maxillary pathways, even when these were available.

Occasional observations of this sort have been made on *in vivo* systems, but there are only a few studies that have attempted to characterize the pathways taken by known groups of neurons following experimental manipulations (Detwiler, 1920; Hamburger, 1939; Taylor, 1944; Castro, 1963; Narayanan, 1964; Hibbard, 1965). In most cases, owing to limitations of the available techniques, these have involved only a characterization of the gross anatomical innervation pattern and not an analysis of how specific cells were behaving. In other cases only the mature innervation pattern was assessed. Thus it was impossible to decide if it was produced by directed axonal growth or by death or retraction of axons that had taken wrong pathways and therefore failed to reach appropriate targets.

It seems likely that a variety of mechanisms will be used in different parts or levels of the nervous system (see also Feldman, 1977) depending on the required task. For example, the mechanism for getting whole groups of processes from one point to another as in the cockroach cercus

(Edwards and Chen, 1979) or moth antenna (Sanes and Hildebrand, 1975) could be much simpler than one that would allow discrete populations of axons to diverge from each other in specific pathways to different targets, as occurs in the vertebrate limb (Lance-Jones and Landmesser, 1980b, 1981a,b). Finally, if one wishes to demonstrate that pathway selection is important in the development of specific connections, it is necessary to disrupt the pathways and document the effect on connectivity. It is not enough simply to show that a precise spatiotemporal pattern, or selective pathway, selection normally occurs and that this would seem sufficient to explain the observed connectivity.

One problem with *in vivo* studies is that it has proved difficult to analyze with sufficient temporal and spatial resolution the outgrowth of identified populations of processes, either normally or following experimental manipulation. On the other hand, axons in tissue culture can be caused to grow under well-defined conditions where their response to specific stimuli can be documented. However, it is often hard to relate these observations to what goes on in normal development. Nonetheless, since they provide some idea of the response capabilities of growing axons, I will first consider some observations from *in vitro* studies that may be of relevance. Next, I will discuss several models that have been proposed to explain pathway guidance and some of the evidence from *in vivo* development supporting each.

WAYS OF DIRECTING AXONAL GROWTH IN TISSUE CULTURE

From the earliest observations of Harrison (1910) and later Weiss (1941), it became apparent that axons had to adhere to some substrate in order to grow. Thus, by providing them with oriented substrates, one could presumably control their direction of growth. This contributed to the general acceptance of the notion that axons were largely guided by non-specific mechanical factors in the substrate, both *in vitro* and during *in vivo* development.

More recently Letourneau (1975a,b) has extended these observations in an interesting way. Given that adhesion of growth-cone filopodia with the substrate is necessary for axon elongation (see Johnson and Wessels, 1979), differences in substrate adhesivity should be able to influence the direction in which an axon grows. Letourneau first showed that there were differences in the strength with which isolated chick dorsal root ganglion (DRG) cells adhered to different substrates. They adhered best to polyornithine, then to collagen, followed by either naked or palladium-coated tissue culture plastic that were essentially equivalent. He next created "pathways" of differing adhesivity by constructing a grid of palladium-coated squares intersected by lanes of either polyorinithine, tissue culture plastic, or other substrates.

Following a certain period in culture during which the axons elongated appreciably, the proportion of growth cones on either the lanes or the squares was scored. When the lanes were composed of tissue culture plastic, there was no difference in the proportion of growth cones on lanes or squares. Thus growth cones appeared to show no preference for palladium or tissue culture plastic. This was substantiated by time-lapse observations in which individual growth cones were found to move from palladium to plastic and back again.

When the lanes were composed of polyornithine, strikingly different results were found; 95 percent of the growth cones were on the lanes, only 5 percent on the squares. The axons grew for long distances along the lanes, sometimes bifurcating at intersections, but they rarely entered the squares. Viewed in time lapse, growth-cone filopodia were seen to "sample" the squares, but this was usually followed by retraction and growth along other filopodia situated on the lanes. These results would not be surprising if palladium were an inadequate substrate for axon elongation. But based on the previous experiment, this explanation is clearly not the case, and it is possible to conclude that nonextreme and therefore possibly physiological differences in adhesivity can result in directed axonal growth.

The preneural pathways alluded to in many classical studies of *in vivo* development could similarly be produced by rather sharp spatial changes in adhesivity. Such tracks would not necessarily be specific in that any axons placed on them would be constrained to trace out the same pathways. They would also not possess any inherent directionality, although this might be conferred by constraints that favor the initial direction of axon outgrowth. Such paths need not have obvious structural correlates (e.g., blood vessels, etc.) but could reflect subtle differences in the chemical composition of cell surfaces or extracellular matrix. If, on the other hand, adhesivity varied continuously as a gradient from some point, axons would not have to follow rigid tracks, but could "home in" on the most adhesive area from anywhere on the gradient. It has, however, not been shown how steep a gradient would be required to influence the direction of axon growth, and therefore whether one would expect adhesive gradients of this magnitude in living systems. Finally, if neurons differed in their cell surface properties such that subclasses adhered preferentially to different substrates, it should be possible to cause specific axons to take different routes and to cease growth at certain targets. Models of this sort have been proposed to explain the topographical projection of retinal ganglion cells onto the tectum (Barbera, 1975; Roth and Marchase, 1976).

Another mechanism for orienting axons was initially championed by

Ramón y Cajal (1909), who proposed that growth cones detected and responded to gradients of diffusible chemical substances perhaps emanating from their targets; in essence, chemotaxis. Until recently this had proven exceedingly difficult to demonstrate experimentally. One problem was that most of the earlier studies failed to distinguish adequately between tropic and trophic effects of chemical agents. For instance, a more robust growth of axons from the side of a ganglion explant facing its presumed target or a chemical source could simply result from the chemical agent stabilizing or enhancing the growth rate of axons that happened to grow near it. This would not be a case of guided growth in the strict sense of the word.

The most convincing evidence for chemotaxis has been obtained on sympathetic or dorsal root ganglion cells, which are dependent on nerve growth factor (NGF) for survival and axon elongation in culture (Chamley and Dowel, 1975; Ebendal and Jacobson, 1977). NGF has also been proposed to play a neurotropic role *in vivo* (Levi-Montalcini et al., 1978). Letourneau (1978) recently showed that it was possible to set up measurable gradients of NGF in an agar matrix in tissue culture. Chick DRG cells plated into this matrix showed a preferential orientation towards the NGF source when it was sufficiently greater than the background NGF concentration, but chick DRG cells showed no orientation when NGF levels were equal throughout the matrix. The orientation was slight (60 percent of the growth cones oriented versus 50 percent in the controls) but consistently demonstrable. The response also showed saturation at high NGF levels, as would be expected of a chemotactic response mediated by specific receptors.

Gunderson and Barrett (1979a) demonstrated chemotaxis in a different way by showing that the growth cones of chick DRG cells rapidly followed a pipette containing 50 times the background NGF concentration, tracing out a horseshoe-shaped path within 9 to 23 minutes. There was no directed growth when the pipette contained the same concentration of NGF as background, or other proteins such as bovine serum albumin, and so on. The actual rate of axon elongation was not affected during a positive response, only the direction of axon growth.

While some might quibble over whether the responses shown both by Letourneau (1978) and by Gunderson and Barrett (1979a) are chemotactic in the strict sense of the word, there is no doubt that the direction of axon growth can be controlled by gradients of diffusible chemical substances of the sort that might occur during *in vivo* development. In fact, at the level of the growth cone, the distinction between tropic and trophic loses its meaning, for it appears that oriented growth of the whole structure can result from the stabilization of filopodia that are extended more or less randomly. Similarly, the distinction between chemical and mechanical guidance becomes blurred. For example, Schubert and Whitlock

(1977) found that NGF increased adhesive interactions between PC12 cells and their substrate, and that this was associated with increases in intracellular cyclic AMP (cAMP) and an influx of calcium. Gunderson and Barrett (1979b) have recently extended their original observations to show that the growth cones of cultured chick DRG cells will also follow pipettes filled with dibutryl-cAMP or 20 nM Ca^{2+} in the presence of a Ca^{2+} ionophore. Thus it appears likely that NGF acts locally by increasing the adhesivity of growth-cone filopodia for the substrate. Campenot (1977) has also found that cultured sympathetic neurons will not extend axons into a chamber unless NGF is present locally.

During normal development it is possible that NGF or other chemical factors could act in a similar manner. It is important to note that such factors may act directly not only on the growth cone but on the substrate as well. Collins (1978) has shown that the enhanced neurite extension of ciliary ganglion cells produced by heart conditioned medium is due to a factor (or to factors) from the conditioned medium bound to the substrate, and thereby apparently altering its adhesivity. Since all cells apparently shed surface proteins into the surrounding medium (Doljanski and Kapellar, 1976; Rosen and Culp, 1977; Schubert, 1977, 1978), the substrate *in vivo* could be modified by glial and other cells as well as by the neurons themselves to produce a complex pattern of adhesiveness. Clearly, the work in culture shows that the growth of neurons can be guided by chemical or mechanical factors (or by both) which alter the adhesive interactions between growth cone and substrate. However, as will be shown below, it is not simple to determine how, or even if, axons are selectively guided during normal *in vivo* development.

NONSPECIFIC "MECHANICAL" GUIDES

Weiss's old concept of contact guidance has recently reemerged in the view that mechanical channels may guide vertebrate axons over relatively long distances during normal development. At one level of analysis it is clear that mechanical structures must play some role, as neurons can not grow through other cells or solid structures. The optic stalk connects the eye to the brain and serves as an obvious structural conduit. Yet Singer and his coworkers (Egar, Simpson, and Singer, 1970; Nordlander and Singer, 1978; Singer, Nordlander, and Egar, 1979) extended this idea to a finer level when they observed that during regeneration of the lizard and newt spinal cord, longitudinally coursing fibers seem to run in "channels or tunnels" of extracellular space between ependymal processes. They proposed that such "oriented" channels of extracellular space provided a "blueprint" for developing axons to follow. Similar observations have been made by Silver and his coworkers in the visual system (Silver and Robb, 1979; Silver and Sidman, 1980).

Such channels would at least seem to play a permissive role in axon elongation. For example, when they are absent in mice with congenital optic nerve aplasia, retinal ganglion cell axons fail to exit through the optic cup and die (Silver and Robb, 1979). Yet it is difficult to prove that axon orientation on any finer level is actually caused by the pattern of such channels. This sort of model could be useful in getting groups of axons from one neural center to another, and possibly in maintaining any topographical arrangement of fibers already present. However, it does not have the capacity to allow the specific sorting out or rearrangement of axons that must occur at some point in many developing systems (Scholes, 1979; Lance-Jones and Landmesser, 1980a,b, 1981a,b; Landmesser, 1980).

PIONEER FIBERS

In some invertebrates bundles of larval or embryonic axons may play a role similar to the nonspecific mechanical guides described above. They often connect two neural structures early in development when the distance between them is extremely small, and then serve as a guide for later added fibers. Two distinctly located bundles of such axons link the cricket cercus to the CNS. Most sensory axons are added later, after substantial cercus elongation, and appear to follow these initial bundles (Edwards and Chen, 1979). Similar observations have been made for the moth antennae (Sanes and Hildebrand, 1975) and the antennae and legs of the locust (Bate, 1976).

Of course, in all cases where a pathfinder or pioneer function is ascribed to a neuron, this must be tested by observing the effect of its early removal on subsequent pathway formation. Edwards, Berns, and Chen (1979) have done this for the cricket cercus and found that sensory axons now course through the cercus in a number of displaced bundles and thus support a pioneer role for the initial bundles. One would also like to know how the pioneer fibers themselves are guided, whether they use specific chemical cues, or still earlier preexisting guides, or trial-and-error sampling. This should be possible to determine by observing the behavior of individual dye-filled pioneer neurons under different conditions (Keshishian, 1979).

Where they have been described, pioneer fibers seem to provide a somewhat nonspecific substrate to link two neural centers or the CNS and periphery. It is not clear that they play an important role in specifically guiding different fiber types to their appropriate targets, although they may get them into a region where other pathway selection mechanisms can take over (e.g., see Shankland, 1979). Pioneer fibers have not been described in vertebrates, although in many different parts of the nervous system nerve fiber links are established very early and could serve as substrates for later originating fibers.

SPATIOTEMPORAL MODELS

A relatively complex pattern of pathways and connectivity between two neural structures need not imply that complex guidance mechanisms were used in its establishment. Such a pattern can also be produced by coordinating in time and space the outgrowth of axons. Any given axon need only follow a simple rule, such as project to the closest available target. Subsequent morphogenetic movements can distort the system, giving it a complicated mature appearance. The most convincing example of a spatiotemporal mechanism has been provided by Macagno on the development of the visual projection in *Daphnia* (Macagno, 1978). Here, the coordinated and sequential differentiation of retinula cells and of their target cells in the lamina seems sufficient to explain the gross wiring diagram.

Similar models have recently been applied to vertebrate development (Horder and Martin, 1978; Horder, 1978). For example, since retinal ganglion cells are added in rings at the circumference of the retina (Gaze et al., 1979) and since some people have observed that retinal axons maintain topographical order within the nerve, (Bunt and Horder, 1978) it has been proposed (Horder and Martin, 1978) that the problem of connecting a whole array of retinal and tectal cells may be less formidable than originally thought. In this view, having axons in the proper order as they reach the tectum may greatly facilitate the establishment of appropriate connections. Furthermore, this order might be achieved by a simple passive process based on the time of axon origin. Nonetheless, other observations of alterations in the ordering of ganglion cell axons along the optic nerve (Horton, Greenwood, and Hubel, 1979), including specific rearrangements (Scholes, 1979) as well as selective selection of the medial branch of the optic nerve by axons of ventral retinal origin (Straznicky, Gaze, and Horder, 1979), are not compatible with an entirely passive mechanism. Neither is the innervation pattern of the chick hindlimb described in the next section. What is needed to prove the importance of such a mechanism is the demonstration that disruption of the spatiotemporal pattern results in a disrupted pathway selection and pattern of connecticity. In fact, some evidence to the contrary exists for the visual system (Beazely and Lamb, 1979; Meyer, 1980).

SPECIFIC CHEMICAL CUES

Specific chemical cues, whether diffusible or laid down in the substrate, have been evoked since the classical studies on specificity (Ramón y Cajal, 1905; Sperry, 1963). Generally, they have been thought to form a spatial coordinate system in which axons home in on their targets by orienting to a gradient of chemical markers. More recently, Katz and Lasek (1979) have suggested that chemical cues may be laid down in distinct tracks or substrate pathways.

The idea comes largely from studies in *Xenopus* in which an eye is transplanted in various atypical locations and the pathways taken by the retinal ganglion cells in the CNS analyzed. Constantine-Paton and Capranica (1976) first observed that when eyes are placed in the position of the ear, ganglion cell axons take a consistent dorsocaudal trajectory growing all the way down the spinal cord, but always in a discrete tract in a characteristic position. Later, Constantine-Paton (1978) and Katz and Lasek (1979) found that when eyes are situated far caudally, the ganglion cell axons could course either rostrally or caudally but again always in the same tract. Katz and Lasek interpreted these results to mean that chemical cues specific for these axons existed as distinct substrate pathways. Similar views have been advanced by Singer and colleagues.

The substrate pathway concept presented by Katz and Lasek (1979) differs from the spatial-coordinate model in several testable ways. First, if axons following substrates are displaced, they will wander about at random until they happen upon a compatible pathway, and they will then grow in a stereotyped fashion down that pathway. Second, substrate pathways lack inherent directionality, so that the displaced axons could grow down them in either direction. The work on displaced eyes cited above is consistent with a substrate pathway model. However, it is difficult to deduce much about the cues normally used by retinal ganglion cells from such extreme displacements. The rest of this paper will therefore be devoted to a consideration of how chick lumbosacral motoneurons behave following a series of more moderate alterations of the environment into which their axons project. After a series of earlier studies (Hamburger, 1939; Castro, 1963; Narayanan, 1969), renewed interest in this system has resulted in a number of studies (Landmesser and Morris, 1975; Hollyday and Farris, 1977; Stirling and Summerbell, 1977, 1979; Landmesser, 1978b; Lance-Jones and Landmesser, 1980a,b, 1981a,b) which provide at least a partial understanding of how these fibers normally find their way.

PATHWAY SELECTION BY CHICK LUMBOSACRAL MOTONEURONS

The nerve pattern of vertebrate limbs is highly stereotyped. In the chick (Fig. 4-1), the first three lumbosacral spinal nerves form the preaxial crural plexus that innervates a number of thigh muscles. Lumbosacral (LS) nerves 4 through 8 form the sciatic plexus, postaxial to the femur, and innervate the remainder of the thigh and all the more distal muscles.

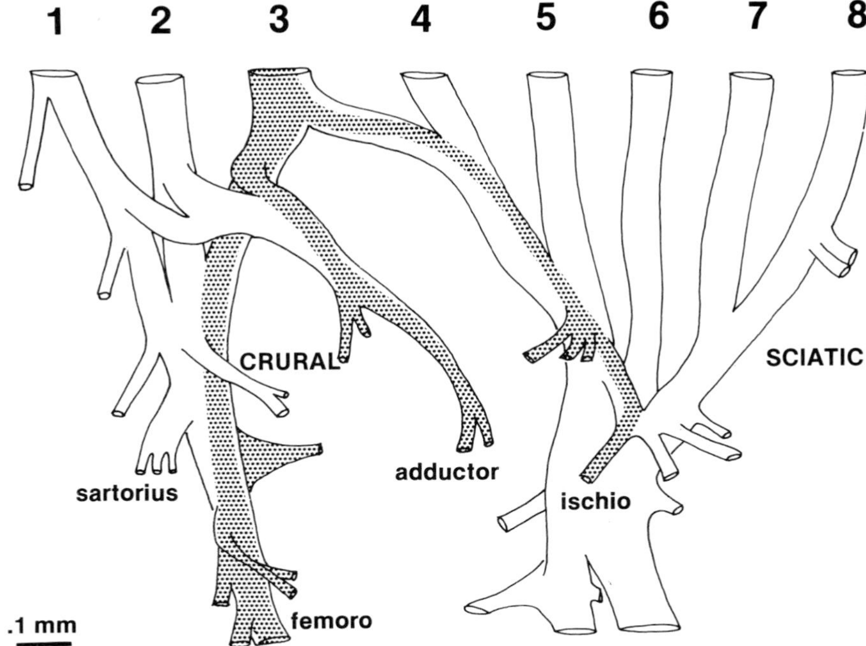

Figure 4.1 The normal innervation pattern of the chick hindlimb. This reconstruction was made from camera-lucida drawings of a transversely sectioned stage 30 hindlimb. Injection of HRP into lumbosacral segment 3 (LS 3) stained many of the axons in the corresponding spinal nerve (stippled area), and these were found to project down a specific set of nerves, appropriate for the mature projection.

The reconstitution of this general pattern following anteroposterior (a-p) limb reversals (Narayanan, 1964; Morris, 1978; Stirling and Summerbell, 1979), or the addition of supernumerary limbs (Hamburger, 1939; Hollyday, Hamburger, and Farris, 1977; Morris, 1978), has led some workers to favor a passive, limb-controlled mechanism for nerve pattern formation (e.g., see Horder, 1978). Thus, whatever spinal nerves entered the limb in the position of LS 1–LS 3 would trace out the innervation pattern of the crural plexus, and so forth. Recent studies in my laboratory (Morris, 1978; Ferguson, 1978 and unpublished observations; Lance-Jones and Landmesser, 1980a,b, 1981a,b) have confirmed some of these observations, but are clearly not compatible with a completely passive model.

At late embryonic stages (stage 36 on) a mature innervation pattern is found with each muscle being innervated by a coherent group or pool of motoneurons situated in a characteristic position in the rostrocaudal and transverse axes of the spinal cord. Further, the axons from each pool project down characteristic nerves (Landmesser and Morris, 1975;

Landmesser, 1978a). In attempting to understand how this pattern was established, we first studied normal development and were able to exclude one mechanism: that of trial-and-error projection of motoneurons down multiple pathways.

Electrophysiological observations indicated that from Stage 28 (5 days), which is prior to the period of normal motoneuron cell death (Hamburger, 1975), errors in pathway selection were extremely rare (i.e., muscle nerves contained almost no segmentally inappropriate axons) (Landmesser and Morris, 1975; Landmesser, 1978b). Cynthia Lance-Jones and I were able to extend these observations to the earliest stages of motoneuron outgrowth into the limb bud (Stage 23) by using an orthograde horseradish peroxidase (HRP) labeling technique. Following injection of HRP into the ventral horn of specific spinal segments, we were able to trace the diffusely stained axons projecting from those segments into the limb. From Stage 26½, when muscle nerves are first formed, we found that axons from a specific segment only projected down appropriate nerves. For example, when axons were labeled in LS 3 (Fig. 4-1), they always projected down the femorotibialis, adductor, and ischioflexorius nerves but never down the sartorius nerve, which receives its innervation from LS 1 and LS 2 (Lance-Jones and Landmesser, 1980a,b, 1981a, b).

At still earlier stages (22 through 24), when axons were just reaching the base of the limb, we were able to visualize growth conelike swellings on their tips (Fig. 4-2a,b). This gave us confidence that we were actually observing the growing tips of fibers, a contention that we are now attempting to confirm at the EM level (K. Tosney, in preparation). While axons from each segment appeared to sample a local region amounting to about one-tenth of the a-p axis of the limb bud (Fig. 4-2a,b,c), at no time did they project diffusely either into the limb or into regions inappropriate for their mature projection (Lance-Jones and Landmesser, 1980c, 1981a). Therefore, in a large number of preparations we could find no evidence of errors at any developmental stage. This led us to conclude that a highly precise form of pathway selection must be occurring. However, based on studies of normal development, it was not possible to determine whether axons were being passively guided by a spatiotemporal mechanism or whether they were playing a more active role in selecting pathways based on specific chemical cues.

To distinguish between these possibilities, we caused motoneurons to enter the limb from abnormal positions by reversing various sized segments of lumbosacral cord about the a-p axis, shortly after neural tube closure (Stage 15 through Stage 16) (Lance-Jones and Landmesser, 1980b, 1981b). When the reversals were of a size that axons entered their original plexus, we consistently found that they projected to the muscles they normally would have innervated. Further, we consistently found that they accomplished this by altering their pathway in the region

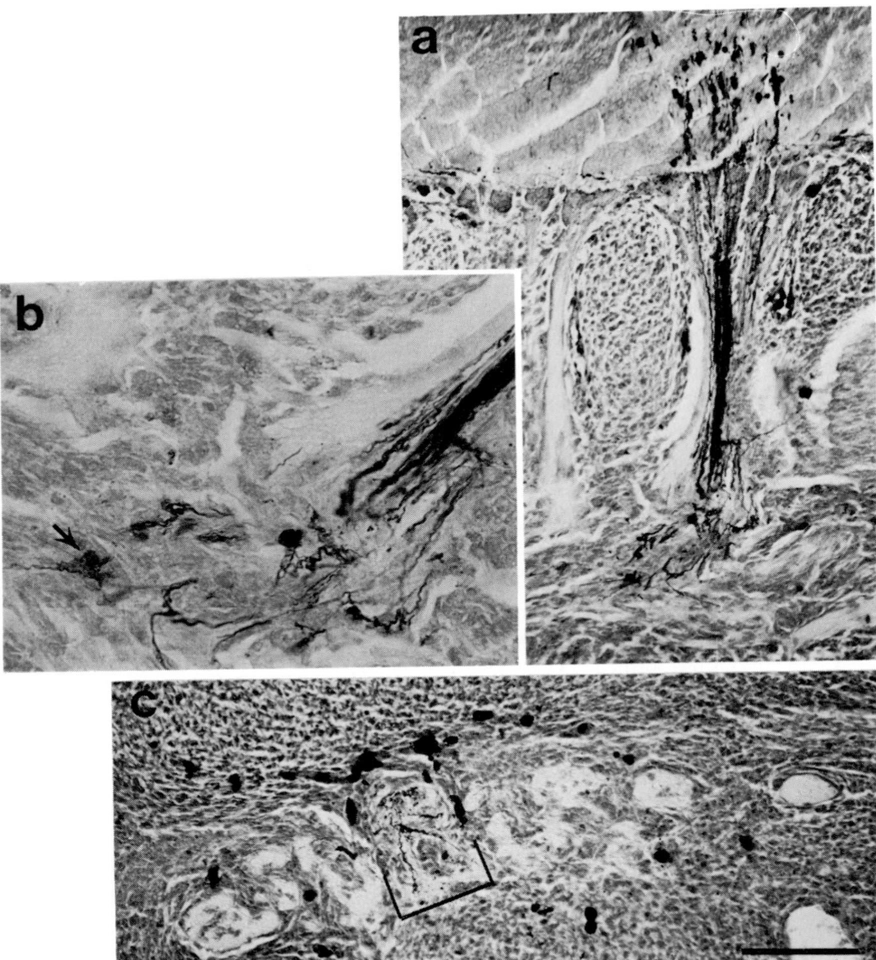

Figure 4.2 The initial outgrowth of axons into the limb bud, visualized by orthograde labeling with HRP. a. In this Stage 23 cord reversal embryo, injection of HRP into the cord (top of the figure) labeled motoneurons within a single segment. Their darkly stained axons can be seen coursing down the adjacent spinal nerve through somitic mesenchyme to the limb base, where a number of growth-conelike structures are seen. This region, which appears to be the growing front of the nerve, is shown at higher magnification (b), where a growth-conelike swelling with thin filopodia is indicated by an arrow. c. Limb bud of a Stage 24 cord reversal embryo. At this stage the spinal nerves (light areas) have just begun to converge at the plexus region. Following HRP injection into a single segment, the stained axons (extent indicated by bar) can be seen to be confined rather locally. They do not project throughout the plexus or a-p axis of the limb. The large, dark-staining structures are red blood cells. Calibration bar: 200 μm for a, c; 150 μm for b.

of the plexus and proximal nerve trunk. For example, in Figure 4-3A the labeled LS 1 axons entered the limb in the position of LS 3 but coursed anteriorly to innervate their normal target, the sartorius muscle. They bypassed the femorotibialis nerve, which would normally be supplied by axons in the position of LS 3 (Compare with Fig. 4-1). These results are clearly inconsistent with a passive spatiotemporal mechanism as proposed by Horder (1978). They demonstrate instead that motoneurons, or rather the regions of neural tube from which they are derived, are specified with respect to peripheral targets as early as Stage 15; and they further demonstrate that they can detect displacements along the a-p axis and respond by altering their path of growth. We assume that they do this by detecting some form of chemical cues contained in the limb bud or in the intervening tissue at the base of the limb.

How, then, can we reconcile these results, in which displaced motoneurons manage to project down their original nerves, with a number of other studies in which motoneurons take pathways in accord with their new position vis-à-vis the limb? (Hamburger, 1939; Narayanan, 1964; Hollyday et al., 1977; Morris, 1978; Stirling and Summerbell, 1979). It would seem that the two sets of data are quite compatible and that it is simply a question of distance. If motoneurons are displaced too far from where they normally enter the limb, they either fail to adequately detect or respond to the cues that they would have originally followed. They then project to a number of inappropriate muscles. This occurred following large cord reversals, the addition of supernumerary limbs, or the shifting of the normal limb a sufficient distance along the a-p axis (Lance-Jones and Landmesser, 1981b). For example, many of the labeled LS 1 and T 7 axons in Figure 4-3A, which entered the sciatic plexus, projected down inappropriate nerves. Normally LS 1 axons never contribute to any nerves emerging from the sciatic plexus.

Surprisingly, however, in all of the above manipulations, we also observed that some of the axons that were channeled into the wrong plexus still managed to reach their correct muscles. They did so by taking aberrant or totally novel paths within the limb (Lance-Jones and Landmesser, 1981b; see Fig. 4-3A,B, arrows). We assume that such paths might have gone undetected in earlier studies where complete nerve pattern reconstructions were not made (Hamburger, 1939; Hollyday et al., 1977; Morris, 1978). The existence of these aberrant paths indicates that pathway selection within the limb cannot be a totally passive process. We have proposed instead that axons are capable of responding only to somewhat local cues. In sufficiently foreign regions, failing to detect these cues, they would be guided in a nonspecific way down a stereotyped set of pathways. These pathways may simply be regions in the developing limb where adhesive interactions (Letourneau, 1975b) favor axon elongation. However, when axons detect cues that are appropriate for them, they are

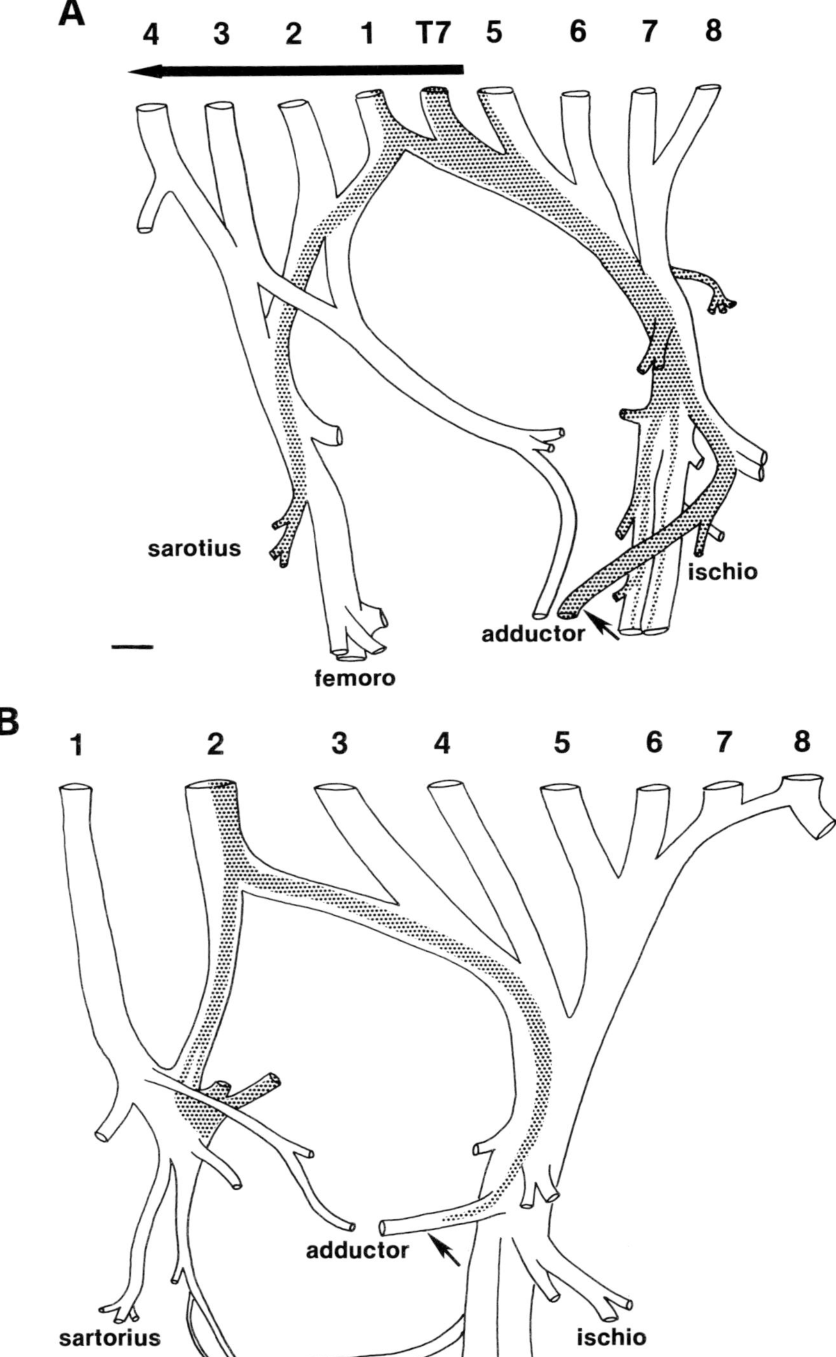
A
4
3
2
1
T7
5
6
7
8
sarotius
ischio
adductor
femoro
B
1
2
3
4
5
6
7
8
adductor
sartorius
ischio
femoro

able to deviate from these stereotyped pathways and form the aberrant paths that we consistently observed. The cues utilized by axons in forming these aberrant paths were apparently quite specific. For example, all of the motoneurons that sent axons down an aberrant path to the sartorius muscle were located in the position of the sartorius motoneuron pool, and therefore, by definition, they were sartorius motoneurons. Similarly, motoneurons projecting to other muscles were located appropriately (Lance-Jones and Landmesser, 1981b).

In all of the experimental manipulations that we carried out, even when motoneurons projected to inappropriate regions along the a-p limb axis, they always projected correctly with respect to the dorsoventral (d-v) axis. Normally, motoneurons that are located medially within the cord project to muscles derived from the ventral muscle mass, those located laterally to muscles of dorsal origin (Landmesser, 1978a). This also occurred following a-p cord reversals, a-p limb shifts, or limb additions and thus confirmed the results of others (Hollyday et al., 1977; Morris, 1978). Similarly, when the limb was shifted about the d-v axis (Ferguson, 1978; and unpublished observations) motoneurons altered their pathways to project to the appropriate dorsal or ventral muscle mass. This suggests that motoneurons also actively follow cues along this axis. They may be able to project appropriately in all cases, simply because none of the experimental manipulations displace them far from their normal position at the center of the d-v axis. Axons may therefore be able to detect appropriate local cues.

What, if anything, can we tell from the behavior of those axons that project inappropriately? In interpreting the behavior of inappropriately projecting motoneurons in supernumerary limbs, Hollyday and her colleagues (1977) proposed a hierarchy of specificities. When motoneurons were unable to reach their original muscle, they were thought to make next-best choices based on either position, function, or embryonic origin of the muscle. In analyzing the inappropriately projecting motoneurons primarily in large cord reversals, but in shifted and supernumerary limbs

Figure 4.3 Nerve patterns following experimental manipulations. A. In this Stage 28 cord reversal embryo, the extent of the cord reversal is indicated by the large arrow at top. The paths of axons stained by HRP injection of the first lumbosacral and last thoracic (T 7) segments are indicated by stipling. The arrow indicates an aberrant path to the adductor. B. In this Stage 29 limb shift embryo, anterior displacement of the limb bud caused all of LS 3 and part of LS 2 to enter the sciatic plexus, posterior to the femur. Compare with Figure 4.1. Aberrant paths to the adductor and femorotibialis muscles (arrows) arise from the sciatic plexus. Electrophysiological observations on this embryo indicated that the aberrant paths contained only segmentally appropriate axons. Calibration bar for both: 100 μm.

as well, we were struck by the lack of a consistent pattern. Except for the general rule described above—that lateral motoneurons projected to dorsally derived muscles and that medial motoneurons projected to ventrally derived ones—the motoneuron pools contributing to inappropriate muscles were widely distributed and lacked a consistent localization in either axis of the cord (Lance-Jones and Landmesser, 1981b). The slight tendency toward a pattern that we observed (i.e., anterior segments tended to innervate more anterior muscles) was probably due to nonspecific geometrical factors. For, in all cases, the diffuseness of the pools was in striking contrast to normal pools or to pools projecting to appropriate muscles via altered paths. Axons appeared to project in an essentially random fashion, and thus they would seem to provide little information on the rules governing normal connectivity.

The results of the different manipulations carried out thus far in the chick limb system do not favor the view that cues exist as rigid substrate pathways as proposed by Katz and Laske (1979). Displaced axons can reach their targets by various alternate routes, including quite proximal changes in their axon trajectories within the plexus or limb base, as well as by a variety of aberrant paths within the limb. Nor does there appear to be a simple gradient of cues distributed through the limb, as has been proposed for the retinotectal system (Roth and Marchase, 1976). If this were the case, we would expect axons to "home in" on their target from anywhere in the limb, or to at least to get as close as possible if they happened to encounter physical constraints. The lack of order observed in inappropriately projecting axons contrasts with the ordered (if altered) projections formed by retinal ganglion cells following a variety of experimental manipulations (Willshaw and Malsberg, 1979). Further, we did not observe any tendency for axons to sort out with each other in any consistent way. This would downplay the importance of mechanisms based on fiber-fiber interactions or the maintenance of neighbor relationships that have been implicated in the visual system (Cook, 1980; see also Lance-Jones and Landmesser, 1980a). Finally, we were unable to demonstrate a hierarchy of specificities, as proposed by Hollyday and her colleagues (1977).

We are left with the conclusion that specific chemical cues are necessary to guide the axons of motoneurons down appropriate pathways, but that these cues must be somewhat local. Alternatively, the capacity of axons to respond to the cues may vary in a discontinuous fashion across the whole population of lumbosacral motoneurons. In fact, we have begun to suspect that motoneurons are labeled in a discontinuous fashion, for we have been unable to detect graded differences in the behavior of motoneurons under various experimental conditions, based on their position in the cord. For example, any motoneuron in the sartorius motoneuron pool was equally likely to behave in a given fashion (i.e., take an aberrant path to reach its muscle), and this differed in an absolute way from the behavior of adjacent adductor or femorotibialis motoneurons. The extent to which the cues are also local, emanating at least in part from the targets themselves, can perhaps be determined by more discrete manipulations involving individual targets or muscles *in vivo. In vitro* experiments using targets or media conditioned by them might also be useful. However, given the findings from tissue culture models described in the introduction, the physicochemical environment actually encountered by axons as they grow into the limb may be extremely complex.

Finally, these experiments attest to the importance of pathway selection in contributing to the specific pattern of motoneuron connectivity. For we found that in all experimental situations in which motoneurons took wrong pathways, and therefore projected to inappropriate muscles, they established functional connections (Lance-Jones and Landmesser, 1981b). Further these incorrect connections were maintained to at least late embryonic stages, which encompassed the normal period of motoneuron cell death. Unless these inappropriately projecting motoneurons had their central connections altered by their new targets (a possibility that we are currently investigating) coordinated function of the limb would be greatly impaired. With no apparent mechanism to remove these wrong peripheral connections once formed, it seems necessary to conclude that pathway selection plays an important role in the generation of specific connectivity normally, and further that it is rather precise.

ACKNOWLEDGMENTS

I would like to thank Cynthia Lance-Jones for many helpful discussions and Mrs. F. Hunihan for secretarial assistance. Supported by NIH grant NS-10666.

REFERENCES

Attardi, D. G., Sperry, R. W. 1963. Preferential selection of central pathways by regenerating optic nerve fibers. *Exp. Neurol.* 7:46–64

Barbara, A. J. 1975. Adhesive recognition between developing retinal cells and the optic tecta of the chick embryo. *Dev. Biol.* 46:167–91

Beazely, L. D., Lamb, A. H. 1979. Rerouted axons in *Xenopus* tadpoles form normal visuotectal projections. *Brain Res.* 179:373–78

Bunt, S. M., Horder, T. J. 1978. Evidence for an orderly arrangement of optic axons in the central pathways of vertebrates and its implications for the formation and regeneration of optic projections. *Neurosci. Abstr.* 4:468

Castro, G. 1963. Effects of reduction of nerve centers on the development of nerve patterns in the wing of the chick embryo. *J. Exp. Zool.* 152:279–95

Chamley, J. H., Dowel, J. J. 1975. Specificity of nerve fiber attraction to autonomic effector organs in tissue culture. *Exp. Cell Res.* 90:1–7

Constantine-Paton, M. 1978. Central projections of anuran optic nerves penetrating hindbrain or spinal cord regions of the neural tube. *Brain Res.* 158:31–43

Constantine-Paton, M., Capranica, R. 1976. Axonal guidance of developing optic nerves in the frog. I. Anatomy of the projection from transplanted eye primordia. *J. Comp. Neurol.* 170:17–32

Cook, J. 1980. Interactions between optic fibres controlling the locations of their terminals in the goldfish optic tectum. *J. Embryol. Exp. Morphol.* 52:89–103

Detwiler, S. R. 1920. Experiments on the transplantation of limbs in *Amblystoma. J. Exp. Zool.* 31:117–69

Doljanski, R., Kapellar, M. 1976. Cell surface shedding—the phenomenon and its possible significance. *J. Theor. Biol.* 62:253–70

Ebendal, T., Jacobson, C. O. 1977. Tissue explants affecting extension and orientation of axons in cultured chick embryo ganglia. *Exp. Cell Res.* 105:379–87

Edwards, J., Chen, S. 1979. Embryonic development of an insect sensory system, the abdominal cerci of *Acheta domesticus. Wilhelm Roux's Archiv., Dev. Biol.* 186:151–78

Edwards, J. S., Berns, M. W., Chen, S. 1979. Laser lesions of embryonic cricket cerci disrupt guidepath role of pioneer fibres. *Neurosci. Abstr.* 5:158

Egar, M., Simpson, S., Singer, M. 1970. The growth and differentiation of the regenerating spinal cord of the lizard. *Anolis carolinensis. J. Morphol.* 131:131–52

Ferguson, B. 1978. Effect of dorsal-ventral limb rotating on the development of motor connections. Neurosci. Abstr. 4:11

Gaze, R. M., Keating, M., Ostberg, A., Chung, S. H. 1979. The relationship between retinal and tectal growth in larval *Xenopus:* Implications for the retino-tectal projection. *J. Embryol. Exp. Morphol.* 53:103–43

Gunderson, R., Barrett, J. 1979a. Neuronal chemotaxis: Chick dorsal-root axons turn toward high concentration of nerve growth factor. *Science* 206:1079–80

Gunderson, R. W., Barrett, J. N. 1979b. Axonal chemotaxis to nerve growth factor and its possible mediation by cyclic nucleotide and calcium. *Neurosci. Abstr.* 5:767

Hamburger, V. 1939. The development and innervation of transplanted limb primordia in chick embryos. *J. Exp. Zool.* 80:347–89.

Hamburger, V. 1961. Experimental analysis of the dual origin of the trigeminal ganglion in the chick embryo. *J. Exp. Zool.* 149:91–120

Hamburger, V. 1975. Cell Death in the development of the lateral motor column of the chick embryo. *J. Comp. Neurol.* 160:535–46

Harrison, R. G. 1908. Embryonic transplantation and development of the nervous system. *Anat. Rec.* 2:367–410

Harrison, R. G. 1910. The outgrowth of the nerve fiber as a mode of protoplasmic movement. *J. Exp. Zool.* 9:787–846

Hibbard, E. 1965. Orientation and directed growth of Mauthner's cell axons from duplicated vestibular nerve roots. *Exp. Neurol.* 13:289–301

Hollyday, M., Hamburger, V. Farris, J. 1977. Localization of motor neuron pools supplying identified muscles in normal and supernumerary legs of chick embryo. *Proc. Natl. Acad. Sci. U.S.A.* 74:3582–86

Hope, R. A., Hammond, B. J., Gaze, R. M. 1976. The arrow model: retino-tectal specificity and map formation in the goldfish visual system. *Proc. R. Soc. London Ser. B* 194:447–66

Horder, T. J. 1978. Functional adaptability and morphogenetic opportunism, the only rules for limb development? *Zoon Suppl.* 6:181–92

Horder, T. J., Martin, K. 1978. Morphogenetics as an alternative to chemospecificity in the formation of nerve connections. In: *Cell-Cell Recognition,* ed. A. S. G. Curtis, pp. 275–358. Cambridge: Cambridge Univ. Press.

Horton, J. C., Greenwood, M. M., Hubel, D. H. 1979. Non-retinotopic arrangement of fibres in cat optic nerve. *Nature* 282:720–22

Johnston, R., Wessels, N. 1980. Regulation of the elongating nerve fiber. In: *Current Topics in Developmental Biology,* vol. 16, Neural Development, Pt. 2: Neural Developments in Model Systems ed. A. A. Moscona, A. Monroy, R. K. Hunt, pp. 165–206. New York: Academic Press.

Katz, M., Lasek, R. 1979. Substrate pathways which guide growing axons in *Xenopus* embryos. 183:817–32

Keshishian, H. 1979. Origin and differentiation of pioneer neurons in the embryonic grasshopper. *Soc. Neurosci.* 5:536

Lance-Jones, C., Landmesser, L. 1980a. Motoneuron projection patterns in embryonic chick limbs following partial deletions of the spinal cord. *J. Physiol.* (London) 302:559–80

Lance-Jones, C., Landmesser, L. 1980b. Motoneuron projection patterns in the chick hindlimb following early partial reversals of the spinal cord. *J. Physiol.* (London) 302:581–602

Lance-Jones, C., Landmesser, L. 1981a. Pathway selection by embryonic chick lumbosacral motoneurons during normal development. (Submitted)

Lance-Jones, C., Landmesser, L. 1981b. Pathway selection by embryonic chick lumbosacral motoneurons in experimentally altered environment. (Submitted)

Landmesser, L. 1978a. The distribution of motoneurones supplying chick hind limb muscles. *J. Physiol.* (London) 284:371–89

Landmesser, L. 1978b. The development of motor projection patterns in the chick hind limb. *J. Physiol.* (London) 284:391–414

Landmesser, L., Morris, D. 1975. The development of functional innervation in the hindlimb of the chick embryo. *J. Physiol.* (London) 249:301–26

Letourneau, P. C. 1975a. Possible roles for cell-to-substratum adhesion in neuronal morphogenesis. *Dev. Biol.* 44:77–91

Letourneau, P. C. 1975b. Cell-to-substratum adhesion and guidance of axonal elongation. *Dev. Biol.* 44:92–101

Letourneau, P. C. 1978. Chemotactic response of nerve fiber elongation to nerve growth factor. *Dev. Biol.* 66:183–96

Levi-Montalcini, R., Menesini Chen, M., Chen, J. 1978. Neurotropic effects of the nerve growth factor in chick embryos and in neonatal rodents. *Zoon Suppl.* 6:201–12

Luduena, M. A., Wessels, N. 1973. Cell locomotion, nerve elongation and microfilaments. *Dev. Biol.* 30:427–40

Macagno, E. 1978. Mechanism for the formation of synaptic projections in the arthropod visual system. *Nature* 275:318–20

Meyer, R. L. 1980. Mapping the normal and regenerating retinotectal projection of goldfish with autoradiographic methods. *J. Comp. Neurol.* 89:273–89

Morris, D. 1978. Development of functional motor innervation in supernumerary hindlimbs of the chick embryo. *J. Neurophysiol.* 41:1450–65

Narayanan, C. H. 1964. An experimental analysis of peripheral nerve pattern development in the chick. *J. Exp. Zool.* 156:49–60

Nordlander, R., Singer, M. 1978. The role of ependyma in regeneration of the spinal cord in the urodele amphibian tail. *J. Comp. Neurol.* 180:349–74

Ramón y Cajal, S. 1905. *Studies on Vertebrate Neurogenesis,* tr. L. Guth, pp. 5–70. Springfield, Ill: Thomas, 1960

Ramón y Cajal, S. 1909. Histogénèse de la Moelle et des ganglions rachidions. In *Histologie du Système Nerveux de'l'homme et de Vertébrés,* vol. 1, p. 599. Madrid: Instituto Ramón y Cajal del C.S.I.C., 1952–1955 (reprint)

Rosen, J. J., Culp, L. A. 1977. Morphology and cellular origins of substrate-attached material from mouse fibroblasts. *Exp. Cell Res.* 107:139–49

Roth, S., Marchase, R. B. 1976. An *in vitro* assay for retinotectal specificity. In *Neuronal Recognition,* ed. S. Barondes, pp. 227–48. New York: Plenum

Rusoff, A., Easter, S. 1979. Order in the optic nerve of goldfish. *Science* 208:311–12

Sanes, J., Hildebrand, J. 1975. Nerves in the antennae of pupal *Manduca sexta. Roux Archiv.* 178:71–78

Scholes, J. H. 1979. Nerve fibre topography in the retinal projection to the tectum. *Nature* 278:620–24

Schubert, D., Whitlock, C. 1977. The alteration of cellular adhesion by nerve growth factor. *Proc. Natl. Acad. Sci. U.S.A.* 74:4055–58

Schubert, D. 1977. The substrate-attached material synthesized by clonal cell lines of nerve, glia and muscle. *Brain Res.* 132:337–46

Schubert, D. 1978. NGF–induced alterations in protein secretion and substrate-attached material of a clonal nerve cell line. *Brain Res.* 155:196–200

Shankland, M. 1979. Development of pioneer and sensory afferent projections in the grasshopper embryo. *Neurosci. Abstr.* 5:585

Silver, J., Robb, R. 1979. Studies on the development of the eye cup and optic nerve in normal mice and in mutants with congenital optic nerve aplasia. *Dev. Biol.* 68:175–90

Silver, J., Sidman, R. L. 1980. A mechanism for the guidance and topographic patterning of retinal ganglion cell axons. *J. Comp. Neurol.* 189:101–11

Singer, M., Nordlander, R., Egar, M. 1979. Axonal guidance during embryogenesis and regeneration in the spinal cord of the newt: The blueprint hypothesis of neuronal pathway patterning. *J. Comp. Neurol.* 185:1–22

Speidel, C. 1942. Studies of living nerves. VII growth adjustments of cutaneous terminal aborizations. *J. Comp. Neurol.* 76:57–69

Sperry, R. 1963. Chemoaffinity in the orderly growth of nerve fiber patterns and connections. *Proc. Natl. Acad. Sci.* U.S.A. 50:703–10

Stirling, R., Summerbell, D. 1977. The development of functional innervation in the chick wing-bud following truncations and deletions of the proximo-distal axis. *J. Embryol. Exp. Morphol.* 41:189–207

Stirling, R., Summerbell, D. 1979. The segmentation of axons from the segmental nerve roots to the chick wing. *Nature* 278:640–42

Straznicky, C., Gaze, R. M., Horder, T. C. 1979. Selection of appropriate medial branch of the optic tract by fibres of ventral retinal origin during development and in regeneration: An autoradiographic study in *Xenopus. J. Embryol. Exp. Morph.* 50:253–67

Taylor, A. C. 1944. Selectivity of nerve fibers from the dorsal and ventral roots in the development of the frog limb. *J. Exp. Zool.* 96:159

Weiss, P. 1941. Nerve patterns: the mechanics of nerve growth. In *Growth,* Suppl. to Vol. 5, pp. 163–203. Third Symposium on Development and Growth

Willshaw, D., Von der Malsberg, C. 1979. A marker induction mechanism for the establishment of ordered neural mappings; its application to the retino-tectal problem. *Phil. Trans. R. Soc. London* 287:203–43

5 / *Neuronal Cell Death and Some Related Regressive Phenomena During Neurogenesis: A Selective Historical Review and Progress Report*

RONALD W. OPPENHEIM

INTRODUCTION

It is both a pleasure and an honor to contribute the present essay on neuronal cell death to a volume marking the 80th birthday of my friend and teacher Viktor Hamburger. I hasten to add at the outset, however, that this pleasure is tempered (but only slightly) by the realization that Viktor was a pioneer in this field and that he has continued to be a major contributor to our understanding of cell death and related issues right up to the present time. Thus, any faint hope that his incisively critical eye might miss factual errors or conceptual flaws owing to unfamiliarity with the subject matter—to say nothing of his always keen ability to detect ambiguities and imperfections in one's prose—is remote. Consequently, my pen has moved more slowly and cautiously than usual across these pages, slackened by the everpresent thought, "Will Viktor consider this to be nonsense?" With that said, and in view of the rather intimate nature of this volume, I wish to begin with a digression that will indicate Viktor's influence on my own decision to study neuronal cell death.

As a student in Viktor's laboratory in the early 1960s, it was my primary task to participate in the still new, but already promising, study of the embryology of behavior in the chick. Following a lifetime devoted to nonbehavioral aspects of neural development, Viktor had begun in the early 1960s to focus instead on the behavioral and neurophysiological development of the embryo. As he recently put it, "The shaping of the nervous system, which has preoccupied us so far, is in a way only a prelude to its functional activation" (1979, p. 22). Despite my own preoccupation during this period with behavioral aspects of neurogenesis, it was impossible not to be influenced by close contact with some of the major participants in the history of the study of neuronal cell death, much of which history had been played out in the same rooms where Viktor and I were now observing the behavioral acrobatics of the embryo. Moreover, the shift of focus by Viktor and his colleagues in St. Louis to problems of behavioral development during this period was never absolute. Down the hall, Rita Levi-Montalcini, who had played such an important role in the study of cell death, continued to pursue the study of related issues with the nerve growth factor (NGF), and Eleanor Wenger, together with Max Cowan at the school of medicine, were beginning their important studies on cell death in the chick visual system. Even some of our neurobehavioral studies on the embryo during this period later proved relevant to certain aspects of neuronal cell death (e.g., see Hamburger, Wenger, and Oppenheim, 1966).

As a result of these various influences, I acquired an early interest in neuronal cell death which, for various reasons, remained latent for several years. It eventually became obvious, however, that my interest in cell death, although latent, was not buried very deep and did not require a major excavation to bring it to the surface. In 1973, while studying the earliest formation of synapses in the ventral horn of the chick spinal cord, my colleagues, Rainer Foelix, I-Wu Chu-Wang, and I observed degenerating spinal ganglion cells and spinal motoneurons between 4 days and 10 days of incubation. The allure of gaining a deeper understanding of neuronal cell death by taking advantage of new techniques (e.g., electron microscopy, thymidine autoradiography, and horseradish peroxidase histochemistry), which were not available to earlier investigators in the 1930s, 1940s, and 1950s, proved overwhelming, and thus we almost immediately set out to study this phenomenon by using the spinal motoneurons of the chick as a model. I suspect an additional motivation for this change in emphasis in my own research, was a vague desire to carry on a tradition begun so many years before by my teachers, Viktor Hamburger and Rita Levi-Montalcini.

EARLY HISTORY

In addition to providing insights into phylogeny, it has often been pointed out that Darwin's theory of evolution by natural selection also provided an impetus for studies of embryological development (Oppenheimer 1967; Gould, 1977). Consequently, since the last half of the 19th century it has been deeply ingrained in embryological thought that ontogeny—like phylogeny—is characterized by a process of change in which new structures, functions, and patterns are gradually added or acquired. That is to say, the prevailing view has long been that embryonic development is a constructive and progressive process. The alternative view—that development could, in any important sense, be destructive or regressive—has always seemed to be counterintuitive and illogical. Despite this prevailing conceptual bias in favor of the general progressive nature of development, a few well-known figures of 19th century biology did, in fact, recognize the potential role of regressive processes in development. Although it may seem that by embracing such a view, one would be in direct opposition to the idea of ontogenetic progress implied in the evolutionary notions of Darwin and his followers, in fact, the proponents of this "regressive" view were not only avowed Darwinists, but they even managed to couch their ontogenetic proposals in language that appeared entirely consistent with the prevailing Darwinian dogma.[1]

Although the presence of transient embryonic and larval features were known to early embryologists (e.g., see von Baer, 1828), the first person to suggest that competition and regression may be a common feature of individual development was Darwin's long time friend and supporter, T. H. Huxley. In 1869, only ten years after the appearance of *The Origin of Species,* Huxley commented that,

> It is a probable hypothesis that what the world is to organisms in general, each organism is to the molecules of which it is composed. Multitudes of these, having diverse tendencies, are competing with one another for opportunity to exist and multiply; and the organism as a whole, is as much the product of the molecules which are victorious as the Fauna and Flora of a country are the product of the victorious organic beings in it (1869, p. 309).

A few years later G. S. Lewes made a similar proposal, arguing that,

> Instead of confining the struggle for existence to the competition of rivals and the antagonism of foes, we must extend it to the competition and antagonism of tissues and organs. The existence of an organism is not only dependent on the external existence of others, and is the outcome of a struggle; but also on the internal conditions which cooperate in the formation of its structure, this structure being the outcome of a struggle (1877, p. 116).

Finally, in 1881 the pioneer German embryologist, Wilhelm Roux, published a book whose central theme is reflected in the title, *Der Kampf der Theile im Organismus (The Struggle of Parts Within the Organism).* Like Huxley and Lewes, Roux attempted to apply the Darwinian leitmotiv of natural selection—"the struggle for existence"—to the molecules, cells, and tissues of the developing organism. Although he had no direct evidence for this proposal, Roux's contribution was unique in at least one respect in that he was the first to suggest explicitly that the notion of struggle, or competition, might apply to neural development.

Despite the fact that Roux's book was widely cited, it had remarkably little influence in arousing an interest in the study of embryonic cell death. This is all the more surprising since its publication coincided with a burgeoning interest in embryology and neuroanatomy (including neural development) throughout the remainder of the 19th century. Part of the reason for the failure of Roux's book to initiate a sustained interest in cell death and other regressive embryological phenomena stems undoubtedly from the deeply ingrained belief that ontogeny *is* progress. Another reason may be that Roux's book had a pronounced Lamarckian bias. By the turn of the century, neo-Lamarckianism was rapidly falling into disrepute, in part because of the persistent criticisms of August Weismann (1894), and in part because of the virtual absence of any convincing evidence for the inheritance of acquired characteristics. Although Roux's ideas about struggle, competition, and regressive processes in cellular development could have stood on their own (i.e., they did not require a Lamarckian underpinning),[2] this fact was probably lost sight of by most supporters and critics in the turn-of-the century haste to rid biology of all traces of Lamarckianism. Thus, by embracing Lamarckianism, Roux may have unwittingly helped delay a serious interest in regressive phenomena in ontogeny by experimental embryologists (Oppenheimer, 1967).

One prominent investigator who may have been influenced by Roux's notions, however, was the great Spanish neuroanatomist, S. Ramón y Cajal.[3] In a paper written in 1919, and included in a reprinted collection of his early studies on neurogenesis (1929), Ramón y Cajal pointed out that,

> To avoid insurmountable mechanical obstacles and adverse contingencies of all kinds, an overproduction of nerve branches . . . would have to occur during development. For the same purpose there may be a superabundance of germinal cells *and even of typical neuroblasts* [italics added]. We must therefore acknowledge that during neurogenesis there is a kind of competitive struggle among the outgrowth (and perhaps even among the nerve cells) for space and nutrition. The victorious neurons . . . would have the privilege of attaining the adult phase and of developing stable dynamic connections (1929, p. 400).

Even earlier, in a discussion of the role of function in establishing specific neuronal connections related to learning and experience, Ramón y Cajal stated that

> during embryonic development dendrites and neuronal processes grow and branch progressively with time and establish relationships with a

> larger and larger number of neurons. But all of these connections which are established early in development are not permanent. In fact, a large number of them disappear by reabsorption of dendritic and axonal branches. Therefore, neuronal connections are not permanent and unchangeable. Tentative connections are created which are destined to be permanent or to be disrupted according to undetermined circumstances (1911, p. 887).

Despite these important insights, it would be misleading to imply that Ramón y Cajal recognized or anticipated the possibility of massive neuronal cell death during normal development. It is obvious from his comments that he was largely concerned with the regression of neuronal processes (axons and dendrites) and synaptic contacts and was only dimly aware of the possibility that large numbers of neurons may be lost during normal ontogeny. For instance, in an apparent reference to Roux, Ramón y Cajal warns that "it is important not to exaggerate, as do certain embryologists, the extent and importance of cellular competition . . . we consider it extremely likely . . . *that the immense majority of the neuroblasts survive to term and succeed in collaborating with the normal structures of the adult nervous system*" (1929, p. 400, italics added).

As the acknowledged leader of developmental neuroanatomy, Ramón y Cajal's failure to observe or even suggest a substantial loss of neurons during normal development may have been another significant factor—along with Roux's Lamarckianism—in delaying the recognition of natural cell death as a fundamental process in neurogenesis. Although there were sporadic reports of cell degeneration in the developing nervous system during the first half of the 20th century (e.g., Collin, 1906; Ernst, 1926; for a review, see Glücksmann, 1951), most embryologists either denied that these were dying cells (Rabl, 1900; Jokl, 1920) or dismissed cell death as a rare and unimportant feature of neurogenesis. One finds no mention of neuronal cell death in the early text books of the period, and even up to the late 1930s, neuroembryologists were apparently completely unaware of (or at least unimpressed by) this phenomenon (e.g., Detwiler, 1936; Weiss, 1939). In retrospect, part of the reason for this failure may also stem from the simple fact that dying embryonic cells degenerate and are phagocytosed exceedingly rapidly. Thus, in the absence of quantitative studies—and if researchers are not explicitly searching for them—degenerating cells often appear to be rare or even nonexistent. Only when neuroembryologists began to carry out systematic cell counts of both healthy and dying neurons at different stages of development, did it become obvious that there was a progressive and massive depletion of neurons. The first such report was the now-classic study of Hamburger and Levi-Montalcini (1949) on the spinal sensory ganglia of the chick embryo.

The study of Hamburger and Levi-Montalcini was motivated by the issue of peripheral or target regulation of neuronal development (i.e., central-peripheral relations), a problem of which the origins can be traced back to the turn of the century (Shorey, 1909; Dürken, 1911). A number of early investigators had observed that surgival removal of structures in the embryo, such as the limb bud or the eye, resulted in a severe depletion of the neurons normally innervating—or innervated by—the removed structure (for a review of this early literature, see Detwiler, 1936). Because many of the issues concerning the function of, and the mechanisms controlling, naturally occurring or spontaneous cell death were foreshadowed in this early work on central-peripheral relations, and because the two phenomena are also related in other respects, I would like to begin by reviewing this literature selectively.

Before doing this, however, I wish to point out that the major focus of this chapter is on what has traditionally been called *histogenetic cell death*. That is, in the case of the nervous system, the death of neurons that have ceased proliferation and have begun to differentiate. Histogenetic cell death has been defined by Glücksmann (1951) as the loss of cells during the differentiation of tissues and organs.[4] In contrast, *phylogenetic cell death* is considered responsible for the regression of vestigal organs or of larval organs during metamorphosis (e.g., the tadpole tail). *Morphogenetic cell death* involves the degeneration of cells during massive changes in morphology such as occur during invaginations and evaginations, and in the bending and folding of tissues and organs during early embryogenesis. Morphogenetic cell death may also be involved in the sculpturing of shape and form (e.g., the digits in the hand or foot). For a detailed discussion of morphogenetic cell death, along with many examples, the reader should consult Glücksmann (1951), Saunders (1966), and Silver (1978).

STUDIES ON CENTRAL-PERIPHERAL RELATIONSHIPS: TOWARD THE DISCOVERY AND UNDERSTANDING OF NATURAL CELL DEATH

Because comprehensive reviews of this literature are readily available (e.g., Hamburger, 1958; Detwiler, 1936; Hughes, 1968; Hughes and Carr, 1978; Jacobson, 1978), my major aim in this section is not to be encyclopedic, but rather to examine the conceptual issues involved and to show how these helped pave the way for the later discovery and interpretation of natural cell death in the nervous system.

The study of central-peripheral relations in the nervous system began as part of the general effort by early experimental embryologists to determine the extent to which cells, tissues, and organs develop by intrinsic mechanisms ("self-differentiation") versus their dependence upon extrinsic interactions ("correlative differentiation). As a result of the experiments, discoveries, and conceptual advances of Roux, Driesch, Weismann, and others in the 1890s (Oppenheimer, 1967; Spemann, 1938), embryology, at the turn of the century, was on the threshold of becoming thoroughly experimental ("causal-analytic"), and the major focus of this new approach was the determination, by experiment, of the chemical, physiological, and mechanical factors controlling development.

One of the early experiments on the nervous system done from this new conceptual approach was by M. Shorey, a student of the American embryologist, F. S. Lillie (Shorey, 1909). Shorey removed the limb buds of early chick and amphibian embryos and found that the spinal motoneurons and sensory ganglion cells were later severely depleted. She interpreted her findings as being consistent with Roux's notions of correlative differentiation and thus suggested that the "missing" neurons had failed to differentiate owing to the absence of a critical, but unknown, influence normally provided by the limb bud.[5]

The next major advance in this field came from the work of Samuel Detwiler, a student of Ross Harrison. In a series of studies nicely summarized in his 1936 book, *Neuroembryology*, Detwiler demonstrated that the transplantation of a limb bud to non limb regions (either spinal or cranial) in salamanders resulted in a numerical hyperplasia (i.e., more cells) in the sensory ganglia innervating the transplant.[6] R. M. May (1933) provided additional support for these results by showing that the transplantation of a supernumerary limb next to the normal limb in frogs also produced a hyperplasia in the sensory ganglia *and* in the motoneurons of the normal limb-innervating segments. Thus, on the basis of these data, it seemed obvious that systematic alterations in the size of peripheral targets of neurons could have either positive or negative effects, resulting in either fewer or more neurons in the affected centers. In amphibians, similar findings were reported for the olfactory system (Burr, 1916), the optic system (Twitty, 1932) and the mesencephalic V neurons that provide sensory innervation for the jaw musculature (Piatt, 1946). The study of congenital defects in mammals, especially those that involve the absence of a limb or the presence of supernumerary limbs or limb parts, showed that similar central-peripheral interactions existed in these forms as well; later, *experimental* studies on mammals verified these reports (Barron, 1945; Hall and Schneiderhan, 1945; Romanes, 1946). And in the chick, Hamburger (1934) confirmed and extended the earlier findings of Shorey on the effects of limb-bud removal.[7]

An important new finding from Hamburger's study was his report of a proportional, almost linear, relationship between the amount of remaining limb musculature—the surgery was not always complete—and the extent of the resulting hypoplasia in the spinal cord. In a later study, Hamburger (1939a,b) showed that the transplantation of a supernumerary wing or leg bud, next to the normal limb, also resulted in a motor and sensory hyperplasia in the limb-innervating segments of the chick spinal cord.

By the 1940s, then, it was widely accepted that interactions between developing neurons and their targets, were responsible for the regulation or adjustment of the size of neural centers relative to the size of the periphery. The major remaining unresolved problems concerned (1) what exactly was being regulated, and (2) how the regulation was being mediated (i.e., the cellular mechanisms involved). The initial attempts to come to grips with the second question involved a return to the ideas of Roux, whereas tentative answers to the first required the creation of novel hypotheses.

Related to his early formulation of two different modes of development (i.e., self-differentiation and correlative differentiation), Roux had also proposed that all cells, tissues, and organs go through two general phases of development. He suggested that there is an initial period when cells, tissues, and organs are formed in anticipation of, and without the participation of, function, and a later period, in which the final stages of differentiation require both endogenous function, and functional activation via extrinsic stimuli for their completion. According to Roux, the two periods were not to be viewed as mutually exclusive, however, for, as he pointed out, "between the two periods lies presumably a transition period . . . in which both classes of causes are concerned" (1905, p. 96).[8]

Following Roux, several of the early papers on central-peripheral relations noted that the onset of the neural changes resulting from alterations in the size of the peripheral targets, were correlated with the emergence of function in the parts concerned. For instance, Detwiler (1936, p. 76) pointed out that if a unilateral limb-bud removal or transplantation is done at the tail bud stage, it is several days before there is a detectable difference between the spinal ganglia of the two sides. He further noted that the time when the neural changes are first observed correlates rather closely with the onset of function in the limbs. Drawing on the related observation of Coghill (1929) that neurons may continue to grow and differentiate while they function, Detwiler and others made the tentative suggestion that the regulation of neural development by peripheral targets might involve functional interactions. Detwiler, for example, pointed out that following limb removal the spinal ganglion neurons would be deprived of the normal flow of impulses from peripheral receptors and that the motor neurons would be deprived of sensory input via reflex arcs. Despite the apparent plausibility of this notion, however, Detwiler and other neuroembryologists of the period apparently tended to discount the role of function.

Based on the classic experiments of Harrison (1904), Carmichael (1926, 1927) and Matthews and Detwiler (1926), in which amphibian embryos were found to develop normally when paralyzed pharmacologically (i.e., when all CNS function was presumably eliminated), it was argued that the critical interaction in regulating the size of neural centers was probably *trophic* and thus not due to neurophysiological function.[9] As Detwiler (1936) summarized the then prevailing view, "we must look for the cause in some characteristic of the growing nerve other than its conduction phase" (p. 134), and "the evidence at hand suggests that the influence is of a trophic nature" (p. 189). Detwiler also suggested that, "chemical substances (neurohumors) given off by the end of the developing nerve constitute the stimulating agent" (p. 134).

Concerning the question of *what* aspects of neural development were being regulated by central-peripheral interactions, the general consensus at the time was that cell proliferation and the recruitment of so-called indifferent cells were the major, if not the sole, processes subject to peripheral control. A critical examination of the early studies cited in support of the notion that cell proliferation is controlled by the size of the peripheral target reveals, however, that most of the data are either weak or nonexistent. In almost all instances, cell counts of differentiated cells made long after the cessation of proliferation were used as the sole criterion for arguing that proliferation was modified following surgical alterations in the size of the peripheral target. In only a few cases (e.g., Burr, 1932) was it directly shown that the number of mitotic figures was increased or decreased relative to the size of the periphery. And even in these cases, it was not possible to determine whether the increased proliferation involved neurons or glia. Later attempts to demonstrate changes in neuronal proliferation as a result of changes in the size of the periphery (e.g., Hamburger and Levi-Montalcini 1949; Kollros, 1953) are also subject to the same drawback. Indeed, recent studies using thymidine autoradiography to label dividing cells have shown rather conclusively that, whereas glial proliferation is, in fact, decreased following peripheral depletion, neuronal proliferation is unaffected (Currie and Cowan, 1974; Carr and Simpson, 1978a,b).[10] (A recent study by Bibb [1978], in which it is argued that neuronal proliferation in frog spinal ganglia is increased following an effective increase in the size of the periphery, is subject to a number of criticisms and thus at present should not be taken as contradicting the current view, as just summarized.)

A related hypothesis for explaining the numerical changes in neurons following modifications in target size was that cells that were initially indifferent, or possibly totipotent, were induced ("recruited") to differentiate into the particular cell types needed—for example, more motoneurons or sensory ganglion cells, as against interneurons or glia—as dictated by the size or demands of the periphery. One of the earliest, if not the first, proponents of this so-called *recruitment hypothesis* was R. L. Carpenter (1932, 1933). In an attempt to explain his findings of hyperplasia in salamander spinal ganglia following peripheral overloading, Carpenter proposed that under the influence of the increased demands of the periphery, indifferent cells are induced to proliferate and differentiate into additional sensory neurons.

This recruitment hypothesis was further defined and elaborated by Hamburger (1934, 1939) and by Hamburger and Keefe (1944). To quote from the latter paper: "pioneer motor fibers . . . which establish the first connection between nerve center and periphery must be instrumental in mediating between the periphery and the motor centers. They report to the central station the extent of the peripheral area to be innervated and control the recruiting of new neurons and motor fibers until the reinforcement is sufficient to satisfy the needs of the periphery" (p. 224). Since the pioneer neurons themselves would not be subjected to hypoplasia or hyperplasia but would merely mediate in this process, Hamburger and Keefe (1944) argued that any putative growth agents must emanate from these differentiating pioneer neurons *within the spinal cord* and recruit indifferent cells in their neighborhood.

The major issue, then, was to determine the nature of this "inductive" influence on indifferent cells. Since it was not recognized at the time that natural cell death was an integral step in the development of either spinal motor centers or of sensory ganglia, Hamburger and Keefe (1944) set out to determine which of the three major known steps in neural development—proliferation, migration, or differentiation—was the primary site of inductive control by the periphery.

Following either unilateral limb-bud removal or transplantation of an additional limb in the chick, Hamburger and Keefe compared the number of mitotic figures in the ependymal layer on the control and operated sides of the brachial spinal cord. Finding no evidence for a significant change in proliferation, they then went on to do cell counts of all motor and nonmotor cells in the ventral half of the spinal cord. They argued that since proliferation is unaffected, the total number of cells, at later stages, should be the same on both sides of the spinal cord. In the case of hypoplasia (after limb removal), there should be more *nonmotor* cells, and in the case of hyperplasia (additional limb), there should be more *motor* cells. In other words, the specific demands (i.e., the size) of the periphery should influence the number of nonmotor or indifferent cells that are induced or recruited to differentiate into motor cells.

The results of Hamburger and Keefe (1944) supported this prediction since they found, for example, in the case of motoneuron hypoplasia following limb removal, that the deficiency of large motoneurons was rather precisely balanced by an increase in smaller, nonmotor cells. They concluded that their findings were consistent with the notion that the periphery controls the recruitment or differentiation, and not the proliferation or migration, of indifferent cells. Although Hamburger and Keefe presented no data on spinal ganglion cells, the possibility that their interpretation might also explain changes in this system was tested a few years later by Hamburger and Levi-Montalcini (1949).

Plausible as their interpretation seemed at the time, the conclusions of Hamburger and Keefe were not the final word on this issue. The later study of Hamburger and Levi-Montalcini (1949), while providing some support for the recruitment hypothesis in the chick spinal ganglia, differed from Hamburger and Keefe's findings on the motor system in that proliferation as well as the induction or recruitment of indifferent cells was apparently affected by the periphery. (As I indicated above, however, the contention of Hamburger and Levi-Montalcini that the size of the periphery controls the proliferation of neurons has been called into question by Carr and Simpson, who argue convincingly [1978a,b] that *glial*, not neuronal, proliferation is affected by limb removal or transplantation.)

However, the most striking and novel finding in Hamburger and Levi-Montalcini's study (1949) was that there is a massive loss of spinal ganglion cells in the nonlimb-innervating segments *during normal development.* This was the first convincing demonstration that naturally occurring cell death is a significant feature in the development of the vertebrate nervous system. And, as they prophetically noted, this finding implies that during normal development an excess of neurons is produced and that some of the cells degenerate owing to the inability of the periphery to support all of them.[11] The inference that the periphery is primarily necessary for the continued *growth and maintenance* of neurons following the cessation of proliferation and the onset of differentiation, forced a reappraisal of the entire earlier literature on central-peripheral interactions.

This important discovery that *differentiating* sensory neurons may subsequently degenerate if not provided with an appropriate peripheral target, raised the question whether a similar situation may also occur in the chick motor system. In 1958, Hamburger reexamined this issue, once again using the limb-bud removal model as a crucial test of the various alternative hypotheses. In contrast to the earlier study of Hamburger and Keefe (1944), Hamburger made detailed observations and quantitative analyses of the motor system from the earliest stage that the lateral motor column was histologically recognizable (i.e., on Day 5 of incubation or 2½ days following limb-bud removal). In doing so, he discovered that the motor system was numerically complete by five days of incubation and that the control and operated sides did not differ in cell number at that time. (See also Hollyday and Hamburger, 1977).

Thus, the earlier contention of Hamburger and Keefe—based on cell counts done at a terminal stage many days after limb removal—that depletion of the motor cells results from a failure of recruitment was shown to be untenable. Furthermore, Hamburger found a significant number of frankly degenerating neurons in the motor column of the operated side after five days. This implied that, as in the case of the sensory ganglion cells, motor cells that had begun differentiation but were prevented from forming contacts with their synaptic target (i.e., the limb musculature) also undergo a secondary degeneration.[12]

Since the interpretation provided by Hamburger (1958) for these findings is now considered valid for a multitude of similar studies in other neural systems and in other species (for a comprehensive review, see Jacobson, 1978), it is only fitting that he be allowed to have the final word on this issue. To quote from the concluding page of his 1958 paper:

> The initial development of the lateral motor column in its qualitative and quantitative aspects is determined by intrinsic factors, and all fibers grow out irrespective of peripheral conditions. The quantitative relationship between the number of motor neurons and the size of the peripheral field of innervation is established by a selective survival of those neurons which find an adequate peripheral milieu, and the degeneration of all others (1958, p. 399).[13]

NATURALLY OCCURRING CELL DEATH: A "NEW" PHENOMENON IN NEUROGENESIS

As noted above, prior to 1950 the few sporadic early reports of neuronal cell death in apparently normal embryos (Collin, 1906; Ernst, 1926) were largely ignored, or met with skepticism and disbelief, or were discounted as rare, even pathological, phenomena. No embryological textbook of the period even mentions neuronal cell death, and neither Detwiler's 1936 book on neuroembryology, nor Paul Weiss's chapter on neurogenesis, in his 1939 text on developmental biology, discuss this issue. Therefore, the report of Hamburger and Levi-Montalcini (1949) served as a watershed, after which virtually all subsequent accounts of neural development took note of their findings and at least acknowledged the existence of naturally occurring neuronal cell death. Despite this newly found—and well-deserved—attention, however, the report of Hamburger and Levi-Montalcini did not elicit (at least not for several years) a sustained interest in neuronal cell death.

This is all the more surprising considering that in this same period Levi-Montalcini (1950) reported finding massive cell death in another region of the chick nervous system and that Glücksmann (1951) had published his comprehensive review on cell death in a variety of developing tissues. Moreover, during this period the report of Bueker (1948), who was a student of Hamburger and Levi-Montalcini, led to the discovery of the NGF, which gave an entirely new and exciting twist to the study of neuronal cell death (for charming personal accounts of these events, see Levi-Montalcini, 1975, and the present volume).

Notwithstanding these events, and with the notable exception of the paper by Hughes (1961) on cell death of spinal motoneurons in the frog, *Xenopus laevis,* it was not until the late 1960s that the study of neuronal cell death finally became a major focus of attention of neuroembryologists (see the reviews by Prestige, 1970, and Cowan, 1973). It is hard to escape the impression that even after 1949 most students of the nervous system continued to be influenced by the deeply ingrained belief that regressive phenomena, such as cell death, were exceptional and thus undeserving of a sustained interest.

EVENTS FROM 1949–1970: A PRELUDE TO THE "MODERN ERA"

To devotees of this field, it is now widely recognized that there is a close relationship between the findings and interpretations derived from the study of central-peripheral interactions in the nervous system on the one hand, and those derived from the study of naturally occurring neuronal cell death on the other. For those less familiar with this literature, I have attempted in the previous sections to summarize the early history of this relationship in order to demonstrate the gradual emergence of concepts that are pertinent to both problems and that we accept today as commonplaces. In doing so, it has been my intention to show that a number of our current concepts about the function of natural cell death, and the mechanisms involved, were foreshadowed in the early literature. Indeed, in some respects we have hardly advanced at all beyond our predecessors. But, as I hope to demonstrate, this is not so much an indication of our own intellectual fossilization as it is a tribute to the sagacity of the pioneers in this field. That they did not discover everything is a welcome consolation and challenge to those of us still struggling to shed new light on these issues.

In addition to the appearance of a number of studies in the older tradition of central-peripheral interactions—that is, without regard to whether or not natural cell death was involved (for a review, see Jacobson, 1978)—the period between 1949 and the early 1970s was also noteworthy, especially after 1960, in that cell death was reported in an increasing number of regions of the nervous system. During this period, naturally occurring cell death, based either on data involving cell counts or on the direct observation of large numbers of degenerating neurons, was reported in the following systems: the cervical motoneurons of the chick (Levi-Montalcini, 1950); the motor and spinal ganglion cells of the frog, *Xenopus laevis* (Hughes, 1961; Prestige, 1965, 1967); motoneurons of the mouse (Romanes, 1946; Harris, 1969); the trochlear and isthmo-optic nuclei of the chick (Cowan and Wenger, 1967, 1968); the mesencephalic nucleus of the trigeminal nerve of the chick (Rogers and Cowan, 1973); the optic tectum of the chick (Cantino and Sisto-Daneo, 1972); cerebellar Golgi cells of the mouse (Larramendi, 1969); the oculomotor nucleus of the mouse (Zilles and Wingert, 1973); the cochlear nuclei of the mouse (Mlonyeni, 1967); the spinal motoneurons of the frog, *Rana pipiens* (Pollack, 1969a); the spinal motoneurons and ganglia of the frog, *Eleutherodactylus martinicensis* (Hughes and Egar, 1972); the spinal motoneurons and sensory ganglia of the opossum (Hughes, 1973); the motoneurons of the fetal monkey spinal cord (Bodian, 1966); and Hofmann's nucleus of the chick embryo spinal cord (Dubey, Kadasne, and Gosavi, 1968).

Beyond extending the original observations of Hamburger and Levi-Montalcini (1949) to other neuronal systems and to other species, however, these reports, with a few notable exceptions, did little to advance our understanding of the function of (i.e., the adaptive role), or the mechanisms controlling, natural cell death. By stating this, I do not wish to imply that these studies were unimportant. In some instances, new light was, in fact, shed on these issues, and in other cases important new theoretical proposals were put forth. Even when this was not the case, the mere documentation of natural cell death in different groups of neurons was still of considerable value, for, as I shall argue later, an important key for understanding natural cell death may be a comparison of the systems exhibiting substantial cell loss during normal ontogeny with those showing little or none.

In Cowan and Wenger's paper (1967) on the trochlear nucleus and in the paper by Prestige (1967) on frog motoneurons, it was shown for the first time in a single population of cells that natural cell loss and cell loss owing to target removal (induced cell death) both occur within the same time frame, thus implying that a common factor is involved. (Hamburger and Levi-Montalcini [1949] had earlier noted that natural cell loss in *nonlimb* ganglia and induced cell loss in *limb* ganglia are also temporally coincident.) In an attempt to explain the *function* of natural cell death, Hughes (1965) was the first to propose that cell death may remove inappropriate or erroneous synaptic connections (i.e., that the adult pattern of neuromuscular connections is accomplished by the loss of inappropriate, and the retention of appropriate, connections). Finally, with regard to the *mechanism* of cell death, Hughes and Prestige (1967) suggested that *functional* interactions between nerve and muscle may be involved, and Prestige (1967) extended the earlier suggestion of Hamburger and Levi-Montalcini (1949) that a trophic or maintenance factor supplied by the peripheral target may regulate cell survival.

In his timely 1973 review on neuronal cell death, Cowan summarized the current state of affairs by commenting that it is "difficult at this time to do more than document the observation that the phenomenon is a real one; much further work will be needed before we can answer most of the critical questions which these descriptive studies have raised" (1973, p. 28). Since then, some fundamental advances have been made, due in no small measure to Cowan's review, which helped focus attention on cell death and the problems yet to be resolved. Accordingly, in the remainder of this essay I shall review the progress that has been made in the last decade and attempt to indicate the types of additional evidence that are still needed to attain a comprehensive understanding of neuronal cell death.

THE DIFFERENTIATION OF NEURONS PRIOR TO CELL DEATH

In attempting to determine the extent to which neurons differentiate prior to naturally occurring cell death, one is immediately confronted with the problem of distinguishing cells that are destined to die from those that will survive, *prior* to the time that the former display frank signs of degeneration. In most, if not all, cell groups showing natural cell death, the neurons that will ultimately die are apparently randomly intermingled with those that survive (i.e., the cells destined to die are not spatially segregated). And although an examination of degenerating cells in the electron microscope can provide some information concerning their prior differentiation, degeneration occurs so rapidly in most populations that even at the earliest stages of the breakdown process, important cytological or cytochemical features that were formally present could be obscured. Detailed ultrastructural studies of entire cell groups prior to the onset of degeneration, while possible, are seriously limited by time and sampling problems inherent in electron microscopy.

Although the study of both frankly degenerating cells and of cell groups exhibiting natural cell death prior to their degeneration can yield valuable information about differentiation, another approach is to examine the differentiation of neurons following removal of their target (induced cell death), which, in many cases, results in the eventual degeneration of virtually all of the cells involved. This approach eliminates some of the drawbacks discussed above.

The use of the induced cell death model rests, however, on the assumption that induced and natural cell death obey similar rules and that induced cell death is merely a quantitative exaggeration of, and not qualitatively different from, natural cell death. If this assumption is correct—and I shall provide information on this point below—then induced cell death provides a powerful model for natural cell death by allowing one to study a particular cell group for which it is known that virtually all the cells will eventually die. Indeed, it was the study of induced cell death which provided the first information on the differentiation capacity of neurons that will ultimately die during embryogenesis.

Based largely on the use of neurofibrillar silver stains, Hamburger and Levi-Montalcini (1949) showed that until the stage when degeneration begins, spinal sensory ganglion cells in the chick may differentiate normally following limb-bud removal. Similarly, Hamburger (1958) demonstrated that spinal motoneurons of the chick appear to differentiate normally, prior to degeneration, following limb-bud removal. Using Nissl and neurofibrillar silver stains, Hamburger found that the initial differentiation of the deprived motoneurons appeared indistinguishable from motoneurons on the contralateral (control) side of the spinal cord. The deprived cells developed axons that grew out to the "periphery" where they formed a neuroma-like tangle of nerve endings some distance from the body wall to which the limb would normally be attached.[14] Judging from the thickness of the ventral roots, on the deprived side, Hamburger inferred that many of the motoneurons had sent out axons prior to cell death. And as noted above, Hamburger made the important observation that prior to the onset of cell death, the number of motoneurons in the ventral horn on the deprived side was similar to that on the control side, thereby indicating that proliferation and migration were unaffected by target removal. And in so doing, he vindicated the earlier contention of Levi-Montalcini and Levi (1942, 1944) that target removal only affects the maintenance of differentiated motoneurons. A number of subsequent studies using the induced cell death model have reported similar findings in several different neuronal systems (for review, see Jacobson, 1978), which confirm the general belief that target removal does not affect initial differentiation.

The first indication that the induced cell death model may not always accurately reflect the differentiation potential of neurons during natural cell death came from the comparison of natural and induced cell death in the chick ciliary ganglion (Pilar and Landmesser, 1976). Although these authors had earlier reported that the ciliary ganglion neurons exhibit a remarkable capacity to differentiate normally when deprived of their target by early eye removal (Landmesser and Pilar, 1974a,b), a more detailed comparison using the electron microscope revealed that the deprived ciliary neurons failed to attain an apparently critical step in differentiation.

Whereas all of the ciliary neurons, including the 50 percent destined to die by natural cell death in control ganglia, developed a well-organized, rough endoplasmic reticulum (RER) and formed polyribosomes (both of which events coincided with the onset of synapse formation in the periphery), none of the peripherally deprived neurons underwent these changes. As a consequence of this failure, when the deprived neurons died, they underwent a different type of cytological degeneration from the control neurons. Natural cell death was heralded by dilation of the RER and cytoplasmic disruption, followed later by nuclear changes, whereas induced cell death in the deprived neurons was mainly characterized by regressive nuclear changes. These different degenerative sequences have been labeled as *Type 1* (nuclear changes) and *Type 2* (cytoplasmic changes) by Chu-Wang and Oppenheim (1978a). Landmesser and Pilar related the difference in differentiation capability between induced and natural cell death to the inability of the deprived neurons to form even transient peripheral synaptic connections.

According to their proposal, this failure to form even provisional synapses deprives the neurons of a critical inductive signal from the periphery which is necessary to trigger the development of the RER and which is ultimately required for the large, normal increase in protein synthesis necessary for survival of the cells; and when deprived of certain proteins, the cells die. Consistent with this suggestion is the independent observation that in normal ciliary ganglia the large increase in the synthesis of the cholinergic enzymes, choline-acetyltransferase (CAT) and acetylcholinesterase (AChE), coincides with the formation of peripheral synapses and with the appearance of RER and polyribosomes (Chiappianelli et al., 1976). It is also of interest that whereas the axons from normal ganglion cells sprout transient axon collaterals following the formation of peripheral contacts, this fails to occur in ganglion cell axons deprived of their synaptic targets (Landmesser and Pilar, 1976). A similar effect of peripheral deprivation has also been noted in the trochlear nucleus (Sohal, Weidman, and Stoney, 1978).

According to Pilar and Landmesser (1976), the sequence of ultrastructural changes observed during natural cell death in the ciliary ganglion are thought to reflect the pathological accumulation of proteins intended for transport to the nerve endings. As a consequence of either forming too few synapses, or of not forming enough *stable* synapses, the materials resulting from increased protein synthesis would not be utilized and their accumulation, as reflected in the swollen RER, would precipitate the Type 2 degenerative process. (See footnote 13.)

In contrast to the observations of Pilar and Landmesser concerning the differences between induced and natural cell death in the ciliary ganglion, the few other systems that have been studied appear indistinguishable in this respect. Spinal motoneurons in the chick (Chu-Wang and Oppenheim, 1978a,b), trochlear motoneurons in the duck (Sohal and Weidman, 1978b), and retinal ganglion cells in the chick (Hughes and McLoon, 1979) all show similar capacities for differentiation and exhibit similar modes of degeneration in both the induced and natural cell death situations. Although the chick spinal motoneurons also exhibit two different types of degeneration, which are similar to the two categories described by Pilar and Landmesser (1976), both types occur during induced *and* natural cell death. And whereas the chick *retinal ganglion* cells, in both the induced and natural situation, appear to degenerate prior to the normal large increase in RER (i.e., similar to induced cell death in the ciliary neurons), many of the chick *spinal motoneurons,* and apparently all of the duck *trochlear neurons,* develop rather extensive RER prior to the onset of either natural or induced degeneration. The situation is further complicated by the finding that some of the chick spinal motoneurons, in both induced and natural cell death, begin to degenerate *prior* to the increase in RER.

Although many more kinds of neurons will have to be examined before any generalizations can be made, at present it seems reasonably clear that neurons differ in the extent to which they may differentiate prior to natural cell death. Furthermore, some neuronal types develop normally (i.e., up to the onset of cell death) in the absence of their targets, whereas others do not. These differences may reflect differing intrinsic capacities for maturation in the absence of target influences, or they may reflect differences in some aspects of synaptogenesis prior to the onset of cell death. For instance, some neurons may establish a greater number of or more stable synaptic connections, which leads to an increased differentiation before cell death begins. The extent of differentiation may, in turn, dictate the type of degenerative changes a neuron exhibits. If the stage of increased RER and polyribosomes is reached, then cells may undergo Type 2 degeneration; if not, then Type 1 may be the primary mode of degeneration.

Despite these variations, the proposal of Pilar and Landmesser (1976) concerning the cytochemical mechanisms underlying the two different modes of cell death (i.e., Types 1 and 2) may still be correct. That is, irrespective of the type of neuron involved (e.g., ciliary versus spinal motoneuron), Type 1 degeneration may reflect cells that are dying due to the *absence* of an inductive signal from the target necessary for increased protein synthesis, whereas, Type 2 may reflect cell death due to an inability of the cell to utilize accumulated proteins following the normal increase in protein synthesis. Because some neurons may degenerate immediately after the cessation of cell division (Hughes & Carr, 1978; Carr, personal communication), or even during migration (e.g., Hughes 1973, Hughes & LaVelle 1975, Chu-Wang, Oppenheim, and Farel, 1980), and because these cells are presumably less differentiated than cells that have completed migration, it will be of interest to determine the type of ultrastructural changes these neurons exhibit during degeneration. There is some evidence to suggest that relatively undifferentiated neurons in the chick thoracic sensory ganglia, which apparently lack neural processes, undergo a type of degeneration which, in some respects, is rather similar to Type 1 (Pannese, 1976).

So far, I have focused primarily on certain specific ultrastructural characteristics for assessing the differentiative capacity of neurons prior to cell death. A number of other anatomical, physiological, and biochemical aspects of differentiation have also been examined in an attempt to characterize better the status of neurons prior to natural and induced cell death.

One modification of the original recruitment hypothesis for explaining the numerical hyperplasia and hypoplasia following modifications in the size of the periphery was made by D. H. Barron, who contended that, when the pioneering fibers reach their peripheral target, they receive a signal that induces dendritic growth in these neurons. The actual recruitment of "indifferent" cells was thought to depend upon a transneuronal signal from the dendrites of these pioneering fibers to adjacent "indifferent" cells with which they were assumed to be in physical contact. According to this proposal, dendrite development should be inhibited or greatly impaired for neurons lacking a peripheral target. (The original hypothesis only required that the *cell bodies* of neurons be in communication and said nothing about dendritic involvement.)

In support of this notion, Barron (1948) reported that spinal motoneurons in the chick fail to develop dendrites following limb-bud removal. My colleagues and I have recently reexamined this question using a variety of anatomical techniques, including silver stains, HRP labeling of neurons, and electron microscopy (Oppenheim, Chu-Wang, and Maderdrut, 1978). In contrast to the report of Barron, we could detect no effects of limb removal on dendritic development. Prior to the onset of induced cell death, the peripherally deprived motoneurons developed extensive dendritic processes. Similar findings have also been reported in other systems (Landmesser and Pilar, 1974a,b; Hughes and LeVelle, 1975; Knyihar, Csillik, and Rakic, 1978).

In the chick spinal motoneurons (Oppenheim et al., 1978) and in the chick ciliary ganglion (Pilar and Landmesser 1974a,b), axodendritic synapses also form on cells lacking a periphery. These synapses in the ciliary ganglion are functional by electrophysiological criteria (Landmesser and Pilar, 1974a,b). Since the cells from which these ciliary afferents arise—that is, the accessory oculomotor nucleus (AON)—are also known to exhibit natural cell death (Cowan and Wenger, 1968; Narayanan and Narayanan, 1978), it is of interest that Landmesser and Pilar (1976) have reported finding degenerating presynaptic profiles on normal as well as on peripherally deprived ciliary neurons. Although some of these degenerating synapses may represent contacts that would have survived in the absence of induced cell death in the ciliary ganglion, in the case of the normal ganglion they must reflect a naturally occurring process. If this is the case, then these findings strongly imply that cells in the AON destined to die by natural cell death can form anatomical and functional synaptic contacts with their normal targets in the ciliary ganglion. I shall discuss this possibility in greater detail in the following sections.

In addition to anatomical, cytological, and physiological criteria for assessing the differentiation potential of cells that will eventually die, neurochemical and histochemical measures also provide a potentially powerful approach to this question. Because neurons may become determined even before they exhibit cytological features characteristic of their mature state, histochemical and neurochemical measures may be the most sensitive early index of determination and initial differentiation (Maderdrut, 1979). Furthermore, since most neurons that have been examined show progressive increases in cell specific substances during development (e.g., neurotransmitters and related enzymes), it is important to determine the extent to which these changes can occur in cells that will subsequently degenerate.

Between three and five days of incubation, spinal motoneurons in normal chick embryos exhibit a severalfold increase in the activity of the cholinergic enzymes choline acetyltransferase (CAT) and acetylcholinesterase (AChE). Following limb-bud removal on Day 2 of incubation, both enzymes show increases similar to that in controls (Oppenheim et al., 1978; Oppenheim and Maderdrut, in preparation). By Day 6, however, when induced cell death is well underway, there is significantly less enzyme activity on the peripherally deprived side, and by Day 10 this difference amounts to a loss of approximately 70 percent. Similarly, on Day 5, the histochemical appearance of AChE in the motoneurons on the deprived side of the spinal cord is indistinguishable from that on the control side. But, by Day 10 there is virtually no histochemically detectable AChE in the ventral horn on the peripherally deprived side. Glutamic acid decarboxylase (GAD) activity (a putative enzyme marker for GABA-secreting interneurons in these same preparations) shows a small but significant decrease at 15 days of incubation but not at earlier stages (Maderdrut and Oppenheim, in preparation). This delayed decrease in GAD activity may reflect a transneuronally mediated loss of interneurons secondary to the loss of motoneurons and sensory ganglion cells which occurs after limb-bud removal (compare with Bueker, 1947). The findings involving the cholinergic enzymes imply that, prior to degeneration, cells destined to die by induced cell death undergo a normal sequence of biochemical differentiation. Moreover, in view of our observation that the cytological differentiation of neurons destined to die by natural cell death appears normal, it seems reasonable to infer from these data on induced cell death that neurochemical differentiation is also normal prior to *naturally* occurring cell death.

Notwithstanding the observations of Pilar and Landmesser (1976) on the ciliary ganglion, the available data—and it is admittedly quite limited—indicates that, for many cell types, most, if not all, aspects of differentiation are normal prior to the onset of neuronal cell death. In all cell types examined so far, including ciliary neurons, there does not appear to be a subpopulation of cells that are programmed to die in the sense that their pathway of differentiation and ultimate fate is already determined at some early stage long before cell death begins. If that were true, the cells could be expected to display some noticeable differences during early differentiation. But, as I have tried to demonstrate here, that does not appear to be true. Even in the case of ciliary neurons, differentiation is normal up to the time that synaptic contacts are formed.

One prominent feature of neuronal differentiation, which I have deliberately avoided discussing so far, is the outgrowth of the axon. Because this phenomenon is such a fundamental aspect of neuronal differentiation, and because, as I discuss below, it appears to be critically involved in the regulation of cell death, I have devoted the following separate section to several issues related to axonal outgrowth and "peripheral" innervation.

AXONAL OUTGROWTH, TARGET INNERVATION, AND CELL DEATH

As noted above, one of the early explanations of naturally occurring neuronal cell death involved the notion of a competition between neurons for synaptic targets or for a trophic substance from the target region (Hamburger and Levi-Montalcini, 1949). Unless one wants to resort to some variation of the early recruitment hypothesis, a central and unavoidable tenet of competition is that all neurons, even those that subsequently die, send axons to their target. For according to this hypothesis, it is interactions at the target that influence the decision leading either to continued growth and maintenance or to regression and ultimate death. Until quite recently, however, the assumption of peripheral or target innervation prior to cell death was unproven.

Circumstantial evidence for target innervation prior to cell death has long been available, from observations showing that target innervation and the onset of cell death are temporally coincident (Hamburger, 1975). More convincing, but still indirect, support for this assumption comes from the observation that at least some peripheral axonal processes that arise from the cervical spinal cord of the chick disappear coincident with the massive loss of cervical motoneurons between four and five days of incubation (Levi-Montalcini, 1950). Similarly, Levi-Montalcini (1964) has also reported the presence of transient nerve processes, in close association with the mesonephric tubules of the chick, whose later disappearance coincides with natural cell death in the thoracic sensory ganglia. Finally, as noted earlier, Hamburger (1958) reported that the ventral roots of peripherally deprived motoneurons appeared to be as thick as those on the unoperated side of the spinal cord, implying that at least some of the deprived motoneurons had sent out axons prior to induced cell death. Since none of these observations unequivocally rule out the possibility that some, or perhaps even most, neurons that undergo natural cell death either fail to sprout an axon or fail to send axons to their target, a new approach was needed to resolve this issue.

The first attempt to address this question directly was made by Prestige and Wilson (1972). Using the electron microscope, they counted all of the axons in one ventral root of the frog, *Xenopus*, at different stages of development. When they compared axon number with cell counts of motoneurons from the same segment at the same stages, they found a consistent 1 : 1 relationship, including during stages prior to the onset of cell death. They concluded that all neurons, including those that later die, had sprouted an axon. Similar findings were subsequently reported for chick spinal motoneurons in both normal and peripherally deprived conditions (Chu-Wang and Oppenheim, 1978b; Oppenheim et al., 1978); in chick ciliary neurons (Pilar and Landmesser, 1976); in duck trochlear neurons (Sohal, Weidman, and Stoney, 1978); and in chick retinal ganglion cells (Rager and Rager, 1978). Although some of these findings could conceivably be explained by assuming that some neurons develop several axonal collateral sprouts whereas others fail to sprout even a single axonal process, this possibility now seems unlikely.

Even when taken at face value, however, these recent findings only indicate that neurons destined to die can sprout an axon; they say nothing whatsoever about whether these neurons actually send axons to their normal targets. Perhaps the axons from cells that die begin to grow but then are prevented from reaching their target. Failure on this account could, in fact, be the major factor determining whether a cell lives or dies. Thus, whereas the findings on correlated cell-axon counts are a decided improvement over the previous circumstantial evidence, they fail to answer the central question of whether neurons innervate their targets prior to cell death. Fortunately, the advent of neuroanatomical techniques for the retrograde labeling of developing neurons with the exogenous enzyme, horseradish peroxidase (HRP), has provided a method for resolving this issue rather unambiguously (Oppenheim and Heaton, 1975; Lamb, 1974).[15]

In a study of neurogenesis and cell death in the isthmo-optic nucleus (ION) of the chick, Clarke and Cowan (1976) demonstrated that HRP injections into the eye—the target of the ION—prior to the onset of natural cell death, labeled virtually all of the neurons in the ION. Since approximately 60 percent of the ION neurons subsequently die, this provides direct proof that even the axons of the neurons that will undergo cell death have reached their synaptic targets in the retina. Using the same approach, my colleague, I. W. Chu-Wang and I have similarly demonstrated that virtually all of the lumbar spinal motoneurons innervate the leg prior to cell death (Chu-Wang and Oppenheim, 1978b). Furthermore, we found HRP-labeled motoneurons that were undergoing degeneration, as defined by ultrastructural criteria, thus indicating that such cells had axons in the limb—the site of HRP injection—prior to the onset of natural cell death (Oppenheim and Chu-Wang, 1977).

Although these findings from the ION and the spinal motoneurons of the chick strongly imply that most neurons may innervate their synaptic targets prior to natural cell death, the reports of degenerating migrating spinal motoneurons in the opossum (Hughes, 1973), of degenerating cells in the neuroepithelium of the chick retina (Hughes and LeVelle, 1975), and of the degeneration of chick sensory ganglion cells immediately (i.e., within 2 hrs) after cell division (Carr and Simpson, 1980), suggest a note of caution. For it is possible in these cases that the dying neurons had not yet sent axons to their targets. Yet, even these findings may not be inconsistent with the notion of target innervation prior to cell death. Migrating neurons may already have sent axons to their targets (Heaton, Moody, and Kosier, 1978; Grobstein, 1979; Farel and Bemelman, 1980) and, unlikely as it seems, the initiation of neuronal differentiation including axonal development may not always be incompatible with continued proliferation (Cone and Cone, 1978; Stillwell, Cone, and Cone, 1978; Scott, 1977; Rothman, Gershon, and Holtzer, 1978).

I hasten to point out that my use of the term innervation in the preceding discussion—consistent with common usage—is only meant to convey the fact that neurons have axonal processes in their target region (i.e., that the targets have a nerve supply). I do not mean to imply the existence of mature synapses, as traditionally defined, nor do I wish to imply that neurons destined to die have formed functional synapses. In the cases involving spinal motoneurons, trochlear motoneurons, and ciliary neurons, however, there is ample evidence for the existence of nascent *anatomical* contacts with target cells prior to, and during, cell death. But it is not known whether the specific neurons making such contacts are the same ones that subsequently die. The same reservation applies to the question of *functional* efferent synapses prior to cell death. Whereas there are electrophysiological and behavioral data to indicate the presence of functional contacts between neurons and their targets prior to, or coincident with, cell death (Hamburger, 1975; Landmesser and Morris, 1974; Stoney and Sohal, 1978), it is at present not known if these represent contacts of the cells that will later die.

Indirect evidence that, in fact, this may be the case, however, comes from the important observations of Landmesser and Pilar on the innervation of the ciliary ganglion by preganglionic afferents from the accessory oculomotor nucleus (AON). They observed that synapses which appear functionally and ultrastructurally normal are formed between the AON neurons and ciliary ganglion cells, prior to the onset of cell death in the AON (Landmesser and Pilar, 1974a,b). Furthermore, since degenerating preganglionic synapses from the AON were also observed on normal ciliary cells (Landmesser and Pilar, 1976), it can be inferred that many AON neurons establish synapses on their normal targets prior to their own natural cell death. (Cell death in the AON occurs between Day 9 and Day 15 and in the ciliary ganglion between Day 8 and Day 12).

COMPETITION AND THE ESTABLISHMENT OF SPECIFIC SYNAPTIC CONNECTIONS AS FACTORS IN THE REGULATION OF CELL DEATH

Having provided compelling evidence that neurons which later die by natural cell death can sprout axons and innervate their targets, one potentially serious objection to the peripheral competition hypothesis has been removed. Yet, this demonstration, while important, is merely consistent with, and not unequivocal proof for, the peripheral competition hypothesis. It is conceivable, for example, that *afferents* also play an important role in the regulation of cell death. Perhaps those neurons that receive an insufficient number of afferents, or afferents of the inappropriate kind, are more likely to undergo natural cell death, irrespective of their efferent connections. Indeed, in the case of induced cell death of spinal motoneurons, considering not only that the targets in the limb-bud are absent but also that a large percentage of *extrinsic* afferents from the dorsal root ganglia are also missing (Hamburger and Levi-Montalcini, 1949), this possibility appears plausible. In addition, the influence of descending fibers, both proprio and supraspinal, must also be considered in any attempt to assess the general role of afferents in natural cell death. In at least one system, however, it would appear that the role of afferents from both short-range and long-range fiber tracts has been ruled out.

In several studies involving a surgical isolation of the brachial or lumbar spinal cord of the chick, prior to the invasion of those regions by longitudinal fiber tracts, motoneuron cell number was found to be totally unaffected (Bueker, 1943; Levi-Montalcini, 1945; Hamburger, 1946). Similarly, the role of short-range afferents from interneurons *within* the limb-innervating segments also appear to have been ruled out in this system. Wenger (1950) determined the number of remaining motoneurons following removal of the contralateral half, the dorsal one-quarter (i.e., one quadrant), or the dorsal half, of the brachial spinal cord in the chick. With the exception of the dorsal half removal, the number of motoneurons was found to be normal at 8 and 11 days of incubation. (The surgery was performed at two days of incubation, and between 8 and 12 days of incubation there is a normal loss of approximately 6,000 brachial motoneurons by natural cell death [Oppenheim and Majors-Willard, 1978].)

Although Wenger did find 40 percent fewer motoneurons following removal of the entire dorsal one half of the spinal cord, she only reported data from one embryo and, in that single case, there was considerable inadvertent damage to the more ventral regions of the spinal cord. Moreover, since she found no effect of *dorsal quadrant* removal on the number of surviving motoneurons, it seems likely that the reduction of motoneurons in the case of *dorsal one-half* removal was, in fact, due to inadvertent damage to the neuroepithelial regions of the basal plate from which the motoneurons arise and not to the loss of afferents.

The negative results of Wenger following dorsal quadrant removal are of additional interest since, by removing the adjacent neural crest, the spinal sensory ganglia in these cases were also absent or reduced unilaterally. Despite the loss of these dorsal root afferents, the motoneurons developed normally on that side. A later study by Hamburger, Wenger, and Oppenheim (1966), involving dorsal one-half spinal cord, and neural crest, removal in the lumbar region also revealed little, if any, effect on cell death or on the initial differentiation of the surviving motoneurons. Although more detailed studies are necessary before one can generalize on this issue, these findings imply that the absence, or reduction, of afferents does not interfere with the development of motoneurons, including the magnitude of natural cell death. In this regard, it is of interest that a substantial *increase* in the number of afferents from spinal sensory neurons produced by chronic treatment of chick embryos with NGF also does not affect the survival of lumbar motoneurons (Bueker, 1948; Levi-Montalcini and Hamburger, 1951; Oppenheim, Maderdrut, and Wells, unpublished observations).

Although the data from studies of chick spinal motoneurons argues against the role of afferents in regulating natural cell death, these findings should not be taken to imply that all neuronal systems are similar in this respect. The removal of peripheral afferents to the chick auditory nuclei in the brainstem (n. magnocellularis and n. angularis) does, in fact, result in increased cell death (Levi-Montalcini, 1949; Parks, 1979). (In contrast, according to Levi-Montalcini [1949], the absence of *descending* fibers has no effect on cell death in these same auditory nuclei.) Removal of tectal afferents to the avian isthmo-optic nucleus (ION) also accentuates the naturally occurring cell loss in this system (Clarke and Cowan, 1976; Sohal, 1976). In the case of the ION, however, it is not clear whether the increased cell death is a direct result of the deafferentation, or whether it reflects an indirect effect resulting from the secondary degeneration of retinal ganglion cells (the targets of the ION), caused by a loss of the ganglion cell targets in the tectum (Clarke and Cowan, 1976). If that is the case, then the increased loss of ION neurons is merely another example of induced cell death following removal of their peripheral targets and thus not evidence for the role of deafferentation per se.

Although Landmesser and Pilar (1974a) have argued that afferents to the chick ciliary ganglion are not involved in the regulation of cell death, embryonic removal of all afferents to this ganglion has been reported to result in a virtually complete loss of neurons (Levi-Montalcini, 1947). Moreover, treatment of chick embryos with the ganglion blocking agent, chlorisondamine, also increases the natural cell loss of ciliary neurons (Wright, 1981a), indicating that receptor mediated interactions may be a critical factor in the role of afferents in this system. Since the removal of ciliary ganglion *targets* also induces cell loss in the ganglion (Landmesser and Pilar, 1974a), cell survival in this system may be under dual control.

Finally, cell number in the rat visual system has been reported to change depending on the extent of afferent input. Cunningham, Huddelston, and Murray (1979) report that following destruction of cortical visual centers at birth, certain lower visual areas such as the ipsilateral nucleus of the optic tract (NOT) and the superior colliculus (SC) become hyperinnervated by collateral sprouts from the optic nerve. Counts in the NOT and SC of experimental animals showed that there was a 39 percent increase in the number of cells in the NOT and 15 percent in the SC. Although the magnitude of *natural* cell death in these regions is not known, degenerating cells have been observed in normal animals. Consequently, Cunningham and colleagues (1979) suggest that their findings represent the prevention of naturally occurring cell death by the afferent hyperinnervation.

Because at present so little unequivocal evidence exists on this issue, the general question as to the role that afferents play in naturally occurring cell death must remain open. It would appear, however, that neurons differ in this respect and that each case will have to be examined on its own merits. Since in the case of chick spinal motoneurons, afferents are probably relatively unimportant, however, one is forced to look to interactions between neurons and their postsynaptic targets for the "causes" of cell death. In attempting to deal with the question of the mechanisms involved in this efferent interaction, one could focus on a number of different levels ranging from the intercellular to the molecular. For the moment, however, I wish to limit my discussion to a more superficial level of analysis. I shall discuss evidence as well as speculations pertinent to the more molecular levels in the following section.

The most popular notion for explaining natural cell death has been some variation of what has come to be known as the "competition hypothesis." The competition hypothesis proposes that during normal development neurons are so overproduced relative to the available synaptic targets that they compete, either for an anatomically defined target (e.g., endplates, receptors, etc.) or for a limited amount of a trophic or maintenance factor contained in the target cells. From the perspective of biological adaptation, the competition hypothesis assumes that cellular redundancy or overproduction is a mechanism for assuring that enough neurons are available to meet normal variations in the number of synaptic targets *prior to* actual innervation. Those neurons that acquire a synaptic target (targets), or those that receive sufficient trophic molecules, survive, whereas those that fail in either of these respects regress and degenerate. The competition hypothesis further assumes that, although under most normal conditions there will always be fewer postsynaptic targets, or less trophic substance, or both than is necessary for the survival of all of the presynaptic neurons, it should nonetheless be possible to increase or decrease natural cell death selectively by manipulating the amount of target tissue (i.e., the number of synaptic sites and/or the amount of trophic substance). The fact that, in most systems, natural cell loss is greatly exaggerated by target depletion or removal (induced cell death) is consistent with this assumption.

Despite the fact that the effects of peripheral depletion are consistent with the competition hypothesis, it would be considerably more satisfying, conceptually, if it could be shown that natural cell death is mitigated or prevented entirely by providing additional synaptic targets or trophic substance. The correctness of the competition hypothesis in fact must stand or fall on the outcome of such experiments. For, although one may subsequently argue over the specific mechanisms involved in mediating competition—until one can demonstrate that natural cell death can be prevented by providing additional targets—such debates are largely gratuitous.

Many of the early studies of central-peripheral interactions, in which increases in the size of the periphery led to neuronal hyperplasia,[16] could be interpreted, in retrospect, as evidence for the prevention of cell death. In the absence of normative data on natural cell death, however, such an inference would be premature. For instance, experimentally produced increases in the size of the jaw musculature in *Ambystoma* embryos of up to 200 percent, result in only a 17 percent increase in cell number in the mature mesencephalic nucleus of the fifth cranial nerve (Piatt, 1946). Until the magnitude of natural cell death in this system is known, there is no way to evaluate the significance of this finding. If natural cell death involves only 20–30 percent of this population, then a 17 percent reduction is significant. On the other hand, if natural cell death amounts to 75 percent, as in fact occurs in the *chick* mesencephalic nucleus (Rogers and Cowan, 1973), then 17 percent represents an exceedingly modest effect of increasing the target size by 200 percent.

One system for which such normative information is now available involves the spinal motoneurons of the chick. Between the fifth and tenth days of incubation there is a loss of 40–50 percent of the neurons in the lumbar region (Hamburger 1975, Chu-Wang, and Oppenheim 1978a,b), and a loss of about 30 percent in the brachial region (Oppenheim and Majors-Willard, 1978). Thus, in retrospect, the early report of Hamburger (1939a) of an average 15 percent increase in cell number over the entire brachial or lumbar region on Day 9, following transplantation of an extra limb, can now be seen to provide support for the competition hypothesis.

In a recent partial replication of that study, Hollyday and Hamburger (1976) found an average 18 percent increase in cell number in the lateral motor column (LMC) of the lumbar region on Day 12, in embryos with a supernumerary leg. In the original study of Hamburger (1939a), it was reported that increased cell number was largely restricted to those rostral limb segments that actually innervated the transplant. In these rostral segments, cell number was greatly increased, averaging almost 40 percent. In apparent contrast, Hollyday and Hamburger (1976) found that the increases in cell number were not restricted to the rostral segments supplying the transplant but occurred throughout the lumbar region. They suggest that since the extra limb is positioned rostrally to the normal limb, a considerable number of rostral motor axons are deflected into the transplant, thereby depriving some muscles in the normal leg of their normal nerve supply. If this is the case, then the peripheral targets from the more caudal segments would also be enlarged, and some of their neurons would have an increased opportunity to survive. This difference in the findings of Hamburger (1939a) and of Hollyday and Hamburger (1976), remains unexplained.

In evaluating experiments that attempt to prevent cell death by an increase in the available synaptic targets, the failure to prevent *all* natural cell death in a neuronal population should not be taken as necessarily inconsistent with the competition hypothesis. As suggested by the experiments of Hamburger (1939a) and of Hollyday and Hamburger (1976), many neurons may never gain access to the extra synaptic targets. Since the competition hypothesis requires that neurons send axons to a target region in order for competition to occur, in order to use evidence in which natural cell death is only partially prevented as an argument against competition, it must be shown that prior to the onset of cell death the axons of all (or most) of the neurons that normally die, do, in fact, gain access to the increased synaptic targets. For instance, in the case of the lumbar spinal motorneurons of the chick, one would have to show that at least 40 percent to 50 percent of the neurons in the lumbar region (the amount of natural cell death) have axons in the supernumerary limb prior to the onset of natural cell death on Day 5. More specifically, it would have to be demonstrated that *all* neurons in a population could be labeled by HRP injections into *either* limb, which would indicate that each neuron had a branch in both limbs. Or, it would have to be shown that half of the population was labeled by injections in one limb and half by injections in the other limb, with no overlap between the two. This, however, has not been demonstrated for the spinal motoneurons of the chick, or, for that matter, in any other system.

If it were possible to prevent all or most natural cell death, then the need to demonstrate such prior innervation would be considerably lessened. At present, this has been shown to occur in only one or two cases. Narayanan and Narayanan (1978) reported a total prevention of cell death in the accessory oculomotor nucleus of the chick following transplantation of an additional eye. This occurred in only one embryo, whereas several other cases showed a considerably smaller effect. In another study, involving the removal of two of the three sensory ganglia normally innervating the frog hindlimb, Bibb (1978) reported finding an increase in cell number in the remaining ganglia equal to that occurring during natural cell death (i.e., 40%). As noted earlier, however, Bibb has interpreted this finding as being consistent with the old recruitment hypothesis in that he attributes the effect to increased cellular proliferation and not to the prevention of natural cell death.

Other studies have shown a considerable—albeit not a total—prevention of cell death by increasing the available synaptic targets. For instance, Boydston and Sohal (1978) found an average numerical increase of 37 percent in the chick trochlear nucleus (where natural cell death accounts for approximately 50% of the cells) and an average increase of 35 percent in the isthmo-optic nucleus (where natural cell death is about 60%) following transplantation of an additional eye. Similarly, Narayanan and Narayanan (1978) found a maximum cellular increase of 27 percent in the chick ciliary ganglion in a few cases (when natural cell death averages 50%) and of 30 percent in the chick trochlear nucleus following transplantation of a supernumerary eye (when natural cell death is 50%). By removing two of the three branches of the ciliary ganglion nerve to the iris and ciliary muscle, and thereby mitigating the need for competition, Pilar, Landmesser, and Burnstein (1980) were able to reduce natural cell death among the neurons normally supplying that branch from 69 percent to 42 percent. Similarly, by cutting one of the three segmental nerves innervating the hindlimb in *Xenopus*, Olek and Edwards (1977) reduced motoneuron cell death from 75 percent to 40 percent. In contrast to these findings, which support the competition hypothesis, Lamb (1979a) has shown that in *Xenopus*, motoneurons which project to knee flexors in the tadpole, but which normally later degenerate, are not saved by removing another separate group of neurons which also project to the knee flexors, but which normally do not degenerate. Thus, in this case, the absence of apparently competing axons does not alter natural cell death. It should be noted, however, that this experiment only rules out heterospecific competition (i.e., competition between different motor pools for a similar target) and says nothing about homospecific competition, or competition *within* a motor pool for synaptic targets. In the chick, the absence of potential competing axons following the removal of several segments of lumbar spinal cord does not alter the *projection patterns* of the remaining neurons; the muscles normally innervated by the removed segments remain uninnervated (Landmesser, 1980; Lance-Jones and Landmesser, 1980). Although this manipulation does not result in a general or widespread modification in the number of surviving motoneurons in the remaining segments, it has been shown that partial depletion of a motoneuron pool for a specific muscle results in an increased number of neurons in the remainder of the pool, apparently by reducing competition, which allows the rescue of cells that would otherwise have undergone natural cell death (Lance-Jones and Landmesser, 1980).

One might expect that in animals with congenitally occurring supernumerary limbs, which are either located close to or merged with the "normal" limb, some of the putative mechanical problems associated with surgically created supernumerary limbs, such as the differential ease of access to the limb of some neurons, would be mitigated. Consequently, it comes as a surprise to find that in at least three well-documented reports, all with frogs, this doesn't appear to be the case. Bueker (1945) found a 22 percent increase in cell number in a *Rana fusca,* that had three well-formed hindlimbs (natural cell death probably amounts to 70% or more in this population). In another case, involving *Rana pipiens,* Pollack (1956b) found only a 15 percent increase in brachial motor cells on the side of the spinal cord which had three, well-formed forelimbs; normal cell loss is about 60 percent in this population (Pollack, 1969a). Finally, Lamb (1969b) recently reported finding a frog *(Xenopus)* with three partially merged hindlimbs, in which the size of the "limb" was increased by 100 percent, but which only had an 18 percent increase in the number of motoneurons (natural cell death of lumbar motoneurons reaches about 75% in *Xenopus*).

Since the 15–20 percent increase in cell number in these three reports rather closely mimics the increase seen following experimentally created supernumerary limbs in frogs (Hollyday and Mendell, 1976), it seems unlikely that the failure to prevent more cell death in these congenital cases is due to some general pathology. Consequently, these consistent failures to prevent more than a small amount of the naturally occurring cell death appear to present serious problems for the current version of the peripheral competition hypothesis (see below).

Additional evidence, which, if taken at face value, also appears to be inconsistent with the competition hypothesis, comes from another study by Lamb (1980), in which he reports that a single frog hindlimb is able to support twice the number of motoneurons that normally innervate one limb. By removing one limb bud in the tadpole and then surgically disrupting extraneural dorsal midline barriers, Lamb was able to foce motoneurons from the contralateral side of the spinal cord to innervate the single remaining limb before the onset of normal cell death. (In some cases, the remaining limb bud was transplanted to the dorsal midline.) If the version of the peripheral competition hypothesis that is favored at present is correct (i.e. that there is a one-to-one relation between muscle mass and neuron survival), then one would expect that the extent of natural cell loss on both sides of the spinal cord would be such that the total number of remaining cells (ipsilateral plus contralateral) would equal that on one side in normal animals (i.e., about 1,500). However, Lamb found a total of up to 2,800 surviving motoneurons on the two sides, which is virtually identical to the total number for both sides found in normal, unoperated frogs. Since Lamb reports that, after dissection, spinal nerves from both sides of the spinal cord were seen to enter the remaining hindlimb, and since HRP injections into the remaining limb (or HRP applied to the cut sciatic nerve) label motoneurons on both sides of the spinal cord, there can be no doubt that at least some motoneurons from both sides innervate the single limb.

A more crucial question, however, and one that was not answered by Lamb's experiment, is whether all of the surviving motoneurons on both sides innervate the limb. If they do not, then the results are not necessarily inconsistent with the peripheral competition hypothesis. Since unilateral limb bud removal usually induces most of the cells on that side to die (e.g., Prestige, 1967), it is, admittedly, rather difficult to understand how, in Lamb's experiment, so many of the "deprived" cells contralateral to the intact limb could survive if they did not innervate that limb. However, the surgical disruption of midline barriers may have created unusual conditions that might facilitate the growth of ipsilateral axons to other abnormal locations besides the contralateral leg (e.g., see Hamburger, 1929). HRP injections into the trunk or tail musculature might help resolve this question, as would removal of the remaining limb; in the latter case one would expect all of the motoneurons on both sides to die. Furthermore, since cell death of spinal motoneurons in *Xenopus* is affected by functional interactions between neurons and muscle (i.e., neuromuscular blockade, resulting in reduced function, prevents cell death [Olek and Edwards, 1978], it is conceivable that the motility in the single remaining limbs in Lamb's study was sufficiently reduced to prevent cell death of the excess innervation. This now seems unlikely, however, as Lamb (personal communication, 1980) has reported that limb movements appear normal. If, however, it is the case that a single, normally functional limb can support twice the usual number of motoneurons, then it seems inescapable that, in the case of *Xenopus* at least, an alternative explanation, other than that now favored must be sought to explain cell death.

This is not to say that the peripheral competition hypothesis has been refuted. Even if Lamb's results are correct, they are not pertinent to the general validity of the competition hypothesis but only to a specific version of that hypothesis. For example, as Purves (1980) has pointed out, if nerves regulate the amount of trophic substance supplied by their targets in a proportional but in a less than one-to-one fashion, then Lamb's results would be consistent with the *general* notion of competition but inconsistent with the *specific* idea that there is a one-to-one relation between muscle mass and motoneuron survival. Although many questions remain to be answered concerning Lamb's results (1980), they may, nonetheless, turn out to be an important contribution to our understanding of neuronal cell death.

So far, my discussion of competition in this section has made the implicit assumption that what is being competed for are synaptic targets of a rather nonspecific nature (i.e., *homospecific* competition), that is, that the neurons that die, as well as those that survive, both innervate their normal synaptic targets. From this assumption, one could argue that those motoneurons which innervate the gastrocnemius muscle but which subsequently die regress because there are a limited number of targets in the gastrocnemius muscle and not because their axons are in the wrong muscle. But, as noted earlier, it has been proposed, in fact, that cell death may be involved in the removal of inappropriate synaptic connections (i.e., *hetrospecific* competition, such as the removal of sartorius motoneurons that have incorrectly innervated the gastrocnemius muscle). And for those cases involving frogs, in which competition has been suggested to be an unlikely cause of cell death, it has been proposed that errors in synaptic connectivity (i.e., heterospecific competition) may be the major cause of cell death (e.g., Lamb, 1980). Furthermore, it has been suggested recently that errors in synaptic connectivity may also explain a large share of the natural cell death in chick brachial spinal motoneurons (Pettigrew, Lindeman, and Bennett, 1979).

Even if one accepts the contention that homospecific competition is not involved in the natural cell death of *frog* motoneurons, it is equally difficult to reconcile much of the frog data with the synaptic specificity hypothesis. Although some neurons in *Xenopus* do, indeed, appear to die for reasons other than competition—presumably due to the formation of inappropriate connections (Lamb, 1977, 1979a)—a considerably greater number die despite having formed apparently correct, adultlike, connections during early limb innervation (Lamb, 1976). Thus, although homospecific competition may not be the whole story in frogs, synaptic specificity appears equally untenable (for a review, see Landmesser, 1980).

Unlike the frog, virtually no motoneurons in the chick lumbar spinal cord appear to die owing to the formation of grossly inappropriate synaptic connections. Based on the HRP labeling of motor pools and on electrophysiological data, lumbar motoneurons form connections with their appropriate muscles at the time of initial innervation and prior to the onset of natural cell death (Landmesser and Morris, 1975; Landmesser, 1978a,b; see also Pettigrew et al., 1979). Furthermore, in some instances, following the manipulation of the spatial position of limb muscles by limb rotation or transplantation, or after the removal of an entire pool of motoneurons, specific muscles may become innervated by "incorrect" motoneurons; yet these "mismatched" motoneurons apparently do not degenerate (Stirling and Summerbell, 1979, 1980; Bennett, Lindeman, and Pettigrew, 1979; Hollyday et al., 1977).

Finally, the prevention of virtually all cell death in the lumbar and brachial spinal cord of the chick by neuromuscular blocking agents (see the following section) does not modify the projection patterns of motoneurons to specific muscles (Oppenheim, 1981a). When HRP is injected into specific wing or leg muscles in these preparations on Day 10 (i.e., after the cessation of almost all natural cell death in normal embryos), the labeled motoneurons are located in virtually the same position in the spinal cord as in control embryos. Since cell death has been prevented in these preparations from an early stage of limb innervation—and prior to the onset of cell death—any significant number of inappropriate connections that exist should have been retained and later revealed, after the HRP injections. That this did not occur provides additional evidence against the synaptic specificity proposal and in favor of peripheral competition. Because of possible technical limitations associated with the identification of motoneuron pools at the earliest stages of innervation,[17] these findings from older embryos, in which such problems are lessened, are particularly significant. Such findings also argue against the role of functional activity as a mechanism in the generation of specific connections as suggested by Changeux and Danchin (1976). Similar results have been found following the pharmacological prevention of cell death in the duck trochlear nucleus (Creazzo and Sohal, 1979). (Since there is some evidence that spinal motoneurons *within* a motor pool may project to particular regions or to specific fiber types within a given muscle [Burke et al., 1977; Burke, 1980], it is conceivable that errors at this level may be corrected by cell loss.) In contrast, in the chick isthmo-optic nucleus, cell death is involved in the removal of neurons that make aberrant connections and of neurons that are located outside the boundaries of the nucleus (Clarke and Cowan, 1976). However, these constitute less than 3 percent of the total population, whereas natural cell death involves 60 percent of the neurons in the nucleus. Thus, the lion's share of cell death in this case also involves neurons that have innervated their correct targets in the contralateral eye.

In summary, judging from the available data, it would seem that the strongest evidence in favor of the competition hypothesis comes from studies in birds, whereas the most compelling evidence against competition comes from studies in frogs: experimental increases in the size of a target region prevents a significant amount of natural cell death in birds, whereas similar manipulations (or congenital increases in target size) in frogs has remarkably little effect on natural cell death. Since it seems exceedingly unlikely that this difference reflects a parochialism on the part of the respective investigators, it may therefore represent a real difference between frogs and birds. Either competition occurs in birds but not frogs, or some more subtle difference, perhaps involving a differential species-specific response to surgical or congenital modifications, and having nothing to do with competition, is responsible.[18] Since motoneuron cell death in frogs occurs during metamorphosis, the absence of this event in avian embryogenesis may represent a fundamental difference in the evolution of mechanisms controlling cell death in frogs and birds. Notwithstanding the present discrepancies, it seems likely that the general mechanisms controlling neuronal cell death in vertebrates (e.g., competition) are similar and that the demonstration of this only awaits the ingenuity of future investigators to devise the appropriate experiments.

FUNCTIONAL AND TROPHIC INTERACTIONS IN THE REGULATION OF CELL DEATH

Although it has long been recognized that natural cell death often occurs during the period when the neurons involved are forming functional synaptic connections with target cells, only recently have functional interactions been considered to play any role in the regulation of cell death (Jacobson, 1974; Changeux and Danchin, 1976). In 1976, my colleague, Randy Pittman and I began a series of experiments to examine this question in the spinal cord of the chick embryo. Specifically, we wanted to see whether the chronic treatment of embryos with neuromuscular blocking agents during the period of normal cell death of lumbar motoneurons would influence the magnitude of cell death.

Based on the working hypothesis that the nicotinic acetylcholine (ACh) receptors in muscle might constitute the target site of developing motoneurons, we assumed that by selectively blocking these receptor sites with substances such as curare or α-bungarotoxin (α-BTX), the motoneurons would not be able to "recognize" their target. We expected that if receptor recognition is, in fact, an important step in the maintenance and survival of motoneurons, then the embryos with a chronic receptor blockade should exhibit enhanced cell death (comparable to that of limb removal) owing to an effective absence of target sites.

Embryos were treated daily with blocking agents in quantities sufficient to induce chronic paralysis as indicated by the virtual absence of overt neuromuscular activity. The duration of treatment typically spanned the period of natural cell death in this population, that is, from embryonic Days 5 or 6 to Day 10. When we later examined the spinal cords of these animals, it was immediately obvious that our prediction of enhanced cell death was incorrect; rather than having fewer motoneurons in the lumbar ventral horn, these embryos had more neurons compared to controls (Pittman and Oppenheim, 1978, 1979). Depending on the duration of treatment, as well as on the extent of the paralysis, the experimental embryos had from 30 percent to 90 percent more motoneurons than controls. Thus a chronic reduction in neuromuscular activity had prevented, rather than enhanced, natural cell death.[19]

In contrast, embryos treated chronically with the same neuromuscular blocking agents after Day 10 were indistinguishable from controls. Furthermore, the induction of paralysis by presynaptic blocking agents, such as botulinum toxin and hemicholinium, also prevents cell death, which implies that a *neuromuscular interaction,* and not merely a *postsynaptic action,* is involved. Additional evidence that the critical factor in the prevention of cell death involves a *peripheral* neuromuscular interaction was the failure of chronic curare treatment to alter the massive cell death induced by limb-bud removal. This finding implies that the drugs do not act *centrally* to prevent cell death.

Although it is conceivable that the chronic drug treatment may have prevented cell death by so inhibiting or retarding the differentiation of motoneurons that they never advanced to the stage necessary for the triggering of regressive changes, the evidence does not support this notion. Detailed histological, morphometric, and ultrastructural analyses of the ventral horn of these preparations, which are now in progress, indicate that the motoneurons attain the same level of differentiation as controls of the same age (Chu-Wang and Oppenheim, in preparation). Despite the fact that the muscles of these preparations are atrophic and contain fewer muscle fibers, axon counts in both the ventral roots and peripheral nerves indicate that all of the additional neurons have sent axons to their normal targets.[20] Indeed, based on histological studies (light and EM) and on the examination of endplates stained for acetylcholinesterase, the muscles of these preparations appear to be *hyperinnervated.* The hyperinnervation may take the form of an increased number of endplates per muscle fiber, an increased number of axons terminating at a single synaptic site (i.e., an increase in polyneuronal innervation), or both (Pittman and Oppenheim, 1979, Sohal, Creazzo, and Oblak, 1979).

Furthermore, biochemical differentiation, as reflected in spinal cord

CAT activity, is also normal in these preparations (i.e., CAT activity is increased proportionately with the increase in cell number; Maderdrut and Oppenheim, in preparation). Finally, when drug treatment is stopped, thereby allowing the resumption of neuromuscular activity, motoneuron cell death quickly ensues, despite the fact that the embryos are now past the stage (i.e., after Day 10) when cell death normally occurs. Thus, the failure of cell death to occur earlier cannot be due to delayed differentiation.

The prevention of cell death in chick lumbar motoneurons by neuromuscular blockade has now been independently confirmed in another laboratory (Laing and Prestige, 1978) and it has also been shown that cell death can be prevented in the *brachial* motoneurons of the chick by these methods (Oppenheim and Majors-Willard, 1978). It has also been reported that natural cell death in the duck trochlear nucleus (Creazzo and Sohal, 1979) and in the frog spinal cord (Olek and Edwards, 1978) can be prevented by chronic treatment with neuromuscular blocking agents. Although it remains to be seen whether functional interactions in other, noncholinergic neurons may play a similar role in the natural cell death, this possibility adds a new dimension to our conceptualization of neuronal cell death. Although at one time it seemed reasonable to conclude that, "neurogenesis does not make use of impulse transmission and its electrical and other correlates as an instrument of differentiation" (Hamburger, 1952, p. 131), this view is gradually giving way to a different perspective, which considers function as simply an additional factor in the epigenetic control of neurogenesis (Changeux and Danchin, 1976; Jacobson, 1974).

Following the pharmacological prevention of cell death in the chick spinal cord, the additional motoneurons are maintained as long as the embryos remain paralyzed, which in some cases persists up to the time of hatching (Pittman and Oppenheim, 1979). Because the blocking agents interfere with the inception of normal pulmonary respiration, however, these embryos do not survive past hatching, and in most instances they die a day or two before hatching. Thus, at present, it is not known whether the additional cells would persist after hatching in the absence of the blocking agents. (In a few embryos that survived for two days past the time of normal hatching on Day 20, but which were unable to escape from the shell, increased cell numbers were maintained; Oppenheim, unpublished observations.) This question can perhaps only be answered by examining other neuronal systems, in which the agents used to prevent cell death do not interfere with the function of such a basic physiological mechanism as respiration.

The observation that the limb muscles are hyperinnervated after motoneuron cell death in the chick spinal cord is prevented raises the question as to what constitutes a "synaptic target" in this sytem. Despite the fact that the muscles in these preparations are smaller than normal, a single limb is able to support almost double the normal number of motoneurons. In contrast to mammals, most muscles in adult birds (and in amphibians) are normally multiply innervated; thus, a single muscle fiber is typically contacted by several axons (Hess, 1970). It is obvious, then, that individual muscle fibers in birds cannot be the unit for which motoneurons compete (i.e., a single fiber is not *the* target). It seems more plausible that a synaptic site or target may be closely associated with, and thus defined by, the quantity, location, or distribution of ACh receptors in muscle. Although it remains to be shown that natural cell death is, in fact, directly related to the number or distribution of ACh receptors, there is a good deal of circumstantial evidence from both developing and adult animals to suggest that this may be the case. Furthermore, some of this evidence indicates that physiological activity is a critical factor in regulating receptor number and distribution.

As the extensive literature on the control of ACh receptors in developing and mature skeletal muscle has been thoughtfully reviewed by Fambrough (1979) and Edwards (1979), there is no need to repeat that material here. Suffice it to say that it has been shown that a high level of extrajunctional ACh receptors is closely correlated with the ability of the *adult* muscle to accept additional innervation. Thus, the presence of extrajunctional ACh receptors may permit innervation by more than one axon, or at least may be characteristic of a particular state of the muscle fiber (e.g., inactive), which allows multiple innervation. The observation that the normal decrease in extrajunctional ACh receptors in certain muscles of the chick embryo following innervation can be prevented by chronic curare-induced paralysis (Burden, 1977) is consistent with this notion. Similarly, chronic paralysis of fetal rats with tetrodotoxin (TTX) maintains the high levels of extrajunctional ACh receptors characteristic of muscle prior to innervation (Braithwaite and Harris, 1979). Therefore, it seems highly likely that the chronic neuromuscular blockade, which results in the prevention of motoneuron cell death in the chick embryo, probably also maintains high levels of extrajunctional ACh receptors. In this regard, it is pertinent that adult muscles exposed to neuromuscular blocking agents are, in fact, capable of accepting additional innervation (Duchen, Heilbronn, and Tonge, 1975).

Electrical stimulation of denervated adult muscle can prevent or reverse the development of increased extrajunctional sensitivity to applied ACh, and the ability of a foreign nerve to innervate a denervated muscle can also be blocked by muscle stimulation (for review, see Fambrough, 1979). TTX blockade of spontaneous muscle contractions *in vitro* increases, and electrical stimulation decreases, ACh receptor synthesis (Shainberg and Burstein, 1976; Hall and Reiness, 1977). A steady electric field applied to immature frog muscle cells *in vitro* leads to a rapid accumulation of ACh receptors at one pole of the cell, which persists even after the electric field is removed (Orida and Poo, 1978). Chronic treatment of spinal cord-muscle cultures from embryonic chick with curare or TTX inhibits the formation of AChE spots in muscle, and either electrical stimulation or treatment with the cyclic nucleotide cyclic GMP (but not cyclic AMP) reverses this effect (Rubin et al., 1979). Treatment of denervated chick myotubes *in vitro* with cyclic GMP also represses ACh receptor synthesis (Betz and Changeux, 1979), whereas cyclic AMP is reported to increase ACh receptor levels in these preparations (Blosser and Appel, 1980).

These findings indicate that cyclic GMP, or cyclic AMP, or both may be an intracellular messenger for the activity-mediated control of both AChE and of ACh receptor synthesis. It has, in fact, been shown that stimulation of the motor nerves to adult frog muscle produces a large increase in cyclic GMP but not in cyclic AMP, and direct muscle stimulation, even in the presence of α-BTX (and thus with the ACh receptors blocked), has a similar effect on cyclic GMP levels (Nestler, Beam, and Greengard, 1978). Thus, depolarization of the muscle membrane (or muscle contraction) appears to be the critical factor in causing an increase in intramuscular cyclic GMP.

In summary, the available evidence suggests the following tentative model for explaining normal development as well as development following neuromuscular blockade. During early synapse formation, one or more axons contact a myotube and begin releasing small amounts of ACh, which leads to a partial depolarization of the muscle membrane. Prior to, and during, this "critical" period of early membrane depolarization, more than one synapse can be maintained by each myotube, since intramuscular cyclic GMP has not yet reached a level sufficient to repress the high numbers of extrajunctional ACh receptors. However, once the frequency and extent of membrane depolarization reach a certain level, cyclic GMP is increased; this leads to a decrease or a redistribution of ACh receptors and thereby produces a change in membrane properties so that only one synapse can be supported. Thus, any additional contacts would be lost and new contacts would be prevented from forming. The axons that are rejected may be at a competitive disadvantage owing to an overextended peripheral territory, or to their smaller size, or owing to their anatomical or functional immaturity relative to the other axons that contact the same target. For muscles that are normally multiply innervated, similar events may also occur. During chronic neuromuscular blockade, the amount of ACh interacting with receptors would decrease considerably, thereby decreasing the membrane depolarization. This would result in the "critical" period being maintained, since the decreased levels of cyclic GMP (and/or increased levels of cyclic AMP) would allow the formation of increased numbers of extrajunctional ACh receptors.

The final effect of the series of events produced by neuromuscular blockade would be the maintenance of additional neurons by the muscle (i.e., the prevention of cell death). It is worth noting that many aspects of this model are consistent with evidence obtained from the study of newly formed neuromuscular junctions following denervation and reinnervation by a foreign nerve (Lømo, 1980).

A critical, but as yet unproved, assumption of this model is that electrical stimulation of either normal or paralyzed embryos should accelerate or enhance natural cell death. Additional predictions from this model are that neuromuscular blockade and the prevention of cell death should be associated with low levels of cyclic GMP (and/or high cyclic AMP levels) in muscle and increased numbers of extrajunctional ACh receptors; that the chronic inhibition of cyclic GMP during the period of natural cell death should mimic the effects of neuromuscular blockade; and that the treatment of curarized embryos with cyclic GMP should reduce ACh receptors and mitigate the effects of curare in preventing cell death. Finally, it should be noted, that although this model depends heavily on the role of physiological interactions between neurons and muscle cells in controlling cell survival, it by no means excludes the possibility that a trophic factor produced by the muscle is also involved. It does, however, assume as a working hypothesis that if such a trophic factor exists, its availability to motoneurons is somehow regulated by muscle activity. Although there is no evidence at present to suggest that either a muscle-associated or an activity-associated trophic factor is, in fact, involved in the control of cell death of spinal motoneurons, there is evidence that muscle-derived trophic factors may enhance neurite growth from spinal neurons *in vitro* (Dribin and Barrett, 1980; Pollack, 1980). Trophic factors have also been implicated in the survival of chick ciliary ganglion neurons *in vitro* (Nishi and Berg, 1979). In contrast to the situation with motoneurons, other cell types, such as those neurons known to be especially sensitive to nerve growth factor (NGF), may rely largely, if not exclusively, on trophic factors for the regulation of cell survival.

Despite the enormous amount of interest shown in NGF since its discovery over 30 years ago, there is still relatively little information available on its role or mode of action in the *embryonic* nervous system. Much of what is known about NGF has been frequently reviewed over the years and thus there is no need for me to duplicate those efforts here. (See, e.g., Levi-Montalcini and Aloe, 1981; Greene and Shooter, 1980; Harper and Thoenen, 1980.) Instead, I wish to focus only on those findings that are pertinent to the possible involvement of NGF in naturally occurring cell death.

Following the transplantation of mouse sarcoma 180 tissue (a source of NGF) into the celomic cavity of two-to-three day-old chick embryos, Levi-Montalcini and Hamburger (1951) found that the sensory and sympathetic ganglia were greatly enlarged when examined from 9 to 15 days of incubation. On the basis of cell counts, they concluded that at least part of this increase was due to the presence of additional neurons in the affected ganglia. Later studies, using both crude tumor NGF and purified preparations of NGF, confirmed the initial report of increased cell numbers in these ganglia following NGF treatment (Levi-Montalcini and Hamburger, 1953; Levi-Montalcini and Cohen, 1956; Levi-Montalcini, 1966). Despite the earlier discovery by Hamburger and Levi-Montalcini (1949) of natural cell death in normal chick sensory ganglia, however, these later findings of increased cell number following NGF treatment were not attributed to the prevention of cell death. Rather, it was generally assumed that increased mitotic activity was responsible. Since it is now known that considerable natural cell death occurs in the sensory ganglia of both limb and nonlimb regions of the chick embryo (Carr and Simpson, 1978a,b; Hamburger, Brunso-Bechtold, and Yip, 1981), it seems reasonable to infer that the increased cell number found in the early NGF experiments does, in large measure, result from the prevention of natural cell death.

Convincing evidence in support of this contention comes from the recent report that the treatment of chick embryos on Day 4 through Day 8 with NGF results in a significant reduction in the number of degenerating neurons in the dorsal root ganglia (Hamburger, Brunso-Bechtold, and Yip, 1981). Hamburger and his associates have found that NGF prevents virtually all cell death among the mediodorsal sensory ganglion cells and considerably reduces cell death in the ventrolateral population. Recent reports that the *in vitro* treatment of chick embryo spinal ganglia with NGF increases their substance-P content may also reflect a reduction of natural cell death among a subpopulation of sensory neurons (Schultzberg et al., 1978; Schwartz and Costa, 1979).

Although it is not known to what extent natural cell death occurs in chick sympathetic ganglia, it seems likely that the early reports of increased cell number in this system following NGF treatment may also reflect the prevention of cell death.[21] Natural cell death has been reported to occur in *rat* sympathetic ganglia postnatally, and NGF treatment appears to reduce some of this cell loss (Hendry and Campbell, 1976). Similarly, injections of mouse embryos *in utero* with NGF increases cell number in the superior cervical ganglion (Kessler and Black, 1980). Although it may seem implausible that cell death in the sensory or sympathetic ganglia, like spinal motoneurons, is associated with functional activity, it is conceivable that physiological activity in these systems may also be involved in the regulation of cell death. For instance, functional interactions between sympathetic neurons and their targets could regulate NGF levels in the target. In any event, our understanding of the mechanisms controlling cell death in some parts of the nervous system is increasing. One may thus reasonably hope that in the near future we may be able to explain both trophic and functional mechanisms of cell death on a molecular basis. An additional key that may be of considerable help in reaching this level of understanding involves the question of whether natural cell death occurs in all neuronal populations or cell types.

HOW WIDESPREAD IS CELL DEATH IN THE VERTEBRATE NERVOUS SYSTEM?

In addition to the cell groups discussed in the earlier sections, naturally occurring neuronal cell death has also been reported in several other regions of the central and peripheral nervous system. In lieu of an exhaustive survey of all of the cell types exhibiting cell death, the following examples are offered merely as an indication of the apparent ubiquity of this phenomenon.[22] Cell death has been reported (or at least, implied to occur) in the human fetal trochlear nucleus (Mustafa and Gamble, 1979); in neonatal mouse cerebral cortex (Leuba and Rabinowicz 1979; Heumann, Reuber, and Rabinowicz, 1978); in the dentate gyrus of the postnatal rat hippocampus (Schlessinger, Cowan, and Gottlieb, 1975); in the postnatal rat cerebellar granule cells (Legrand, 1979; Heinsen, 1978); in the neonatal rat superior colliculus (Arees and Astrom, 1977), in the postnatal mouse lateral geniculate nucleus (Heumann and Rabinowicz, 1980); in the cerebral hemispheres of chick embryos (Hyndeman and Zamenhof, 1978); in the postnatal mouse and posthatching chick cerebellar Purkinje cells (Fritzsch, 1979); in the chick embryo inferior olive (Armstrong and Clark, 1979); in the chick embryo dorsal motor nucleus of the vagus nerve (Wright, 1981b); in the mesencephalic nucleus of the fifth nerve in hamsters (Alley, 1974); in several visual nuclei in the tree shrew (Zilles, 1978); in the chick auditory relay nuclei (Rubel, Smith, and Miller, 1976); in the hamster optic tectum (Berg and Finlay, 1979); and in the guinea pig retina (Fry and Spira, 1979).

The neurons that exhibit natural cell death include such a wide variety of cells, involving practically every conceivable type of neuron, that it is impossible at this time to argue that any particular feature is characteristic of all neurons exhibiting cell death. Although it has been suggested that cell death may only occur among neurons that have greatly restricted or limited numbers of synaptic targets (Jacobson, 1978), the variety of cell types represented in even the incomplete list cited above, would appear to contradict such a suggestion. Admittedly, however, a much more careful analysis of the efferent projections as well as of the afferent inputs, of each of these systems is essential before this possibility can be ruled out. Indeed, apparent support for this notion comes from the report of the absence of natural cell death in the pontine nuclei of the chick embryo (Armstrong and Clarke, 1979). Cell numbers in the medial and lateral pontine nuclei apparently remain constant (ca. 3,000 cells) from embryonic Day 12 to 3 weeks posthatching.[23] The cells that compose these nuclei cease dividing after six to six-and-a-half days and are clearly identifiable as a recognizable cell group by Day 12. Armstrong and Clarke (1979) estimate that each neuron in the pontine nuclei has an average of 9,000 potential synaptic targets available among the cerebellar granule cells, and they suggest that only 1 percent or so of these need to be contacted to maintain or prevent the presynaptic cell from dying. They contrast this with the situation in the chick inferior olive, where they find that 30 percent of the neurons die between embryonic Day 16 and hatching, and in which they estimate that each neuron has only 20 to 50 potential synaptic targets available among the Purkinje cells. If this should prove to be correct, the contention that the presence or absence of cell death is related to the number of available synaptic targets provides additional support for the peripheral competition hypothesis.

In addition to the pontine nuclei of the chick, a preliminary unpublished investigation of the noradrenergic nucleus locus ceruleus (LC) and the nucleus ruber (NR) in the chick also indicates an apparent absence of cell death in these populations. Dr. Linda Wright has found in my laboratory that between embryonic Day 10 and hatching the number of cells in the LC remains virtually constant at 3,500, and the same is true of the NR, which contains about 3,000 neurons. Although it is conceivable that some cell death may occur after hatching in these two populations—and we are currently investigating this possibility—the LC, which is known to project to many different regions of the brain and spinal cord (Ikeda and Gotah, 1971, 1974; Amaral and Sinnamon, 1977), may be similar to the pontine nuclei in that the number of available synaptic targets may be sufficient to preclude the occurrence of natural cell death. Cell death has also been reported to be minimal or absent in the avian auditory nucleus magnocellularis (Parks, 1979; Rubel, Smith, and Miller, 1976). (Despite the apparent absence of cell death in the pontine nuclei, n. magnocellularis the LC and the NR of the chick, such reports must be interpreted cautiously. If natural cell death in these regions is of a small magnitude [e.g., $< 15\%$], then large sample sizes may be needed to detect cell death if normal variations in cell number, or if counting errors, approach 15%.) As has been true of so much of the history of the study of natural cell death, the spinal motoneurons may provide a particularly important model for testing the validity of this notion. By comparing the magnitude of cell death in specific pools of motoneurons with the size of the mature motor units in the relevant muscles, both within and between species, it may be possible to provide compelling—albeit correlative—evidence for the role of synaptic targets in controlling the magnitude of natural cell death. It should be noted that this rationale has been previously applied to natural cell death in chick spinal ganglia by Hamburger and Levi-Montalcini (1949), when they pointed out that there was considerably more natural cell death in nonlimb-innervating, as opposed to limb-innervating, ganglia (also see Carr and Simpson, 1978). Hamburger and Levi-Montalcini recognized that this difference might reflect the greater availability of synaptic targets in the limb regions due to the normally larger periphery provided by the limb.

REGRESSIVE PROCESSES NOT INVOLVING NEURONAL CELL DEATH

Space does not permit a complete discussion of other kinds of regressive processes involved in normal neurogenesis (for reviews of this literature see Purves and Lichtman, 1980; Oppenheim, 1981b). However, one group of phenomena deserves at least brief consideration in the present context; this involves the transient polyinnervation or hyperinnervation of synaptic targets or of neurons early in development. Although first observed in the neuromuscular system (Redfern, 1970; Bagust, Lewis, and Westerman, 1973), it now appears that this phenomenon occurs in several other systems as well (e.g., Crepel, Mariani, and Delhaye-Bouchard, 1967; Lichtman, 1977; Ronnevi, 1977; Ivy, Akers, and Killackey, 1979; Innocenti and Frost, 1979; Speidel, 1941). Of particular interest is the demonstration in some systems that, similar to cell death, the normal loss of the excess or polyneuronal innervation apparently involves competition between neurons as well as functional interactions between the presynaptic and postsynaptic cells.

The polyneuronal innervation of single synaptic sites (i.e., of a neuromuscular endplate) in skeletal muscle is perhaps the most extensively studied example of this phenomenon (Jansen, Thompson, and Kuffler, 1978). In all vertebrates that have been studied, endplates, which in the mature animal are innervated by a single axon, are contacted by several axons at embryonic, or early postnatal, stages. In the rat (Brown, Jansen, and Van Essen, 1976) and chick (Oppenheim and Majors-Willard, 1978) the transition from polyneuronal innervation to the mature condition is independent of motoneuron cell death (i.e., cell death is virtually over before the major loss of polyneuronal innervation occurs). Moreover, whereas cell death involves the cytological degeneration of peripheral axons (e.g., Chu-Wang and Oppenheim, 1978b), the loss of polyneuronal innervation apparently only involves the retraction of collaterals back to the parent axon without actual degeneration (e.g., Korneliussen and Jansen, 1976; Riley, 1977). Partial denervation of specific limb muscles in the newborn rat leads to a delay or inhibition of the normal loss of polyneuronal innervation, indicating that competition between motoneurons plays a role in this process (Betz, Caldwell, and Ribchester, 1979; but see Thompson and Jansen, 1977). Furthermore, the reduction or abolition of neuromuscular activity by tenotomy (Benoit and Changeux, 1975), chronic curarization (Sirhari and Vrbová, 1978) or deafferentation (Zelena, Vyskočil, and Jiranová, 1979) delays, whereas electrical stimulation (O'Brien, Ostberg, and Vrbová, 1978) and functional muscle hypertrophy (Zelena et al., 1979) hastens, the loss of polyneuronal innervation.

Additional evidence for the role of competition in eliminating polyneuronal innervation comes from the study of climbing fiber (CF)-Purkinje cell (P-cell) contacts in mice and rats. In the mature rodent, each P-cell is contacted synaptically by a single CF. In contrast, neonatal mice and rats are found to have several functional CF synapses on each P-cell (Crepel et al., 1976). Adult mice or rats lacking granule cells retain the multiple innervation of P-cells of CFs (e.g., Woodward, Hoffer, and Altman, 1974; Crepel et al., 1976; Crepel and Mariani, 1976), indicating that the formation of synapses between granule cells and P-cells somehow eliminates (competitively?) the excess of CFs during normal ontogeny. That this interaction between the granule and P-cells involves a *synaptic* mechanism was shown by examining the genetic mouse mutant *staggerer* (Crepel et al., 1980). In *staggerer,* granule cells are present during the normal loss of the excess CFs, but they fail to make synaptic contact with P-cells. The mere presence of granule cells is not sufficient to bring about the normal reduction in the excess number of climbing fibers; in the adult *staggerer* mouse, each P-cell retains the multiple CF contacts. At present, however, it is not known to what extent the critical factor in eliminating the multiple CFs involves functional or trophic synaptic interactions between the granule and P-cells.

In view of the fact that polyneuronal innervation is a relatively new discovery and has so far only been reported in a few systems, it remains to be seen just how extensive this phenomenon is in the developing nervous system. Nevertheless, it is becoming increasingly evident that, in general, regressive processes, often involving competition and functional mechanisms, are an important feature in the otogeny of the nervous system. Various aspects of neurogenesis, ranging from the control of cell number to synaptogenesis to the more subtle regulation of physiological input to neurons (including behavioral mechanisms), appear to depend as much on the elimination, loss, or suppression of previously formed characteristics as on the attainment or formation of new characteristics. (For detailed discussions of this view, see Jacobson, 1974; Hirsch and Jacobson, 1975; Mark, 1974, 1980, Changeux and Danchin, 1976; Oppenheim, 1981b.)

Regressive processes such as cell death and the loss of synaptic connections may have arisen phylogenetically as necessary epigenetic mechanisms to compensate for individual nonheritable variations in the number of presynaptic and postsynaptic components of specific neuronal systems (e.g., motoneurons and muscle fibers) during ontogeny. Alternatively, or additionally, these phenomena may also serve to compensate for heritable mutations in either the presynaptic or postsynaptic components. As Jacobson (1974), and more recently Katz and Lasek (1978), have pointed out, in the absence of such epigenetic canalizing mechanisms, a mutation involving quantitative changes in only the postsynaptic target (e.g., fewer muscle fibers) would be maladaptive, in that the result would be a quantitative mismatch between neurons and their targets. For instance, mutations or developmental errors leading to the formation of supernumarary limb parts may not involve direct or primary neurogenetic changes, yet the extra parts become innervated, apparently due to secondary epigenetic regulations such as increased cell survival or axonal branching.

Although our predecessors were probably only dimly aware of the full significance of regressive processes in evolution and development, this may have been what Huxley, Lewes, Roux, Ramón y Cajal, and others were striving toward in their own nascent attempts to come to grips with the difficult and counterintuitive notion that the appearance of developmental progress on one level may often involve regression and loss on other levels. Despite the fact that many of the details of this view still remain to be worked out, it is already evident that our present recognition of regressive processes represents a fundamentally different perspective from most previous conceptions of neural development in which progress was thought to be the major, if not the sole, leitmotiv on all levels.

ACKNOWLEDGMENTS

My own research reported in this chapter was supported by the National Science Foundation and by the North Carolina Department of Mental Health. I thank many of my colleagues who were willing to share their thoughts and unpublished data with me. I owe a special debt of gratitude to Viktor Hamburger, Lanny Haverkamp, Alan Lamb, and Jerome Maderdrut for valuable comments and discussions. Rubenia Daniels and Ann Sterling provided prompt and competent clerical support, which made it possible to meet the submission deadline. For that, I am most grateful.

NOTES

1. The long history of attempts to relate evolution (phylogeny) and individual development (ontogeny) has been brilliantly reviewed recently by Gould

(1977). The most well known of these attempts, the biogenetic law or recapitulation theory of E. Haeckel, is only one of the many made at drawing parallels between these two levels of phenomena.

2. For instance, A. Weismann, an ardent opponent of neo-Lamarckianism, adopted Roux's notion of cellular and molecular "struggle" and applied it to the competition between specific genes and chromosomes in the cell nucleus of the zygote following fertilization (Weismann, 1896).
3. Part of the reason for Ramón y Cajal's acceptance of the notion of regressive processes and functional adaptations may stem from his belief in the correctness of recapitulation theory for neural development (e.g., Ramón y Cajal, 1937, pp. 458–59) and from his apparent inclination towards Lamarckianism (Ramón y Cajal, 1937, pp. 458, 576).
4. This definition of histogenetic cell death is admittedly quite vague, and I would agree with Jacobson (1978) that we should be prepared to abandon such terms as our knowledge of neuronal cell death increases.
5. The first experiments involving the peripheral depletion of targets were by Steinmetz (1906) and Braus (1906) with frogs. They also interpreted their findings as being consistent with Roux's notion of correlative differentiation.
6. In apparent contrast to the situation in birds, however, Detwiler (1936) claimed that in salamanders the only changes in *motoneurons* following these manipulations were an increase in the size and the peripheral branching of neurons and not a numerical hypoplasia or hyperplasia. He attributed this to the fact that salamanders lack a clearly defined lateral motor column (for a detailed discussion of this issue, see Hamburger & Keefe, 1944).
7. It was later shown by Simmler (1949) that wing-bud removal in the chick also results in induced cell death in the brachial sympathetic ganglia (see also Levi-Montalcini and Levi, 1944).
8. I have recently discussed the early ideas of Roux and others with regard to the role of function in neural development (Oppenheim, 1979).
9. It is important to note, however, that none of these studies actually investigated neuroanatomical development. They merely inferred that since the behavior (i.e., swimming) appeared normal, neural development must also have been normal.
10. Glial proliferation typically occurs after neuronal proliferation in a given region (Jacobson, 1978). Thus, the lower rates of mitosis following these manipulations most likely reflect a secondary effect on gliogenesis following induced neuronal cell death.
11. A similar view was expressed earlier by Ernst (1926) and Levi-Montalcini and Levi (1942, 1944). In a paper also published in 1949, Levi-Montalcini presented evidence that removal of the otocyst in the chick resulted in the failure of migration of neurons into the vestibular nucleus tangentialis. Although long cited as a paradigm example of the peripheral control of migration, these findings have recently been disputed (Peusner and Morest, 1977). It is now thought that after otocyst removal the tangentialis neurons migrate normally, but in the absence of vestibular fibers, they are not maintained and undergo induced cell death.
12. It is important to point out that in all experiments with chicks that involved limb-bud removal, the surgery was performed at least two days prior to limb

innervation. Thus, induced cell death of either sensory or motoneurons cannot reflect degeneration owing to axotomy.

13. Although Hamburger (1958) considered several alternatives to explain induced cell death, such as the absence of a trophic factor and the failure to form anatomical connections with the target, he favored the explanation that an impediment to continued axoplasmic flow caused by the neuroma was responsible. Pilar and Landmesser (1976) have suggested a similar—albeit more detailed—explanation for induced cell death in the ciliary ganglion.
14. In view of recent proposals suggesting that cues from the limb guide axons to their correct targets (for review, see Landmesser, 1980), it is of historical interest that both Detwiler (1923) and Hamburger (1929, 1939b) made similar proposals many years ago.
15. There are some potential pitfalls in using HRP retrograde labeling in developing animals. It is often difficult to rule out diffusion into adjacent muscles. Moreover, since it is known that growth cones can take up HRP (e.g., Chu-Wang and Oppenheim, 1980), it is possible that the neurons that take up HRP at early stages of innervation, following injection of a specific muscle, are still in the process of extending axons to other muscles. If so, then the labeled neurons may not represent the same cells that project to that muscle in the mature animal.
16. In the early literature, the terms *hyper*plasia and *hypo*plasia sometimes were used to designate numerical changes and sometimes to refer to volumetric changes in neural centers following modifications in the size of the periphery. And at times the terms *atrophy* and *hypertrophy* were used interchangeably with hypoplasia and hyperplasia. In view of the fact that virtually all cases of hypoplasia and hyperplasia resulting from changes in target size probably reflect increased or reduced natural cell death—and not changes in proliferation—Hollyday and Hamburger (1976) have suggested that if one designates natural cell death as *neurothanasia,* then the terms *hypothanasia* and *hyperthanasia* are more appropriate for describing the actual events (but see note 23). Hyperplasia and hypoplasia would then be restricted to numerical changes in a neuronal population resulting from modifications of proliferation, while atrophy and hypertrophy would refer to changes in *cell size* owing to peripheral manipulations.
17. As indicated in note 15, HRP diffusion between muscles may be a possible source of error, and electrophysiological recordings or stimulation at the earliest stages of innervation may not be sufficiently precise, owing to the small size of the preparations, to resolve specific synaptic connections.
18. As already noted, at one time it was believed that there were differences between salamanders and chicks with regard to the responsiveness of spinal motoneurons to changes in the size of the periphery. Later studies, however, showed that this was not the case (Hamburger & Keefe, 1944).
19. At present, it appears that *behavioral* immobilization, per se, is a necessary but not a sufficient condition for preventing motoneuron cell death. Chronic treatment of embryos from Day 5 through Day 10 with concentrations of eserine, nicotine, carbachol, or hexacarbacholine sufficient to produce behavioral immobilization equal to, or greater than, that resulting from curare or α-BTX, does not prevent cell death; motoneuron numbers on Day 10 are

comparable to, or in the case of eserine even less than, controls (Oppenheim and Maderdrut, in preparation; also see Olek and Edwards, 1980). Furthermore, the observation that some pharmacological agents produce behavioral paralysis without altering the magnitude of cell death is additional evidence against the role of afferent sensory input in cell death. Movement-associated sensory feedback is greatly altered in this situation, yet cell death is unaffected.

20. The degree of atrophy and histological immaturity in the muscles of embryos in which cell death has been prevented appears to depend upon the duration and extent of prior paralysis. We are currently attempting to determine whether some aspect of muscle differentiation—other than, or in addition to, ACh receptors—is critical in the control of motoneuron cell death.
21. Levi-Montalcini and Aloe (1981) have raised the possibility that the increased cell number in the superior cervical ganglia of rats treated with NGF since birth may reflect more than the prevention of natural cell death. They point out that the increase in cell number following such treatment is far greater than the relatively modest loss of cells by natural cell death. Therefore, they argue that, in addition to the prevention of cell death, NGF treatment may also increase proliferation or even alter the phenotype of neural crest cells that would not normally form sympathetic neurons (e.g., by inducing the recruitment of "indifferent" crest cells to assume the "sympathetic" phenotype).
22. These examples represent estimates or claims of cell death that rely upon evidence with differing degrees of validity, including direct cell counts, axon counts, the observation of pyknotic cells, and estimates of total DNA content in a given neural region. Direct cell counts of either healthy or degenerating cells is the most conclusive method for proving cell death.
23. Harkmark (1954) has shown that *induced* cell death occurs in these same pontine nuclei between embryonic Day 16 through Day 19, following removal of the anlage of the cerebellum on embryonic Day 5. Thus, the ability of a neuronal population to exhibit induced cell death cannot be taken as evidence that natural cell death also occurs in that population.

REFERENCES

Alley, K. E. 1974. Morphogenesis of the trigeminal mesencephalic nucleus in the hamster: Cytogenesis and neurone death. *J. Embryol. Exp. Morphol.* 31:99–121

Amaral, D. G., Sinnamon, H. M. 1977. The locus coeruleus: Neurobiology of a central noradrenergic nucleus. *Prog. Neurobiol.* 9:147–96

Arees, E. A., Aström, K. E. 1977. Cell death in the optic tectum of the developing rat. *Anat. Embryol.* 151:29–34

Armstrong, R. C., Clarke, P. G. H. 1979. Neuronal death and the development of the pontine nuclei and inferior olive in the chick. *Neuroscience* 4:1635–47

Baer, K. E. von 1828. *Entwicklungsgeschichte der Thiere: Beobachtung und Reflexion.* Königsberg: Bornträger.

Bagust, J., Lewis, D. M., Westerman, R. A. 1973. Polyneuronal innervation of kitten skeletal muscle. *J. Physiol.* (London) 229:241–55

Barron, D. H. 1945. The role of sensory fibers in the differentiation of the spinal cord in sheep. *J. Exp. Zool.* 100:431–43

Barron, D. H. 1948. Some effects of amputation of the chick wing bud on early differentiation of the motor neuroblasts in the associated segments of the spinal cord. *J. Comp. Neurol.* 88:93–127

Bennett, M. R., Lindeman, R., Pettigrew, A. G. 1979. Segmental innervation of the chick forelimb following embryonic manipulation. *J. Embryol. Exp. Morphol.* 54:141–54

Benoit, P., Changeux, J. -P. 1975. Consequences of tenotomy on the evolution of multinnervation in developing rat soleus muscle. *Brain Res.* 99:345–58

Berg, A. T., Finlay, B. L. 1979. Cell death in the developing hamster optic tectum. *Neurosci. Abstr.* 5:153

Betz, W. J., Caldwell, J. H., Ribchester, R. R. 1979. The effects of partial denervation at birth on the development of motor units and muscle fibers in rat lumbrical muscle. *Neurosci. Abstr.* 5:475

Betz, H., Changeux, J. -P. 1979. Regulation of muscle acetylcholine receptor synthesis *in vitro* by cyclic nucleotide derivatives. *Nature* 278:749–52

Bibb, H. D. 1978. Neuronal death in the development of normal and hyperplastic spinal ganglia. *J. Exp. Zool.* 206:65–72

Blosser, J. C., Appel, S. H. 1980. Regulation of acetylcholine receptor by cyclic AMP. *J. Biol. Chem.* 255:1235–38

Bodian, D. 1966. Spontaneous degeneration in the spinal cord of monkey foetuses. *Bull. Johns Hopkins Hosp.* 119:217–34

Boydston, W. R., Sohal, G. S. 1979. Grafting of additional periphery reduces embryonic loss of neurons. *Brain Res.* 178:403–10

Braithwaite, A. W., Harris, A. J. 1979. Neural influence on acetylcholine receptor clusters in embryonic development of skeletal muscles. *Nature* 270:549–51

Braus, H. 1906. Vordere Extermität und Operculum bei Bombinatorlarven. Ein Beitrag zur Kenntnis morphogener Correlation und Regulation. *Morphol. Jahrb.* 35:139–220

Brown, M. C., Jansen, J. K. S., Van Essen, D. 1976. Polyneuronal innervation of skeletal muscle in newborn rats and its elimination during maturation. *J. Physiol.* (London) 261:387–422

Bueker, E. D. 1943. Intracentral and peripheral factors in the differentiation of motor neurons in transplanted lumbosacral spinal cords of chick embryos. *J. Exp. Zool.* 93:99–127

Bueker, E. D. 1945. Hyperplastic changes in the nervous system of a frog *(Rana)* as associated with multiple functional limbs. *Anat. Rec.* 93:323–31

Bueker, E. D. 1948. Implantation of tumors in the hindlimb field of the embryonic chick and the developmental response of the lumbosacral nervous system. *Anat. Rec.* 102:369–89

Burden, S. 1977. Development of the neuromuscular junction in the chick embryo: The number, distribution, and stability of acetylcholine receptors. *Dev. Biol.* 57:317–29

Burke, R. E. 1980. Motor unit types: functional specializations in motor control. *Trends Neurosci.* 3:255–258

Burke, R. E., Strick, P. L., Kanda, C., and Walmsley, B. 1977. Anatomy of medical gastrochemius and soleus motor nuclei in cat spinal cord. *J. Neurophysiol.* 40:667–680

Burr, H. S. 1916. The effects of the removal of the nasal pits in *Am-*

blystoma embryos. *J. Exp. Zool.* 20:27–57

Burr, H. S. 1932. An electro-dynamic theory of development suggested by studies of proliferation rates in the brain of *Amblystoma. J. Comp. Neurol.* 56:347–71

Cantino, D., Sisto-Daneo, L. 1972. Cell death in the developing optic tectum. *Brain Res.* 38:13–25

Carmichael, L. 1926. The development of behavior in vertebrates experimentally removed from the influence of external stimulation. *Psychol. Rev.* 33:51–58

Carmichael, L. 1927. A further study of the development of behavior in vertebrates experimentally removed from the influence of external stimuli. *Psychol. Rev.* 34:34–47

Carpenter, R. L. 1932. Spinal ganglion responses to the transplantation of differentiated limbs in *Amblystoma* larvae. *J. Exp. Zool.* 61:149–73

Carpenter, R. L. 1933. Spinal ganglion responses to the transplantation of limbs after metamorphosis in *Amblystoma punctatum. J. Exp. Zool.* 64:287–301

Carr, V. M., Simpson, S. B. 1978a. Proliferative and degenerative events in the early development of chick dorsal root ganglia. I. Normal development. *J. Comp. Neurol.* 182:727–40

Carr, V. M., Simpson, S. B. 1978b. Proliferative and degenerative events in the early development of chick dorsal root ganglia. II. Responses to altered peripheral fields. *J. Comp. Neurol.* 182:741–56

Changeux, J-P., Danchin, A. 1976. Selective stabilization of developing synapses as a mechanism for the specification of neuronal networks. *Nature* 264:705–11

Chiappinelli, V., Giacobini, E., Pilar, G., Uchimura, H. 1976. Induction of cholinergic enzymes in chick ciliary ganglion and iris muscle cells during synapse formation. *J. Physiol.* (London) 257:746–66

Chu-Wang, I-W., Oppenheim, R. W. 1978a. Cell death of motoneurons in the chick embryo spinal cord. I. A light and electron microscopic study of naturally occurring and induced cell loss during development. *J. Comp. Neurol.* 177:33–58

Chu-Wang, I-W., Oppenheim, R. W. 1978b. Cell death of motoneurons in the chick embryo spinal cord. II. A quantitative and qualitative analysis of degeneration in the ventral root, including evidence for axon outgrowth and limb innervation prior to cell death. *J. Comp. Neurol.* 177:59–86

Chu-Wang, I-W., Oppenheim, R. W. 1980. Uptake, intraaxonal transport and fate of horseradish peroxidase in embryonic spinal neurons of the chick. *J. Comp. Neurol.* 193:753–776

Chu-Wang, I-W., Oppenheim, R. W., Farel, P. B. 1981. Ultrastructural characteristics of migrating motoneurons. *Brain Res.* (in press)

Clarke, P. G. H., Cowan, W. M. 1976. The development of the isthmo-optic tract in the chick, with special reference to the occurrence and correction of developmental errors in the location and connections of isthmo-optic neurons. *J. Comp. Neurol.* 167:143–64

Clarke, P. G. H., Rogers, L. A., Cowan, W. M. 1976. The time of origin and the pattern of survival of neurons in the isthmo-optic nucleus of the chick. *J. Comp. Neurol.* 167:125–42

Coghill, G. E. 1929 *Anatomy and the Problem of Behavior.* Cambridge: University Press

Collin, R. 1906. Recherches cytologigues sur le développement de la cellule nerveuse. *Névraxe* 8:181–308

Cone, C. D., Cone, C. M. 1976. Induction of mitosis in mature neurons in central nervous system by sustained depolarization. *Science* 192:155–58

Cone, C. D., Cone, C. M. 1978. Evidence of normal mitosis with complete cytokinesis in central nervous system neurons during sustained depolarization with ouabain. *Exp. Neurol.* 60:41–55

Cowan, W. M. 1973. Neuronal death as a regulative mechanism in the control of cell number in the nervous system. In *Development and Aging in the Nervous System*, ed. M. Rockstein, pp. 19–41. New York: Adademic

Cowan, W M., Wenger, E. 1967. Cell loss in the trochlear nucleus of the chick during normal development and after radical extirpation of the optic vesicle. *J. Exp. Zool.* 164:265–80

Cowan, W. M., Wenger, E. 1968. Degeneration in the nucleus of origin of the preganglionic fibers to the chick ciliary ganglion following early removal of the optic vesicle. *J. Exp. Zool.* 168:105–24

Creazzo, T. L., Sohal, G. S. 1979. Effects of chronic injections of α-Bungarotoxin on embryonic cell death. *Exp. Neurol.* 66:135–45

Crepel, F., Delhaye-Bouchaud, N., Guastavino, J. M., Sampaio, I. 1980. Multiple innervation of cerebellar Purkinje cells by climbing fibers in *staggerer* mutant mouse. *Nature* 283:483–84

Crepel, F., Mariani, J. 1976. Multiple innervation of Purkinje cells by climbing fibers in the cerebellum of the weaver mutant mouse. *J. Neurobiol.* 7:579–82

Crepel, F., Mariani, J., Delhaye-Bouchaud, N. 1976. Evidence for a multiple innervation of Purkinje cells by climbing fibers in the immature rat cerebellum. *J. Neurobiol.* 7:567–78

Cunningham, T. J., Huddelston, C., Murray, M. 1979. Modification of neuron numbers in the visual system of the rat. *J. Comp. Neurol.* 184:423–34

Currie, J., Cowan, W. M. 1974. Some observations on the early development of the optic tectum in the frog *(Rana pipiens)*, with special reference to the effects of early eye removal on mitotic activity in the larval tectum. *J. Comp. Neurol.* 156:123–42

Detwiler, S. R. 1923. Experiments on the transplantation of the spinal cord in *Amblystoma* and their bearing upon the stimuli involved in differentiation of nerve cells. *J. Exp. Zool.* 37:339–93

Detwiler, S. R. 1936. *Neuroembryology, An Experimental Study.* New York: Macmillan

Dribin, L. B., Barrett, J. N. 1980. Conditioned medium enhances neurite outgrowth from rat spinal cord explants. *Dev. Biol.* 74:184–95

Dubey, P. N., Kadasne, D. K., Gosavi, V. S. 1968. The influence of the peripheral field on the morphogenesis of Hofmann's nucleus major of chick spinal cord. *J. Anat.* 102:407–14

Duchen, L. W., Heilbronn, E., Tonge, D. H. 1975. Functional denervation of skeletal muscle in the mouse after the local injection of a postsynaptic blocking fraction of *Naja siamensis* venom. *J. Physiol.* (London) 250:26–27P

Dürken, B. 1911. Über einseitige Augenextirpation bei jungen Froschlarven. *Z. Wiss. Zool. Abt.* 105:192–242

Edwards, C. 1979. The effects of innervation on the properties of acetylcholine receptors in muscle. *Neuroscience,* 4:565–84

Ernst, M. 1926. Über Untergang von Zellen während der normalen Entwicklung bei Wirbeltieren. *Z. Anat. Entwicklungsgesch.* 79:228–62

Farmbrough, D. M. 1979. Control of acetylcholine receptors in skeletal muscle. *Physiol. Rev.* 59:165–27

Farel, P. B., Bemelmans, S. E. 1980.

Retrograde labelling of migrating spinal motoneurons in Bullfrog larvae. *Neurosci. Lett.* 18:133–136

Fritzsch, B. 1979. Observations on degenerative changes of Purkinje cells during early development in mice and in normal and otocyst deprived chickens. *Anat. Embryol.* 158:95–102

Fry, K. R., Spira, A. W. 1979. Quantitative analysis of neuronal development in guinea pig retina. *Neurosci. Abstr.* 5:160

Glücksmann, A. 1951. Cell deaths in normal vertebrate ontogeny. *Biol. Rev.* 26:59–86

Gould, S. J. 1977. *Ontogeny and Phylogeny.* Cambridge, Mass.: Belknap

Greene, L. A., Shooter, E. M. 1980. The nerve growth factor: Biochemistry, synthesis and mechanism of action. *Ann. Rev. Neurosci.* 3:353–402

Grobstein, C. S. 1979. The development of afferents and motoneurons in rat spinal cord. *Neurosci. Abstr.* 5:161

Hall, E. K., Schneiderhan, M. A. 1945. Spinal ganglion hypoplasia after limb amputation in the fetal rat. *J. Comp. Neurol.* 82:19–34

Hall, Z. W., Reiness, C. G. 1977. Electrical stimulation of denervated muscles reduces incorporation of methionine into the ACh receptor. *Nature* 268:655–57

Hamburger, V. 1929. Experimentelle Beiträge zur Entwicklungsphysiologie der Nervenbahnen in der Froschextremität. *Arch. Entwicklungsmech. Org.* 119:47–99

Hamburger, V. 1934. The effects of wing bud extirpation on the development of the central nervous system in chick embryos. *J. Exp. Zool.* 68:449–94

Hamburger, V. 1939a. Motor and sensory hyperplasia following limbbud transplantation in chick embryos. *Physiol. Zool.* 12:268–84

Hamburger, V. 1939b. The development and innervation of transplanted limb primordia of chick embryos. *J. Exp. Zool.* 80:347–89

Hamburger, V. 1946. Isolation of the brachial segments of the spinal cord of the chick embryo by means of tantalum foil blocks. *J. Exp. Zool.* 103:113–42

Hamburger, V. 1952. Development of the nervous system. *Ann. N. Y. Acad. Sci.* 55:117–32

Hamburger, V. 1958. Regression versus peripheral control of differentiation in motor hypoplasia. *Am. J. Anat.* 102:365–410

Hamburger, V. 1975. Cell death in the development of the lateral motor column of the chick embryo. *J. Comp. Neurol.* 160:535–46

Hamburger, V. 1976. The developmental history of the motor neuron. *Neurosci. Res. Program Bull.* 15:1–37

Hamburger, V. 1980. Trophic interactions in neurogenesis: A personal historical account. *Ann. Rev. Neurosci.* 3:269–78

Hamburger, V., Brunso-Bechtold, J. K., Yip, J. 1981. Neuronal death in the spinal ganglia of the chick embryo and its reduction by nerve growth factor. *J. Neurosci.* 1:60–71

Hamburger, V., Keefe, E. L. 1944. The effects of peripheral factors on the proliferation and differentiation in the spinal cord of chick embryos. *J. Exp. Zool.* 96:223–42

Hamburger, V., Levi-Montalcini, R. 1949. Proliferation, differentiation and degeneration in the spinal ganglia of the chick embryo under normal and experimental conditions. *J. Exp. Zool.* 111:457–502

Hamburger, V., Wenger, E., Oppenheim, R. W. 1966. Motility in the chick embryo in the absence of sensory input. *J. Exp. Zool.* 162:133–60

Harkmark, W. 1954. The influence of the cerebellum on development and maintenance of the inferior olive and the pons. *J. Comp. Neurol.* 100:333–71

Harper, G. P., Thoenen, H. 1980. Nerve growth factor: Biological significance, measurement and distribution. *J. Neurochem.* 34:5–16

Harris, A. E. 1969. Differentiation and degeneration in the motor horn of foetal mouse. *J. Morphol.* 129:281–305

Harrison, R. G. 1904. An experimental study of the relation of the nervous system to the developing musculature in the embryo of the frog. *Am. J. Anat.* 3:197–220

Heaton, M. B., Moody, S. A., Kosier, M. E. 1978. Peripheral innervation by migrating neuroblasts in the chick embryo. *Neurosci. Lett.* 10:55–59

Heinsen, H. 1978. Postnatal quantitative changes in the cerebellar uvula of albino rats. *Anat. Embryol.* 154:285–304

Hendry, I. A., Campbell, J. 1976. Morphometric analysis of rat superior cervical ganglion after axotomy and nerve growth factor treatment. *J. Neurocytol.* 5:351–60

Heumann, D., Rabinowicz, T. 1980. Postnatal development of the dorsal lateral geniculate nucleus in the normal and enucleated albino mouse. *Exp. Brain Res.* 38:75–85

Heumann, D., Reuber, G., Rabinowicz, T. 1978. Postnatal development of the mouse cerebral cortex. IV. Evolution of the total cortical volume, of the population of neurons and glial cells. *J. Hirnforsch.* 19:385–93

Hess, A. 1970. Vertebrate slow muscle fibers. *Physiol. Rev.* 50:40–62

Hirsch, H., Jacobson, M. 1975. The perfectible brain: Principles of neuronal development. In *Handbook of Psychobiology,* ed. M. S. Gazzaniga, C. Blakemore, pp. 107–37. New York: Academic

Hollyday, M., Hamburger, V. 1976. Reduction of the naturally occurring motor neuron loss by enlargement of the periphery. *J. Comp. Neurol.* 170:311–20

Hollyday, M., Hamburger, V. 1977. An autoradiographic study of the formation of the lateral motor column in the chick embryo. *Brain Res.* 132:197–208

Hollyday, M., Hamburger, V., Farris, J. M. G. 1977. Localization of motor neuron pools supplying identified muscles in normal and supernumerary legs of chick embryos. *Proc. Nat. Acad. Sci. U.S.A.* 74:3582–86

Hollyday, M., Mendell, L. 1976. Analysis of moving supernumerary limbs of *Xenopus laevis. Exp. Neurol.* 51:316–24

Hughes, A. 1961. Cell degeneration in the larval ventral horn of *Xenopus laevis. J. Embryol. Exp. Morphol.* 9:269–84

Hughes, A. 1965. A quantitative study of the development of the nerves in the hind-limb of *Eleutherodactylus martinicensis. J. Embryol. Exp. Morphol.* 13:9–34

Hughes, A. 1968. *Aspects of Neural Development.* London: Logos

Hughes, A. 1973. The development of dorsal root ganglia and ventral horns in the opossum. A quantitative study. *J. Embryol. Exp. Morphol.* 30:359–76

Hughes, A., Carr, V. M. 1978. The interaction of periphery and center in the development of dorsal root ganglia. In *Handbook of Sensory Physiology,* vol. 9, ed. M. Jacobson, pp. 85–114. Berlin: Springer

Hughes, A., Egar, M. 1972. The innervation of the hind limb of *Eleutherodactylus martinicensis:* Further comparison of cell and fiber numbers during development. *J. Embryol. Exp. Morphol.* 27:389–412

Hughes, W. F., LaVelle, A. 1975. The effects of early tectal lesions on development in the retinal ganglion cell layer of chick embryos. *J. Comp. Neurol.* 163:265–84

Hughes, W. F., McLoon, S. C. 1979. Ganglion cell death during normal retinal development in the chick: Comparisons with cell death induced by early target field destruction. *Exp. Neurol* 66:587–601

Huxley, T. H. 1869. The geneology of animals. In *Critiques and Addresses of T. H. Huxley* (1873). London: Macmillan

Hyndman, A. G., Zamenhof, S. 1978. Cell proliferation and cell death in the cerebral hemispheres of developing chick embryos. *Dev. Neurosci.* 1:216–25

Ikeda, H., Gotoh, J. 1971. Distribution of monoamine-containing cells in the central nervous system of the chicken. *Jpn. J. Pharmacol.* 21:763–84

Ikeda, H., Gotoh, J. 1974. Distribution of monoamine-containing terminals and fibers in the central nervous system of the chicken. *Jpn. J. Pharmacol.* 24:831–41

Innocenti, G. M., Frost, D. O. 1979. Effects of visual experience on the maturation of the efferent system to the corpus callosum. *Nature* 280:231–34

Ivy, G. O., Akers, R. M., Killackey, H. P. 1979. Differential distribution of callosal projection neurons in the neonatal and adult rat. *Brain Res.* 173:532–37

Jacobson, M. 1974. A plentitude of neurons. In *Aspects of Neurogenesis* ed. G. Gottlieb, pp. 151–66. New York: Academic

Jacobson, M. 1978. *Developmental Neurobiology.* New York: Plenum

Jansen, J. K. S., Thompson, W., Kuffler, D. P. 1978. The formation and maintenance of synaptic connections as illustrated by studies of the neuromuscular junction. *Prog. Brain Res.* 48:3–18

Jokl, A. 1920. Zur Entwicklung des Anurenauges. *Anat. Hefte* 49: 217–41

Katz, M. J., Lasek, R. J. 1978. Evolution of the nervous system: role of ontogenetic mechanisms in the evolution of matching populations. *Proc. Natl. Acad. Sci. U.S.A.* 75:1349–52

Kessler, J. A., Black, I. B. 1980. The effects of nerve growth factor (NGF) and antiserum to NGF on the development of embryonic sympathetic neurons *in vivo.* *Brain Res.* 189:157–68

Knyihar, E., Csillik, B., Rakic, P. 1978. Transient synapses in the embryonic primate spinal cord. *Science* 202:1206–09

Kollros, J. J. 1953. The development of the optic lobes in the frog. I. The effects of unilateral enucleation in embryonic stages. *J. Exp. Zool.* 123:153–87

Korneliussen, H., Jansen, J. K. S. 1976. Morphological aspects of the elimination of polyneuronal innervation of skeletal muscle fibers in newborn rats. *J. Neurocytol.* 5:591–604

Laing, N. G., Prestige, M. C. 1978. Prevention of spontaneous motoneurone death in chick embryos. *J. Physiol.* (London) 282:33–34P

Lamb, A. H. 1974. The timing of the earliest motor innervation to the hind limb bud in the *Xenopus* tadpole. *Brain Res.* 67:527–30

Lamb, A. H. 1976. The projection patterns of the ventral horn in the hind limb during development. *Dev. Biol.* 54:82–99

Lamb, A. H. 1977. Neuronal death in the development of the somatotopic projections of the ventral horn in *Xenopus.* *Brain Res.* 134:145–50

Lamb, A. H. 1979a. Evidence that some developing limb motoneurons die for reasons other than peripheral competition. *Dev. Biol.* 71:8–21

Lamb, A. H. 1979b. Ventral horn cell counts in a *Xenopus* with naturally occurring supernumerary hindlimbs. *J. Embryol. Exp. Morph.* 49:13–16

Lamb, A. H. 1980. Motoneurone counts in *Xenopus* frogs reared with one bilaterally-innervated hindlimb. *Nature* 284:347–50

Lance-Jones, C., Landmesser, L. 1980. Motoneurone projection patterns in embryonic chick limbs following partial deletions of the spinal cord. *J. Physiol.* 302:559–80

Landmesser, L. 1978a. The distribution of motoneurones supplying chick hind limb muscles. *J. Physiol.* (London) 284:371–89

Landmesser, L. 1978b. The development of motor projection patterns in the chick hind limb. *J. Physiol.* (London) 284:391–14

Landmesser, L. 1980. The generation of neuromuscular specificity. *Ann. Rev. Neurosci.* 3:279–302

Landmesser, L., Morris, D. G. 1975. The development of functional innervation in the hind limb of the chick embryo. *J. Physiol.* (London) 249:301–26

Landmesser, L., Pilar, G. 1974a. Synapse formation during embryogenesis on ganglion cells lacking a periphery. *J. Physiol.* (London) 241:715–36

Landmesser, L., Pilar, G. 1974b. Synaptic transmission and cell death during normal ganglionic development. *J. Physiol.* (London) 241:737–49

Landmesser, L., Pilar, G. 1976. Fate of ganglionic synapses and ganglion cell axons during normal and induced cell death. *J. Cell Biol.* 68:357–74

Larramendi, L. M. 1969. Analysis of synaptogenesis in the cerebellum of the mouse. In *Neurobiology of Cerebellar Evolution and Development,* ed. R. Llinas, pp. 803–43. Chicago: A.M.A. Research Foundation

Legrand, J. 1979. Morphogenetic action of hormones. *Trends Neurosci.* 2:234–36

Levi-Montalcini, R. 1945. Correlations dans le développement des différentes parties du systéme nerveux. II. Correlations entre le développment de l'encephale et celui de la moelle épinière dans l'embryon de poulet. *Arch Biol.* 56:71–93

Levi-Montalcini, R. 1947. Regressione secondaria del ganglio ciliare dopo aspotazione della resicola mesencefalica in embrione di pollo. *R. Acad. Naz. Linncie* 3:144–146

Levi-Montalcini, R. 1949. The development of the acoustico-vestibular centers in the chick embryo in the absence of the afferent root fibers and of descending tracts. *J. Comp. Neurol.* 91:209–42

Levi-Montalcini, R. 1950. The origin and development of the visceral system in the spinal cord of the chick embryo. *J. Morphol.* 86:253–83

Levi-Montalcini, R. 1964. Events in the developing nervous system. *Prog. Brain Res.* 4:1–29

Levi-Montalcini, R. 1966. The nerve growth factor: Its mode of action on sensory and sympathetic nerve cells. *Harvey Lect.* 60:217–59

Levi-Montalcini, R. 1975. NGF: An uncharted route. In *Neurosciences: Paths of Discovery,* ed. F. G. Worden, J. P. Swazey, G. Adelman, pp. 245–65. Cambridge: M.I.T.

Levi-Montalcini, R., Aloe, L. 1981. Mechanism(s) of action of nerve growth factor in intact and lethally injured sympathetic nerve cells in neonatal rodents. In *Cell Death in Biology and Pathology,* eds. I. D. Bowen and R. A. Lockshin, pp. 295–327. London: Chapman and Hall

Levi-Montalcini, R., Cohen, S. 1956. *In vitro* and *in vivo* effects of a

nerve growth stimulating agent isolated from snake venom. *Proc. Nat. Acad. Sci. U.S.A.* 42:695–99
Levi-Montalcini, R., Hamburger, V. 1951. Selective growth stimulating effects of mouse sarcoma on the sensory and sympathetic nervous system of the chick embryo. *J. Exp. Zool.* 116:321–51
Levi-Montalcini, R., Hamburger, V. 1953. A diffusable agent of mouse sarcoma, producing hyperplasia of sympathetic ganglia and hyperneurotization of viscera in the chick embryo. *J. Exp. Zool.* 123:233–78
Levi-Montalcini, R., Levi. G. 1942. Les conséquences de la destruction d'un territoire d'innervation périphérique sur le développment des centres nerveux correspondants dans l'embryon de poulet. *Arch. Biol.* 53:537–45
Levi-Montalcini, R., Levi, G. 1944. Correlazioni nello sviluppo tra varie partie del sistema nervoso. I. Consequenze della demolizione dell'abbozzo die un arto sui centri nervosi nell'embrione di pollo. *Comment. Pontif, Acad. Sci.* 8:527–68
Leuba, G., Rabinowicz, T. 1979. Long-term effects of postnatal undernutrition and maternal malnutrition on mouse cerebral cortex. I. Cellular densities, cortical volume and total number of cells. *Exp. Brain Res.* 37:283–98
Lewes, G. H. 1877. *The Physical Basis of Mind, Problems of Life and Mind* (2nd series). Boston: Osgood
Lichtman, J. W. 1977. The reorganization of synaptic connections in the rat submandibular ganglion during postnatal development. *J. Physiol.* (London) 273:155–77
Lømo, T. 1980. What controls the development of neuromuscular junctions? *Trends Neurosci.* 3:126–29
Maderdrut, J. L. 1979. A radiometric microassay for glutamic acid decarboxylase activity. *Neuroscience* 4:995–1005
Mark, R. F. 1974. *Memory and Nerve Cell Connections.* Oxford: Claredon
Mark, R. F. 1980. Synaptic repression at neuromuscular junctions. *Physiol. Rev.* 60:355–95
Matthews, S. A., Detwiler, S. R. 1926. The reaction of *Amblystoma* embryos following prolonged treatment with chloretone. *J. Exp. Zool.* 45:279–92
May, R. M. 1933. Reactions neurogéniques de la moelle à la greffe en surnombre, ou à l'ablation d'une ébauche de patte posterieure chez l'embryon de l'anoure, *Discoglossus pictus. Bull. Biol.* (Liège) 67:327–49
Mlonyeni, M. 1967. The late stages of the development of the primary cochlear nuclei in mice. *Brain Res.* 4:334–44
Mustafa, G. Y., Gamble, H. J. 1979. Changes in axonal numbers in developing human trochlear nerve. *J. Anat.* 128:323–30
Narayanan, C. H., Narayanan, Y. 1978. Neuronal adjustments in developing nuclear centers of the chick embryo following transplantation of an additional optic primordium. *J. Embryol. Exp. Morphol.* 44:53–70
Nestler, E. J., Beam, K. G., Greengard, P. 1978. Nicotinic cholinergic stimulation increases cyclic GMP levels in vertebrate skeletal muscle. *Nature* 275:451–53
Nishi, R., Berg, D. K. 1979. Survival and development of ciliary ganglion neurones grown alone in cell culture. *Nature* 277:232–34
O'Brien, R. A. D., Ostberg, A. J., Vrbová, G. 1978. Observations on the elimination of polyneuronal innervation in developing mammalian skeletal muscle. *J. Physiol.* (London) 282:571–83
Olek, A. J., Edwards, C. 1977. The effect of spinal nerve section on motor neuron loss during development

in *Xenopus. Neurosci. Abstr.* 3:115

Olek, A. J., Edwards, C. 1978. Effect of alpha and beta bungarotoxin on the naturally occuring motoneuron loss in *Xenopus* larvae. *Neurosci. Abstr.* 4:122

Olek, A. J., Edwards, C. 1978. Effects of anesthetic treatment on motor neuron death in *Xenopus. Brain Res.* 191:483–88

Oppenheim, R. W. 1979. Laura Bridgeman's brain: An early consideration of functional adaptations in neural development. *Dev. Psychobiol.* 12:533–37

Oppenheim, R. W. 1981a. Cell death of motoneurons in the chick embryo spinal cord. V. Evidence on the role of cell death and neuromuscular function in the formation of specific peripheral connections. *J. Neurosci.* 1:141–151

Oppenheim, R. W. 1981b. Ontogenetic adaptations and retrogressive processes in the development of the nervous system and behavior. In *Maturation and Development* eds. H. Prechtl and K. Connolly (in press). Lippincott: Philadelphia

Oppenheim, R. W., Chu-Wang, I-W. 1977. Spontaneous cell death of spinal motoneurons following peripheral innervation in the chick embryo. *Brain Res.* 125:154–60

Oppenheim, R. W., Chu-Wang, I-W., Maderdrut, J. L. 1978. Cell death of motoneurons in the chick embryo spinal cord. III. The differentiation of motoneurons prior to their induced degeneration following limb-bud removal. *J. Comp. Neurol.* 177:87–112

Oppenheim, R. W., Heaton, M. B. 1975. The retrograde transport of horseradish peroxidase from the developing limb of the chick embryo. *Brain Res.* 98:291–302

Oppenheim, R. W., Majors-Willard, C. 1978. Neuronal cell death in the brachial spinal cord of the chick is unrelated to the loss of polyneuronal innervation in wing muscle. *Brain Res.* 154:148–52

Oppenheimer, J. 1967. *Essays in the History of Embryology and Biology.* Cambridge: M.I.T.

Orida, N., Poo. M-M. 1978. Electrophoretic movement and localization of acetylcholine receptors in the embryonic muscle cell membrane. *Nature* 275:31–35

Pannese, E. 1976. An electron microscopic study of cell degeneration in chick embryo spinal ganglia. *Neuropathol. Appl. Neurobiol.* 2:247–67

Parks, T. N. 1979. Afferent influences on the development of the brain stem auditory nuclei of the chicken: Otocyst ablation. *J. Comp. Neurol.* 183:665–78

Pettigrew, A. G., Lindeman, R., Bennett, M. R. 1979. Development of the segmantal innervation of the chick forelimb. *J. Embryol. Exp. Morphol.* 49:115–37

Peusner, K. D., Morest, D. K. 1977. Neurogenesis in the nucleus vestibularis tangentialis of the chick embryo in the absence of the primary afferent fibers. *Neuroscience* 2:253–70

Piatt, J. 1946. The influence of the peripheral field on the development of the mesencephalic V nucleus in *Amblystoma. J. Exp. Zool.* 101:109–41

Pilar, G., Landmesser, L. 1976. Ultrastructural differences during embryonic cell death in normal and peripherally deprived ciliary ganglia. *J. Cell Biol.* 68:339–56

Pilar, G., Landmesser, L., Burstein, L. 1980. Competition for survival among developing ciliary ganglion cells, *J. Neurophysiol.* 43:233–54

Pittman, R., Oppenheim, R. W. 1978. Neuromuscular blockade increases motoneurone survival during normal cell death in the chick embryo. *Nature* 271:364–66

Pittman, R., Oppenheim, R. W. 1979.

Cell death of motoneurons in the chick embryo spinal cord. IV. Evidence that a functional neuromuscular interaction is involved in the regulation of naturally occurring cell death and the stabilization of synapses. *J. Comp. Neurol.* 187:425–46

Pollack, E. D. 1969a. Normal development of the lateral motor column in the brachial cord in *Rana pipiens. Antat. Rec.* 163:111–20

Pollack, E. D. 1969b. Response of the lateral motor column in multiple forelimbs in *Rana pipiens. Teratol.* 2:159–62

Pollack, E. D. 1980. Target-dependent survival of tadpole spinal cord neurites in tissue culture. *Neurosci. Lett.* 16:269–74

Prestige, M. C. 1965. Cell turnover in the spinal ganglia of *Xenopus laevis tadpoles. J. Embryol. Exp. Morphol.* 13:63–72

Prestige, M. C. 1967. The control of cell number in the lumbar ventral horn during the development of *Xenopus laevis* tadpoles. *J. Embryol. Exp. Morphol.* 18:359–387

Prestige, M. C. 1970. Differentiation, degeneration and the role of the periphery: quantitative considerations. In *The Neurosciences: Second Study Program,* ed. F. O. Schmitt, pp. 73–82. New York: Rockefeller

Prestige. M. C., Wilson, M. A. 1972. Loss of axons from ventral roots during development. *Brain Res.* 41:467–70

Purves, D. 1980. Neuronal competition. *Nature* 287:585–586

Purves, D. & Lichtman, J. 1980. Elimination of synapses in the developing nervous system. *Science* 210:153–57

Rabl, C. 1900. *Ueber den Bau und die Entwicklung der Linse.* Leipzig; Grieben

Rager, C., Rager, U. 1978. Systems matching by degeneration. I. A quantitative electron microscopic study of the generation and degeneration of retinal ganglion cells in the chick. *Exp. Brain Res.* 33:65–78

Ramon y Cajal, S. 1911. *Histologie du système nerveux de l'homme et des Vertébrés,* vol. 2. Madrid: Instituto Ramón y Cajal

Ramón y Cajal, S. 1929. *Studies on Vertebrate Neurogenesis.* Springfield, Ill.: Thomas

Ramón y Cajal, S. 1937. *Recollections of My Life.* Cambridge: M.I.T. (reprint)

Redfern, P. S. 1970. Neuromuscular transmission in newborn rats. *J. Physiol.* (London) 209:701–9

Rogers, L. A., Cowan, W. M. 1973. The development of the mesencephalic nucleus of the trigeminal nerve in the chick. *J. Comp. Neurol.* 147:291–320

Romanes, G. J. 1946. Motor localization and the effects of nerve injury on the ventral horn cells of the spinal cord. *J. Anat.* 80:117–31

Ronnevi, L-O. 1977. Spontaneous phagocytosis of boutons on spinal motoneurons during early postnatal development. An electron microscopical study in the cat. *J. Neurocytol.* 6:487–504

Rothman, T. P., Gershon, M. D., Holtzer, H. 1978. The relationship of cell division to the acquisition of adrenergic characteristics by developing sympathetic ganglion cell precursors. *Dev. Biol.* 65:322–41

Roux, W. 1881. *Der Kampf der Theile im Organismus.* Leipzig: Englemann

Roux, W. 1905. Die Entwickelungsmechanik, ein neuer Zweig der biologischen Wissenschaft. In *Vorträge und Aufsatze uber Entwickelungsmechanik der Organismen,* ed. W. Roux, pp. 1–105. Leipzig: Englemann

Rubel, E. W., Smith, D. J., Miller, L. C. 1976. Organization and development of brain stem auditory nuclei of the chicken: Ontogeny of N. magnocellularis and N. laminaris. *J. Comp. Neurol.* 166:469–90

Rubin, L. L., Schuetze, S. M., Weill, C. L., Fischbach, G. D. 1980. Regulation of acetylcholinesterase appearance at neuromuscular junctions *in vitro*. *Nature* 283:264–67

Saunders, J. W. 1966. Death in embryonic systems. *Science* 154:604–12

Schlessinger, A. R., Cowan, W. M., Gottlieb, D. I. 1975. An autoradiographic study of the time of origin and the pattern of granule cell migration in the dentate gyrus of the rat. *J. Comp. Neurol.* 159:149–76

Schultzberg, M., Ebendal, T., Hökfelt, T., Nilsson, G., Pfenninger, K. 1978. Substance P-like immunoreactivity in cultured spinal ganglia from chick embryos. *J. Neurocytol.* 7:107–17

Schwartz, J. P., Costa, E. 1979. Nerve growth factor-mediated increase of the substance P content of chick embryo dorsal root ganglia. *Brain Res.* 170:198–202

Scott, B. S. 1977. The effect of elevated potassium on the time course of neuron survival in cultures of dissociated dorsal root ganglia. *J. Cell Physiol.* 91:305–16

Shainberg, A., Burstein, M. 1976. Decrease of acetylcholine receptor synthesis in muscle cultures by electrical stimulation. *Nature* 264:368–69

Shorey, M. L. 1909. The effect of the destruction of peripheral areas on the differentiation of the neuroblasts. *J. Exp. Zool.* 7:25–64

Silver, J. 1978. Cell death during development of the nervous system. In *Handbook of Sensory Physiology*, vol. 9, ed. M. Jacobson, pp. 419–36. Berlin: Springer

Simmler, G. M. 1949. The effects of wing bud extirpation on the brachial sympathetic ganglia of the chick embryo. *J. Exp. Zool.* 110:247–57

Sirhari, T., Vrbová, G. 1978. The role of muscle activity in the differtiation of neuromuscular junctions in slow and fast chick muscles. *J. Neurocytol.* 7:529–40

Sohal, G. S. 1976. An experimental study of cell death in the developing trochlear nucleus. *Exp. Neurol.* 51:684–98

Sohal, G. S. 1976. Effects of deafferentation on the development of the isthmo-optic nucleus in the duck. *Exp. Neurol.* 50:161–73

Sohal, G. S., Weidman, T. A. 1978a. Development of the trochlear nerve: Loss of axons during normal ontogeny. *Brain Res.* 142:455–65

Sohal, G. S., Weidman, T. 1978b. Ultrastructural sequence of embryonic cell death in normal and peripherally deprived trochlear nucleus. *Exp. Neurol.* 61:53–64

Sohal, G. S., Creazzo, T. L., Oblak, T. G. 1979. Effects of chronic paralysis with α-Bungarotoxin on development of innervation. *Exp. Neurol.* 66:619–28

Sohal, G. S., Weidman, T. A., Stoney, S. D. 1978. Development of the trochlear nerve: Effects of early removal of periphery. *Exp. Neurol.* 59:331–41

Speidel, C. C. 1941. Adjustments of nerve endings. *Harvey Lect.* 36:126–58

Spemann, H. 1938. *Embryonic Development and Induction.* New Haven: Yale Univ. Press

Steinitz, E. 1906. Über den Einfluss der Elimination der embryonalen Augenblasen auf die Entwicklung des Gesamtorganismus beim Frosche. *Arch. Entwicklungsmech. Org.* 20:537–68

Stillwell, E. F., Cone, C. M., Cone, C. D. 1978. Stimulation of DNA synthesis in CNS neurones by sustained depolarization. *Nature* 246:110–11

Stirling, R. V., Summerbell, D. 1979. The segmentation of axons from the segmental nerve roots to the chick wing. *Nature* 278:640–42

Stirling, R. V., Summerbell, D. 1980. Evidence for non-selective motor innervation of rotated chick limbs. *J. Physiol.* (London) 300:7–8P

Stoney, S. D., Sohal, G. S. 1978. Development of the trochlear nerve: Neuromuscular transmission and electrophysiologic properties. *Exp. Neurol.* 62:798–803

Thompson, W., Jansen, J. K. S. 1977. The extent of sprouting of remaining motor units in partly denervated immature and adult rat soleus muscle. *Neuroscience* 2:523–35

Twitty, V. C. 1932. Influence of the eye on the growth of its associated structures, studied by means of heteroplastic transplantation. *J. Exp. Zool.* 61:333–74

Weismann, A. 1894. *The Effect of External Influences on Development.* Oxford: Clarendon

Weismann, A. 1896. *On Germinal Selection as a Source of Definite Variation.* Chicago: Open Court

Weiss, P. A. 1939. *Principles of Development, a Text on Experimental Embryology.* New York: Hafner (1969 reprint)

Wenger, E. L. 1950. An experimental analysis of relations between parts of the brachial spinal cord of the embryonic chick. *J. Exp. Zool.* 114:51–85

Woodward, D. J., Hoffer, B. J., Altman, J. 1974. Physiological and pharmacological properties of Purkinje cells in rat cerebellum degranulated by postnatal x-irradiation. *J. Neurobiol.* 5:283–304

Wright, L. 1981b. Time of cell origin and cell death in the dorsal motor nucleus of the vagus. *J. Comp. Neurol.* (in press).

Wright, L. 1981a. Cell survival in chick embryo ciliary ganglion is reduced by chronic ganglionic blockade. *Brain Res.* (in press).

Zelena, J., Vyskočil, F., Jiranová, I. 1979. In *The Cholinergic Synapse,* ed. S. Tuček, pp. 365–72. Amsterdam: Elsevier

Zilles, K. 1978. *Ontogenesis of the Visual System.* Berlin, New York: Springer

Zilles, K., Wingert, F. 1973. Quantitative studies of the development of the fresh volumes and the number of neurones of the N. oculomotorii of white mice during ontogenesis. *Brain Res.* 56:63–75

6 / *Embryonic Development of the Neural Circuitry Underlying Motor Coordination*

ANNE BEKOFF

INTRODUCTION

Motor coordination requires the selective activation of specific muscles in a defined temporal sequence. Muscle contractions are, of course, initiated by motoneuron impulses. Therefore, motor coordination depends on neural circuits that can activate pools of motoneurons in specific patterns. In recent years it has become clear that the neural circuitry underlying motor coordination may develop during embryonic stages despite the fact that embryonic motor behavior often appears jerky and uncoordinated, and bears little obvious resemblance to coordinated adult behaviors. It therefore seems appropriate at this time to review the information available on the embryonic development of motor behavior and to examine critically our current understanding of the underlying neural mechanisms. In order to lay the groundwork for a discussion of recent research on the ontogeny of motor coordination, it will be useful first to review briefly the historical origins of this rapidly progressing field of neuroembryology.

The first extensive comparative study of the development of motor behavior in a wide variety of vertebrate, as well as some invertebrate, embryos was conducted by Preyer (1885). In addition to providing an exhaustive survey of the literature and very careful descriptions of embryonic movements based on his own behavioral observations, Preyer made perhaps his most important contribution in showing that experimental techniques, such as carefully delivered mechanical, electrical, or thermal stimulation, could be applied to the study of embryonic behavior in an attempt to determine the underlying physiological mechanisms. One significant result that came out of this work, the importance of which was not fully appreciated until over 75 years later when Viktor Hamburger began his studies of embryonic behavior (Hamburger, 1963), was the discovery that "impulsive" or spontaneous movements occurred before reflexogenic movements could be elicited in the chick embryo. This was the first evidence that centrally generated behavior (behavior produced by the central nervous system in the absence of sensory input) could, and did, occur during development.

After the publication of Preyer's work in 1885, there followed a hiatus of nearly 20 years before the next major period of interest in embryonic movement began. The literature of this period has been reviewed many times (e.g., Hamburger, 1963, 1973; Carmichael, 1970) and therefore only a few salient points need be discussed here. Between 1902 and 1941 Coghill published his classic studies of the salamander, *Ambystoma.* Many of these are summarized in his book, *Anatomy and the Problem of Behavior* (1929). Coghill's most important contribution was his demonstration that steps in the development of behavior could be correlated with specific anatomical changes in the nervous system. For example, each new behavioral capability, from head bending to coiling to S-flexure, immediately followed the appearance of new connections in the spinal cord.

In addition, based on his work on *Ambystoma,* Coghill developed a theory which holds that particular behaviors, for example, the independent limb movements observed during walking, are developed by individuation ("emancipation") from a total pattern that is continuously integrated throughout life. However, while Coghill's theory was important during the 1920s and 1930s in generating a great deal of research, it did not prove to be a general description of the development of motor behavior in vertebrates since amniotes (reptiles, birds, and mammals) do not show continuously integrated behavior (Hamburger, 1963, 1973).

Rephrasing Coghill's theory in terms of our current understanding of motor control, it could imply (1) that, at least in anamniotes (fish and amphibians), the neural pattern generating circuitry for coordinated limb movements is activated obligatorily with the pattern generator(s) for the trunk movements of swimming at early developmental stages and (2) that only at later stages of development can the limb pattern generating circuitry be turned on and off separately. The idea that there may be separate pattern generators for trunk and limbs is suggested by data from adult fish where it seems likely (though it is not proven) that each spinal segment has its own pattern generator for alternating trunk movements (Grillner and Kashin, 1976) and that each fin has its own separate pattern generator (Holst, 1939). Despite the fact that in amniotes coordinated trunk and limb movements do not continue throughout embryonic life (and therefore behavior is not continuously integrated as originally suggested by Coghill, 1929), early leg movements are synchronous with trunk movements (for review, see Hamburger, 1963). Thus early coupling of the activity of trunk and limb pattern generators may be a common phenomenon. This issue deserves further research using currently available techniques.

An opposing theory, which held that behaviors develop by the integration of local reflexes into coordinated action patterns, was first proposed by Swenson (1929) and strongly supported by Windle (1940, 1944) and his coworkers. While Coghill had made it clear that the neural mechanisms underlying the generation of the coordinated movements involved in walking, for example, were laid down before the embryo could respond to sensory input (Coghill, 1929, pp. 86–87), most other workers in the 1930s and 1940s were strongly under the influence of reflexology; they accepted that reflexes are the functional units of behavior. In consequence, the onset and development of behavior were defined purely in terms of responses to stimulation. While spontaneous movements were occasionally mentioned, effort was concentrated on the description of evoked, or reflexogenic, movements and on anatomical studies of the development of neural connections that might underlie them, such as the closing of reflex arcs (e.g., *chick:* Kuo, 1932; Orr and Windle, 1934; Windle and Orr, 1934; Visintini and Levi-Montalcini, 1939; *guinea pig:* Bridgman and Carmichael, 1935; *rat:* Swenson, 1929; Angulo y Gonzalez, 1932; Windle and Baxter, 1936; *cat:* Windle, O'Donnell, and Glasshagle, 1933; *sheep:* Barcroft and Barron, 1939; *human:* Minkowski, 1920, 1921; Hooker, 1952). The light microscopic studies of silver-stained sections revealed that the first appearance of reflex responses coincided with the approach of afferent fibers to interneurons, which had already established connections with motoneurons.

However, limiting the study of the development of behavior only to the study of reflexes eventually led to a standstill in the progress toward understanding the neural mechanisms underlying the development of coordinated behavior. The evidence did suggest that the closing of the reflex arcs was correlated with the onset of reflex sensitivity. But the underlying assumption that the development of behavior could be understood solely through the study of the development of the form and pattern of reflexes imposed too great a limitation on the search for the physiological mechanisms (Hamburger, 1963).

During this period the first attempt to come to terms with the neural basis for the development of *coordinated* movement in the embryo was made (Weiss, 1941b). While others (with the exception of Coghill; see Oppenheim, 1978) had thought in terms of "diffuse" or "local" responses and reflexes, Paul Weiss was the first to state explicitly that the problem of the development of behavior was the problem of how the nervous system could arrange to selectively activate "definite groups of units in such combinations that their united action will result in an organized peripheral effect that makes sense" (Weiss, 1941b, p. 7).

Weiss did not deal with the description of the onset and gradual development of coordinated movements during ontogeny. He did, however, perform important experimental manipulations that shed light on some of the mechanisms by which the neural organization underlying coordinated movements developed. In one set of experiments on frog tadpoles, Weiss (1941a) deafferented the region of the spinal cord that innervates the legs, before the legs were functional. No impairment of coordinated leg movements was observed after the legs began to move. In addition, Weiss (1941b) transplanted limb buds in reversed anteroposterior polarity in embryonic salamanders. These legs moved in a coordinated manner, but in reverse, and never changed their pattern. The necessarily distorted sensory feedback did not disrupt the normal sequential activation of muscles. Weiss concluded, therefore, that (1) sensory pathways are not necessary for the development of normally patterned locomotor movements and (2) abnormal sensory input cannot alter the development of such movements. This is not to say that Weiss failed to recognize the importance of sensory information as it affects motor output. However, he felt that it was important to:

> abandon such inarticulate utterances about sensory control as that it is "of paramount importance," "dominant," "essential," "vital," or, on the other side of the picture, "irrelevant," "practically insignificant," etc., and replace them with precise statements as to what phases of motor activity depend on the integrity of sensory innervation, in what respect, and to what degree (Weiss, 1941b, p. 74).

Although conclusions about motor coordination based on purely behavioral observations have been shown to be open to question (for review, see Bekoff, 1978), Weiss's work is remarkable in that it foreshadowed by over 20 years work that would be done by neurophysiologists using more sophisticated techniques—first on invertebrates and later on vertebrates—to finally resolve the controversy of central versus peripheral control of coordinated movement in adult animals. For reviews of this controversy, see Evarts et al. (1971) and Grillner (1975).

Although some work continued on the development of embryonic behaviors (e.g., Hooker, 1952; Humphrey, 1952) in the late 1940s and the 1950s, no substantial progress was made in understanding the mechanisms involved. Furthermore, little attention was paid to the implications of Weiss's results for the study of development of behavior.

However, in the early 1960s a new era was ushered in by the elegant experiments of Viktor Hamburger and his coworkers. This work focused on the development of spontaneous behavior for the first time since Preyer's studies in 1885. Hamburger and his colleagues began an extensive series of studies on the development of normal, spontaneous behavior in the chick embryo. First they described the onset and development of the embryonic movements from three and one-half days through hatching at 20 to 21 days (Hamburger and Balaban, 1963; Hamburger et al., 1965; Hamburger and Oppenheim, 1967). Using behavioral observations, they characterized the movements at various ages and, in addition, obtained measures of the periodicity of the movements. Similar data have since been collected on amphibians (Hughes, 1966; Hughes and Prestige, 1967); reptiles (Decker, 1967; Hughes, Bryant, and Bellairs, 1967); mammals (Edwards and Edwards, 1970; Narayanan, Fox, and Hamburger 1971) and a few invertebrates (Berrill, 1973; Provine, 1977; Kammer and Kinnamon, 1979). Using the baseline of behavioral data collected on the chick embryo, Hamburger and his coworkers applied the techniques of experimental embryology to try to elucidate some of the mechanisms involved in the production of motility. In a very important set of experiments, Hamburger, Wenger, and Oppenheim, (1966) removed the entire dorsal half of the lumbosacral spinal cord at two to two and one-half days and, in addition, performed a thoracic cord transection. This effectively removed all spinal ganglia from the region of the cord that innervates the leg; thus, both sensory input from the lumbosacral region and all descending input were eliminated. The spontaneous movements were normal as judged by behavioral criteria until 15 days of incubation. Degeneration of the lateral motor column was thought to be responsible for the deterioration seen after 15 days. This experiment produced unequivocal proof that normal, spontaneous embryonic motility in the chick was centrally generated and, furthermore, that it could develop in the absence of both sensory and descending input.

At the same time that Hamburger and his colleagues were revitalizing the study of the embryonic development of vertebrate behavior, neurophysiologists were developing methods for analyzing the neural circuitry responsible for generating the motor control of coordinated behaviors in unanaesthetized, behaving adult animals. This led to the very important experimental demonstration that sensory input was not necessary for the production of coordinated flight behavior in the adult locust (Wilson, 1961; Wilson and Gettrup, 1963). Moreover, it made it clear that subsequent studies of the neural mechanisms involved in producing coordinated adult behaviors would have to explore the role of central pattern generating circuits, in addition to considering the role of sensory input. In recent years a wealth of information has been obtained detailing the organization of the neural circuitry underlying behaviors as diverse as the crayfish tail flip (Wine, 1978), lobster stomach movements (Mulloney and Selverston, 1974), *Aplysia* gill withdrawal (Kupferman, Carew, and Kandel, 1974) and leech swimming (Friesen, Poon, and Stent, 1978; Poon,

Friesen, and Stent, 1978). Substantial progress has also been made in the analysis of the pattern generating circuitry underlying adult locomotion in both vertebrates and invertebrates (for reviews, see Grillner, 1975; Herman et al., 1976; Wetzel and Stuart, 1976; Stein, 1978).

Progress in the study of the development of the neural circuits underlying coordinated behavior has not been as rapid, in part because of technical difficulties encountered in dealing with developing organisms. Nevertheless, as a result of the two independent lines of research discussed above, the primary thrust of the research in the last decade or so has been to attempt to determine how the neural circuitry that underlies *normal, spontaneous* behavior is assembled. That is, the emphasis has moved away from a consideration of embryonic reflexes as the major determinant of behavior to a concentration on the normal behavioral repertoire of the undisturbed embryo, which may include reflex as well as centrally programmed components.

The emphasis of the following sections of this review is on neural mechanisms underlying the development of motor coordination in vertebrates. Nevertheless, a great deal of relevant information is available from recent studies of invertebrates and I shall make extensive use of this material. Much of the work I shall discuss has focused on the development of limb coordination, mainly because limbs are easy to see and are involved in relatively stereotyped, and obviously coordinated, behaviors such as walking, flying, and swimming in the adult. As will become clear in the following discussion, my views of the development of motor coordination are strongly influenced by recent work on the neural control of adult locomotion in both vertebrates and invertebrates (for reviews, see Grillner, 1975; Herman et al., 1976; Wetzel and Stuart, 1976; Stein, 1978). The aim of this review is to put into perspective the advances that have been made in recent years and to point out areas of research that deserve further attention.

METHODOLOGICAL CONSIDERATIONS

In order to study the development of neural circuitry underlying coordinated behavior, it is first necessary to recognize its earliest manifestations. In theory, one expects to be able to detect early stages of motor coordination simply by observing normal behavior. However, in practice, at least three factors may render this technique, by itself, inadequate. The first is that some motor sequences may not appear to be coordinated despite the fact that they are produced by a coordinated sequence of muscle contractions. For example, hatching in chick embryos is clearly recognizable as a coordinated behavior. It consists of smooth, relatively large amplitude movements of various body parts arranged in a relatively stereotyped sequence (Hamburger and Oppenheim, 1967; Kovach, 1970; Oppenheim, 1972c; Bekoff, 1976; Bakhuis, 1974). However, during early stages of development, movements are produced that bear little resemblance to coordinated behavior. They are characterized as "random, jerky, showing no coordination between body parts" (Hamburger and Oppenheim, 1967). Nevertheless, electromyogram (EMG) recordings from identified leg muscles during spontaneous motility in early chick embryos show clearly that these movements are the result of a coordinated sequence of muscle contractions (Bekoff, Stein, and Hamburger, 1975; Bekoff, 1976). Moreover, as will be discussed later, the gradual development of the neural circuitry underlying the coordinated leg movements of hatching can be followed from 7 to 21 days of incubation using EMG recordings despite the fact that behavioral observations fail to suggest gradual changes in motor coordination (Bekoff, 1976). In the chick embryo, the jerkiness of the movements and the apparent absence of coordination between head and trunk and between trunk and limbs prevent the human observer from recognizing patterns of movement within a limb which can be detected when EMG records are analyzed. Thus, in looking for early manifestations of motor coordination it is important to be aware of the strengths and limitations of the techniques used (Bekoff, 1978).

Another problem is that some coordinated behaviors may be expressed only under permissive conditions. That is, the neural circuitry may be present at a particular stage of development but the behavior may be produced only if neural inhibition or specific physical barriers are removed (see pp. 145; 159 ff). Thus, it must be kept in mind that the absence of a behavior at one stage does not necessarily indicate the absence of the underlying neural circuitry.

A third factor to be considered is the possibility that a neural circuit which is developed to produce one type of behavior at one stage of development may be used to produce another at a later stage. Abrupt transitions (discontinuities) in behavior during development need not necessarily be the result of the turning on of newly constructed neural circuits. For example, it has been suggested that the neural circuit which generates walking in the hatched chick may be the same pattern generating circuit which produces the leg movements of hatching in the chick embryo (Bekoff, 1978). If this proves to be the case, then the study of the development of the circuit for the leg movements of hatching would also be a study of the development of the neural circuit for walking, despite the fact that the two behaviors are quite distinct. Hamburger (1973) has also discussed the issue of continuity at the neural level despite discontinuity at the behavioral level.

LEVELS OF COORDINATION

In considering the development of motor circuits it will be useful to discuss separately three levels of coordination (Bekoff, 1978): (1) intrajoint coordination—coordination between muscles acting on a common joint; (2) interjoint coordination—coordination between muscles of the same limb (or other structure) acting at different joints; and (3) interlimb coordination—coordination between muscles of different limbs. This separation is based on obvious anatomical characteristics of limbs and the need for any model of a pattern-generating circuit for limb movements to account for coordination both within and between limbs. One conception of the organization of the neural circuitry underlying adult locomotion is that it may consist of separate pattern generating circuits for each limb which are coordinated with each other to produce interlimb coordination. Limb pattern generators may, in turn, be made up of subunits ("unit generators") for each joint. Each unit generator would produce intrajoint coordination, and these would be coordinated with each other to produce interjoint coordination. Alternative possibilities exist (Edgerton et al., 1976), and there is as yet only suggestive evidence for the existence of separate pattern generators for each joint (Grillner and Zangger, 1979). However, many studies have found evidence for separate pattern generators for each limb (for reviews, see Stein, 1974, 1976; Grillner, 1975).

In this section I will deal with a description of the development of the three levels of coordination. Subsequent sections will cover information on neural mechanisms underlying the developmental changes observed.

Intrajoint coordination

Significantly more information is available on the development of intrajoint coordination than on the other levels of coordination. I shall present the information available on intrajoint coordination during spontaneous behavior first and then during reflex responses to stimulation (reflexogenic behavior). It is important to remember that spontaneous behavior and reflexogenic behavior are not necessarily mutually exclusive. "Spontaneous" refers to behavior occurring in the absence of *obvious* peripheral stimulation. The actual movements seen during spontaneous behavior may be entirely centrally programmed. In this case, central pattern generating circuits produce coordinated (patterned) muscle contractions and sensory feedback plays no role in patterning the movements. On the other hand, spontaneous behavior may be initiated centrally, but sensory feedback may pattern the movements. Finally, spontaneous behavior may reflect a combination of central programming and peripheral feedback as is found in most adult locomotory systems (for reviews, see Grillner, 1975; Herman et al., 1976). Only when sensory input is either naturally absent, as occurs early in ontogeny in many animals, or has been experimentally removed can spontaneous activity be clearly identified as centrally programmed.

Spontaneous behavior is usually studied by interfering as little as possible with normal embryonic conditions and analyzing ongoing behavior in the absence of applied stimulation. However, some developing animals (e.g., crickets, locusts) are not spontaneously active and permissive conditions must be identified in order to study the development of neural mechanisms underlying motor coordination. In these cases it is often possible to elicit behavior by applying a tonic stimulus. For example, a steady windstream over the head elicits the motor output of flying in larval crickets (Bentley and Hoy, 1970). Technically, of course, this is not spontaneous behavior since it is not initiated centrally. Nevertheless, the presumed action of the tonic stimulus is to raise the pattern generating circuitry to threshold and not to participate in the pattern generating mechanism itself. Thus, although the specific mechanisms by which spontaneous and tonically stimulated behaviors are initiated differ, both methods allow investigation of the development of the pattern generating mechanisms involved in producing intrajoint coordination.

In almost every case in which the development of intrajoint coordination has been examined using electromyographic (EMG) recordings, gradual refinement of the patterned motor output has been observed. In several species of insects, the first activity of pairs of antagonist muscles is uncoordinated or may even show a tendency toward synchrony. Later, short periods of somewhat irregular alternation are observed. And, finally, the consistent alternating pattern of the adult is seen (Bentley and Hoy, 1970; Kutsch, 1971; Weber, 1972; Bentley, 1973; Altman, 1975; Kammer and Rheuben, 1976). In the one-day to four-month-old postnatal cat a gradual emergence of coordinated activation of ankle antagonists during spontaneous walking has also been reported (Scheibel and Scheibel, 1970).

A more detailed study of the ontogeny of antagonist alternation has been carried out in spontaneously active chick embryos (Bekoff et al., 1975; Bekoff, 1976). A tendency toward alternation of the ankle antagonists is observed as early as seven days of incubation, but this early alternation is not symmetrically reciprocal. While ankle flexor activity rarely begins before ankle extensor activity terminates, the reverse is not true. By nine days of incubation, the alternation is reciprocal. From 17 days to hatching (20 to 21 days), a gradual increase in the tightness of coupling of the activities of the two antagonists is observed. That is, over this period the consistency with which one muscle is activated just at the termination of activity in the other increases.

In all cases that have been analyzed so far, intrajoint coordination has been found prior to the time at which coordinated behavior is recognized behaviorally. In the chick embryo, intrajoint coordination is already present at the time the first active leg movements are observed. Thus the neural circuitry underlying intrajoint coordination is assembled very early in development.

Evidence that spontaneous intrajoint coordination is centrally programmed is available in several systems. In crickets (Bentley and Hoy, 1970), locusts (Altman, 1975; Kutsch, 1971, 1974) and moths (Kammer and Rheuben, 1976; Kammer and Kinnamon, 1979), the pattern generating circuit for flight develops in the absence of normal sensory input. In the chick embryo, intrajoint coordination of the ankle muscles is observed at seven days of incubation, prior to the time at which reflex responses can first be elicited (Bekoff et al., 1975). Further discussion of these results will be found below (pp. 153–157).

In contrast to spontaneous behavior, reflexogenic behavior is generally studied by applying specific stimuli to various parts of the embryo during intervals between spontaneous movements (e.g., Oppenheim, 1972a; Narayanan, Fox, and Hamburger, 1971). As first pointed out by Preyer (1885), reflex responses appear after the onset of spontaneous activity in most vertebrates. In mammals, however, reflex sensitivity and spontaneous motility appear synchronously. Because behavior appearing after reflex responses are present is not necessarily reflexogenic, the role of sensory input must be experimentally determined in each case. Numerous studies have dealt with the development of reflexes; however, none has yet shown that reflexes actually play a role in patterning embryonic motor behavior (see pp. 153–157 below; Oppenheim, 1972b).

As mentioned earlier, the onset of reflex sensitivity correlates well with the anatomically observed closure of reflex arcs. All anatomical studies (both light and electron microscopic) which examined both motor and association neuropil have found that the reflex arcs close in a retrograde sequence (e.g., Windle et al., 1933; Windle and Orr, 1934; Windle and Baxter, 1936; Visintini and Levi-Montalcini, 1939; Vaughn and Grieshaber, 1973; Singer, Skoff, and Price, 1978). That is, synapses are seen between interneurons and motoneurons before they are seen between sensory neurons and interneurons. Recently these results have been confirmed in an electrophysiological study of the isolated rat spinal cord *in vitro* (Saito, 1979).

A wide variety of behavioral studies has shown that the earliest response to tactile sensory stimulation is widespread movement of the embryo ("total response") and then later the response becomes more and more restricted until finally a specific, predictable reflex response ("local response") occurs (Carmichael, 1934; Barcroft and Barron, 1939; Hamburger and Narayanan, 1969; Oppenheim, 1972a,b).

In agreement with the behavioral studies, EMG recordings from the hindlimb muscles of sheep and guinea pig fetuses have shown that the earliest response to stretch of the gastrocnemius muscle is excitation of both gastrocnemius and its antagonists (Änggard, Bergström, and Bernhard, 1961; Bergström, Hellström, and Stenberg, 1962). At later stages the response becomes limited to the stretched muscle. A similar spread in excitation and its later restriction was observed in response to plantar skin stimulation in the fetuses. Thus both monosynaptic and polysynaptic reflexes show this phenomenon. Gatev (1972) found similar results for elbow muscles of human infants between one month and three years. In his *in vitro* study of spinal reflexes in the isolated fetal rat spinal cord, Saito (1979) also found widespread excitation at early stages, which later appeared to be suppressed by inhibition and was finally eliminated.

Interjoint Coordination

Little is known about the mechanisms involved in interjoint coordination in either adult or developing organisms. Studies of the development of interjoint coordination within a single limb can contribute valuable information on the organization of the neural circuitry involved in producing coordinated behavior. That is, if there are separate pattern generating networks for each joint, they may be uncoupled early in development. In the only such study available, an EMG analysis of leg muscle coordination in chick embryos, Bekoff (1976) found that as early as nine days of incubation interjoint coordination was present as evidenced by a consistent pattern of coactivation of knee and ankle extensors. There is a gradual improvement in the precision with which these muscles are coactivated between nine days and hatching, which parallels the improvement seen in the intrajoint coordination of antagonists (Bekoff, 1974). Unfortunately, it has not yet been possible to obtain EMG recordings from identified knee and ankle muscles at earlier stages, so it is not known when interjoint coordination first appears. More detailed studies of the early development of interjoint coordination in this system will undoubtedly be profitable.

Interlimb Coordination

Several studies of locomotion in adult animals have indicated that there is a pattern generating circuit for each limb (e.g., Hughes and Wiersma, 1960; Wilson, 1961; Kulagin and Shik, 1970; Holst, 1973; Lennard and Stein, 1977). The neural mechanisms involved in coordinating the activity of the separate limb pattern generators to produce the adult pattern of interlimb coordination have been reviewed by Stein (1974, 1976) and Grillner (1975).

Considerably less is known about the development of interlimb coordination. Circuitry for activation of right and left legs in the alternating pattern typical of walking or in-phase activation of the two wings as during flying is evidently specified in the lumbrosacral and brachial regions of the chick embryo spinal cord, respectively, at early stages (Straznicky, 1963; Narayanan and Hamburger, 1971). This is illustrated by the observation that, if the lumbosacral cord is transplanted to the brachial region in two-and-one-half-day embryos and allowed to innervate the wings, the posthatching chick will move its wings alternately in a stepping-like pattern. Moreover, if brachial cord is transplanted to the lumbosacral region, the legs of the posthatching chick will move in-phase, as if flapping.

These experiments do not provide information about the embryonic stages of development of interlimb coordination, and this has not yet been examined in the chick embryo with techniques adequate to determine when interlimb coordination first appears and whether or not it develops gradually. Provine (1980) has reported a gradual increase in simultaneous movements of right and left wings. However, since simultaneous movements are not necessarily coordinated, these results are only suggestive. It should be possible to resolve this issue with EMG recordings.

In mammals, there is evidence for prenatal development of interlimb coordination. For example, kitten fetuses have been reported to show interlimb coordination patterns typical of postnatal walking (Graham Brown, 1915; Windle and Griffin, 1931) and recently, Bekoff and Lau (1980) have found evidence for interlimb coordination in 20-day-old rat fetuses through the use of frame-by-frame analysis of videotape records. One interesting finding in this study was that coordination was observed between homologous limbs (forelimb-forelimb or hindlimb-hindlimb), but coordination between homolateral limbs (right or left forelimb-hindlimb pair) was not. This suggests that neural mechanisms for coordinating homologous limb pairs develop before those for coordinating homolateral pairs. On the day after birth, interlimb coordination of all four limbs can be seen during swimming. This shows that coordinating pathways among the four limb pattern generators are established by this time (Bekoff and Trainer, 1979).

However, in both rats (Bekoff and Trainer, 1979) and mice (Fentress, 1972) fewer than four limbs may participate in swimming during the first few days after birth. Often one or more will remain stationary while the other limbs perform coordinated swimming movements. This suggests that the excitability thresholds of the individual limb pattern generators or the excitatory inputs to them continue to develop gradually for several days after birth. An important point to remember here is that the early appearance of postnatal interlimb coordination in mice and rats was not appreciated until permissive conditions (the buoyant medium of water) were identifed (Fentress, 1972; Bekoff and Trainer, 1979).

Only in insects have the relative times of first appearance of interlimb versus intralimb (specifically intrajoint) coordination been worked out. Interlimb coordination between forewings and hindwings in crickets and locusts develops after coordination of muscles within each wing is established (Bentley and Hoy, 1970; Kutsch, 1971; Altman, 1975).

Based on the very limited data available, a general scheme for the development of interlimb coordination could be proposed as follows: (1) After neural mechanisms for intralimb coordination (intrajoint and interjoint) are established, coordinating pathways for interlimb coordination develop. (2) Coordination between homologous limbs develops before coordination between homolateral limbs. And (3) these developmental events occur gradually. Further studies are needed to test the validity of this scheme.

MECHANISMS INVOLVED IN THE ASSEMBLY OF THE NEURAL CIRCUITRY FOR MOTOR COORDINATION

Basic Mechanisms of Neural Development

While we do not yet have a detailed understanding of the neural mechanisms involved in the assembly of any one neural circuit for motor coordination, data from a variety of developing systems provide insight into some of the mechanisms that may be used. The available data also suggest potentially useful directions for future research.

Internal Neuronal Properties It has been suggested that the gradual development of motor coordination could be, in part, the result of changes in internal properties of neurons such as threshold, refractory period, and responsiveness (Bentley, 1973). At present, there is little evidence for functionally important changes in thresholds of neurons during ontogeny. However, studies of embryonic muscle fibers show that such factors as ionic conductances (Kano, 1975) and transmitter receptor distribution (Diamond and Miledi, 1962; Dennis and Ort, 1975; Letinsky, 1975) change during development.

Developmental changes in the size and form of the action potential (e.g., Eaton et al., 1977; Spitzer and Baccaglini, 1976; Goodman and Spitzer, 1979) and in its ionic basis (Spitzer and Baccaglini, 1976; Llinás and Sugimori, 1980; Goodman and Spitzer, 1979) have been shown. In developing neuromuscular synapses in bullfrog tadpoles, Letinsky (1974a,b) found evidence for an initially low quantal content that increases with age, and an apparent rapid depletion of transmitter stores during repetitive stimulation at early stages. Developmental changes in these parameters and others, such as size of synaptic area and surface area of postsynaptic neurons, would affect the efficiency of synaptic transmission. Nevertheless, no study has yet linked a normal developmental change in any of the internal neuronal properties discussed above with a specific change in the functioning of a neural circuit involved in generating patterned motor output.

It has become apparent in recent years that the size and shape of a neuron may profoundly influence its physiological properties (Rall, 1964; Grossman, Spira, and Parnas, 1973; Graubard, 1975; Wong and Pearson, 1975) and especially its responsiveness to synaptic input (i.e., its input-output relationship). Changes in length, diameter, and branching complexity of neuronal processes are common during development (e.g., Scheibel and Scheibel, 1970, 1971; Altman and Tyrer, 1975; Truman and Reiss, 1976; Goodman and Spitzer, 1979). Although the effect of these changes on neuronal properties during normal development is not known, Murphey and coworkers (1975; Murphey and Levine, 1980) have documented a decrease in size in specific giant interneuron dendrites following partial deafferentation during postembryonic development in the cricket. They suggest that this would substantially decrease the surface area and therefore increase the overall input resistance of the interneuron, thus accounting in part for the observed increase in the effectiveness of the remaining inputs (Palka and Edwards, 1974). In addition, Peretz and Lukowiak (1976) have shown that the smaller size of neuron R_2 in young *Aplysia* is correlated with the expected increased input resistance and increased effectiveness of a known synaptic input when compared to the larger R_2 in older animals.

Experiments designed to test whether developmental changes in particular internal neuronal properties are causally related to specific changes in pattern generator function are needed. This will probably be done first in invertebrate embryos owing to the presence of large, identified neurons from which intracellular recordings can be obtained (e.g., Goodman and Spitzer, 1979). As discussed below, somewhat more is known about the role of changes in synaptic connections in the development of motor coordination.

Peripheral Synaptic Connections The development of peripheral neuromuscular synapses has so far been studied primarily in vertebrates. This topic has been reviewed recently by Landmesser (1980). In general, the results to date from studies of the chick hindlimb, which provide the most complete information now available, suggest that initial axon outgrowth and peripheral synapse formation are quite specific so that the overall pattern of functional peripheral connections does not change during development. Thus, although there is evidence for a transient period of hyperinnervation of motor endplates during embryonic development in the chick (Bennett and Pettigrew, 1974), the elimination of this hyperinnervation does not appear to contribute to changes in motor coordination since it occurs after specific peripheral connections have formed (Landmesser, 1978). Furthermore, EMG recordings from leg muscles of seven-day-old chick embryos show discrete bursts with few motor units firing between bursts. These EMG recordings suggest that most motoneurons that have made functional peripheral connections have made appropriate central connections (Bekoff et al., 1975; Bekoff, unpublished observations). It is not known whether the units that fire out of phase (between bursts) represent motoneurons with inappropriate central connections or immature, randomly firing units. In either case, these units are in the minority and do not affect the overall conclusion that most motoneurons make specific, appropriate connections very early in development and that changes in the peripheral connections of motoneurons do not explain the gradual changes in motor coordination seen during development. In addition, because the adult pattern of innervation is established so early, the massive motoneuron cell death that occurs during development (Hamburger, 1975) does not appear to be responsible for establishing coordinated motor output by removing motoneurons which have made functionally inappropriate synapses.

In contrast to the situation in the chick hindlimb, errors in peripheral motoneuron connectivity have been described in axolotl hindlimb (McGrath and Bennett, 1979), chick wing (Pettigrew, Lindeman, and Bennett, 1979), and rat intercostal muscle (Harris and Dennis, 1977) development. Neither the reason for the discrepancy in results nor the functional significance for motor coordination is known because EMG recordings have not been made in these cases.

Recent data on sensory neuron projection patterns in chick embryos also suggest that axon outgrowth is highly specific and that peripheral connections do not change during development (Honig, 1979).

Central Synaptic Connections Because existing evidence suggests that changes in peripheral motor and sensory neuron synapses are not responsible for the initial establishment or subsequent changes in motor coordination in chicks, and because the situation is still unclear in other vertebrates, attention must be focused on central synaptic connections. Unfortunately, much less is known about the mechanisms by which specific central synapses are made and maintained during the development of neural circuits for motor coordination.

There is substantial evidence for an increase in the number of synapses within the central nervous system during development (e.g., Bentley, 1973; Oppenheim, Chu-Wang, and Foelix, 1975). Moreover, there are marked changes in the number of synapses that can be classified into the various morphological types, for example, flattened versus spherical vesicles, symmetric versus asymmetric, and so on (Oppenheim et al., 1975).

It is reasonable to suppose that, in addition to the formation of new synapses, the loss of previously formed synapses may play a role in the development of organized neural circuits. A variety of electrophysiological and anatomical studies have shown losses of synapses during development (e.g., *Purkinje cells:* Crepel, Mariana, and Delhaye-Bouchard, 1976; *submandibular gland:* Lichtman, 1977; *superior cervical ganglion:* Smolen and Raisman, 1980). The loss of up to 50 percent of the synapses formed on cat lumbosacral motoneuron cell bodies and dendrites between birth and the second postnatal week has been documented in EMG studies (Ronnevi and Conradi, 1974; Conradi and Ronnevi, 1975). The functional significance of this loss of synapses on motoneurons is unknown. Scheibel and Scheibel (1970) found a progressive improvement in intrajoint coordination during this period. However, improvement continues after synapse loss has apparently ceased. One possible source of synapses which are eventually lost might be Ia inputs from inappropriate sensory neurons. However, studies on other mammals have shown that these are lost before birth (Änggard et al., 1961; Bergström et al., 1962; Saito, 1979) and Mendell and Scott (1975) have found that the adult pattern of Ia projections to motoneurons in cats is already present at birth.

The studies on the ontogeny of insect flight and chick motility also suggest that there is a determinate order of synapse development. That is, in both crickets (Bentley and Hoy, 1970) and locusts (Altman, 1975) the depressor units consistently become activated earlier than elevator units, suggesting that depressor motoneurons develop connections first. In the seven-day-old chick embryo, both antagonist muscles are active and fire in alternation, but there is a consistent asymmetry in the firing pattern, which is lost by nine days. This suggests that the order in which synapses are established is predetermined, and it has been proposed that this gradual increase in the symmetry of the alternation may represent a gradual increase in the efficiency or numbers of specific inhibitory synapses (Bekoff, 1976). On the other hand, it is also possible that the ability of the neurons to respond to synaptic input changes in a determinate order.

Despite a great deal of interest in the relative times of appearance of excitatory and inhibitory synapses during development, relatively little is known with certainty (for review, see Oppenheim and Reitzel, 1975). The results of the behavioral and electrophysiological studies of reflexes discussed earlier have been used to suggest that excitatory synapses develop first to produce the generalized response in which both agonist and antagonist muscles are activated and that the subsequent restriction is due at least in part to the development of functional inhibitory connections to antagonist muscles (e.g., Carmichael, 1934; Barcroft and Barron, 1939; Änggard et al., 1961; Bergström et al., 1962; Saito, 1979). Nevertheless, in the chick embryo, for which the most complete information is available, inhibitory synapses develop very early. Organized reflexes that are known to involve inhibitory connections, such as flexor withdrawal and crossed excitation, are not observed in response to leg stimulation until nine days of incubation (Oppenheim, 1972a). However, pharmacological and anatomical evidence indicates that inhibitory as well as excitatory synaptic connections develop prior to nine days (Oppenheim and Reitzel, 1975; Oppenheim et al., 1975; Reitzel, Maderdrut, and Oppenheim, 1979; Reitzel and Oppenheim, in press). Moreover, EMG recordings from leg muscles during spontaneous motility in seven-day-old embryos have shown clear evidence of antagonist alternation, which strongly suggests that both excitatory and inhibitory synapses are present at this very early stage (Bekoff et al., 1975).

It seems useful to consider explanations other than the development of inhibition for the restriction of reflexes since, at least in the chick embryo, coordinated motor output is present during spontaneous activity prior to restriction of reflexes. Furthermore, by analogy to adult vertebrate locomotion, the inhibitory mechanisms involved in both spontaneous embryonic motor coordination and reflexes are likely to involve Ia inhibitory interneurons (Hultborn, 1972; Jankowska and Roberts, 1972; Feldman and Orlovsky, 1975; Hultborn, Illert, and Santini, 1976). For example, it may be that sensory neurons form excess inappropriate branches, which are functional before they are eliminated. Retraction of excess branches seems to be a common event in the development of both central and peripheral processes of neurons (e.g., Redfern, 1970; Brown, Jansen, and Van Essen, 1976; Crepel, et al., 1976; Lichtman, 1977; Goodman and Spitzer, 1979). However, it is interesting that, using HRP histochemistry at the light microscopic level, Grobstein (1979) found no anatomical evidence of excessive branching of sensory neurons in rat fetuses. This issue clearly deserves further study.

Intracellular recordings from motoneurons in anesthetized cat fetuses have also suggested that inhibitory as well as excitatory synapses involved in reflex pathways develop early in ontogeny, at least as early as three weeks before birth (Naka, 1964a,b; Maksimova, 1971). However, recordings were not obtained from younger fetuses owing to the difficulty encountered in keeping the animals alive during experimentation. Since, in the cat, both spontaneous and reflex leg movements begin five weeks before birth (Windle and Griffin, 1931; Windle, 1940), it is not known whether the excitatory input develops in advance of the inhibitory input in the reflex pathways tested.

While Atwood and Kwan (1976) found that both excitatory and inhibitory neuromuscular synapses were present at very early stages of synapse development in the crayfish opener muscle, several other studies of invertebrates suggest that excitatory synapses may develop earlier than inhibitory synapses. For example, in the cricket and locust flight systems, antagonist alternation develops gradually out of a period of uncoordinated activity or a period during which antagonists are often coactivated (Bentley and Hoy, 1970; Kutsch, 1971; Altman, 1975). In the zebrafish embryo the Mauthner cell fires spontaneously more frequently than in larvae, which possibly suggests that inhibitory control develops after the onset of excitatory activation (Eaton et al., 1977). Further, in the Mauthner cell system habituationlike response decrements do not develop until after hatching.

Despite the suggestive evidence in favor of the early development of excitatory activation, it is clear that at present there is insufficient evidence to make a general case for temporal priority for either excitatory or inhibitory synapses.

Does the gradual emergence of coordinated motor output reflect the gradual withdrawal of inhibition from previously constructed circuits (Pollack and Crain, 1972; Bentley, 1973; Crain, 1974)? In this case, overt behavior and recorded peripheral nerve or EMG activity might be misleading as assays for the development of central neural circuitry. The only evidence in favor of this in a developing animal is the report that strychnine releases various stages of swimming behavior from one to two days prior to their normal appearance in fish embryos (Pollack and Crain, 1972). However, these results are difficult to interpret since the pattern generating circuit for coordinated swimming undoubtedly involves postsynaptic inhibition to account for alternation of the right and left sides and the constant phase-coupling between segments (Grillner, 1974; Cohen and Wallén, 1979). Perhaps strychnine selectively affects postsynaptic inhibitory synapses involved in inhibiting pattern generator function and not those that are actually part of the pattern generator function itself (see also Hart, 1971). That this does happen, in fact, is indicated by EMG recordings which show that coordinated swimming typical of later stages results after the application of strychnine to young fish embryos. Alternatively, strychnine may have a direct excitatory effect (Klee, Faber, and Heis, 1973). Until it is shown otherwise, it seems most economical to assume that the gradual emergence of coordinated motor output reflects the actual construction of functional neural circuitry rather than the gradual withdrawal of inhibition from previously constructed circuits.

While the gradual improvement in motor coordination seen during ontogeny must be explained, it is undoubtedly too simplistic to consider this improvement to be a reflection of the changes in only one class of synapses (e.g., inhibitory). More likely the construction of neural circuits for motor coordination results from the development of a variety of types of synapses. Ultimately, of course, it will be necessary to make intracellular recordings during the earliest stages of development to resolve the question of when inhibitory synapses develop and what role they play in the gradual development of coordinated movement. Recent developments in microelectrode technology (e.g., Corson, Goodman, and Fein,

1979) should help to overcome some of the problems encountered in earlier attempts to record intracellularly from motoneurons in embryos (Naka, 1964a,b). Moreover, the recent study of fetal reflexes in the isolated rat spinal cord *in vitro* (Saito, 1979) introduces a promising new approach.

Spontaneous Activity

Spontaneous activity seems to be characteristic of developing motor circuits (Hamburger, 1963; Corner, 1977; Berrill, 1973; Kammer and Kinnamon, 1979). This property has been extensively studied in the chick embryo by Hamburger and his coworkers. As discussed earlier, they showed that embryonic movements continue even in chronically deafferented, spinal embryos (Hamburger et al., 1966). This suggests that motoneurons and/or interneurons are spontaneously active. In support of this, electrophysiological work has shown that unit activity is greatest in the ventral two-thirds of the spinal cord in 15-day to 19-day embryos (Provine et al., 1970). Furthermore, the lumbosacral region of the spinal cord still showed spontaneously active units when it was isolated from both descending and afferent input (Sharma et al., 1970).

Spontaneous neural activity has also been noted in the developing moth motor system (Kammer and Rheuben, 1976; Kammer and Kinnamon, 1979) and in a developing cricket motoneuron (Clark and Cohen, 1979). When single unit recordings have been obtained from interneurons that are spontaneously active in developing mammals, the frequency of firing has been found to increase with age (Purpura, Shofer, and Scarff, 1965; Spear et al., 1972; Romand and Marty, 1975; Lannou, Precht, and Cazin, 1979). However, in an insect motoneuron, spontaneous firing decreased and eventually disappeared (Clark and Cohen, 1979). Sharma and coworkers found that the number of spontaneously active neurons in the isolated chick spinal cord decreased with age (Sharma et al., 1970). During normal development in the zebrafish embryo, Eaton and coworkers (1977) found that the frequency of spontaneous firing of the Mauthner cell increased during embryonic stages and then decreased in larval animals.

It is not known whether the spontaneous activity is a property of individual neurons or of neural networks. In addition, the importance of such spontaneous activity for the development and/or maintenance of organized neural circuitry is not known. However, it has been suggested that electrical activity may be important in the selective stabilization of synapses during development (Changeux and Danchin, 1976). Spontaneous activity would provide a mechanism for insuring adequate neural activity in developing systems despite the reduced sensory stimulation available to embryos (Reynolds, 1962; Oppenheim, 1972a,b; Bradley and Mistretta, 1975).

In cases in which it has been examined, spontaneous motor activity has been shown to be centrally generated, that is, it is nonreflexogenic (Hamburger et al., 1966; Sharma et al., 1970; Kammer and Rheuben, 1976; Kammer and Kinnamon, 1979). Moreover, in the seven-day-old chick embryo (Bekoff et al., 1975) and the developing moth (Kammer and Rheuben, 1976; Kammer and Kinnamon, 1979), centrally generated spontaneous activity results in coordinated motor output. This indicates that the activity is produced by a central pattern generating circuit rather than just by random firing of unorganized neural circuits. Although this needs to be determined experimentally, spontaneous motor activity in older chick embryos and other developing animals may also be produced by central pattern generating circuits. As mentioned above, the advantage of having neural activity be centrally generated may be to insure adequate amounts of activity. The reason why very early movements should be organized or coordinated may simply be one of economy. Teleologically, if a pattern generating circuit must be built eventually and synaptic connections and activity are important to neural development at early stages, then, rather than establishing one set of connections at early stages and later changing them, the first connections might as well be those involved in assembling the pattern generating circuit.

Role of Sensory Input

Although locomotion is produced by a central pattern generating mechanism in a wide variety of animals, sensory input appears to play an important role in enabling an animal to react to, and compensate for, various kinds of environmental variables and perturbations (e.g., Wilson, 1972; Forssberg et al., 1976; Pearson and Duysens, 1976; Grillner and Wallén, 1977). For example, afferent input can serve to modulate the centrally generated pattern. At the same time, the effect of a specific stimulus may be dependent on the phase of the step cycle. It has been shown that in both cats and dogfish specific reflexes are gated or reversed so as to insure the best possible matching between reflex responses and the ongoing phase of locomotion (Duysens and Pearson, 1976; Forssberg et al., 1976). In addition, sensory input resulting from postural factors may turn central pattern generators off (Pearson and Duysens, 1976; Berkinblit et al., 1978).

As discussed earlier, probably a good deal of embryonic motor activity is centrally patterned. Nevertheless sensory input is presumably available throughout much of embryonic development. Its role during ongoing embryonic movements is unknown. In developing moths, sensory input is present but is apparently unpatterned (i.e., is not phasically related to the motor output; Kammer and Athey, cited in Kammer and Kinnamon, 1979). It may be that the pattern generating circuitry in developing animals has decreased sensitivity to sensory input. Little is known about this, although Oppenheim (1972a) has found that chick embryos cease to respond to sensory stimulation very rapidly. Overall we know almost nothing about what sensory input does during ongoing behavior when it is present in developing animals and this entire issue deserves further investigation.

A separate issue is whether or not (or to what extent) sensory input is important in the development of pattern generating circuits. One approach is to consider cases in which movements begin before sensory input is available. Many recent studies, in a variety of developing organisms, have provided evidence for the normal development of motor circuits in the absence of sensory input. For example, EMG analysis of the activity of identified leg muscles of the chick embryo has shown that synergist coactivation and antagonist alternation at the ankle joint are present at an age (7 days) when responses to sensory stimulation cannot be detected behaviorally (Hamburger and Balaban, 1963; Oppenheim, 1972a; Narayanan and Malloy, 1974) or by using EMG recordings (Bekoff et al., 1975). Anatomical (silver-stain) evidence suggests that reflex arcs are only beginning to close at this stage (Windle and Orr, 1934; Visintini and Levi-Montalcini, 1939). Thus it might be that the neural organization responsible for coordinated contraction of ankle muscles is established independently of sensory input.

Similar results have been obtained from several insect larvae. Bentley and Hoy (1970) have demonstrated that the neural circuits responsible for producing flight and singing are completely functional prior to the final moult in crickets. Since the wing pads do not move in a flapping pattern at this stage of development, sensory feedback from stretch receptors in the wing does not appear to contribute to the establishment of the underlying neural organization. The flight motor pattern of locusts and moths has also been shown to develop in the absence of normal phasic sensory input from the wings (Altman, 1975; Kammer and Kinnamon, 1979). These results suggest that some neural circuits which are responsible for programming coordinated motor behaviors normally develop in the absence of phasic sensory feedback.

In many cases, however, sensory input is present during important stages of the development of pattern generating circuits. To try to determine whether the sensory input plays an important role in the assembly of the circuits during these stages, deafferentation experiments are often employed. For example, as discussed earlier, embryonic chick leg movements were found to continue after complete, chronic deafferentation of the leg (Hamburger et al., 1966). This shows clearly that embryonic movements can develop in the absence of sensory input. However, EMG recordings will be necessary to determine whether or not the deafferented embryos produce normally coordinated movements during embryonic development.

Several studies have examined the effect of early deafferentation on later production of coordinated adult motor behavior. For example, in one case it has been shown that a motor program for coordinated movement can develop in the complete absence of sensory feedback although in the normal situation sensory feedback might have been available. Prior to differentiation of muscle and sense organs, presumptive swimmerets (locomotor appendages) were extirpated from larval lobsters (Davis, 1973; Davis and Davis, 1973). Two weeks later, electrophysiological recordings from swimmeret motoneurons showed that normal coordinated patterns of rhythmic locomotor output were present despite the absence of target muscles and sense organs. Thus normal differentiation of the neural circuitry responsible for producing coordinated swimmeret movement was not dependent on the presence of sensory feedback from the swimmerets.

Other suggestive data come from experimental studies of fetal monkeys. Bilateral deafferentation of the forelimbs of monkey fetuses near term (Berman and Berman, 1973), or even as early as two-thirds of the way through gestation (Taub, 1976), did not prevent the monkeys from using their forelimbs in coordinated locomotion after birth. In other studies, complex face-grooming sequences were observed in mice after neonatal forelimb amputation (Fentress, 1973) and wing flapping developed in chicks after wing amputation at hatching (Provine, 1979) despite the absence of normal sensory feedback. These studies suggest the importance of central pattern generators in the production of behaviors in developing animals. However, in each case the operations were performed *after* sensory neurons had formed functional connections and at a time when the animals had already had extensive motor experience. Thus, although the operations were performed prior to the time at which the animals would normally perform the particular behavior assayed for, it is possible that some elements of the neural program for that behavior had been established previously.

In a study with results in apparent contradiction to those discussed above, Narayanan and Malloy (1974) chronically deafferented the legs of chick embryos and raised the chicks through hatching. Behavioral observations of the posthatching chicks showed that they did not perform normal alternating walking. Because behavioral observation may not be adequate in this case to assess the presence of coordinated motor output (Bekoff, 1978), it would be useful to repeat these experiments using EMG recordings. Nevertheless, if their interpretation is correct and chronic deafferentation does prevent the normal development of pattern generating circuitry for walking, the question of what specific role sensory input normally plays is still open. Deafferentation, which in this case was accomplished by destruction of the neural crest precursors of sensory neurons, involves more than simply removing patterned sensory feedback. It removes all input, whether it is due to trophic or electrical activity, which would normally have come from the large population of sensory neurons.

The neurons involved in the circuitry that produces normal walking may need input, but it is not yet clear whether the required input is in the form of trophic substances or action potentials, or both. It is also possible that the input need not be from primary sensory neurons or represent feedback of information about events in the periphery. If this were the case, the input would be required to serve a maintenance rather than an organizing function.

It is now clear that the specific question to be addressed is this: does sensory input provide information that directs the organization of the circuitry, or is it only permissive, that is, does it simply provide nonspecific conditions that allow the construction of the circuitry, the organization of which is genetically programmed? At present, there are no reported studies that distinguish between these two alternatives during the initial development of the neural circuitry underlying coordinated movement in cases in which sensory input appears to be required.

Another approach, less drastic than deafferentation, is to alter the sensory input and determine the effect of this manipulation. Unfortunately, no study has yet examined the effect of the altered input during development. Nevertheless, several studies have suggested that altered sensory input does not modify previously established motor connections. For example, rotation of the eye of the frog, at any time after retinotectal connections have been programmed, results in maladaptive motor responses (Sperry, 1951; Jacobson, 1967). A similar result is obtained if a salamander limb with its anteroposterior axis reversed is transplanted in place of the normal limb (Weiss, 1937). The transplanted limb always moves "backwards." Moreover, in a neurophysiological study of cross-reinnvervation in newborn cats, Mendell and Scott (1975) found that even as long as one year after the medial gastrocnemius and lateral gastrocnemius-soleus nerves were crossed, no change was observed in the distribution of Ia afferent terminals to the relevant motoneurons.

At present, the available data suggest that, in many cases, the neural circuitry involved in the production of rhythmic, relatively stereotyped behaviors, such as locomotion, does not depend on peripheral feedback to determine its functional organization during ontogeny. One reason that the development of the neural circuitry underlying such basic behaviors as locomotion, hatching, grooming, and so on is not dependent on sensory input may be that appropriate sensory stimulation in the embryonic or larval environment is limited or absent (Reynolds, 1962). For example, the wing pads of crickets and locusts are so small and virtually immobile throughout most of development that the relevant sensory feedback may not be available. For a bird floating in a pool of amniotic fluid encased in a hard-shelled egg, or for a mammal *in utero*, the level of sensory stimulation might very well fall below that required to stimulate leg movements. Moreover, Oppenheim (1972a) has provided data which suggest that sensory receptors rapidly cease to respond in embryonic chicks. Similar suggestions have been made for sensory receptors in mammalian fetuses (Scherrer, 1968; Sedlacek, 1971; Bradley and Mistretta, 1973, 1975) and for locust wind-receptor hairs (Sviderskii, 1970). Alternatively, of course, it could be argued that because the neural circuitry is capable of self-assembly, there is no need for the provision of adequate stimulus situations.

Role of Descending Control

The pattern generating circuitry for locomotor movements of each limb has been shown to be located in the spinal cord in adult vertebrates (for reviews, see Grillner, 1975; Stein, 1976, 1978). The descending control of the spinal circuits generating locomotion has been extensively studied in cats and dogfish (Grillner, 1976; Grillner and Kashin, 1976). In these experiments a "mesencephalic locomotor region" located in the midbrain tegmentum has been found which, when stimulated electrically, can elicit stereotyped locomotion. Although the identity and precise location of the fibers descending from the mesencephalic locomotor region have not been well worked out, there is suggestive evidence that they descend in the dorsolateral funiculus of the spinal cord (Roaf and Sherrington, 1910; Lennard and Stein, 1977; Mori, Shik, and Yagodnitsyn, 1977) and that they may be noradrenergic (Forssberg and Grillner, 1973; Grillner and Zangger, 1975). However, there are data that contradict this view (Steeves et al., 1980).

Even less is known about the development of these descending pathways. None of the experimental studies of vertebrate embryos have used EMG or other electrophysiological recording techniques to monitor motor coordination. Nevertheless, behavioral observations have provided suggestive results. Several experiments on mammalian and chick embryos have shown that spinal embryos (both chronic and acute) continue to produce behavior that appear to be qualitatively normal (Hooker and Nicholas, 1930; Barcroft and Barron, 1939; Barron, 1941; Hamburger et al., 1965; Oppenheim, 1975). These experiments indicate that the circuits that generate limb movements are located in the spinal cord and that they can develop in the absence of descending input. Furthermore, there is good evidence that pattern generating circuits for locomotion are specified very early in embryonic development in newts and chicks and that they are located specifically in the brachial and lumbosacral regions of the spinal cord (e.g., Narayanan and Hamburger, 1971; for review, see Székely, 1976).

The nature of the descending influences during development has been examined closely only in the chick embryo. It has been found that while leg movements produced in cervical spinal chick embryos appear qualitatively normal between 4 and 19 days of incubation, some quantitative differences are observed. Specifically, despite the fact that frequency of movements per minute is the same as in controls, after ten days of incubation cervical spinal chicks have shorter activity periods (Oppenheim, 1975). This suggests that the descending input is not simply exciting the spinal pattern generating circuitry to produce more movements, but instead is modulating the activity of the circuits so that they are active less often, but for a longer period each time. Whether this is done via a tonic or a phasic descending input is not known. Nor has the identity of the descending path(s) been determined. Since Oppenheim (1975) found that the earliest observable effect of either chronic or acute spinal transection occurs at ten days of incubation, it would be interesting to know what descending tracts reach the lumbosacral cord on or around ten days. In agreement with the behavioral results, multiunit recordings from spinal cord neurons in cervical spinal chick embryos were quite similar to controls up to 13 days (Provine and Rogers, 1977). The cause of the differences after 13 days is not known.

In an attempt to assess the roles of various brain regions in the descending influences on embryonic chick movement, extirpations have been carried out at very early stages (Decker and Hamburger, 1967; Oppenheim, 1972d). These studies have shown that removal of the forebrain (telencephalon and diencephalon) interferes neither with the timing or organization of embryonic movements nor with prehatching behavior. However, hatching (climax) behavior is not initiated in these embryos. Removal either of the cerebral hemispheres only or of the midbrain tectum has no detectable effect at any embryonic stage, including hatching. Removal of the entire midbrain (including the tegmentum) had the most serious effect. Prehatching and hatching were never seen after such lesions and, while the amount of embryonic activity was normal, convulsion-like movements occured after 17 days. These are basically the same defects seen in cervical spinal chick embryos after 17 days.

On the basis of these studies, and on the results of acute brain transections (Corner and Bakhuis, 1969; Oppenheim, 1972d), Oppenheim has suggested that a midbrain center actually organizes the hatching behavior while the forebrain simply provides a triggering input. However, these experiments do not rule out the possibility that, as in locomotion, the function of the midbrain center is to raise the excitability level of pattern generating circuits located in the spinal cord. Experiments designed to examine this issue are currently in progress (Bekoff and Kauer, unpublished data).

In addition to descending excitatory influences, descending inhibition may also play a role during development. It has been suggested that inhibition near the time of hatching or birth may be necessary to prevent excessive activity that might result in damage to the embryo or to its fragile environment (Blest, 1960; Crain, 1974). For example, Blest (1960) has found that adult behaviors such as shivering and wing flapping could be elicited on Day 21 in moths prematurely removed from the pupal cuticle, but not on Day 22, the day before eclosion. Since he observed that flightlike movements of the wing bases help to rupture the pupal cuticle during normal eclosion, Blest (1960) has suggested that the apparent inhibition of flight behavior immediately prior to the moult prevents premature rupture of the weakened pupal cuticle. In birds, however, this is probably not the explanation since chick embryos have never been found to damage themselves or the egg during massive strychnine or picrotoxin-induced convulsions even just before hatching (Oppenheim, personal communication).

In birds and mammals a decrease in spontaneous activity is normally seen toward the end of incubation or gestation (Barcroft and Barron, 1939; Barron, 1941; Hamburger and Balaban, 1963; Narayanan, Fox, and Hamburger, 1971; Edwards and Edwards, 1970). In sheep fetuses, spinal transection experiments indicate that the lowered level of activity near the end of normal fetal life is due to supraspinal inhibition (Barcroft and Barron, 1939; Barron, 1941). In chick embryos, although spinal cord transection does not increase the level of spontaneous activity above the level of controls, addition of strychnine does increase the activity as early as ten days of incubation. Moreover, this increase in response to strychnine is not seen in chicks with chronic or acute cervical transections (Oppenheim, 1975), which suggests that there may be strychnine-sensitive descending inhibitory pathways.

Inhibition of specific behavior patterns has been recognized in several invertebrate larvae and pupae. In the developing cricket, descending inhibition has been shown to selectively prevent the expression of a specific behavior, stridulation, until after the final moult (Bentley and Hoy, 1970). The neural circuitry responsible for producing stridulation in crickets is fully assembled in the last instar nymph. Nevertheless, the behavior is never expressed unless lesions are made in the mushroom bodies in the brain. Reingold and Camhi (1977) have also shown that adult-type abdominal grooming reflexes can be elicited from late-instar cockroach nymphs, but only after decapitation. The adult motor patterns of flight and walking can be elicited from normally quiescent, pharate moths by canaline and by picrotoxin, a known antagonist of GABA, which suggests that these behavior patterns may normally be inhibited at this stage (Truman, 1976; Kammer, Dahlman, and Rosenthal, 1978).

Inhibition of a specific behavior pattern has also been recognized in a vertebrate. Newborn to four-week-old kittens have an excitatory somatovesical reflex that elicits micturition and defecation when the mother licks the perineal regions. DeGroat and coworkers (1975) have shown that this reflex becomes suppressed by supraspinal inhibitory controls after four weeks, when the normal supraspinal pathways for micturition develop and this reflex is no longer necessary. That the circuitry for the excitatory reflex is still intact can be shown by transecting the spinal cord. Within a short period the animals begin responding to the perineal stimulation again.

Thus, while as yet there is no conclusive evidence that gradual withdrawal of inhibition from previously constructed circuits is responsible for the gradual appearance of coordinated behaviors, it is clear that some circuits can be completely inhibited during normal development. For this reason, it is important to consider that the absence of a particular behavior from the repertoire of an organism at any stage of development does not necessarily mean absence of the neural circuitry which underlies the production of that behavior. Hamburger (1973), for example, has emphasized that it may not always be possible to analyze the steps in the gradual assembly of neural circuitry from observations of overt behavior. Nevertheless, in cases that have been examined using electrophysiological techniques, it has been experimentally demonstrated that the *de novo* appearance of various behaviors was preceded by a period during which the neural circuitry was gradually assembled (Bentley and Hoy, 1970; Kutsch, 1971; Altman, 1975; Bekoff, 1976).

CONCLUSION

It is clear that we do not yet have a complete understanding of the neural mechanisms underlying the development of the pattern generating circuits for any coordinated behavior. Nevertheless, an enormous amount of information relevant to this issue has been accumulated. The availability of new techniques for reanalyzing old issues and the development of new conceptual approaches to the study of embryonic motor behavior, such as those pioneered by Viktor Hamburger, have made this an appropriate time to review the available data and to rephrase some of the unanswered questions in light of current understanding of pattern generating mechanisms in adults.

ACKNOWLEDGMENTS

I would like to thank Drs. R. Eaton, R. Oppenheim, P. S. G. Stein, and Ms. J. Kauer for valuable comments on an earlier draft of this chapter. I would also like to thank Ms. P. Holman for typing the manuscript. Support during the preparation of this review was provided by a grant from the National Science Foundation (BNS79-13826) and a fellowship from the Alfred P. Sloan Foundation.

REFERENCES

Altman, J. S. 1975. Changes in the flight motor pattern during the development of the Australian plague locust, *Chortoicetes terminifera*. *J. Comp. Physiol.* 97:127–42

Altman, J. S., Tyrer, N. M. 1974. Insect flight as a system for the study of the development of neuronal connections. In *Experimental Analysis of Insect Behaviour*, (ed. L. Barton, pp. 159–79. New York: Springer

Änggard, L., Bergström, R., Bernhard, C. G. 1961. Analysis of prenatal spinal reflex activity in sheep. *Acta Physiol. Scand.* 53:128–36

Angulo y Gonzalez, A. W. 1932. The prenatal development of behavior in the albino rat. *J. Comp. Neurol.* 55:395–442

Atwood, H. L., Kwan, I. 1976. Synaptic development in the crayfish opener muscle. *J. Neurobiol.* 7:289–312

Bakhuis, W. J. 1974. Observations on hatching movements in the chick *(Gallus domesticus)*. *J. Comp. Physiol. Psychol.* 87:997–1003

Barcroft, J., Barron, D. H. 1939. The development of behavior in foetal sheep. *J. Comp. Neurol.* 70:477–502

Barron, D. H. 1941. The functional development of some mammalian neuromuscular mechanisms. *Biol. Rev.* 16:1–33

Bekoff, A. 1974. Development of coordinated motor output in the hindlimb of the chick embryo. Ph. D. Thesis. St. Louis: Washington University

Bekoff, A. 1976. Ontogeny of leg motor output in the chick embryo: A neural analysis. *Brain Res.* 106:271–91

Bekoff, A. 1978. A neuroethological approach to the study of the ontogeny of coordinated behavior. In *The Development of Behavior: Comparative and Evolutionary Aspects*, ed. G. M. Burghardt, M. Bekoff, pp. 19–41. New York: Garland

Bekoff, A., Lau, B. 1980. Interlimb coordination in 20-day old rat fetuses. *J. Exp. Zool.* 214:173–75

Bekoff, A., Stein, P. S. G., Hamburger, V. 1975. Coordinated motor output in the hindlimb of the 7-day chick embryo. *Proc. Natl. Acad. Sci.* U.S.A. 72:1245–48

Bekoff, A., Trainer, W. 1979. Development of interlimb coordination during swimming in postnatal rats. *J. Exp. Biol.* 83:1–11

Bennett, M. R., Pettigrew, A. G. 1974. The formation of synapses in striated muscle during development. *J. Physiol.* (London) 241:515–45

Bentley, D. 1973. Postembryonic development in insect motor systems. In *Developmental Neurobiology of Arthropods*, ed. D. Young, pp. 147–78. Cambridge: Cambridge Univ. Press

Bentley, D. R., Hoy, R. R. 1970. Postembryonic development of adult motor patterns in crickets: A neural analysis. *Science* 170: 1409–11

Bergström, R. M., Hellström, P.-E., Stenberg, D. 1962. Studies in reflex irradiation in the foetal guinea-pig. *Ann. Chir. Gynaecol. Fenn.* 51:171–78

Berkinblit, M. B., Deliagina, T. G., Feldman, A. G., Gelfand, I. M., Orlovsky, G. N. 1978. Generation of scratching. II. Nonregular regimes of generation. *J. Neurophysiol.* 41:1058–69

Berman, A. J., Berman, D. 1973. Fetal deafferentation, the ontogenesis of movement in the absence of peripheral sensory feedback. *Exp. Neurol.* 38:170–76

Berrill, M. 1973. The embryonic behavior of certain crustaceans. In *Behavioral Embryology*, ed. G. Gottlieb, pp. 141–58. New York: Academic

Blest, A. D. 1960. The evolution, ontogeny and quantitative control of the settling movements of some new world saturniid moths, with some comments on distance communication by honey-bees. *Behaviour* 16:188–253

Bradley, R. M., Mistretta, C. M. 1975. Fetal sensory receptors. *Physiol. Rev.* 55:352–82

Bridgman, C. S., Carmichael, L. 1935. An experimental study of the onset of behavior in the fetal guinea-pig. *J. Genet. Psychol.* 47:247–67

Brown, M. C., Jansen, J. K. S. Van Essen, D. 1976. Polyneuronal innervation of skeletal muscle in newborn rats and its elimination during maturation. *J. Physiol.* (London) 261:387–422

Brown, T. G. 1915. On the activities of the central nervous system of the unborn foetus of the cat; with a discussion of the question whether progression (walking, etc.) is a "learnt" complex. *J. Physiol.* (London) 49:208–15

Carmichael, L. 1934. An experimental study in the prenatal guinea pig of the origin and development of reflexes and patterns of behavior in relation to the stimulation of specific receptor areas during the period of active fetal life. *Genet. Psychol. Monogr.* 16:337–491

Carmichael, L. 1970. The onset and early development of behavior. In *Carmichael's Manual of Child Psychology,* vol. 1, ed. P. Mussen, pp. 447–563. New York: John Wiley & Sons

Changeux, J.-P., Danchin, A. 1976. Selective stabilisation of developing synapses as a mechanism for the specification of neuronal networks. *Nature* (London) 264:705–12

Clark, R. D., Cohen, M. J. 1979. Changes in firing behaviour of an insect motor neurone during development and following experimental surgery. *J. Insect Physiol.* 25:725–31

Coghill, G. E. 1929. *Anatomy and the Problem of Behavior.* New York: Hafner. 110 pp.

Cohen, A. H., Wallén, P. 1979. A simple vertebrate locomotor system: "Fictive swimming" in the *in vitro* lamprey spinal cord. *Neurosci. Abstr.* 5:493

Conradi, S., Ronnevi, L.-O. 1975. Spontaneous elimination of synapses on cat motoneurons after birth· Do half of the synapses on the cell bodies disappear? *Brain Res.* 92:505–10

Corner, M. A. 1977. Sleep and the beginnings of behavior in the animal kingdom—Studies of ultradian motility cycles in early life. *Prog. Neurobiol.* (N.Y.) 8:279–95

Corner, M. A., Bakhuis, W. L. 1969. Developmental patterns in the central nervous system of birds. V. Cerebral electrical activity, forebrain function and behavior in the chick at the time of hatching. *Brain Res.* 13:541–55

Corson, D. W., Goodman, S., Fein, A. 1979. An adaptation of the jet stream microelectrode beveler. *Science* 205:1302

Crain, S. M. 1974. Tissue culture models of developing brain functions. In *Aspects of Neurogenesis,* ed. G. Gottlieb, pp. 69–114. New York: Academic

Crepel, F., Mariana, J., Delhaye-Bouchard, N. 1976. Evidence for a multiple innervation of Purkinje cells by climbing fibers in the immature rat cerebellum. *J. Neurobiol.* 7:567–78

Davis, W. J. 1973. Development of locomotor patterns in the absence of peripheral sense organs and muscles. *Proc. Natl. Acad. Sci. U.S.A.* 70:954–58

Davis, W. J., Davis, K. B. 1973. Ontogeny of a simple locomotor system: Role of the periphery in the development of central nervous circuitry. *Am. Zool.* 13:409–25

Decker, J. D. 1967. Motility of the turtle embryo, *Chelyda serpentina* (Linné). *Science* 157:952–54

Decker, J. D., Hamburger, V. 1967. The influence of different brain regions on periodic motility of the chick embryo. *J. Exp. Zool.* 165:371–80

DeGroat, W. C., Douglas, J. W., Glass, J., Simonds, W., Weimer, B., Werner, P. 1975. Changes in somato-vesical reflexes during postnatal development in the kitten. *Brain Res.* 94:150–54

Dennis, M. J., Ort, C. A. 1975. Physiological properties of nerve-muscle junctions developing *in vivo*. *Cold Spring Harbor Symp. Quant. Biol.* 40:435–42

Diamond, J., Miledi, R. 1962. A study of foetal and new-born rat muscle fibres. *J. Physiol.* (London) 162:393–403

Duysens, J., Pearson, K. G. 1976. The role of cutaneous afferents from the distal hindlimb in the regulation of the step cycle of thalamic cats. *Exp. Brain Res.* 24:245–55

Eaton, R. C., Farley, R. D., Kimmel, C. B., Schabtach, E. 1977. Functional development in the Mauthner cell system of embryos and larvae of the zebra fish. *J. Neurobiol.* 8:151–72

Edgerton, V. R., Grillner, S., Sjöström, A., Zangger, P. 1976. Central generation of locomotion in vertebrates. In *Neural Control of Locomotion*, ed. R. M. Herman, S. Grillner, P. S. G. Stein, D. Stuart, pp. 439–64. New York: Plenum

Edwards, D. D., Edwards, J. S. 1970. Fetal movement: Development and time course. *Science* 169:95–97

Evarts, E. V., Bizzi, E., Burke, R. E., Delong, M., Thach, W. T. 1971. Central control of movement. *Neurosci. Res. Programs Bull.* 9:1–170

Feldman, A. G., Orlovsky, G. N. 1975. Activity of interneurons mediating reciprocal Ia inhibition during locomotion. *Brain Res.* 84:181–94

Fentress, J. C. 1972. Development and patterning of movement sequences in inbred mice. In *The Biology of Behavior*, ed. J. A. Kiger, pp. 83–132.Eugene, Ore.: Oregon State Univ. Press

Fentress, J. C. 1973. Development of grooming in mice with amputated forelimbs. *Science* 179:704–5

Forssberg, H., Grillner, S. 1973. The locomotion of the acute spinal cat injected with clonidine i.v. *Brain Res.* 50:184–86

Forssberg, H., Grillner, S., Rossignol, S., Wallén, P. 1976. Phasic control of reflexes during locomotion in vertebrates. In *Neural Control of Locomotion*, ed. R. M. Herman, S. Grillner, P. S. G. Stein, D. G. Stuart, pp. 647–74. New York: Plenum

Friesen, W. O., Poon, M., Stent, G. S. 1978. Neuronal control of swimming in the medicinal leech. IV. Identification of a network of oscillatory interneurons. *J. Exp. Biol.* 75:25–44

Gatev, V. 1972. Role of inhibition in the development of motor coordination in early childhood. *Dev. Med. Child Neurol.* 14:336–41

Goodman, C. S., Spitzer, N. C. 1979. Embryonic development of identified neurones: Differentiation from neuroblast to neurone. *Nature* 280:208–14

Graubard, K. 1975. Voltage attenuation within *Aplysia* neurons: The effect of branching pattern. *Brain Res.* 88:325–32

Grillner, S. 1974. On the generation of locomotion in the spinal dogfish. *Exp. Brain Res.* 20:459–70

Grillner, S. 1975. Locomotion in vertebrates: Central mechanisms and reflex interaction. *Physiol. Rev.* 55:247–304

Grillner, S. 1976. Some aspects of the descending control of the spinal circuits generating locomotor movements. In *Neural Control of Locomotion,* ed. R. M. Herman, S. Grillner, P. S. G. Stein, D. G. Stuart, pp. 351–75. New York: Plenum

Grillner, S., Kashin, S. 1976. On the generation and performance of swimming in fish. In *Neural Control of Locomotion,* ed. R. M. Herman, S. Grillner, P. S. G. Stein, D. Stuart, pp. 181–201. New York: Plenum

Grillner, S., Wallén, P. 1977. Is there a peripheral control of the pattern generators for swimming in dogfish? *Brain Res.* 127:291–95

Grillner, S., Zangger, P. 1975. How detailed is the central pattern generation for locomotion? *Brain Res.* 88:367–71

Grillner, S., Zangger, P. 1979. On the central generation of locomotion in the low spinal cat. *Exp. Brain Res.* 34:241–61

Grobstein, C. S. 1979. The development of afferents and motoneurons in rat spinal cord. *Neurosci. Abstr.* 5:161

Grossman, Y., Spira, M. E., Parnas, I. 1973. Differential flow of information into branches of a single neuron. *Brain Res.* 64:379–86

Hamburger, V. 1963. Some aspects of the embryology of behavior. *Q. Rev. Biol.* 38:342–65

Hamburger, V. 1973. Anatomical and physiological basis of embryonic motility in birds and mammals. In *Behavioral Embryology,* ed. G. Grottlieb, pp. 52–76. New York: Academic

Hamburger, V. 1975. Cell death in the development of the lateral motor column of the chick embryo. *J. Comp. Neurol.* 160:535–46

Hamburger, V., Balaban, M. 1963. Observations and experiments on spontaneous rhythmical behavior in the chick embryo. *Dev. Biol.* 7:533–45

Hamburger, V., Balaban, M., Oppenheim, R., Wenger, E. 1965. Periodic motility of normal and spinal chick embryos between 8 and 17 days of incubation. *J. Exp. Zool.* 159:1–14

Hamburger, V., Narayanan, C. H. 1969. Effects of the deafferentation of the trigeminal area on the motility of the chick embryo. *J. Exp. Zool.* 170:411–26

Hamburger, V., Wenger, E., Oppenheim, R. 1966. Motility in the chick embryo in the absence of sensory input. *J. Exp. Zool.* 162:133–60

Harris, A. J., Dennis, M. J. 1977. Deletion of "mistakes" in nerve muscle connectivity during development of rat embryos. *Neurosci. Abstr.* 3:107

Hart, B. L. 1971. Facilitation by strychnine of reflex walking in spinal dogs. *Physiol. Behav.* 6:627–28

Herman, R. M., Grillner, S., Stein, P. S. G., Stuart, D. G. (eds.) 1976. *Neural Control of Locomotion.* New York: Plenum. 822 pp.

Holst, E. von 1939. Relative coordination as a phenomenon and as a method of analysis of central nervous functions. (Tr. from German) In *The Behavioural Physiology of Animals and Man,* tr. R. Martin, vol. 1, 1973, pp. 33–135. Coral Gables, Fla.: Univ. of Miami Press

Honig, M. G. 1979. Development of sensory neuron projection patterns under normal and experimental conditions in the chick hindlimb. *Neurosci. Abstr.* 5:163

Hooker, D. 1952. *The Prenatal Origin of Behavior.* Univ. of Kansas Press

Hooker, D., Nicholas, J. S. 1930. Spinal cord section in rat fetuses. *J. Comp. Neurol.* 50:413–67

Hughes, A. 1966. Spontaneous movements in the embryo of *Eleutherodactylus martinicensis. Nature* (London) 211:51–53

Hughes, A., Bryant, S. V., Bellairs,

A. d'A. 1967. Embryonic behaviour in the lizard, *Lacerta vivipara*. *J. Zool.* 153:139–52

Hughes, A., Prestige, M. C. 1967. Development of behaviour in the hindlimb of *Xenopus laevis*. *J. Zool.* 152:347–59

Hughes, G. M., Wiersma, C. A. G. 1960. The coordination of swimmeret movements in the crayfish *Procambarus clarkii* (Girard). *J. Exp. Biol.* 37:657–70

Hultborn, H. 1972. Convergence on interneurons in the reciprocal Ia inhibitory pathway to motoneurons. *Acta Physiol. Scand. Suppl.* 375:1–42

Hultborn, H., Illert, M., Santini, M. 1976. Convergence on interneurones mediating the reciprocal Ia inhibition of motoneurones. I. Disynaptic Ia inhibition of Ia inhibitory interneurones. *Acta Physiol. Scand.* 96:193–201

Humphrey, T. 1952. The spinal tract of the trigeminal nerve in human embryos between 7½ and 8½ weeks of menstrual age and its relation to early fetal behavior. *J. Comp. Neurol.* 97:143–210

Jacobson, M. 1967. Retinal ganglion cells: Specification of central connections in larval *Xenopus laevis*. *Science* 155:1106–8

Jankowska, E., Roberts, W. J. 1972. Synaptic action of single interneurons mediating reciprocal Ia inhibition of motoneurones. *J. Physiol.* (London) 222:623–42

Kammer, A. E., Dahlman, D. L., Rosenthal, G. A. 1978. Effects of the non-protein amino acids L-canavanine and L-canaline on the nervous system of the moth *Manduca sexta* (L.). *J. Exp. Biol.* 75:123–32

Kammer, A. E., Kinnamon, S. C. 1979. Maturation of the flight motor pattern without movement in *Manduca sexta*. *J. Comp. Physiol.* 130:29–37

Kammer, A. E., Rheuben, M. B. 1976. Adult motor patterns produced by moth pupae during development. *J. Exp. Biol.* 65:65–84

Kano, M. 1975. Development of excitability in embryonic chick skeletal muscle cells. *J. Cell Physiol.* 86:503–10

Klee, M. R., Faber, D. S., Heis, W. D. 1973. Strychnine- and pentylenetetrazol-induced changes of excitability in *Aplysia* neurons. *Science* 179:1133–36

Kovach, J. K. 1970. Development and mechanisms of behavior in the chick embryo during the last five days of incubation. *J. Comp. Physiol. Psychol.* 73:392–406

Kuo, Y.-Z. 1932. Ontogeny of embryonic behavior in Aves. I. The chronology and general nature of the behavior of the chick embryo. *J. Exp. Zool.* 61:395–430

Kulagin, A. S., Shik, M. L. 1970. Interaction of symmetrical limbs during controlled locomotion. *Biophysics* 15:171–77

Kupferman, I., Carew, T. J., Kandel, E. R. 1974. Local, reflex and central commands controlling gill and siphon movements in *Aplysia*. *J. Neurophysiol.* 37:996–1019

Kutsch, W. 1971. The development of the flight pattern in the desert locust, *Schistocerca gregaria*. *Z. Vergl. Physiol.* 74:156–68

Kutsch, W. 1974. The influence of the wing sense organs on the flight motor pattern in maturing adult locusts. *J. Comp. Physiol.* 88:413–24

Landmesser, L. T. 1978. The development of motor projection patterns in the chick hind limb. *J. Physiol.* (London) 284:391–414

Landmesser, L. T. 1980. The generation of neuromuscular specificity. *Ann. Rev. Neurosci.* 3:279–302

Lannou, J., Precht, W., Cazin, L. 1979. The postnatal development of functional properties of central vestibular neurons in the rat. *Brain Res.* 175:219–32

Lennard, P. R., Stein, P. S. G. 1977. Swimming movements elicited by electrical stimulation of turtle spinal cord. I. Low-spinal and intact preparations. *J. Neurophysiol.* 40:768–78

Letinsky, M. S. 1974a. The development of nerve-muscle junctions in *Rana catesbeiana* tadpoles. *Dev. Biol.* 40:129–53

Letinsky, M. S. 1974b. Physiological properties of developing frog tadpole nerve-muscle junctions during repetitive stimulation. *Dev. Biol.* 40:154–61

Letinsky, M. S. 1975. Acetylcholine sensitivity changes in tadpole tail muscle fibers innervated by developing motor neurons. *J. Neurobiol.* 6:609–17

Lichtman, J. W. 1977. The reorganization of synaptic connexions in the rat submandibular ganglion during postnatal development. *J. Physiol.* (London) 273:155–77

Llinás, R., Sugimori, M. 1980. Calcium conductances in Purkinje cell dendrites: Their role in development and integration. *Prog. Brain Res.* 51:323–34

Maksimova, E. V. 1971. Inhibition of spinal reflex responses in the last period of embryogenesis. *Neurophysiology* (U.S.S.R.) 3:51–56

Mendell, L. M., Scott, J. G. 1975. The effect of peripheral nerve cross-union on connections of single Ia fibers to motoneurons. *Exp. Brain Res.* 22:221–34

Minkowski, M. 1920. Réflexes et mouvements de la tête, du tronc et des extrémités du foetus humain, pendant la premier moitié de la grossesse. *C. R. Soc. Biol.* 81:1202–4

Mori, S., Shik, M. L., Yagodnitsyn, A. S. 1977. Role of pontine tegmentum for locomotor control in mesencephalic cat. *J. Neurophysiol.* 40:284–95

Mulloney, B., Selverston, A. I. 1974. Organization of the stomatogastric ganglion of the spiny lobster. III. Coordination of the two subsets of the gastric system. *J. Comp. Physiol.* 91:53–78

Murphey, R. K., Levine, R. B., 1980. Mechanisms responsible for changes observed in response properties of partially deafferented insect interneurons. *J. Neurophysiol.* 43:367–82

Murphey, R. K. Mendenhall, B., Palka, J., Edwards, J. S. 1975. Deafferentation slows the growth of specific dendrites of identified giant interneurons. *J. Comp. Neurol.* 159:407–18

Naka, K. I. 1964a. Electrophysiology of the fetal spinal cord. I. Action potentials of the motoneuron. *J. Gen. Physiol.* 47:1003–22

Naka, K. I. 1964b. Electrophysiology of the fetal spinal cord. II. Interaction among peripheral inputs and recurrent inhibition. *J. Gen. Physiol.* 47:1023–38

Narayanan, C. H., Fox, M. W., Hamburger, V. 1971. Prenatal development of spontaneous and evoked activity in the rat *(Rattus norvegicus albinus). Behaviour* 40:100–34

Narayanan, C. H., Hamburger, V. 1971. Motility in chick embryos with substitution of lumbosacral by brachial and brachial by lumbosacral spinal cord segments. *J. Exp. Zool.* 178:415–32

Narayanan, C. H., Malloy, R. B. 1974. Deafferentation studies on motor activity in the chick. I. Activity pattern of hindlimbs. *J. Exp. Zool.* 189:163–76

Oppenheim, R. W. 1972a. An experimental investigation of the possible role of tactile and proprioceptive stimulation in certain aspects of embryonic behavior in the chick. *Dev. Psychobiol.* 5:71–91

Oppenheim, R. W. 1972b. Embryology of behavior in birds: A critical review of the role of sensory stimulation in embryonic development. *Proc. 15th Int. Ornithol. Congr.*, 283–302

Oppenheim, R. W. 1972c. Prehatching and hatching behaviour in birds: A comparative study of altricial and precocial species. *Anim. Behav.* 20:644–55

Oppenheim, R. W. 1972d. Experimental studies on hatching behavior in the chick. III. The role of the midbrain and forebrain. *J. Comp. Neurol.* 146:479–506

Oppenheim, R. W. 1975. The role of supraspinal input in embryonic motility: A re-examination in the chick. *J. Comp. Neurol.* 160:37–50

Oppenheim, R. W. 1978. G. E. Coghill (1872–1941): Pioneer neuroembryologist and developmental psychobiologist. *Persp. Biol. Med.* 22:45–64

Oppenheim, R. W., Chu-Wang, I.-W., Foelix, R. F. 1975. Some aspects of synaptogenesis in the spinal cord of the chick embryo: A quantitative electron microscopic study. *J. Comp. Neurol.* 161:383–418

Oppenheim, R. W., Reitzel, J. 1975. Ontogeny of behavioral sensitivity to strychnine in the chick embryo: Evidence for the early onset of CNS inhibition. *Brain Behav. Evol.* 11:130–59

Orr, D. W., Windle, W. F. 1934. The development of behavior in chick embryos; the appearance of somatic movements. *J. Comp. Neurol.* 60:271–85

Palka, J., Edwards, J. S. 1974. The cerci and abdominal giant fibers of the house cricket, *Acheta domesticus*. II. Regeneration and effects of chronic deprivation. *Proc. R. Soc. London Ser. B.* 185:105–21

Pearson, K. G., Duysens, J. 1976. Function of segmental reflexes in the control of stepping in cockroaches and cats. In *Neural Control of Locomotion*, ed. R. M. Herman, S. Grillner, P. S. G. Stein, D. G. Stuart, pp. 519–37. New York: Plenum

Pettigrew, A. G., Lindeman, R., Bennett, M. R. 1979. Development of the segmental innervation of the chick forelimb. *J. Embryol. Exp. Morphol.* 49:115–37

Pollack, E. D., Crain, S. M. 1972. Development of motility in fish embryos in relation to release from early CNS inhibition. *J. Neurobiol.* 3:381–85

Poon, M., Friesen, W. O., Stent, G. S. 1978. Neuronal control of swimming in the medicinal leech. V. Connexions between the oscillatory interneurones and the motor neurones. *J. Exp. Biol.* 75:45–64

Preyer, W. 1885. *Specielle Physiologie des Embryo.* Leipzig: Grieben

Provine, R. R. 1977. Behavioral development of the cockroach *(Periplaneta americana). J. Insect Physiol.* 23:213–20

Provine, R. R. 1979. "Wing-flapping" develops in wingless chicks. *Behav. Neural. Biol.* 27:233–37

Provine, R. R. 1980. Development of between-limb movement synchronization in the chick embryo. *Dev. Psychobiol.* 13:151–63

Provine, R. R., Rogers, L. 1977. Development of spinal cord bioelectric activity in spinal chick embryos and its behavioral implications. *J. Neurobiol.* 8:217–28

Provine, R. R., Sharma, S. C., Sandel, T. T., Hamburger, V. 1970. Electrical activity in the spinal cord of the chick embryo *in situ. Proc. Natl. Acad. Sci., U.S.A.* 65:508–15

Purpura, D. P., Shofer, R. J., Scarff, T. 1967. Properties of synaptic activities and spike potentials of neurons in immature neocortex. *J. Neurophysiol.* 28:925–42

Rall, W. 1964. Theoretical significance of dendritic trees for neuronal input-output relations. In *Neuronal Theory and Modeling*, ed.

R. F. Reiss, pp. 73–97. Palo Alto, Calif.: Stanford Univ. Press

Redfern, P. A. 1970. Neuromuscular transmission in new-born rats. *J. Physiol.* (London) 209:701–9

Reingold, S. C., Camhi, J. M. 1978. Abdominal grooming in the cockroach: Development of an adult behavior. *J. Insect Physiol.* 24:101–10

Reitzel, J. L., Maderdrut, J. L., Oppenheim, R. W. 1979. Behavioral and biochemical analysis of GABA-mediated inhibition in the early chich embryo. *Brain Res.* 172:487–504

Reitzel, J. L., Oppenheim, R. W. 19 Further studies on strychnine-mediated inhibition in chick embryo spinal cord. *Dev. Psychobiol.*

Reynolds, S. R. M. 1962. Nature of fetal adaptation to the uterine environment: A problem of sensory deprivation. *Am. J. Obstet. Gynecol.* 83:800–8

Roaf, H. E., Sherrington, C. S. 1910. Further remarks on the mammalian spinal preparation. *Q. J. Exp. Physiol.* 3:209–11

Romand, R., Marty, R. 1975. Postnatal maturation of the cochlear nuclei in the cat: A neurophysiological study. *Brain Res.* 83:225–33

Ronnevi, L.-O., Conradi, S. 1974. Ultrastructural evidence for spontaneous elimination of synaptic terminals on spinal motoneurons in the kitten. *Brain Res.* 80:335–39

Saito, K. 1979. Development of reflexes in the rat fetus studied *in vitro*. *J. Physiol.* (London) 294:591–94

Scheibel, M. E., Scheibel, A. B., 1970. Developmental relationship between spinal motoneuron dendrite bundles and patterned activity in the hind limb of cats. *Exp. Neurol.* 29:328–35

Scheibel, M. E., Scheibel, A. B. 1971. Developmental relationship between spinal motoneuron dendrite bundles and patterned activity in the forelimb of cats. *Exp. Neurol.* 30:367–73

Scherrer, J. 1968. Electrophysiological aspects of cortical development. *Prog. Brain Res.* 22:480–89

Sedláček, J. 1971. Cortical responses to visual stimulation in the developing guinea pig during prenatal and perinatal period. *Physiol. Bohemoslov.* 20:213–20

Sharma, S. C., Provine, R. R., Hamburger, V., Sandel, T. T. 1970. Unit activity in the isolated spinal cord of chick embryo, *in situ*. *Proc. Natl. Acad. Sci.* U.S.A. 66:40–47

Singer, H. S., Skoff, R. P., Price, D. L. 1978. The development of intersegmental connections in embryonic spinal cord: An anatomic substrate for early embryonic motility. *Brain Res.* 141:197–209

Smolen, A., Raisman, G. 1980. Synapse formation in the rat superior cervical ganglion during normal development and after neonatal deafferentation. *Brain Res.* 181:315–23

Spear, P. K., Chow, K. L., Masland, R. H., Murphey, E. H. 1972. Ontogenesis of receptive field characteristics of superior colliculus neurons in the rabbit. *Brain Res.* 45:67–86

Sperry, R. W. 1951. Regulative factors in the orderly growth of neural circuits. *Growth Symp.* 10:63–87

Spitzer, N. C., Baccaglini, P. I. 1976. Development of the action potential in amphibian neurons *in vivo*. *Brain Res.* 107:610–16

Steeves, J. D., Schmidt, B. J., Skovgaard, B. J., Jordan, L. M. 1980. Effect of noradrenaline and 5-hydroxytryptamine depletion on locomotion in the cat. *Brain Res.* 185:349–62

Stein, P. 1974. Neural control of interappendage phase during locomotion. *Am. Zool.* 14:1003–16

Stein, P. S. G. 1976. Mechanisms of interlimb phase control. In *Neural Control of Locomotion*, ed. R. M.

Herman, S. Grillner, P. S. G. Stein, D. G. Stuart, pp. 465–87. New York: Plenum

Stein, P. S. G. 1978. Motor systems, with specific reference to the control of locomotion. *Ann. Rev. Neurosci.* 1:61–81

Straznicky, K. 1963. Function of heterotopic spinal cord segments investigated in the chick. *Acta Biol. Acad. Sci. Hung.* 14:143–53

Sviderskii, V. L. 1970. Receptors of the forehead of *Locusta migratoria*. *J. Evol. Biochem. Physiol.* 5:392–98

Swenson, E. A. 1926. The development of movement of the albino rat before birth. Ph.D. Thesis. Univ. of Kansas, 195 pp.

Székely, G. 1976. Developmental aspects of locomotion. In *Neural Control of Locomotion*, ed. R. M. Herman, S. Grillner, P. S. G. Stein, D. G. Stuart, pp. 735–57. New York: Plenum

Taub, E. 1976. Motor behavior following deafferentation in the developing and motorically mature monkey. In *Neural Control of Locomotion*, ed. R. M. Herman, S. Grillner, P. S. G. Stein, D. G. Stuart, pp. 675–705. New York: Plenum

Truman, J. W. 1976. Development and hormonal release of adult behavior patterns in silkmoths. *J. Comp. Physiol.* 107:39–48

Truman, J. W. Reiss, S. E. 1976. Dendritic reorganization of an identified motoneuron during metamorphosis of the tobacco hornworm moth. *Science* 192:477–79

Vaughn, J. E., Grieshaber, J. A. 1973. A morphological investigation of an early reflex pathway in developing rat spinal cord. *J. Comp. Neurol.* 148:177–210

Visintini, F., Levi-Montalcini, R. 1939. Relazione tra differenziazione strutturale e funzionale dei centri e delle vie nervose nell'embrione di pollo. *Schweiz. Archiv. Neurol. Psychol.* 43:1–45

Weber, T. 1972. Stabilizierung des Flugrhythmus durch "Erfahrung" bei der Feldgrille. *Naturwissenschaften* 59:366

Weiss, P. 1937. Further experimental investigations on the phenomenon of homologous response in transplanted amphibian limbs. II. Nerve regeneration and the innervation of transplanted limbs. *J. Comp. Neurol.* 66:481–535

Weiss, P. 1941a. Does sensory control play a constructive role in the development of motor coordination? *Schweiz. Med. Wochenschr.* 12:591–95

Weiss, P. 1941b. Self-differentiation of the basic patterns of coordination. *Comp. Psychol. Monogr.* 17:1–96

Wetzel, M. C., Stuart, D. G. 1976. Ensemble characteristics of cat locomotion and its neural control. *Progr. Neurobiol.* 7:1–98

Wilson, D. M. 1961. The central nervous control of flight in a locust. *J. Exp. Biol.* 38:471–90

Wilson, D. M. 1972. Genetic and sensory mechanisms for locomotion and orientation in animals. *Am. Sci.* 60:358–65

Wilson, D. M., Gettrup, E. 1963. A stretch reflex controlling wingbeat frequency in grasshoppers. *J. Exp. Biol.* 40:171–85

Windle, W. F. 1940. *Physiology of the Fetus: Origin and Extent of Function in Prenatal Life.* Philadelphia: W. B. Saunders

Windle, W. F. 1944. Genesis of somatic motor function in mammalian embryos: A synthesizing article. *Physiol. Zool.* 17:247–60

Windle, W. F., Baxter, R. E. 1936. Development of reflex mechanisms in the spinal cord of albino rat embryos. Correlations between structure and function, and comparisons with the cat and the chick. *J. Comp. Neurol.* 63:189–204

Windle, W. F., Griffin, A. M. 1931. Observations on embryonic and fetal movements of the cat. *J. Comp. Neurol.* 523:149–88

Windle, W. F., O'Donnell, J. E., Glasshagle, E. E. 1933. The early development of spontaneous and reflex behavior in cat embryos and fetuses. *Physiol. Zool.* 4:521–41

Windle, W. F., Orr, D. W. 1934. The development of behavior in chick embryos: Spinal cord structure correlated with early somatic motility. *J. Comp. Neurol.* 60:287–308

Wine, J. J. 1978. Crayfish escape behavior. III. Monosynaptic and polysynaptic sensory pathways involved in phasic extension. *J. Comp. Physiol.* 121:187–203

Wong, R. K. S., Pearson, K. G. 1975. Limitations on impulse conduction in the terminal branches of insect sensory nerve fibers. *Brain Res.* 100:431–36

7 / *The First Stage in Myoblast Development: Skeletal Muscles, Myocardium, and Iris*

GUIDO FILOGAMO

INTRODUCTION

There are two sequences in the development of the peripheral neuromuscular system. On the one hand, there is the *neural sequence* that starts with the appearance of the neural plate and groove; continues through the migration of neural crest cells, the genesis and growth of motoneurons, and the sprouting of motor axons from the neural tube and their ramification within the developing muscles; and finally concludes with the conjunction of the terminal arborizations of the axons with their target cells. On the other hand, there is the *muscle sequence* that commences with the determination of myoblasts; continues through the differentiation of muscle fibers; and ends with the innervation of their junctional regions by the terminal arborizations of the motor nerves.

In this essay, attention will be paid to the points of interaction of these two sequences. We shall not deal with the final stabilization of the nerve-muscle synapses, with the problems of their localization and specificity, or with the intermediate stages during which the nerve and the muscle fibers interact with, and regulate, each other. Our attention will be confined to the first steps in the process and with the first points of contact between the future partners in the peripheral motor system. Our purpose is to analyze the *in vivo* responsibilities of each partner in the epigenetic history of the motor system.

At the outset it must be stated that there is no single developmental model, any more than there is a single pattern, that is typical of the mature motor system. But in what follows we shall deal with three models for the innervation of striated skeletal musculature.

The First Stage in the Development of Skeletal Muscle Myoblasts

Skeletal muscle tissue makes its appearance at a very early stage in development; myoblasts can already be detected in the more cranial segments of the mesoderm of the chick at the nine-somite stage after about 35

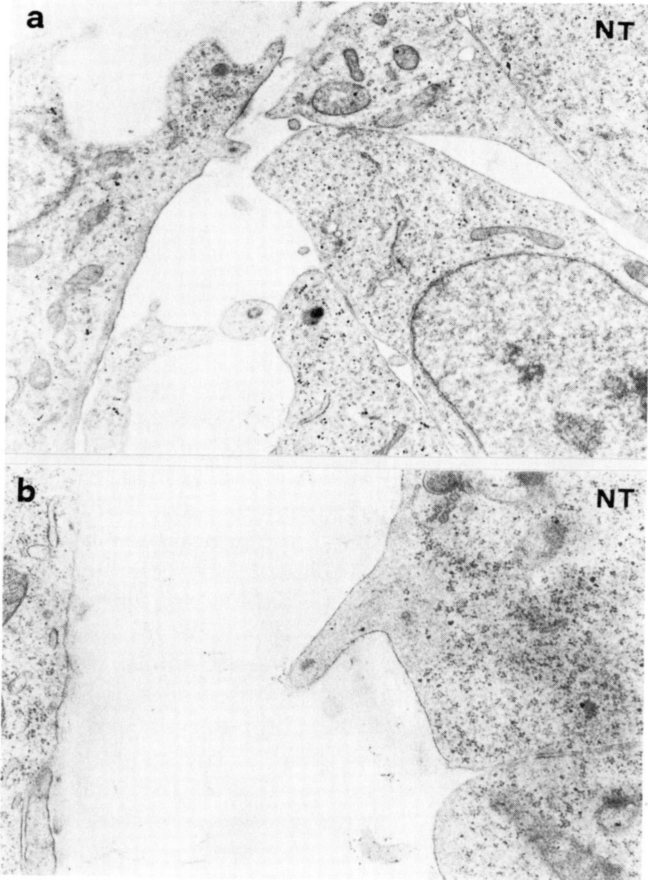

Figure 7.1 (a) Cells of the medial somite wall are contacted by processes of cells in the neural tube (NT). (From a Stage 10 embryo, at 35 hours of incubation, ×17,500.) (b) A filopodium that projects from the neural tube (NT) towards the medial wall of the somite. (×12000)

hours of incubation (Sisto Daneo and Filogamo, 1973). Subsequently there is a distinct rostrocaudal gradient in the time of appearance of further myogenic centers, similar to that which characterizes the proliferation of neurons in the spinal cord (Hamburger, 1977). Toward the end of the third day of incubation, myoblasts can also be observed in the limbs (Filogamo and Gabella, 1967).

The crucial moment in the determination of the myogenic cell line occurs at the level of the first somite, when numerous filopodia project into the perimedullary space, from the cells that lie toward the side of the neural tube (Fig. 7-1) and establish contact with those put out by the cells along the edge of the somite (Filogamo, 1976). At this time, the neural crest cells are migrating through the "neurosomitic corridor," but they do not appear to make contact with the two sets of filopodia.

Three-dimensional models reconstructed from serial sections through the thoracic somites (Fig. 7-2) have shown that 14 percent of the total somite volume is occupied by myoblasts that are packed into its dorsomedial sector, a few hours after the filopodia have made contact with each other (Filogamo, Peirone, and Sisto Daneo, 1978; Panattoni and Sisto Daneo, 1979). Presumptive myoblasts, identified as large cells with clear cytoplasm, rosettelike ribosomes, a few cisternae of agranular cytoplasmic reticulum, large Golgi complexes, and loose webs of thin filaments, can

Figure 7.2 A reconstructed model of a thoracolumbar somite, seen from its medial aspect. The region that is contacted by filopodia from the neural tube cells is the area of the primitive myotome here labeled "M."

be clearly distinguished from the other three components into which the somite is divided, that is, the sclerotome, dermatome, and undifferentiated cells. The first two are formed at about the same time, as a result of chordal and ectodermal interactions.

It is well-known that the first neuroblasts in the lateral motor column arise during the second day of incubation. Between the 35th and the 40th hours of incubation, a large number of flattened growth cones with many large vesicles, abundant filopodial projections and occasional undulating membranes invade the perimedullary space and make contact with the medial wall of the somite (Filogamo and Sisto Daneo, 1977).

As soon as they take shape, the presumed myoblasts migrate outward, either singly or in small groups, in a direction toward the dermatome and away from the neural tube. This migration takes place when the sclerotomal cells migrate toward the notochord, that is, when the core of the somite is being emptied. The presumed muscle cells cross the path of the sclerotomal cells, which are moving in a different direction, and possibly also the paths of other undifferentiated cells. The growth cones of the nerve fibers remain in contact with the myoblasts and travel with them toward the deep surface of the dermatome; the latter is rapidly being covered by the myoblasts, which arrange themselves to form the myotome. Nerve fibers do not penetrate the myotome for another two days. During this time they confine themselves to flanking the myotome, particularly toward its rostral tip; here there are many close contacts between the axons and the undifferentiated cells and myoblasts that are being actively transformed.

At this point, consideration must be given to the origin of the limb-bud myoblasts, since their derivation from the somites has long been disputed. Our recent ultrastructural studies (Sisto Daneo and Filogamo, 1976; Filogamo et al., 1978) have indicated that between Stage 21 and Stage 23 of the Hamburger and Hamilton series, thin bundles of fibers from the anterior ramus of the spinal nerve separate into single axons at the base of the wing bud, where neither chondrogenic nor myogenic blastemas can be seen in semithin sections. The axons make close contact with the mesodermal cells and myoblasts either singly or in small groups. Throughout the growth period of the wing bud (up to the 5th day of incubation), the advancing front of axons is always located further forward than that of the myoblasts. This spatial relationship again suggests that myogenic determination is induced by nerve fibers. However, against this view must be set the fact that the wing muscles appear when the first myoblasts have already formed in the somitic region. It is possible that the myoblasts move from the anterior borders of the somites into the base of the wing bud as myogenic blastemal pioneers (Christ, Jacob, and Jacob, 1977; Chevallier, Kieny, and Mauger, 1978). Even though this possibility in no way contradicts our hypothesis of primary neural determination, a search has been made for myoblast pioneers that might determine myogenic evolution.

The crucial influence of the exploratory nerve fibers upon the presumptive muscle cells is apparent when one examines the development of the acetylcholine system. Zacks (1954) has found a pronounced cholinesterase reaction in the perimedullary space during the first 48 hours of development. This finding has since been confirmed (Filogamo, 1964) and more recently Robecchi and Filogamo (1980) have observed a positive Karnovsky reaction for acetylcholinesterase in some cells as they migrate from the somite to the myotome in contact with the exploratory nerve fibers (Fig. 7-3). Giacobini (1972) has similarly reported the appearance of, and a rapid rise in, cholineacetyltransferase in the somites on the third day of incubation. The appearance of acetylcholine in the spinal cord on the second day has been reported by Szepsenwol and Caretti

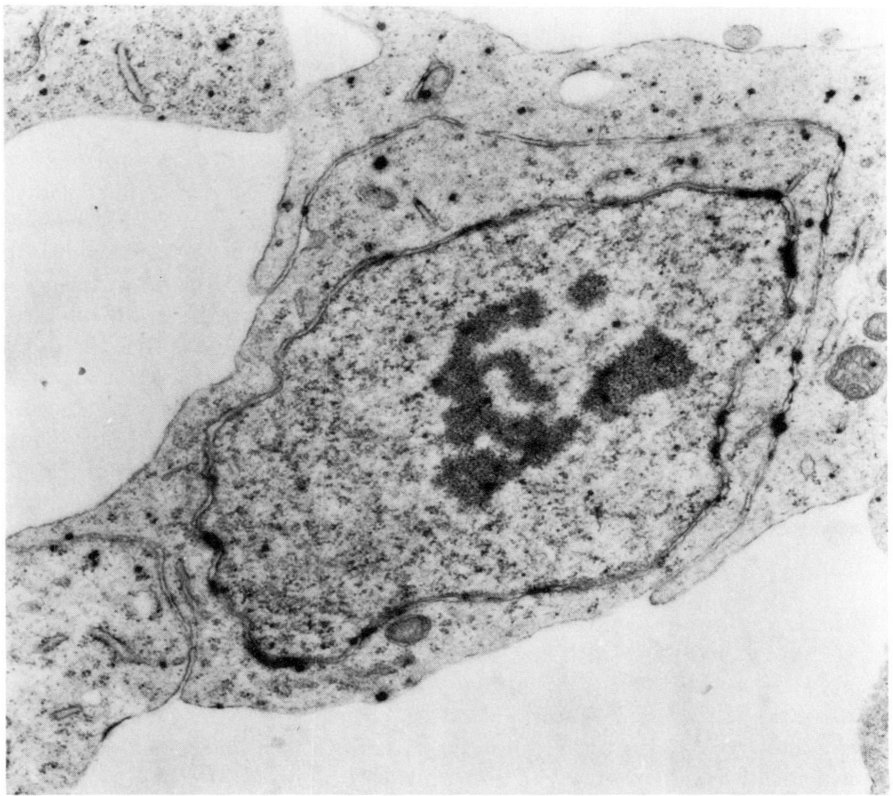

Figure 7.3 The Karnovsky reaction for AChE is positive in this section of a migrating myoblast, en route towards the myotome. (From an embryo after 60 hours of incubation, ×18,000)

(1942), but Kuo (1939) is of the opinion that it may be present even earlier. Data on the time of appearance of actylcholine receptors are lacking.

These findings lead one to conclude that the motoneurons possess the "chronological and topographical qualifications" required of agents for myoblast determination. But the question remains: do they uniquely "create" the myogenic line, whch constitutes their own target?

Mention should be made here of the important work of Strudel (1955) and of Sandor and Amels (1970), who have shown that the myotomes appear to follow their usual sequence of differentiation even after excision of the axial organs. But it is not clear from their work that the excision was not performed at a relatively late stage, when neural contact had already been established; or alternatively that the excised segments of the neural tube were long enough to prevent exploratory fibers that arise in other segments from making contact with the somites that were deprived of their usual innervation. One answer to this question has come from the *in vitro* study of Peirone, Sisto Daneo, and Filogamo (1977), who cultured early somites that had not yet been reached by nerve fibers, either alone or in the presence of neural tube fragments. This was possible because Ellison, Ambrose, and Easty (1969) had earlier shown that at Stage 11 (but, significantly, not at Stage 9 or Stage 10) the somites give rise to muscle cells *in vitro.*

Peirone and his colleagues (1977) prepared two groups of cultures; in the first, six pairs of somites, from Stage 10 embryos, were cultured with or without neural tube explants; in the second, somites from Stage 9 embryos were cultured in the same way. After about one week *in vitro* myoblasts and myotubes were only observed in the cultures that included neural tube explants, and especially where the nerve fibers had grown most strongly. Replacement of the neural tube by explanted spinal ganglia failed to induce myogenic cells, even though there was an active outgrowth of nerve fibers.

The conclusion to be drawn from these experiments is that *in vitro* myogenic determination of mesodermal cells requires the influence of nerve fibers from motoneurons. In addition they serve to strengthen the view that *in vivo* motoneurons are not merely bystanders, but probably play a comparable determining role. (The possible influence of the crest cells, however, is still uncertain.)

Peirone and Filogamo (1980) have recently cultured the six most caudal somites from Stage 9 embryos (i.e., before they were contacted by axonal filopodia) with or without fragments of muscle or tendon from nine-day-old quail embryos. When cultured alone or with tendon fragments, only chick somites gave rise to fibroblasts. But when chick somites were cultured with muscle fragments from quail embryos, they produced muscle cells and hybrid (quail/chick) myotubes (Fig. 7-4). That is to say,

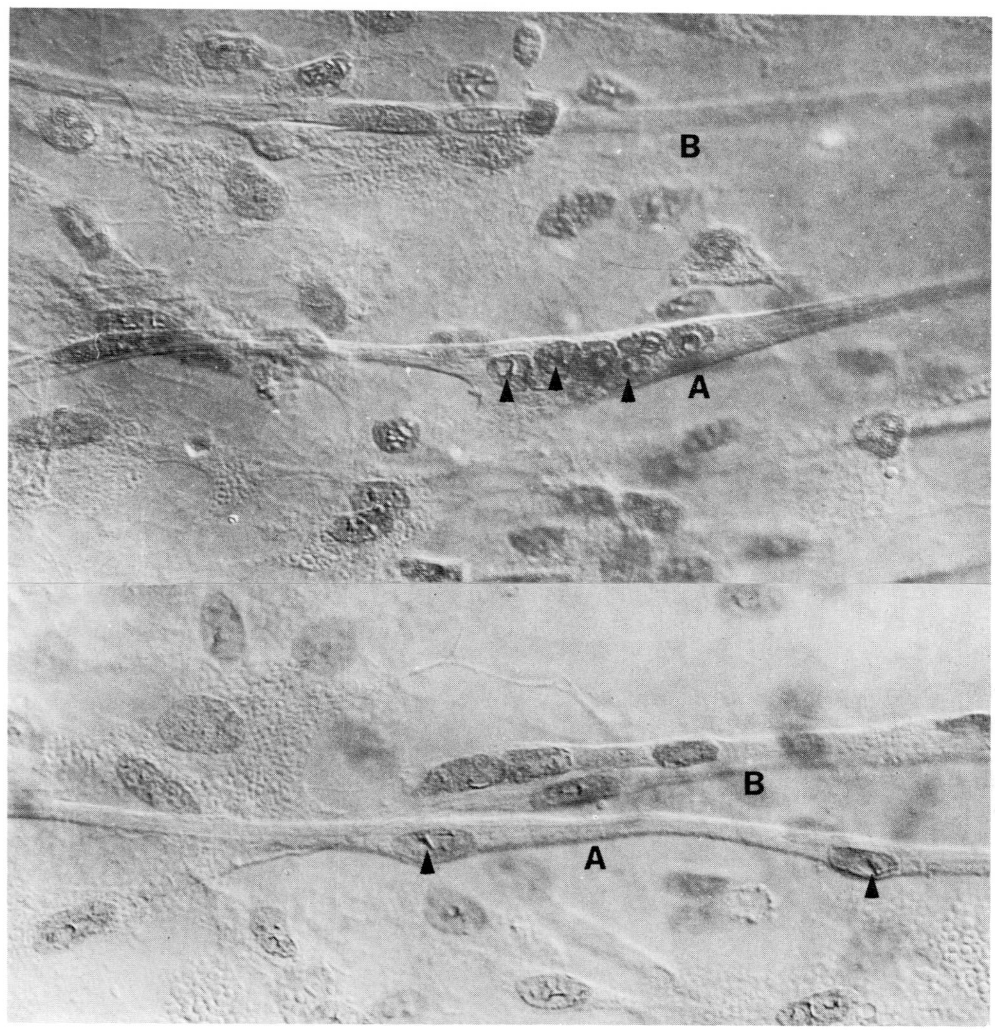

Figure 7.4 Coculture of somites from a Stage 9 chick embryo and cells from the pectoral muscle of a 10-day-old quail embryo after seven days in culture. Four myotubes are shown; the difference in their origin is suggested by the different arrangement of the nuclear chromatin. (A = quail myotube, B = chick myotube). The arrowheads point to a chromatic mass in the quail nuclei. (Feulgen stain–interference contrast microscopy, ×800)

in vitro muscle cells can induce myogenesis in somitic mesodermal cells.

We are thus of the opinion that *in vivo* the first muscle cells are determined by some neural influence, both in the somites and in the wing bud. Other nonneural agents, such as myoblasts and myotubes, may be secondarily involved as myogenic determinants at a later stage (Filogamo, 1980a) so that the first muscle cells may set in motion a "neuroindependent" mechanism, whose appearance considerably enhances myogenesis. Initially, neural and nonneural determination would thus form a sequence, whereas later they may overlap and interfere with each other. It is possible that muscle growth is modeled both by central influences naturally supplied by the motoneurons and by various peripheral influences, including possibly the reciprocal interaction of the contractile activity of the agonist and antagonist muscles (Filogamo, 1980a).

The First Stage in the Development of Myocardial Myoblasts

That motor root fibers emerging from the neural tube come into direct contact with somitic mesodermal masses seems to be beyond question. There is little doubt, too, that they exert an inductive influence on the determination of myoblasts. However, the position is not so clear with respect to the myocardium. The heart-bud muscle cells are both morphologically and functionally precocious: the heart is already beating at the eight or nine somite stage in chick embryos (Olivo, 1925). Set against this is the apparent lateness of the interaction between the cardial myocytes and their related nerve fibers (96 hours of incubation, Abel, 1912).

We have recently re-examined the relationship between the mesodermic structures responsible for the development of the myocardium and the related neural structures, including the neurites that grow out from the neural tube and neural crest cells (Sisto Daneo and Filogamo, 1980). At stage 6 (24 hours of incubation), the mesodermal cells in the presumptive cardiac area extend forwards and backwards across the primitive node on either side of the midline in contact with the cells of the neural groove (Rudnick, 1937). The contact consists of distinct filopodia between the neural groove and mesodermal cells; the Koelle reaction for acetylcholinesterase is positive here but the "contacted" and "noncontacted" cells are morphologically similar. Three hours later, the endothelial cardiac tubes are formed cranial to the first somite and ventral to the foregut endoderm. They are covered by the mesothelium of the splanchnopleure, which we think is derived from the presumptive cardiac area. Starting caudally from the angle between the splanchnopleure and the somatopleure, a long section of the mesothelium becomes thickened. Manasek (1968) and Viragh and Challice (1973) have shown that this thickened region contains cells that resemble myoblasts, especially toward the cau-

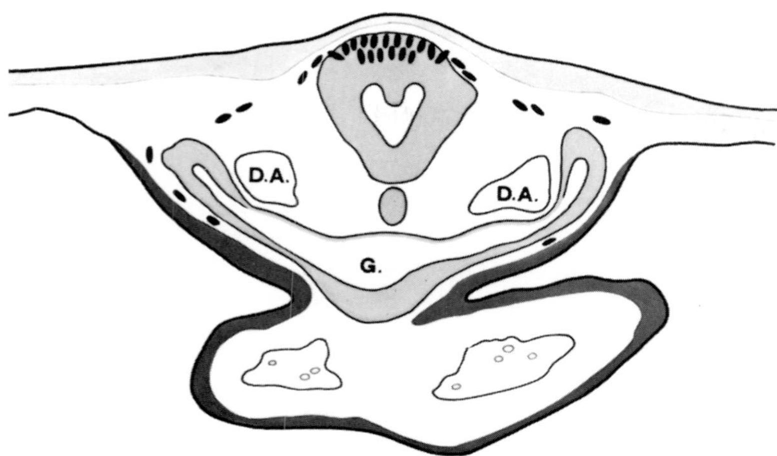

Figure 7.5 A diagram to show the course taken by neural crest cells as they migrate through the corridor between the endoderm and spanchnopleure. Based on observations of semithin sections from a Stage 8 chick embryo.

dal end of the heart tube. These cells migrate ventrally to cover the endothelial tubes. We have observed that they react positively with the Karnovsky reaction for acetylcholinesterase.

Nerve fibers are first observed in the dorsal part of the dorsal aorta at the 50th hour of incubation (Cajal, 1929). The ingrowth of nerve fibers would thus appear to be, both temporally and spatially, incapable of determining the cardiac myogenic line. On the other hand, crest cells are well placed to exert such an influence. The narrow interval between the foregut endoderm and the thickened splanchnopleure contains elongated cells, either singly or in groups with a dorsoventral orientation, as early as Stage 8 (Fig. 7-5). These cells look like those found in the corridor between the neural tube and the medial edge of the somite (Fig. 7-6), which are migrating towards the dorsal aorta; some of the cells will remain there to form the periaortic ganglia, while others will proceed toward the root of the primitive mesocardium. Both sets of cells display a positive reaction for acetylcholinesterase, and it seems reasonable to regard them as equivalent to the neural crest derivatives demonstrated experimentally by Noden (1975).

While the data are not yet sufficient to provide a final answer to the question of the time and manner of determination of the myocardial myogenic line, we can identify two points at which neural cells and the mesodermal cells that are destined to become myocytes interact. The first is when contact is established between the filopodia of the cells in the neural groove and those in the presumptive cardiac area, and the

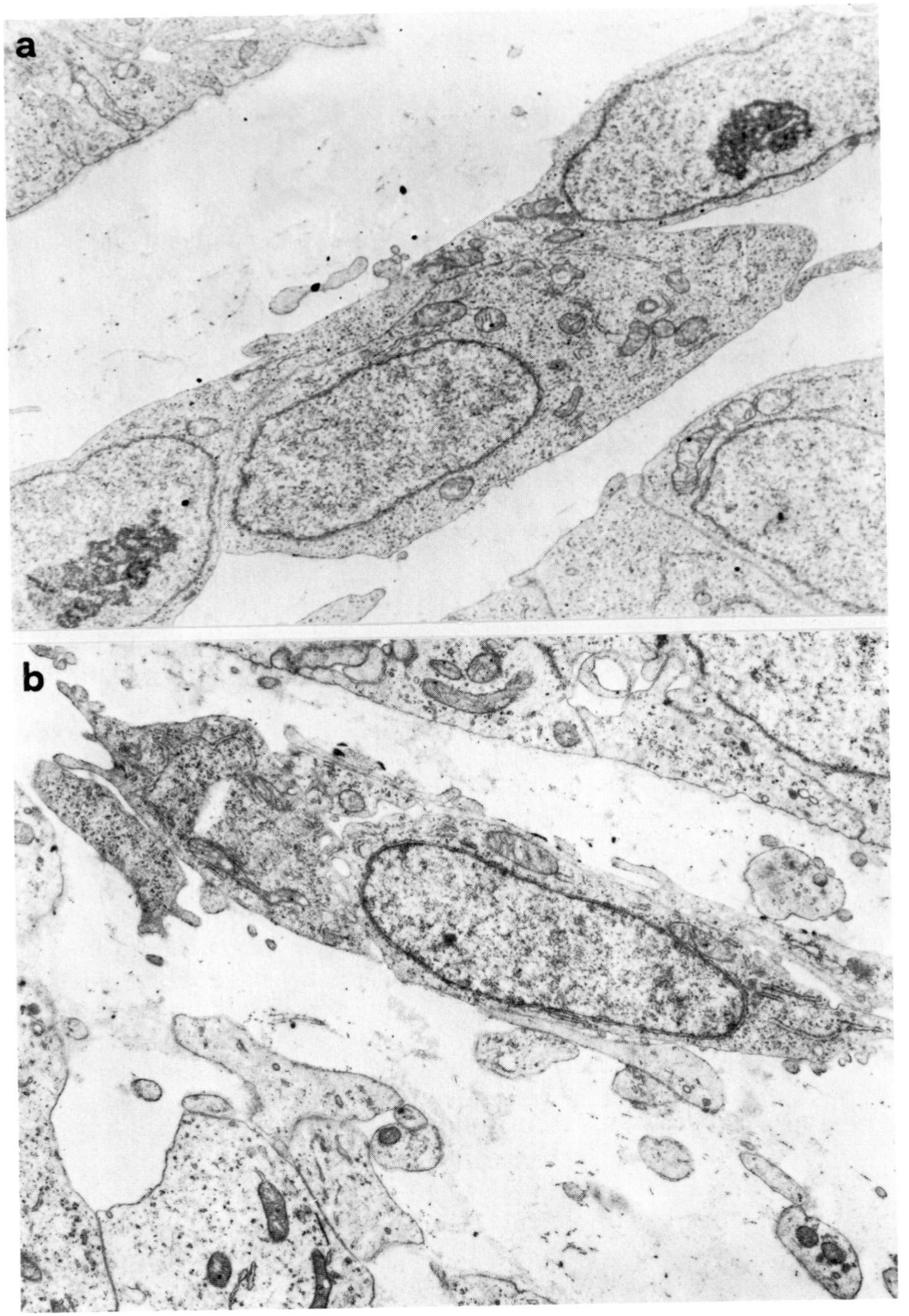

Figure 7.6 Chick embryo, stage 13. (a) Neural crest cells in the perimedullary space. 12000× (b) One cell, probably a neural crest cell, in the corridor between the endoderm and splanchnopleure. 10000×

second is when the migrating crest cells come into close topographic relation with the splanchnopleure. At least, some of these crest derivatives are destined to form the postganglionic neurons that lie scattered throughout most of the myocardium by the third or fourth day of incubation.

Both of these contacts may be regarded as potential initiations of myogenic differentiation in the heart bud, and, of course, their influences may not be mutually exclusive. An initial influence by the cells of the neural groove, followed by one from the crest cells, may reasonably be postulated. Certainly the long-lasting interplay between spinal motoneurons and their skeletal muscle targets (Sisto Daneo and Filogamo, 1976) suggests that one cannot rule out the possibility that certain unknown myocardial-inducing factors are released by two different neural elements at different stages in development. It should also be pointed out that the "cholinergicity" of neuroblasts, of crest cells, and of their respective targets is evident at such an early stage in development that it cannot be regarded as being critically linked only to synaptogenesis (Filgamo and Marchisio, 1971). Rather, one might postulate that the acetylcholine system plays some *cooperative* role in conjunction with the true inducers of the myogenic line.

A further point should be made with respect to the crest cells. Our preliminary work suggests that they may also become transformed into sinoatrial node cells. Nanot and Le Douarin (1977) have put forward the notion that the node cells arise from a cell population other than that which later gives rise to the myocardial cells. But in this context, we may note that Lemon, Peterson, and Schubert (1979) have proposed that neurectodermal cells may be transformed into muscle.

The First Stage in the Development of the Cells in the Intrinsic Eye Muscles

Earlier descriptions of the development of the intrinsic eye muscles have suggested that they do not appear in the chick embryo until as late as the eighth day of incubation (Laplat, 1912; Ramanoff, 1960), when groups of cells (either of retinal or of mesenchymal origin) are contacted by neurites from the ciliary ganglion, and form the sphincter pupillae and ciliary muscles. It has also been suggested that the determination of these two groups of muscles is under neural influence, a conclusion that is consonant with our earlier work on the somites (Aloisi and Mussini, 1973; Lucchi, Bortohami, and Callegari, 1974).

However, the eight-day stage is clearly too late. Tello (1923) maintains that both the oculomotor nerve and certain neural crest cells reach the posterior surface of the optic vescicle as early as the 70th hour of incubation. At this time the optic vesicle is only surrounded by a thin layer of mesenchyme that contains capillaries. (Their thickening around the optic cup to form the choroid and sclera will not occur until later.) Tello has also described groups of myoblasts and a few nerve cells as already being present at 70 hours, and is of the opinion that these form the primordium of the ciliary body and the ciliary ganglion, respectively. Van Campenhout (1936) followed the oculomotor nerve from the mesencephalon, to a mass of poorly differentiated myoblastic tissue, and to the ciliary ganglion in embryos as early as 60 to 90 hours of incubation; at this time thin nerve bundles were said to pass from the ganglion to the dorsal and ventral aspects of the optic vesicle. By 96 hours, long nerve fibers (which are evidently the ciliary nerves) run from the ganglion, around the eye, to reach the margin of the iris. Noden (1978) using orthotopic transplantations of ^{3}H-thymidine-labeled neural crest tissue, has obtained a radioauto-

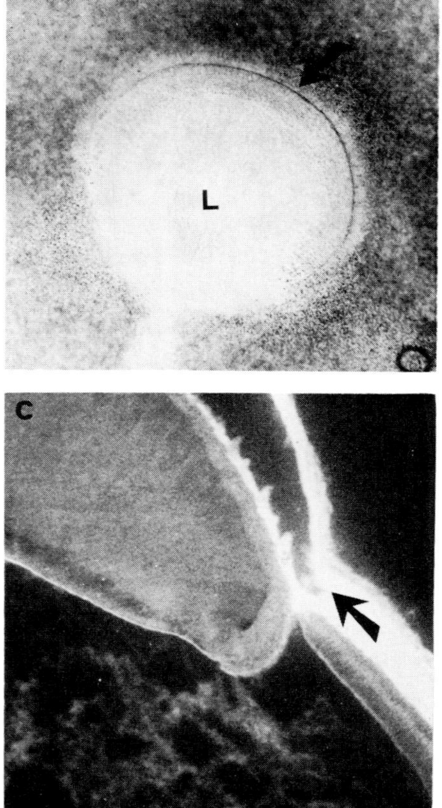

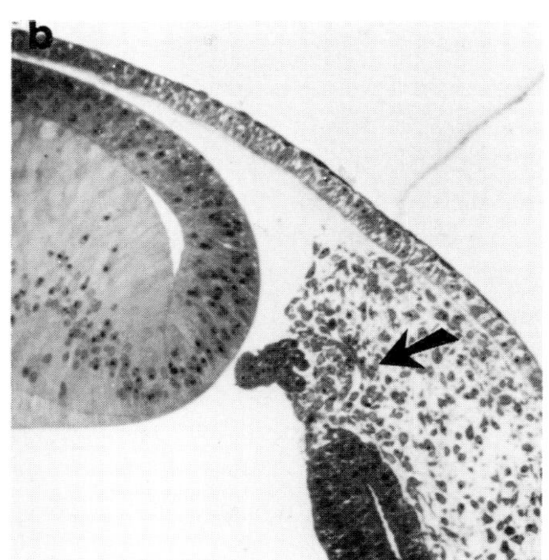

Figure 7.7 Chick embryo, stage 23. (a) The arciform distribution of the Koelle reaction for AChE is indicated by the arrow (L = lens). 70× (b) The reaction for AChE corresponds to the site within the presumptive iris indicated by the arrow. (c) A fibronectin bundle (indicated by the arrow) bridges the pupillary rim between the irideal region and the circumferential border of the lens.

graphic picture of the distribution of the neural crest cells. He reported that the oculomotor and ophthalmic nerves had reached the optic vesicle by the third day of incubation, and that at this stage they are surrounded by large numbers of neural crest cells. According to Noden the crest cells are also grouped medially and caudally to the vesicle to form the primordium of the ciliary ganglion. By Stage 11 and Stage 12 of the Hamburger and Hamilton series, they had spread around the optic stalk from the rostral mesencephalon and diencephalon and were often associated with vascular endothelia. Noden has also stressed the ability of the crest cells to disperse into the corneal endothelium and the presumptive iris.

Stimulated by these observations, and by our own findings on the early determination of skeletal and cardiac muscle, we have re-examined the early development of the optic vesicle. Our preliminary findings indicate that, as early as the fourth day of incubation, myoblasts are located both on the future anulus irideus at the edge of the pupillary opening and in the mesenchyme surrounding the anterior part of the optic vesicle. The presumptive myoblasts can be identified in the electron microscope, along an arc parallel to the pupillary margin and to the edge of the lens (Fig. 7-7); they show an intensely positive Koelle reaction for AChE. The lens itself is already connected to the presumptive iris by means of amorphous material; immunofluorescence microscopy has revealed the presence of a thick bundle of fibronectin bridging the pupillary rim and the circumferential borders of both the corneal and lens rudiments (Kurkinen et al., 1979; Di Renzo et al., 1980). We have not yet observed nerve fibers further rostral than the caudal pole of the eye, and we have nothing to add to Noden's findings on the presence of neural crest cells on the outer surface of the optic cup.

We have observed, however, many cells that penetrate the choroid fissure and invade the interior of the optic cup by moving from the caudal pole to the pupillary margin, to the iris epithelium, and to the edge of the lens (Fig. 7-8). El-Hifnawi (1977) has pointed out that mesenchymal cells reach the pupillary margin prior to the onset of melanogenesis. But, in my view the cells in question include neural crest derivatives and presumptive myoblasts. The Koelle reaction for AChE, which is clearly positive within the ciliary ganglion from the third day of incubation, is also positive in these cells (Fig. 7-9).

In summary, our data, taken together with those of other workers who have recognized the early presence of myogenic cells in contact with the optic vesicle as early as the third day of incubation, indicate that the intrinsic eye muscles are present earlier than has hitherto been supposed. They also suggest that crest cells may serve as inducers of myogenesis in the intrinsic eye muscles.

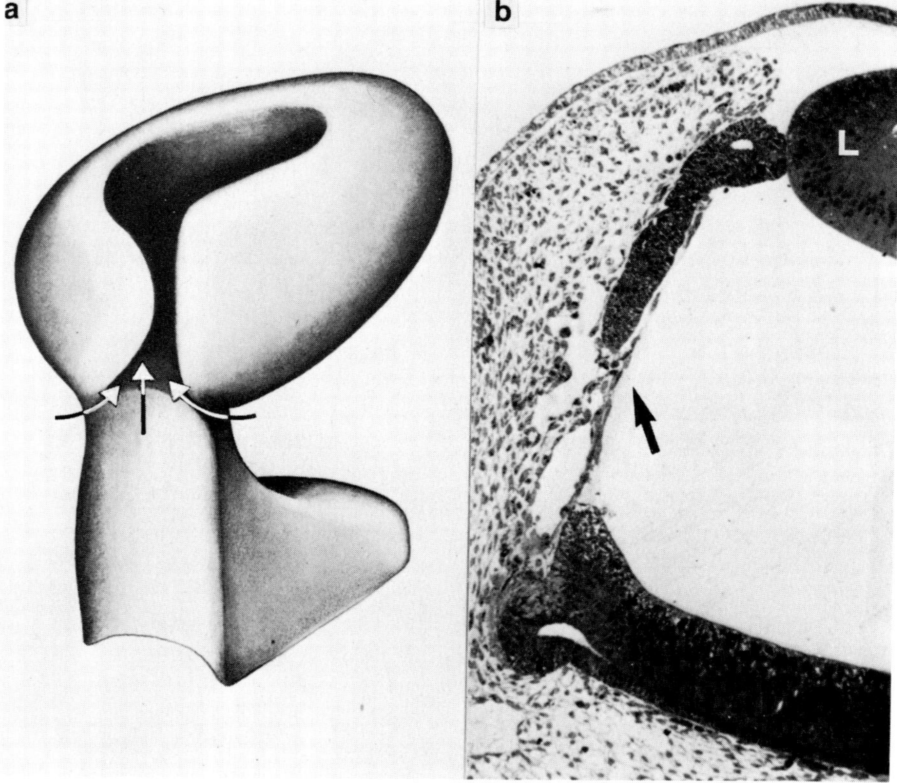

Figure 7.8 The optic cup in a Stage 23 embryo. (a) The arrows point to the choroid fissure. (b) Small cell clusters migrate through the choroid fissure (arrow) into the primitive eye cup and can be followed towards the pupillary margin.

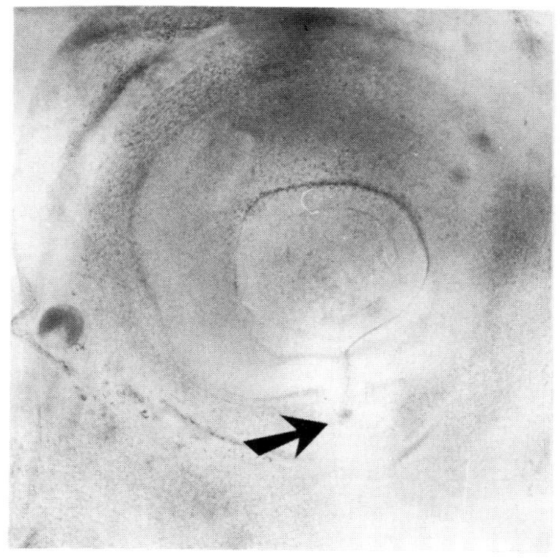

Figure 7.9 An eye cup for a Stage 23 embryo treated with the Koelle method for AChE and photographed from its anterior aspect. Note that the reaction product surrounds the lens and extends toward the primitive rudiment of the ciliary ganglion (arrow), which is also positive.

CONCLUSIONS

The nervous system has long been regarded as unnecessary for the *determination* of skeletal muscle, as opposed to the *maintenance* of its differentiated state. However, more recent morphological data from chick embryos suggest that the processes of motoneurons and certain neural crest cells may exert a determining influence on myogenesis (Couteaux, 1978). The contractile elements do not exist as "preformed targets" awaiting the arrival of nerve fibers. The targets are, in fact, created *de novo* at early embryonic stages. They do not attract nerve fibers, but, rather, are determined by their association with these fibers, or neural crest cells. The "creation" of the targets is followed by their "capture" by the neurons that innervate them, at first loosely and then definitively. Thus is initiated the long process of interplay that constitutes reciprocal neuromuscular regulation (Filgamo, 1980b). The primary determination of myogenic differentiation, which I regard as neural, in every case, may be followed by secondary, nonneural, processes, (e.g., other mesodermal cells may be recruited by the cells that are already determined). And, of course, this too would involve the passage of "inductive information."

REFERENCES

Abel, W. 1912. Further observations on the development of the sympathetic nervous system in the chick. *J. Anat.* 47:37–72

Aloisi, M., Mussini, I. 1973. Neuromyoblastic junctions in the differentiation of muscle of epithelial origin. 7th Symposium on Physiomorphology of Neuromuscular Junction, Krakow, Poland.

Chevallier, A., Kieny, M., Mauger, A. 1978. Limb somite relationship effect of removal of somitic mesoderm on the wing musculature. *J. Embryol. Exp. Morphol.* 43:263–78

Christ, B. H., Jacob, J., Jacob, M. 1977. Experimental analysis of the origin of the wing musculature in avian embryos. *Anat. Embryol.* 150:171–86

Couteaux, R. 1978. *Recherches morphologiques et cytochimiques sur l'organisation des tissus excitables.* Paris: Robin et Marenge, 225 pp.

Di Renzo, F., Tarone, G., Corvetti, G., Sisto Daneo, L. 1980. The development of fibronectin in the eye. Submitted for publication.

El-Hifhawi, E. 1977. Interaction between mesenchimal cells and the posterior iris epithelium in chicken embryos. *Anat. Embryol.* 151:109–18

Ellison, M. L., Ambrose, E. J., Easty, G. C. 1969. Myogenesis in chick embryo somites in vitro. *J. Embryol. Exp. Morphol.* 21:341–46

Filogamo, G. 1964. Activité AChE au niveau des junctions neuro-musculaires. *C. R. Assoc. Anat.* 121:115–22

Filogamo, G., Gabella, G. 1967. The development of neuro-muscular correlations, in vertebrates. *Arch. Biol.* 78:9–60

Filogamo, G. 1976. Neurogenic control

versus autonomous determination of muscle cell development. (Gif Lectures in Neurobiology) Naturalia et Biologia, publ. B.P.3, 78350 Jouj en Josas, France

Filogamo, G. 1980a. Neuromuscular interactions during development. In *Multidisciplinary Approach to Brain Development*, ed. Di Benedetta et al. New York: Elsevier/North Holland

Filogamo, G. 1980b. Maturational changes in the development of motor innervation in chick embryo. *Acta Acad. Pontif. Sci.* in press.

Filogamo, G., Marchisio, P. C. 1971. Acetylcholine system and neural development. *Neurosci. Res.* 4:29–64

Filogamo, G., Peirone, S., Sisto Daneo, L. 1978. How early do myoblast determination and fast and slow muscle fiber differentiation occur? In Maturation of neurotransmitters, ed. A. Vernadakis, E. Giacobini, G. Filogamo, Basel: Karger

Filogamo, G., Sisto Daneo, L. 1977. Nervous projection and myotome maturation. *J. Submicrosc. Cytol.* 9:307–10

Giacobini, G. 1972. Embryonic and postnatal developmental of choline acetyltransferase activity in muscles and sciatic nerve of the chick. *J. Neurochem.* 19:1401–3

Hamburger, V. The developmental history of the motor neuron. *Neurosci. Res. Program Bull. Supp.* 15:1–37

Hamburger, V., Hamilton, L. 1951. A series of normal stages in the development of the chick embryo. *J. Morphol.* 88:49–92

Kuo, Z. Y. 1939. Development of acetylcholine in the chick embryo. *J. Neurophysiol.* 2:488–93

Kurkinen, M., Alitalo, K., Vaheri, A., Stenman, S., Saxen, L. 1979. Fibronectin in the development of embryonic chick eye. *Dev. Biol.* 69:589–600

Laplat, S. 1912. Recherches sur le développment et al structure de la membrane vasculaire de l'oeil des oiseaux. *Arch. Biol.* 27:403–523

Lemon, V. A., Peterson, S., Schubert, D. 1979. Neuroectoderm markers retained in phenotypical skeletal muscle cells arising from a glial cell line. *Nature* 28:586–88

Lucchi, L. M., Bortolami, B., Callegari, E. 1974. Fine structure of intrinsic eye muscles of Birds: Development and postnatal changes. *J. Submicrosc. Cytol.* 6:205–18

Manasek, F. J. 1968. Embryonic development of the heart. A light and electron microscopic study of myocardial development in the early chick embryo. *J. Morphol.* 125:329–66

Nanot, J., Le Douarin, G. 1977. Ultrastructure du noeud sino-atrial de l'embryon de souris aux jeunes stades du développement. *J. Embryol. Exp. Morphol.* 37:13–147.

Noden, D. M. 1975. An analysis of the migratory behavior of avian cephalic neural crest cells. *Dev. Biol.* 42:106–30

Noden, D. M. 1978. The control of avian cephalic neural crest cytodifferentiation. *Dev. Biol.* 67:296–312

Olivo, O. M. 1925. Sull'inizio della funzione contrattile del cuore e dei miotomi dell'embrione di pollo in rapporto alla loro differenziazione morfologica e strutturale. *Arch. Exp. Zellforsch. Besonders Gewebezuech.* 1:427–500

Panattoni, G. L., Sisto Daneo, L. 1979. Three dimensional models of chick embryo somites obtained by mathematical methods applied to ultrastructural data. *Wilhelm Roux's Arch. Dev. Biol.* 186:273–95

Peirone, S., Filogamo, G. 1980. Non-neural influences in determination of chick skeletal muscle.

Arch. Anat. Microsc. Morphol. Exp.

Peirone, S., Sisto Daneo, L., Filogamo, G. 1977. Myogenic imprinting of early somites by nerve fibres "in vitro." *J. Submicrosc. Cytol.* 9:311–14

Ramón y Cajal, S. 1929. Etudes sur la neurogenèse de quelques Vertébrés. Madrid, pp. 19–69

Robecchi, M. G., Filogamo, G. 1980. AChE in the myoblasts immediately after determination. Submitted for publication.

Romanoff, A. 1960. *The Avian Embryo.* New York: Macmillan

Rudnick, D. 1937. Differentiation in culture of pieces of the early chick blastoderm. *Anat. Rec.* 70:351–68

Sandor, S., Amels, D. 1970. Researchers on the development of axial organs. *Rev. Roum. Embryol. Cytol. Ser.* 7:49–57

Strudel, G. 1955. L'action morphogène du tube nerveux et de la corde sur la différenciation des vertèbres et des muscles vertèbraux chez l'embryon de poulet. *Arch. Anat. Microsc. Morphol. Exp.* 44:209–35

Szepsenwol, J., Caretti, J. A. 1942. Acetilcolina en embryones de pollo. *Rev. Arg. Biol.* 18:532–38

Sisto Daneo, L., Filogamo, G. 1973. Ultrastructure of early neuromuscular contacts in the chick embryo. *J. Submicrosc. Cytol.* 9:219–25

Sisto Daneo, L., Filogamo, G. 1976. Neurotisation in muscles anlages of chick embryo: Myoneural contacts in a period preceding synapse formation. *J. Submicrosc. Cytol.* 8:303–18

Sisto Daneo, L., Filogamo, G. 1980. Morphological analysis of chick embryo heart development. In *International Symposium on Cholinergic Mechanism,* pp. Florence: 1980

Tello, J. F. 1923. Les différenciations neuronales dans l'embryon de poulet, pendant les premiers jours de l'incubation. *Travaux* 21:1–95

Viragh, S., Challice, C. E. 1973. Origin and differentiation of cardiac muscle cells in the mouse. *J. Ultrastruct. Res.* 42:1–24

Van Campenhout, E. 1936. Le développement du système nerveux cranien chez le poulet. *Arch. Biol.* 68:611–65

Zacks, J. T. 1954. Esterases in the early chick embryo. *Anat. Rec.* 118:455–87

8 / *Of Limbs and Eyes and Neuronal Connectivity*

MARGARET HOLLYDAY
PAUL GROBSTEIN

AN UNDERSTANDING OF THE DEVELOPMENTAL MECHANISMS that operate to produce functionally appropriate patterns of neuronal connections is one of the principal objectives of developmental neurobiology. Work directed towards this objective has been done in a variety of systems and has yielded a host of candidate mechanisms. In this essay, we will focus on the mechanisms that experimental evidence suggests participate in patterning connections in two of these systems, vertebrate limb innervation and nonmammalian retinotectal projections. Studies on limb innervation played an important early role in the elaboration of the concept of "neuronal specificity" (Sperry, 1950a; Weiss, 1950); a wealth of new information on the genesis of limb innervation patterns has appeared in recent years (see reviews by Landmesser, 1980; Hollyday, 1980a; Mark, 1980). A series of classical studies on retinotectal projections led to the enunciation of a general hypothesis to account for the formation of specific patterns of neuronal connections (Sperry, 1963). Reviewing these, Sperry wrote, "it seemed a necessary conclusion that the cells and fibers of the brain and spinal cord must carry some kind of individual identification tags, presumably cytochemical in nature . . . and further that the growing fibers are extremely particular when it comes to establishing synaptic connections." Testing and refinement of the hypothesis that neurons recognize their targets in the case of retinotectal projections has continued to the present (for reviews, see Jacobson, 1978, pp. 345–413; Gaze, 1978; Horder and Martin, 1978; Fraser and Hunt, 1980).

Some years ago, Viktor Hamburger, after discussing evidence from retinotectal projections on the existence of target recognition, posed the question, "To what extent are the findings on the visual system paradigmatic for other systems?" (Hamburger, 1975b). His feeling, at the time, was that developmental neurobiology was not yet itself sufficiently well developed to yield "a single 'law' or even a broad generalization" about the genesis of specific connectivity. Our objective here is most assuredly not to argue that there is a single law which accounts for the formation of functionally appropriate patterns of connections. Our objective is, however, to suggest that some broad generalizations about the genesis of specific connectivity are beginning to emerge, at least in the sense that experimental evidence on retinotectal connections and limb innervation is pointing in similar directions. In particular, we will argue not only that target recognition does seem to be present in both cases but also that in both cases target recognition should be understood as only one among several mechanisms influencing patterns of connections. Others include substrate guidance, fiber sorting, competition, and cell death.

We will first discuss the developmental mechanisms that influence patterns of connections between retina and tectum. A comprehensive review of the extensive and contentious literature in this area is far beyond the scope of the present essay. Rather, we want to focus on the logic of the experimentation necessary to document the existence of target recognition and of other mechanisms influencing neuronal connectivity and on the evidence for the existence of multiple mechanisms influencing retinotectal patterns. This includes both studies of regeneration and of initial development. Four points emerge from this: (1) experimental evidence showing abnormal patterns of connections does not preclude the existence of target recognition as a significant developmental mechanism; (2) the target recognition process is not sufficiently stringent to prevent abnormal patterns of connections—outgrowing axons are intrinsically somewhat less particular than Sperry might have imagined; (3) several other mechanisms, such as fiber sorting and competition, interact with target recognition to yield the observed retinotectal patterns; (4) the variety of results in experiments on the retinotectal system can be understood in terms of the likelihood that experimental manipulations may enhance the influence of one of these mechanisms relative to the others. In the second section, we will treat experimental evidence on the development of limb innervation in somewhat greater detail, but we shall do so in the same logical framework and with particular reference to the parallels between observations on limb innervation and retinotectal connections. In this case, as with retinotectal projections, we will conclude that several different mechanisms, including target recognition, interact in determining connectional patterns.

Our principal general conclusion, that there is not a single mechanism but rather a complex of mechanisms which is involved in establishing specific connectivity, seems to us a significant generalization not only in making sense of the existing experimental evidence but also in clarifying the experimental logic and in defining the issues to which further experiments should be addressed. One important point concerning experimental design is that a particular set of results can be used to infer the existence of a given developmental mechanism but not to exclude the existence of others. A second point is that given the existence of multiple developmental mechanisms, one needs to be cautious in attributing a particular connectivity pattern, either normally or experimentally produced, to one or another of these mechanisms exclusively. We will briefly discuss some of the wider issues related to multiple mechanisms in the final section of this essay. We are uncertain whether Viktor will accept our generalization as a sign of incipient maturity in developmental neurobiology. However, we are quite certain of our indebtedness to Viktor not only for, as he puts it, "not being smart enough to solve all of the problems" but for being smart enough to pose the kind of questions we seek to answer and to pursue the phenomena from which mature generalizations must ultimately emerge.

RETINOTECTAL PROJECTIONS

Target Recognition

The hypothesis that target recognition plays an important role in the patterning of retinotectal connections was developed by Roger Sperry as a consequence of behavioral observations following optic nerve regeneration in amphibians (Sperry, 1951). Sperry inferred from these observations that the axons of the retinal ganglion cells returned to their original tectal targets following optic nerve section, an inference that was subsequently confirmed physiologically (Gaze, 1959; Maturana et al., 1959) and anatomically (Meyer, 1980). Sperry's behavioral observations allowed him to test and exclude the involvement of experience in reestablishing the original retinotectal projection pattern. Since the fibers were scrambled at the crush site and yet each ultimately returned to a distinct (and correct) location, Sperry argued that the ganglion cells must differ from one another in some intrinsic way such as by a "cytochemical label." Although there have been some successes in trying to analyze these intrinsic differences at the cellular and molecular levels (Barbera, Marchase, and Roth, 1973; Marchase, 1977; see also Gottlieb and Glaser, 1981 in this volume), the differential behavior of the regenerating axons starting from the same (or at least a randomized) point seems to us still one of the strongest lines of evidence for the existence of "labels" on ganglion cells that influence their formation of connections.

Confirmation of the existence of a target recognition mechanism requires the demonstration of intrinsic differences not only among ganglion cells but also among the tectal target loci. The latter has been more difficult to establish than the existence of labels on the ganglion cells. As noted by Gaze (1970) and by others, there is nothing in the regeneration experiments that requires that differences among tectal loci exist, if one permits the possibility that the ganglion cells—by virtue of their labels—might be capable of recognizing and sorting among themselves. Similarly, there need be no actual target differences if there are cues along the growth path so precise that a given ganglion cell axon cannot but come to the right place. These two possibilities suggest two additional mechanisms that may influence retinotectal patterning: fiber sorting and selective substrate guidance. By *fiber sorting,* we mean the ability of axons to recognize their neighbors and to organize themselves in an orderly way. By *substrate guidance* we mean specific interactions between individual axonal growth cones and cues in the environment into which they grow. We will discuss some of the evidence for the existence of these and other mechanisms below. We mention them here to make it clear why evidence, over and above the reestablishment of a normal pattern of connections in regeneration, is necessary to show that tectal targets are distinct from one another. It should be noted that the conclusion from the regeneration experiment (i.e., that *ganglion cells* differ from one another) is not called into question by the possible existence of these additional mechanisms. The reestablishment of normal order, despite the fiber scrambling associated with optic nerve section, need not necessarily depend on distinctiveness of tectal targets, but the differential behavior of ganglion cells, whether brought about by target recognition, fiber sorting, or substrate guidance, must be taken as evidence for their distinctiveness from one another.

The demonstration of differences between tectal loci requires that it be shown that ganglion cells that are free to choose between tectal loci, actually discriminate among them. The first effort to obtain such evidence was that of Attardi and Sperry (1963). Anatomical observations on regeneration from eyes with partial retinal lesions in the goldfish indicated that ganglion cell axons would bypass other available loci and return to their normal tectal loci. Subsequently, however, abnormal projections were observed in a variety of similar experimental paradigms. It is now clear that depending on the circumstances, either normal or abnormal patterns of connections can be formed. The significance of these abnormal projection patterns will be discussed in connection with the existence of multiple mechanisms in the next section. A more serious problem with the interpretation of the Attardi and Sperry experiment was that it involved *regeneration.* This left open the possibility that what the ganglion cells were recognizing were not intrinsic differences among tectal loci but remnants of the previous innervation (Schmidt, 1978). Strong evidence for recognition of remnants is present in several invertebrates (Hoy, Bittner, and Kennedy, 1967; Muller and Carbonetto, 1979).

Evidence for intrinsic differences between tectal loci that are present before innervation comes from three sources. In the chick, adhesion of dissociated retinal cells to virgin tecta suggests at least some degree of appropriate distinctiveness in different tectal regions (Barbera et al., 1973). Furthermore, Crossland and coworkers (1974) performed experiments in chick embryos, like those of Attardi and Sperry, but early in development before tectal innervation. They made partial eye cup lesions before optic nerve outgrowth and found that the remaining retina projected exclusively to the normal tectal loci, leaving significant areas uninnervated. In the frog, evidence for the independent development of intrinsic differences among tectal loci comes from the recent work of Straznicky (1978), who rotated tectal fragments in virgin tecta and then allowed them to become innervated. His finding that the projection onto the rotated fragment was correspondingly rotated suggests the existence of distinctive properties of different tectal loci. However, as shown in modeling studies by Hope, Hammond, and Gaze (1976), under some circumstances, a rotated projection might be obtained even in the absence of distinct differences between tectal loci. Because of this it would be desirable to repeat Straznicky's experiments using transplantations of virgin tectal fragments instead of rotations. This experiment has been done at later stages, in the frog (Jacobson and Levine, 1975) and goldfish (Hope et al., 1976), but since it was done in combination with optic nerve regeneration, definitive conclusions about the existence of independently developed tectal labels cannot be drawn. Although the evidence is better in some animals than in others, collectively these studies make it highly likely that some distinctive differences are present among tectal loci before innervation. This does not preclude the possibility that such differences may be altered by subsequent innervation (Schmidt and Easter, 1978).

In summary, evidence does exist for distinctiveness (or "labeling") of both ganglion cells and tectal loci. Evidence for the former comes directly from the differential behavior of ganglion cells following optic nerve scrambling; evidence for the latter comes from observations that ganglion cells can distinguish among tectal loci. Together, these two lines of evidence indicate that a target recognition capability is present in the retinotectal system. It is important to note that the experimental results indicate the *existence* of a target recognition mechanism and in no way answer the question of whether in normal (or in other experimental) circumstances the observed patterns are caused by this mechanism. Similarly, the finding of abnormal patterns of connectivity in other experimental circumstances cannot be used to rule out the existence of a target recognition mechanism. We will consider the significance of abnormal connection patterns in the following section.

Multiple Developmental Mechanisms

In considering the relation between the regeneration experiments and target recognition, we noted two other mechanisms, both based on intrinsic ganglion cell differences, which might influence the patterning of the retinotectal projection: substrate guidance and fiber sorting. Here we discuss the involvement of these two mechanisms as well as a third, competition, in retinotectal pattern formation.

Either fiber sorting or substrate guidance might affect the growth path of ganglion cell axons and, in doing so, influence the ultimate pattern of connections by constraining the possible tectal targets with which the axons come into proximity. It has been found, however, that as long as ganglion cell axons reach the tectum, neither their growth trajectory nor their neighbor relations seem to be critical for the patterning of retinotectal connections. Normal patterning is seen even when experimental perturbations produce highly abnormal growth paths (e.g., Hibbard, 1967; Gaze and Grant, 1978). Indeed, even in regeneration, the evidence suggests that the details of the growth paths of regenerating axons are abnormal (Udin, 1978; Meyer, 1980), which supports Sperry's assumption that considerable scrambling accompanies optic nerve section. The relative independence of the retinotectal projection pattern from the details of the growth path should not be taken, however, as evidence that during normal development neither substrate guidance nor fiber sorting are operative. It may simply be that processes occuring at the tectum are adequate to correct disorder produced by experiments which disturb the normal organization of axons along their growth path.

Positive evidence for the existence of developmental mechanisms in addition to target recognition comes from the large literature referred to above showing that in many cases experimentally produced mismatches between retinal and tectal size result in ganglion cells terminating at abnormal locations. (Early observations included Gaze, Jacobson, and Székely, 1963; Gaze and Sharma, 1970; Yoon, 1971, 1972. For more current lists of references see the reviews by Jacobson, 1978; Gaze, 1978; Horder and Martin, 1978; and Fraser and Hunt, 1980.) An early ambiguity about the interpretation of these findings was the possibility that surgical removal of parts of the tectum or retina resulted in a process of regulation such as to recreate the normal complement of labels across the remaining cells (Yoon, 1971; Meyer and Sperry, 1976). More recent observations make it clear, however, that ganglion cells can be caused to terminate at abnormal locations, even under circumstances for which an explanation in terms of altered retinal or tectal labels is unlikely. One indication of this is the finding that ganglion cells in a single retinal region will, under some circumstances, project to different locations in the two tectal lobes (Feldman, Keating, and Gaze, 1975; Fraser, 1979). A similar observation was made by Meyer (1978), who found that a small portion of one retina displayed an expanded projection over a denervated ipsilateral tectal lobe while the remainder of the retina mapped normally over the contralateral tectal lobe. Schmidt (1978) tested for alteration of retinal markers associated with an expanded projection from a half retina by deflecting its optic nerve into a normally innervated tectum. The finding that this resulted in a normally restricted projection suggests that altered retinal markers were not responsible for the expansion. Further evidence that abnormal projections do not require altered labels comes from additional experiments discussed below in connection with competition.

The conclusion that ganglion cells with unchanged labels will terminate at abnormal locations has two important implications. First, it indicates that the target recognition process does not absolutely determine the target of a given ganglion cell. Second, it implies that additional developmental mechanisms are present that may interact with the target recognition process in patterning the normal retinotectal projections. Evidence exists for at least two mechanisms, over and above target recognition, which can influence these patterns. One is fiber sorting, not as a determinant of growth path but as a determinant of target location. The other is competition. Together these two mechanisms provide a possible straightforward explanation of the finding that, for example, the retina will project in an orderly and continuous manner onto both normal tecta and surgically reduced tecta. Fiber sorting could account for the maintenance of the continuous projection, while competition could account for the displacement of the ganglion cells from their normal targets.

The direct evidence for fiber sorting, that is, for an ability of ganglion cell axons to interact among themselves so they will yield a continuous retinotopic representation, is suggestive but not definitive. One line of evidence is that retinotopy in the projection pattern is almost invariably maintained in the large number of experiments involving removal of regions of either retina or tectum. This indicates that retinotopy can be achieved relatively independently of specific tectal markers. A second line of evidence is the increasing information suggesting some organization of ganglion cell axons in the optic nerve (Scalia and Fite, 1974; Bunt and Horder, 1978; Sharma, 1979). This organization is not necessarily retinotopic (Scholes, 1979; Rusoff and Easter, 1980) and could simply be a consequence of the normal growth pattern of optic nerve fibers; it may, indicate, however, some ability of the fibers to organize themselves. A third line of evidence for fiber sorting comes from the observation that during development in the chick, optic nerve fibers organize themselves into a retinotopic projection on the surface of the tectum before descending to contact terminal sites (Crossland, Cowan, and Rogers, 1975). This finding indicates that retinotopy can be attained without reference to the actual sites of synapse formation but does not preclude the possibility that fibers are ordered by other tectal cues. Recently, Meyer (1979) has reported a situation in which fibers project to the tectum retinotopically but with reversed polarity. This reversed polarity is most easily understood as resulting from fiber sorting. Whether this is in fact the explanation, the presence of a retinotopic projection with reversed polarity strongly suggests that retinotopy can be achieved independently of tectal cues. Ideally, what is required to establish fiber sorting is the demonstration that in the absence of any external organizing cue, growing optic nerve fibers will establish a retinotopic order among themselves and that they are capable of reestablishing such an order after being scrambled. Such a definitive experiment, perhaps involving long-term maintenance of the developing eye in culture, and hence away from any possible external organizing cues, has not been reported. The evidence described, however, strongly suggests that optic nerve fibers do in fact have the ability to interact with one another in such a way as to produce a continuous retinotopic organization.

The second mechanism suggested by the experiments involving size disparities between the retina and tectum, is competition; that is, the ability of ganglion cell axons to displace one another from tectal sites. A mechanism of this kind is implied by the finding that the axons normally projecting to a removed caudal tectum will terminate in the caudal part of the remaining tectum and cause a reorganization of the terminations of undamaged axons so that the entire retinotectal projection is evenly compressed onto the tectal fragment (Yoon, 1971). It is also suggested by considerations of the normal developmental process (Gaze, Keating, and Chung, 1974). Cell proliferation in the retina and tectum of both goldfish (Johns and Easter, 1978; Meyer, 1978) and *Xenopus* (Straznicky and Gaze, 1971, 1972) continues well beyond the initial appearance of a retinotopic projection. The patterns of cell addition in the two structures would seem to require a caudal displacement of older ganglion cell axons by younger ones. Evidence for such a process has been described for *Xenopus* (Scott and Lazer, 1976; Longley, 1978; Gaze et al., 1979; but see also Jacobson, 1976, 1977).

The strongest direct evidence for competition comes from experiments in which it is shown that an established retinotectal pattern is altered by the addition of a new afferent population. Law and Constantine-Paton (1980) have, for example, recently reported that in the frog superinnervation of a normally innervated tectal lobe by an additional optic nerve results in segregated bands of afferent terminals from the two eyes. Cook (1979) has reported that the compression of the entire retina onto a rostral tectal fragment is retarded by delaying the ingrowth of fibers normally terminating in caudal tectum. The existence of competitive interactions is also implied by the finding that the projection from a smaller than normal retinal complement will terminate over a greater or lesser extent of tectum depending on the presence of other fibers (Meyer, 1978; Schmidt, 1978; see also Fraser and Hunt, 1980, pp. 334–35).

In sum, there is evidence for at least two mechanisms, in addition to target recognition, which can influence the retinotectal projection pattern: fiber sorting and competition. As in the case of target recognition, the evidence for each of these two mechanisms is better in some organisms than in others. Our feeling, though, is that with properly designed experiments, of the kind described here, evidence for all three mechanisms can probably be found in any organism. Strong evidence for target selectivity exists for both chick and frog; the necessary perturbations of virgin tecta have not yet been possible in goldfish. Strong evidence for fiber sorting and competition comes from experiments on goldfish. Adjustments for size disparity, the basic phenomenon suggesting the existence of fiber sorting and competition, has, however, been documented in the frog (Udin, 1977). Direct evidence for competition in the frog comes from observations of afferent segregation in dually innervated tecta. Whether adjustments for size disparities will occur in the chick is unknown.

Conclusions

This brief survey obviously barely scratches the surface of the enormous literature of the development of retinotectal projections. It suffices, however, to illustrate what we believe to be the major conclusions with respect to the narrow question of how the connection pattern is established. The most important conclusion is that there appear to be at least three different mechanisms that can influence where a given ganglion cell terminates: target recognition by the ganglion cell; competition among ganglion cells for targets; and sorting among ganglion cells themselves.

Of the three mechanisms we have mentioned, the first corresponds to the differential cytochemical labeling and resulting chemoaffinity hypothesized by Roger Sperry; evidence for the existence of the latter two has emerged from experimental efforts directed at confirming or refuting Sperry's hypothesis. The upshot of these efforts, as we view them, has been to do neither. Rather, they indicate that there is some ability of retinal cells to recognize tectal cells, as Sperry suggested, but that this recognition process is not the sole determinant of the target of a retinal cell. Additional mechanisms may also influence the target choice under particular experimental circumstances.

It is worth making more explicit some of the implications of the existence of several mechanisms for experimental strategy. An important point is that no single experimental observation can rule out the existence of any of the mechanisms discussed; the patterns resulting from a given experimental perturbation may simply reflect the relative strength of the various mechanisms under the particular experimental circumstances. For example, from size disparity experiments in the goldfish, it now seems clear that at short times after surgery the target recognition mechanism dominates the observed pattern; at later times, the competition mechanism, perhaps coupled with some fiber sorting dominates (Yoon, 1976; Schmidt, Cicerone, and Easter, 1978). A second point is that it is very difficult to deduce the characteristics of any one mechanism from experiments in which there is an unknown influence from several other mechanisms. It would be nice to know, for example, whether the target recognition mechanism is such as to make one target preferable for a given ganglion cell and all others equal or, instead, such as to produce a graded series of target preferences for a given ganglion cell. In the absence of evidence as to whether abnormal target choices are wholly determined by one mechanism (e.g., fiber sorting) or instead represent a compromise influenced by several mechanisms, we see no way to answer this question. Finally, it is worth again making the point that, just as we cannot be sure of the relative strengths of these various mechanisms in any given experimental situation, we cannot, from the evidence described, say anything certain about the involvement of each in normal development. We will return to this issue in the final section.

LIMB INNERVATION

As described in the previous section, the hypothesis that retinal and tectal cells are so individuated from one another that some degree of target recognition plays a role in the establishment of connections between them has experimental support. At the same time, it is clear that additional mechanisms exist which can influence connection patterns. In the following pages, we will consider whether motoneurons, like ganglion cells, display some ability to recognize their targets and whether additional mechanisms are involved in patterning their connections. First, however, because the relationship between the target and the innervating cell population is not so straightforward as it is in the retinotectal system, we will briefly discuss the best way to characterize the normal organization of the limb innervation pattern.

Normal Organization of Limb Innervation Pattern

In the case of retinotectal connections there is an obvious relationship between the anatomical projection pattern and the functional requirements of the map. The input to a ganglion cell is determined by the cell's position in the retina. The ganglion cells project in a continuous and topographic way to the tectum. Both the function and the appropriate tectal target is clearly related to ganglion cell soma location.

For limb innervation, the relationship between the anatomical pattern of connections and the functional requirements of the system is less clear. What is obviously important is that a muscle receive connections from motoneurons which are active in such a way as to produce behaviorally appropriate contractions in that muscle. The activity of motoneurons depends upon the pattern of inputs they receive. Both classical and more detailed modern evidence indicates that individual muscles are consistently innervated by motoneurons found in characteristic locations in the spinal cord (Romanes, 1951, 1964; Sharrard, 1955; Cruce, 1974; Lamb, 1976; Landmesser, 1978a; Hollyday, 1980b). There is some evidence for a correlation between motoneuron position and function (Romanes, 1964; Luiten, 1976; Jacobson, 1979). We presume then that at least certain types of presynaptic inputs are correlated with motoneuron cell body position, although we recognize that the distribution of the motoneuron dendritic tree may, in fact, be more directly related to the pattern of inputs.

Despite these reservations, the position of the motoneuron cell body seems to us the best available identifying criterion for motoneurons. Exclusive use of this rigorous criterion will not be possible, however, in this essay since it would preclude discussion of much of the older literature in which motoneurons were identified in terms of the peripheral nerve in which their axons were running. In considering these older experiments it is important to bear in mind the untested assumption that the set of motoneurons whose axons are distributed within a particular peripheral nerve is not changed by the experimental manipulation. It is also important to realize that a given nerve may contain axons going to several different muscles and that a given muscle may receive branches from several different nerves. These possible ambiguities illustrate why it is preferable to design experiments in which motoneurons are identified by cell body location rather than peripheral axonal trajectory.

An additional important characteristic of normal limb innervation is that the projection to a given muscle comes not from a single motoneuron but from a *group* of motoneurons, the "motor pool." It is the location of this group that we use to name motoneurons. It should be pointed out that the degree of motor-pool discreteness within the lateral motor column varies for different muscles and species (for a review, see Romanes, 1964; Székely and Czéh, 1967; Burke et al, 1977; Hollyday, 1980a). Some muscles are supplied by motor pools located in discrete, unique positions whereas for other groups of muscles there may be a substantial degree of overlap of the locations of the motoneurons supplying them. We also recognize that there are both morphological and physiological differences among the motoneurons of a pool and among the fibers of the muscles they innervate. We will not consider in this essay the mechanisms by which these microheterogeneities might be brought into functional correspondence, but will focus instead on the problems of matching motor pools and muscles.

In general, it is possible to characterize the normal relation between motor pools and the limb muscles rather simply. The motor pools of the lateral motor column are organized in a continuous, somatotopic way that is related to the position of the muscle precursor tissue in the embryonic limb. All vertebrate limb muscles differentiate from two primary sheets of mesoderm: the dorsal and ventral premuscle masses (Romer, 1927, 1964). A central condensation of cartilage forms the precursors of the skeletal elements. Limb muscles derived from the ventral muscle mass are innervated by medially positioned motor pools; muscles derived from the dorsal mass are supplied by lateral pools (Landmesser, 1978a). Limb muscles derived from adjacent portions of one or the other muscle mass are innervated by adjacent motor pools (Hollyday, 1980b).

Motoneurons innervating particular muscles are characterized not only by their location in the spinal cord but also by their birthdates (Prestige, 1973; Hollyday and Hamburger, 1977). The topographic relation between motoneuron position and target location in the embryonic muscle mass, coupled with the regular order of motoneuron birthdates, has suggested the hypothesis that normal innervation patterns may be a consequence of the spatiotemporal pattern of motoneuron axon outgrowth into the developing limb bud (Horder, 1978). From a detailed analysis of the normal innervation pattern itself (Hollyday, 1980b), this seems unlikely since there are some discontinuities between motor pools innervating neighboring muscles derived from different muscle masses. More importantly, we present evidence in the following section that indicates that motoneurons display a significant ability to discriminate targets. The orderly spatiotemporal pattern of motoneuron production may be an important element in the acquisition of the distinctive properties necessary for target recognition.

Target Selectivity

As described in connection with the retinotectal pathways, the initial evidence for the involvement of target selectivity in patterning neuronal connections came from experiments which indicate that ganglion cells can reestablish their normal connections during regeneration. This indicates that the ganglion cells themselves, though not necessarily their targets, are intrinsically individuated from one another.

With respect to limb innervation, regeneration experiments have yielded variable results. In the ichthyopsidans (fish and amphibians) good functional recovery following peripheral nerve cuts and crushes has been reported (Weiss, 1936; Sperry, 1950b; Sperry and Deupree, 1956; Arora and Sperry, 1957; Mark, 1965). In mammals, functional recovery is poor in those cases in which significant disturbance of the normal innervation pattern of the limb is produced (for a review, see Sperry, 1945). For the moment, we defer the issue of why there are phylogenetic differences in the extent of functional restitution, but let us note here that those animals in which good functional recovery of limb movements is observed are, in general, the same as those in which successful optic nerve regeneration occurs. Functional recovery of limb movements following regeneration was at one time interpreted as indicating that motoneurons make random contacts with muscles and then had their central connections adjusted to match the muscle in which they terminated (Sperry, 1950b; Arora and Sperry, 1957). At the present time evidence for such a "peripheral respecification" is nonexistent (Eccles, Eccles, and Magni, 1960; Eccles et al., 1962; Sperry and Arora, 1965; Mark, 1969; Grimm, 1971; Mendell and J. Scott, 1975; S. Scott, 1977). It seems to us that this classical evidence in nonmammalian species is now most easily understood as indicating that the axons of motoneurons do in fact largely return to their original targets and, in parallel with the interpretation of regeneration in the retinotectal system, that motoneurons have some intrinsic differences that affect their target choice. It would be worth redoing such regeneration experiments using modern tracing techniques to show that the location of motoneurons innervating particular muscles is, in fact, the same in regenerated and normal animals.

More dramatic evidence for the distinctive behavior of motoneurons with respect to their targets comes from experiments in which peripheral nerves are only cut but so intentionally deflected so as to favor entry of their axons into foreign muscles. While some degree of abnormal innervation is observed under such circumstances, there is a dramatic ability of motoneurons to return to their normal muscle (Sperry and Arora, 1965; Mark, 1965; Grimm, 1971; Cass and Mark, 1975). This not only indicates that motoneurons behave differently from one another but also that the muscles must have distinctive labels, since the axons of the motoneurons, despite the ready availability of one muscle, display a tendency to return to a different one. The ability of motoneuron axons during development to find the appropriate target, even when growing from an abnormal spinal cord location, has recently been reported (Lance-Jones and Landmesser, 1980b).

As argued above for the retinotectal projection, studies which indicate the reestablishment of normal connections following regeneration do not preclude the possibility that fibers are recognizing remnants of the previous innervation rather than intrinsic labels on the target structures. Manipulations prior to initial innervation are necessary to determine whether there are intrinsic target labels. Such experiments have now been done for limb innervation in the chick.

Deletion of several spinal cord segments prior to axon outgrowth results in a predictable absence of innervation to various muscles which indicates that the surviving motoneurons have distinguished between their normal targets and the available uninnervated muscle (Lamb, 1979; Lance-Jones and Landmesser, 1980a). Embryonic surgery resulting in specific limb segment deletions yields normal innervation of the surviving limb segments; the motoneurons in positions that normally supply the muscles of the deleted limb segments apparently die (Whitelaw and Hollyday, 1980). If neuromuscular connections were formed solely on the basis of sequentially arriving axons growing passively to the nearest target site, then one would predict that the same early-born population of motoneurons would innervate whatever leg segment was closest to the body, either calf or thigh muscles, depending on the operation. This result has not been obtained. Rather, different motor pools supply the muscles of the manipulated legs, depending on which muscles are present. Both the normal innervation of the remaining muscles and the distinctive pattern of cell death in cases of limb segment deletions indicate the existence of some specific recognition of their normal targets by motoneurons.

The evidence for target selectivity described so far relates to the ability of motoneurons to form their normal connections despite perturbations in the spatiotemporal outgrowth patterns or in the size of the outgrowing population or their target structures or both. Additional evidence for target selectivity stems from the observations of Arora and Sperry (1965) on competitive interactions between motoneurons. In these experiments, abnormal innervation of a muscle was established, and subsequently the axons that normally innervate the muscle were given access to it. Behavioral observations suggested that the normal innervation had displaced or suppressed the abnormal innervation (see also Marotte and Mark, 1970; and Mark and Marotte, 1972). Further investigations of this phenomenon have given somewhat conflicting results. In some systems, both native and foreign synapses can be maintained apparently indefinitely (Scott, 1975, 1977; Frank and Jansen, 1976; Haimann, Mallart and Zilber-Gachelin, 1976). In others, the suppression phenomenon has been confirmed either physiologically (Yip and Dennis, 1976; Bennett and Raftos, 1977; Dennis and Yip, 1978; Grinnell, Letinsky, and Rheuben, 1979) or anatomically (Fangboner and Vanable, 1974; Dennis and Yip, 1978). Whether the suppression involves actual physical displacement or simply physiological inactiviation of the abnormal innervation has not been completely resolved (Mark, 1980). However, the evidence clearly indicates that there is some distinctive relation between motoneurons and their normal target.

Multiple Developmental Mechanisms

In the previous section we summarized the evidence which indicates that, at least in the ichthyopsidans, motoneurons display some ability to distinguish among target muscles. It is well known, however, that, even in fish and amphibians, motoneurons will under many circumstances form synapses on abnormal muscles. This indicates that the target recognition mechanism that operates in limb innervation is not sufficiently strict to preclude the formation of abnormal connections, and it suggests furthermore the existence of additional developmental mechanisms.

In the case of limb innervation, some further information about the nature of the target recognition process comes from studies on the innervation of supernumerary limbs in chick embryos. When, in these experiments, muscles are innervated by abnormal motoneurons, the innervation is not random but displays characteristic patterns (Hollyday, Hamburger, and Farris, 1977; Hollyday, 1978, 1981; Morris, 1978). Abnormally innervated muscles invariably are supplied by motor pools whose normal target is derived from the same embryonic premuscle mass as the muscle in question. Furthermore, abnormally innervated muscles derived from superficial or deep subdivisons of the premuscle mass always receive innervation from motor pools normally projecting to muscles that are derived from the same subdivisions. These results suggest that the distinguishing characteristics of muscles should perhaps be thought of as a series of features related to the cleavage divisions of the embryonic muscle mass and that the distinguishing characteristics of motoneurons consist of a similar series of features. When motoneurons are confronted with abnormal muscles, they match these sets of features to the extent possible. The choices made by a given motoneuron in a variety of experimental circumstances corresponds to the hierarchy of selective chemoaffinities previously suggested (Hollyday et al., 1977).

In the case of the retinotectal projection, we argued that in addition to target recognition other mechanisms also influence the pattern of neuronal connectivity. A similar argument can be made for limb innervation. As described, in fish and amphibians, motoneurons that innervate abnormal muscles can be inactivated or displaced by the normally innervating motoneurons. This suggests the existence of a competitive interaction between motoneurons of different motor pools. (The possible existence of a competitive interaction between motoneurons of a single motor pool will be considered below). Competitive interactions between motoneurons of different motor pools are of particular interest in the case of limb innervation since a difference in the potency of such interactions may account for the previously noted dichotomy between mammals and the icthyopsida with respect to the extent of functional recovery during regeneration. Given that target recognition is inadequate to prevent the formation of abnormal connections, the essential element in functional recovery, as previously proposed by Mark (1969), may be the ease with which abnormal connections are displaced by normal ones. Mark hypothesized that this replacement was facilitated by the multiterminal innervation pattern of icthyopsidan muscle fibers. An alternative hypothesis is motivated by the suggestion that there is normally a significant turnover of neuromuscular junctions (Barker and Ip, 1966). Such turnover may be more rapid in the Icthyopsida than in mammals since growth in these species (unlike that in mammals) is continuous and may involve the ongoing addition of muscle fibers. If synapses can be made and broken rapidly, this may facilitate adjustments of the motor supply to this growth. It might also facilitate the displacement of abnormal by normal innervation since it would increase the opportunity for competition between correct and incorrect motoneurons.

This line of argument suggests a way to test the existence of some degree of target selectivity in mammalian limb innervation. Bixby and Van Essen (1979) have reported that foreign motoneurons will acquire synaptic targets in normally innervated muscle in rats. This is consistent with some normal degree of turnover of neuromuscular junctions. An interesting question is whether the rate of target acquisition, or the number of synapses made, is higher for a normal nerve taking over targets in an abnormally innervated muscle than it is for an abnormal nerve acquiring targets in a normally innervated one. A positive answer would indicate that at least during competitive interactions, mammalian motoneurons do display some ability to recognize their normal targets.

In the case of retinotectal connections, we considered the importance of the growth path in determining the projection pattern and concluded that it was not a serious constraint on the possible target choices of retinal axons. In limb innervation, where axons from different spinal segments sort out in the limb plexuses and follow different paths to nonadjacent targets, it is likely that mechanisms for assuring that axons reach the correct general location of their targets are more important. The idea that a series of developmental mechanisms may be responsible in turn for plexus formation, for producing the typical branching patterns of the peripheral nerves, and for establishing specific terminal connections was clearly stated by Viktor Hamburger in 1929 in a paper written to honor an earlier distinguished embyrologist, Hans Spemann.

Recent evidence demonstrates that under normal conditions the outgrowth paths of motor axons are organized before axons reach their targets. Axons emerging from a particular ventral root reach the base of the limb and distribute within the limb bud in an orderly way (Lance-Jones and Landmesser, 1979; Hollyday, unpublished observations). A recognizable innervation pattern is formed in the limb bud before the target muscles have separated completely from the dorsal and ventral premuscle masses (Taylor; 1942, Roncali, 1970; Fouvet, 1973; Bennett, Davey, and Uebel, 1980). These patterns could result from the timing of axon outgrowth and passive mechanical guidance, from specific substrate guidance, or from fiber sorting. To distinguish the first from the latter two possibilities, what is necessary is to determine whether typical nerve patterns are produced when the normal timing of limb innervation is disturbed. The classic experiments of Piatt (1942), in which limbs were allowed to develop to advanced stages in the absence of innervation and subsequently allowed to become innervated, showed that normal peripheral branching patterns were formed. This finding suggests that such peripheral nerve patterns are not critically dependent on the time of innervation and hence probably result from selective substrate guidance, from fiber sorting, or from both. Piatt had no information about the identity of the axons in the peripheral nerves in his material. It would be worth repeating such delayed innervation experiments using modern tracing techniques to determine whether or not the distribution of motoneuron axons in the peripheral nerves is normal.

Further evidence for the active selection of axon outgrowth pathways comes from the experiments of Lance-Jones and Landmesser (1980b) who rotated a few spinal cord segments in young chick embryos. Motor axons emerging from an unusual level of the spinal cord do not follow the path through the limb plexus characteristic of the axons which normally emerge from that segment; instead, they traverse the limb plexus in such a way as to emerge to innervate their appropriate target muscle. This result differs from the findings of Stirling and Summerbell (1979, 1980), who rotated limb buds rather than spinal cord segments and found that under these circumstances motor axons innervated incorrect limb muscles. One possible explanation for this difference is that the motor axons had already sorted in the limb plexus at the limb base before entering the rotated tissue. This suggests that once axons choose particular peripheral pathways, they may be forced to innervate abnormal targets.

It is not yet possible to evaluate the relative importance of specific substrate cues and fiber sorting in the genesis of axon outgrowth patterns in limb innervation. It would be worth exploring explicitly the question of fiber sorting by asking whether the motor axons in spinal cord explant cultures exhibit any tendency to recognize one another and segregate. It would also be interesting to know whether axons segregate according to motor pool classifications and whether larger aggregations typical of those in various peripheral nerves would occur.

Cell Death Among Motoneurons

In this discussion of the mechanisms influencing connectivity patterns in limb innervation, we have ignored one of the most dramatic phenomena occurring during the development of limb innervation, namely, naturally occurring cell death among motoneurons. This phenomenon has been reported for anuran amphibians (Hughes, 1965; Prestige, 1967) for birds (Hamburger, 1975a; Oppenheim and Willard-Majors, 1978) and for mammals (Harris Flanagan, 1969). The analysis of the phenomenon has been greatly advanced by Viktor Hamburger's work for he was one of the first embryologists to relate cell death to interactions with target tissues. The magnitude of normal motoneuron loss varies in different species, but in the chick embryo, at least 40 percent of the motoneurons generated die rapidly during the time when the limb muscles separate from the two embryonic premuscle masses (Hamburger, 1975a). This is also when the limbs first begin to move (Hamburger and Balaban, 1963). The vast majority of motoneurons send an axon into the limb bud (Prestige and Wilson, 1973; Oppenheim and Chu-Wang, 1977). The amount of motoneuron loss is related to the size of the periphery. Partial and total limb-bud removals increase cell death in direct proportion to the amount of tissue removed (Hamburger, 1934) while enlargement of the periphery by the addition of a supernumery limb rescues some of the cells that would normally die (Hamburger, 1939; Hollyday and Hamburger, 1976).

A few years ago, Viktor Hamburger wrote concerning cell death: "The consensus among developmental neurobiologists seem to be that we are dealing with a competition process among axons at the site of their projection fields . . . but the question 'competition for what?' is still not settled." The issue remains unsettled. In the present context, the question is whether the competition is among motoneurons that can be regarded, by virtue of their positions within the lateral motor columns, as belonging to different motor pools or among motoneurons of single motor pools. If the former, cell death may play a signficant role in the patterning of neuronal connections. In the chick embryo, the evidence available at present indicates that at stages prior to the onset of normal cell death, motoneurons project into the vicinity of their adult targets (Landmesser, 1978b; Hollyday et al., 1977; but see Pettigrew, Lindeman, and Bennett, 1979). Because motoneuron axons grow into and distribute themselves within largely undifferentiated tissue, it is difficult to determine the degree of accuracy of the initial projection pattern; the targets themselves are poorly defined. While it is clear that motoneurons belonging to widely separated motor pools do not project to the same limb region before cell death, there may be some overlap in the projections to muscles derived from neighboring portions of the muscle masses. Although these considerations make it likely that a substantial amount of the competition is between motoneurons of a single motor pool, they leave open the possibility of some competition between motoneurons of pools that innervate adjacent muscles. Further evidence that the relevant competition accounting for cell death is not between motoneurons belonging to widely separated motor pools comes from the observation that the normal cell death in the intact spinal cord segments is not reduced following partial spinal cord removals (Lamb, 1979).

In the frog, some evidence for the innervation of embryonic muscle by abnormal motor pools and the subsequent death of such cells has been provided. Lamb (1976) has found that during the earliest stages of limb innervation in *Xenopus,* motoneurons in abnormal positions project to one region of the presumptive thigh. These abnormally projecting motoneurons do not survive past the period of normal cell death. Thus, while it is possible that cell death plays some role in patterning limb innervation, such a role would seem to be restricted to the removal of small projection errors.

In general, the numbers of naturally dying cells seems much greater than is called for by the observed incidence of abnormal innervation. Hence the major part of cell death seems to result from competition among motoneurons with similar labels. If so, the significance of cell death for development must be sought in terms of problems other than those addressed in this essay. One hypothesis is that the overproduction of motoneurons and their subsequent death provides a means for matching numbers of motoneurons to the size of the periphery (Cowan, 1973). A possible approach to evaluating this is suggested by the fact that homologous muscles in closely related species can vary substantially in size. A corresponding variation in the amount of cell death in the homologous motor pools would provide support for the hypothesis that cell death is involved in matching motoneuron number to muscle size. A somewhat different hypothesis about cell death is related to the assumption, mentioned earlier, that there is a correspondence between motoneuron location and the pattern of inputs it receives. The mechanisms controlling the development of this pattern are unknown but, as we noted, the critical issue in the development of limb innervation is not the relation between motoneuron location and muscle innervated but rather the relation between the functional pattern of inputs to a motoneuron and the muscle innervated. Electromyographic recordings made from identified muscles in spontaneously moving embryos before the completion of cell death indicate that the basic adultlike patterns of muscle activity are present from the beginning (Bekoff, Stein, and Hamburger, 1975; Bekoff, 1976). It is conceivable, however, that there is some variability in the pattern of inputs onto motoneurons at a given location. If so, some of these may be inappropriate for the muscles with which these motoneurons formed connections. Cell death might be involved in eliminating these inappropriately functioning motoneurons.

In summary, further experiments are necessary before the developmental significance of normal cell death among limb motoneurons can be established with certainty. Although we did not discuss cell death in the context of retinotectal connections, there is in fact evidence for cell death among retinal ganglion cells during development (Rager and Rager, 1976), and for variations in the amount of cell death consequent on manipulations of their tectal target (Hughes and LaVelle, 1975; Hughes and McLoon, 1979). Whether or not the significance of cell death in both systems will prove to be similar remains unclear.

CONCLUSIONS

The evidence reviewed indicates that, in limb innervation as in retinotectal projections, a target recognition mechanism can be demonstrated experimentally but that several additional mechanisms may influence the pattern of connections. The existence of multiple mechanisms in the case of limb innervation has the same implications for experimental strategy as those previously discussed in connection with the retinotectal projection. From a single observation one cannot preclude the existence of any given mechanism. Nor can one easily evaluate the contribution of any given mechanism to an observed connection pattern.

In our comparison of retinotectal connections and limb innervation, we have focused on the common feature that several developmental mechanisms are present in both systems. At the same time, it is clear from the brief surveys given here that additional insights into development might emerge from more detailed comparisons of the two systems than we have undertaken in this essay. As we remarked in passing, in addition to matching motor pools and muscles, there is, in limb innervation, a finer level of matching, that of motoneuron type and muscle fiber type. There is substantial evidence that at least part of this matching involves induction of muscle fiber type by different motoneurons. While some dependence of tectal histogenesis on optic nerve innervation is known (Kelly and Cowan, 1977), there is at present no evidence for differential effects of different ganglion cell types. We also discussed briefly the issue of cell death and left unanswered the question whether or not it plays the same developmental role in the two systems.

A third point on which more detailed comparison of the two systems might be instructive relates to the importance of developmental mechanisms that act to organize axons along their growth paths. We noted that such organization is probably of more significance in limb innervation than in the retinotectal projection. It may, however, prove that a more appropriate comparison is that between limb innervation and the projection pattern of the retina in its entirety. The various target structures of the optic nerve receive different sets of ganglion cell projections, and the axons of these sets run in different tracts after the optic chiasm. Organizing mechanisms along the growth path may be quite important in bringing the correct complement of optic fibers to each target structure. This would be analogous to the importance of such mechanisms operating in the limb plexuses to bring motoneuron axons into the vicinity of the appropriate muscle. This perspective also suggests that retinotectal projections could usefully be compared with the behavior of motoneurons running in a particular peripheral nerve. Evidence from the retinotectal system tends to emphasize mechanisms that are important in patterning fibers once they are in the vicinity of their targets. Similar processes in limb innervation may be most important after the limb plexus, when fibers in one peripheral nerve have to make choices among the muscles available to them. For example, it may be that significant competition is normally only between motoneurons projecting to nearby regions of the muscle mass.

MULTIPLE DEVELOPMENTAL MECHANISMS: GENERAL CONSIDERATIONS

In the previous sections, we have drawn attention to the fact that studies of limb innervation and of retinotectal projections have indicated that in each there are multiple developmental mechanisms which can influence patterns of connections. We have also discussed some of the implications of multiple mechanisms for the interpretation and design of experiments. In this section, we want to discuss two more general and related issues raised by the demonstration of multiple possible influences on connectional patterns.

The two issues are: why are there multiple mechanisms? And what is the relative significance of each in normal development? One possible answer to these questions is that in the normal development of any given system, only one of the mechanisms we have discussed is really significant and that the others ought properly to be regarded almost as experimental artifacts, that is, as influences operating only in highly unusual cases associated with experimental perturbation. A second possible answer is that the multiple mechanisms create an error-correcting redundancy, that is, that each of the mechanisms by itself would produce the same result, but with some probability of mistakes. Collectively they assure a highly ordered connectivity. A third, and perhaps the most interesting possibility, is that the various mechanisms exist in order to cope with a series of somewhat distinct problems in elaborating functionally appropriate neuronal circuitry.

In earlier sections, we have noted some possible instances of the latter. In the case of limb innervation, for example, we noted that the outgrowth path taken by the outgrowing axons of motoneurons may be quite important and that an incorrect path may preclude correct termination. In this situation both a pathway guidance mechanism (either by fiber sorting or by substrate guidance) and a target recognition mechanism may operate sequentially to deal first with the outgrowth path and then with the determination of target within a local area. It should be noted that the importance of each mechanism, and perhaps its potency, may vary from situation to situation. As we have discussed, pathway guidance may be more important and target recognition less so in the case of limb innervation as opposed to the retinotectal projection.

A second instance in limb innervation in which different mechanisms may serve slightly different purposes was discussed in connection with the possible difference between mammalian and ichthyopsidan muscle innervation. Competitive interactions may be an important mechanism for assuring correct connectivity in cases in which continued turnover of synapses is necessary to maintain functional innervation during growth. One might predict less effective competition and more reliance on pathway guidance in mammalian limb innervation, where the necessary connection pattern is more stable.

In the case of the retinotectal connections, an example in which different mechanisms serve slightly different purposes was mentioned in connection with the discussion of competition and afferent sorting. We noted that together these two mechanisms would be capable of adjusting the retinotectal projection to changes in size of either the afferent or the target structure. Again, one might speculate that such mechanisms may be more potent in those animals that have to cope with continuous growth than in animals that have stable connectional patterns which are functionally adequate.

It is not possible at the present time to provide definitive answers to the questions of why multiple mechanisms that influence connectivity exist and of what the relative importance of each is; indeed, these may not really be experimentally approachable questions in their general form. It seems worth raising these issues, however, if for no other reason than to emphasize that in analyzing developmental processes it is important to keep in mind the required functional outcome and the circumstances within which the process must operate. Such considerations both promote a desirable humility in approaching the problems of development and suggest ways for unraveling complex processes experimentally. A "connection pattern" is often thought of as a well-defined and invariant outcome; arguments about how it is brought about are often couched in terms of "the most simple mechanism" or "the mechanism requiring least genetic information." In general, the functionally required outcome may be neither well defined nor invariant; the means by which it is accomplished are unlikely to be deduced by naive conceptions of design constraints. In the future, it may be possible to arrive at some generalizations about the mechanisms responsible for specific connectivity broader than those we have reached in this essay. We suspect, however, that they will have to await a larger catalogue of examples and will have to include a clearer understanding of the actual functional requirements placed on a developing system. Depending on one's mind set, one can be depressed at the prospect that developmental neurobiology is doomed to long-term immaturity or elated at the vision that the processes responsible for the production of functionally appropriate patterns of connections are subtler than we have thus far appreciated. For some time to come there need be no fear that any of us will be "smart enough to solve all the problems."

ACKNOWLEDGMENTS

M. H. was a postdoctoral fellow in the laboratory of Viktor Hamburger. The importance of interactions with him, then and since, for the development of the ideas presented here cannot be overestimated. This essay grew out of interactions with a number of students in a course in developmental neurobiology taught by the two of us for the past several years and out of ongoing discussions with colleagues in our two laboratories and in the Committee on Neurobiology. We are grateful to all these people and in particular to C. S. Grobstein, S. Hoskins, R. Jacobson, and V. Whitelaw. The research program of M. H. is supported by the Spencer Foundation and PHS NS 14066; that of P. G. by PHS EY 01658 and RCDA EY 00057, NSF BNS 794122, and an Alfred P. Sloan Fellowship.

REFERENCES

Arora, H. L., Sperry, R. W. 1957. Myotypic respecification of regenerated nerve fibers in cichlid fishes. *J. Embryol. Exp. Morphol. 5:256–63*

Attardi, D. G., Sperry, R. W. 1963. Preferential selection of central pathways by regenerating optic nerve fibers. *Exp. Neurol.* 7:46–64

Barbera, A. S., Marchase, R. B., Roth, S. 1973. Adhesive recognition and retino-tectal specificity. *Proc. Natl. Acad. Sci. U.S.A.* 70:2482–86

Barker, D., Ip, M. C. 1966. Sprouting and degeneration of mammalian motor axons in normal and deafferented skeletal muscle. *Proc. R. Soc. London Ser. B* 163:538–54

Bekoff, A. 1976. Ontogeny of leg motor output in the chick embryo: A neural analysis. *Brain Res.* 106:271–91

Bekoff, A., Stein, P. S. G., Hamburger, V. 1975. Coordinated motor output in the hindlimb of the 7-day chick embryo. *Proc. Natl. Acad. Sci. U.S.A.* 72:1245–8

Bennett, M. R., Davey, D. F., Uebel, K. E. 1980. The growth of the segmental nerves from the brachial myotomes into the proximal muscle of the chick prelimb during development. *J. Comp. Neurol.* 189:335–58

Bennett, M. R., Raftos, J. 1977. Formation and elimination of foreign synapses of adult salamander muscle. *J. Physiol. (London)* 265:261–95

Bixby, J., Van Essen, D. 1979. Competition between foreign and original nerves in adult mammalian skeletal muscle. *Nature (London)* 282:726–28

Bunt, S. M., Horder, T. J. 1978. Evidence for an orderly arrangement of optic axons in the central pathways of vertebrates and its implications for the formation and regeneration of optic projections. *Neurosci. Abstr.* 4:468.

Burke, R. E., Strick, P. L., Kanda, K., Kim, C. C., Walmsley, B. 1977. Anatomy of medial gastrocnemius and soleus motor nuclei in cat spinal cord. *J. Neurophysiol.* 40:667–80

Cass, D. T., Mark, R. F. 1975. Re-innervation of axolotl limbs I. Motor nerves. *Proc. R. Soc. London Ser. B* 190:45–58

Cook, J. E. 1979. Interactions between optic fibers controlling the locations of their terminals in the goldfish optic tectum. *J. Embryol. Exp. Morphol.* 52:89–103

Cowan, W. M. 1973. Neuronal death as a regulative mechanism in the control of cell number in the nervous system. In *Development and Aging in the Nervous System,* ed. M. Rockstein, pp. 19–41. New York: Academic

Crossland, W. J., Cowan, W. M., Rogers, L. A. 1975. Studies on the development of the chick optic tectum. IV An autoradiographic study of the development of retino-tectal connections. *Brain Res.* 91:1–23

Crossland, W. J., Cowan, W. M., Rogers, L. A., Kelly, J. P. 1974. The specification of the retino-tectal projection in the chick. *J. Comp. Neurol.* 155:127–64

Cruce, W. L. R. 1974. The anatomical organization of hindlimb motoneurons in the lumbar spinal cord of the frog, *Rana catesbiana. J. Comp. Neurol.* 153:59–76

Dennis, M. J., Yip, J. W. 1978. Formation and elimination of foreign synapses on adult salamander muscle. *J. Physiol. (London)* 274:299–310

Eccles, J. C., Eccles, M., Magni, F. 1960. Monosynaptic excitatory action on motoneurons regenerated to antagonistic muscles. *J. Physiol. (London)* 154:68–88

Eccles, J. C., Eccles, R. M., Shealy, C. N., Willis, W. D. 1962. Experiments utilizing monosynaptic excitatory action on motoneurons for testing hypotheses relating to specificity of neuronal connections. *J. Neurophysiol.* 25:559–80

Fangboner, R. F., Vanable, J. W., Jr. 1974. Formation and regression of inappropriate nerve sprouts during trochlear nerve regeneration in *Xenopus laevis. J. Comp. Neurol.* 157:391–406

Feldman, J. D., Keating, M. J., Gaze, R. M. 1975. Retino-tectal mismatch: A serendipitous result. *Nature* (London) 253:445–46

Fouvet, B. 1973. Innervation et Morphogènese de la patte chez l'embryon de Poulet I. Mise en place de l'innervation normale. *Arch. 'Anat. Microsc. Morphol. Exp.* 62:269–80

Frank, E., Jansen, J. K. S. 1976. Interaction between foreign and original nerves innervating gill muscles in fish. *J. Neurophysiol.* 39:84–90

Fraser, S. E., Hunt, R. K. 1980. Retinotectal specificity. *Ann. Rev. Neurosci.* 3:319–52

Fraser, S. E. 1979. Late LEO: A new system for the study of neuroplasticity in *Xenopus.* In *Developmental Neurobiology of Vision,* ed. W. Singer, J. Freeman. New York: Plenum

Gaze, R. M. 1959. Regeneration of the

optic nerve in *Xenopus laevis. Q. J. Exp. Physiol.* 44:290–308

Gaze, R. M. 1970. *The Formation of Nerve Connections*, New York: Academic

Gaze, R. M. 1978. The problem of specificity in the formation of nerve connections. In *Specificity of Embryological Interactions*, ed. D. Garard, pp. 53–93. London: Chapman and Hall

Gaze, R. M., Grant, P. 1978. The diencephalic course of regenerating retino-tectal fibers in *Xenopus laevis. J. Embryol. Exp. Morphol.* 44:201–16

Gaze, R. M., Jacobson, M., Székely, G. 1963. The retino-tectal projection in *Xenopus* with compound eyes. *J. Physiol. (London)* 165:484–99

Gaze, R. M., Keating, M. J., Ostberg, A., Chung, S.-H. 1979. The relationship between retinal and tectal growth in larval *Xenopus:* Implications for the development of the retino-tectal projection *J. Embryol. Exp. Morphol.* 53:103–43

Gaze, R. M., Keating, M. J., Chung, S.-H. 1974. The evolution of the retino-tectal map during development in *Xenopus. Proc. R. Soc. London Ser. B.* 194:447–66

Gaze, R. M., Sharma, S. C. 1970. Axial differences in the reinnervation of the goldfish tectum by regenerating optic nerve fibers. *Exp. Brain Res.* 10:171–81

Grimm, L. M. 1971. An evaluation of myotypic respecification in Axolotls. *J. Exp. Zool.* 178:479–96

Grinnell, A. D., Letinsky, M. S., Rheuben, M. 1979. Competitive interaction between foreign nerves innervating frog skeletal muscle. *J. Physiol. (London)* 289:241–62

Haimann, C., Mallart, A., Zilber–Gachelin, N. F. 1976. Competition between motor nerves in the establishment of neuromuscular junctions in striated muscles of *Xenopus laevis. Neurosci. Lett.* 3:15–20

Hamburger, V. 1929. Experimentelle Beiträge zur Entwicklungsphysiologie der Nervenbuhnen in der Froschextremität. *Wilhelm Roux' Archiv. Entwicklungsmech. Org.* 119:47–99

Hamburger, V. 1934. The effects of wing bud extirpation on the development of the central nervous system in chick embryos. *J. Exp. Zool.* 68:449–94

Hamburger, V. 1939. Motor and sensory hyperplasia following limbbud transplantations in chick embryos. *Physiol. Zool.* 12:268–84

Hamburger, V. 1975. Cell death in the development of the lateral motor column of the chick embryo. *J. Comp. Neurol.* 160:535–46

Hamburger, V. 1975. Changing concepts in developmental neurobiology. *Perspect. Biol. Med.* 18:162–78

Hamburger, V. 1976. The developmental history of the motor neuron. *Neurosci. Res. Program Bull. Suppl.* 15:1–36

Hamburger, V., Balaban, M. 1963. Observations and experiments on spontaneous rhythmical behavior in the chick embryo. *Dev. Biol.* 7:533–45

Harris Flanagan, A. E. 1969. Differentiation and degeneration in the motor horn of fetal mouse. *J. Morphol.* 129:281–305

Hibbard, E. 1967. Visual recovery following regeneration of the optic nerve through the oculomotor nerve root in *Xenopus. Exp. Neurol.* 19:350–56

Hollyday, M. 1978. Target selectivity of motor pools in chick embryos. *Neurosci. Abstr.* 4:115

Hollyday, M. 1980a. Motoneuron histogenesis and development of limb innervation. *Curr. Top. Dev. Biol.* 15:181–215

Hollyday, M. 1980b. Organization of motor pools in chick lumbar lateral motor column. *J. Comp. Neurol.* 194:143–70

Hollyday, M. 1981. Rules of motor innervation in chick embryos with supernumerary limbs. Submitted for publication

Hollyday, M., Hamburger, V. 1976. Reduction of the naturally occurring motor neuron loss by enlargement of the periphery. *J. Comp. Neurol.* 170:311–20

Hollyday, M., Hamburger, V. 1977. An autoradiographic study of the formation of the lateral motor column in the chick embryo. *Brain Res.* 132:197–208

Hollyday, M., Hamburger, V., Farris, J. M. G. 1977. Localization of motor neuron pools supplying identified muscles in normal and supernumerary legs of chick embryo. *Proc. Natl. Acad. Sci. U.S.A.* 74:3582–86

Hope, R. A., Hammond, B. J., Gaze, R. M. 1976. The arrow model: Retino-tectal specificity and map formation in the goldfish visual system. *Proc. R. Soc. London Ser. B* 194:447–66

Horder, T. J. 1978. Functional adaptability and morphogenetic opportunism, the only rules for limb development? *Zoon* 6:181–92

Horder, T. J., Martin, K. A. C. 1978. Morphogenetics as an alternative to chemospecificity in the formation of nerve connections. *Symp. Soc. Exp. Biol.* 32:275–358

Hoy, R. R., Bittner, G. D., Kennedy, D. 1967. Regeneration in crustacean motoneurons: Evidence for axonal fusion. *Science* 156:251–52

Hughes, A. F. 1961. Cell degeneration in the larval ventral horn of *Xenopus laevis* (Daudin). *J. Embryol. Exp. Morphol.* 9:269–84

Hughes, W. F., LaVelle, A. 1975. The effects of early tectal lesions on development in the retinal ganglion cell layer of chick embryos. *J. Comp. Neurol.* 163:265–84

Hughes, W. F., McLoon, S. C. 1979. Ganglion cell death during normal retinal development in the chick: Comparisons with cell death induced by early target field destruction. *Exp. Neurol.* 66:587–601

Jacobson, M. 1976. Histogenesis of retina in the clawed frog with implications for the pattern of development of retino-tectal connections. *Brain Res.* 103:541–45

Jacobson, M. 1977. Mapping the developing retino-tectal projection in frog tadpoles by a double label autoradiographic technique. *Brain Res.* 127:55–67

Jacobson, M. 1978. *Developmental Neurobiology* New York: Plenum

Jacobson, M., Levine, R. L. 1975. Stability of implanted duplicate tectal positional markers serving as targets for optic axons in adult frogs. *Brain Res.* 92:468–71

Jacobson, R. 1979. Limb muscle activity in chick during locomotion. *Neurosci. Abstr.* 5:374

Johns. P. R., Easter, S. S., Jr. 1977. Growth of the adult goldfish eye II. Increase in cell number. *J. Comp. Neurol.* 176:331–41

Kelly, J. P., Cowan, W. M. 1972. Studies on the development of the chick optic tectum III. Effects of early eye removal. *Brain Res.* 42:263–88

Lamb, A. H. 1976. The projection patterns of the ventral horn to the hind limb during development. *Dev. Biol.* 54:82–99

Lamb, A. H. 1977. Neuronal death in the development of the somatotopic projections to the ventral horn in *Xenopus. Brain Res.* 134:145–50

Lamb, A. H. 1979. Evidence that some developing limb motoneurons die for reasons other than peripheral competition. *Dev. Biol.* 71:8–21

Lance-Jones, C., Landmesser, L. 1979. Pathway selection by embryonic chick lumbosacral motoneurons." *Neurosci. Abstr.* 5:166

Lance-Jones, C., Landmesser, L. 1980a. Motoneuron projection patterns in the chick hind limb following partial deletions of the spinal cord. *J. Physiol. (London)* 302:559–80

Lance-Jones, C., Landmesser, L. 1980b. Motoneuron projection patterns in the chick hind limb following early partial reversal of the spinal cord. *J. Physiol (London)* 302:581–602

Landmesser, L. 1978a. The distribution of motoneurons supplying chick hind limb muscles. *J. Physiol. (London)* 284:371–89

Landmesser, L. 1978b. The development of motor projection patterns in the chick hind limb. *J. Physiol. (London)* 284:391–414

Landmesser, L. 1980. The generation of neuromuscular specificity. *Ann. Rev. Neurosci.* 3:279–307

Law, M. I., Constantine-Paton, M. 1980. Right and left eye bands in frogs with unilateral tectal ablations. *Proc. Natl. Acad. Sci. U.S.A.* 77:2314–18

Longley, A. 1978. Anatomical mapping of retino-tectal connexions in developing and metamorphosed *Xenopus:* Evidence for changing connexions. *J. Embryol. Exp. Morphol.* 45:249–70

Luiten, P. G. M. 1976. A somatotopic and functional representation of the respiratory muscles in the trigeminal and facial motor nuclei of the carp (Cyprinns carpio L.). *J. Comp. Neurol.* 166:191–200

Marchase, R. B. 1977. Biochemical studies of retino-tectal specificity. *J. Cell. Biol.* 75:237–57

Mark, R. F. 1965. Fin movement after regeneration of neuromuscular connections: An investigation of myotypic specificity. *Exp. Neurol.* 12:292–302

Mark, R. F. 1969. Matching muscles and motoneurons. A review of some experiments on motor nerve regeneration. *Brain Res.* 14:245–54

Mark, R. 1980. Synaptic repression at neuromuscular junctions. *Physiol. Rev.* 60:355–95

Mark, R. F., Marotte, L. R. 1972 The mechanism of selective reinnervation of fish eye muscles III. Functional electrophysiological and anatomical analysis of recovery from section of the third and fifth nerves. *Brain Res.* 46:131–48

Marotte, L. R., Mark, R. F. 1970 The mechanism of selective reinnervation of fish eye muscle. I. Evidence from muscle function during recovery. *Brain Res.* 19:41–51

Maturana, H. R., Lettvin, J. L., McCulloch, W. S., Pitts, W. H. 1959. Evidence that cut optic nerve fibers in a frog regenerate to their proper places in the tectum. *Science* 130:1709–10

Mendell, L. M., Munson, J. B., Scott, J. G. 1974. Connectivity changes of Ia afferents on axotomized motoneurons. *Brain Res.* 73:338–42

Mendell, L. M., Scott, J. G. 1975. The effect of peripheral nerve cross-union on connections of a single Ia fibers to motoneurons. *Exp. Brain Res.* 22:221–34

Meyer, R. L. 1978. Deflection of selected optic fibers into a denervated tectum in goldfish. *Brain Res.* 155:213–27

Meyer, R. L. 1978. Evidence from thymidine labelling for continuing growth of retina and tectum in juvenile goldfish. *Exp. Neurol.* 59:99–111.

Meyer, R. L. 1979. Retino-tectal projection in goldfish to an inappropriate region with a reversal in polarity. *Science* 205:819–21

Meyer, R. L. 1980. Mapping the normal and regenerating retino-tectal projection of goldfish with autoradiographic methods. *J. Comp. Neurol.* 189:273–89

Meyer, R. L., Sperry, R. W. 1976. Retino-tectal specificity: Chemoaffinity theory. In *Neural and Behavioral Specificity,* ed. G. Gottlieb, pp. 111–48. New York: Academic

Morris, D. G. 1978. Development of functional motor innervation in supernumerary hindlimbs of the chick embryo. *J. Neurophysiol.* 41:1450–65

Muller, K. J., Carbonetto, S. 1979. The morphological and physiological properties of a regenerating synapse in the C.N.S. of the leech. *J. Comp. Neurol.* 185:485–516

Oppenheim, R. W., Chu-Wang, I. W. 1977. Spontaneous cell death of spinal motoneurons following peripheral innervation in the chick embryo. *Brain Res.* 125:154–60

Oppenheim, R., Willard–Majors, C. 1978. Neuronal cell death in the brachial spinal cord of the chick is unrelated to the loss of polyneuronal innervation in wing muscle. *Brain Res.* 154:148–52

Pettigrew, A., Lindeman, R., Bennett, M. R. 1979. Development of the segmental innervation of the chick forelimb. *J. Embryol. Exp. Morphol.* 49:115–37

Piatt, J. 1942. Transplantation of aneurogenic forelimbs in *Amblystoma punctatum. J. Exp. Zool.* 91:79–101

Prestige, M. C. 1967. The control of cell number in the lumbar ventral horns during the development of *Xenopus laevis* tadpoles. *J. Embryol. Exp. Morphol.* 18:359–87

Prestige, M. C. 1973. Gradients in time of origin of tadpole motoneurons. *Brain Res.* 59:400–404

Prestige, M. C., Wilson, M. A. 1972. Loss of axons from ventral roots during development. *Brain Res.* 41:467–70

Rager, G., Rager, U. 1976. Generation and degeneration of retinal ganglion cells in the chicken. *Exp. Brain Res.* 25:551–53

Roncali, L. 1970. The brachial plexus and the wing nerve pattern during early developmental phases in chicken embryos. *Monit. Zool. Ital. N. S.* 4:81–98

Romanes, G. J. 1951. The motor cell columns of the lumbo-sacral spinal cord of the cat. *J. Comp. Neurol.* 94:313–63

Romanes, G. J. 1964. The motor pools of the spinal cord. *Prog. Brain Res.* 11:93–119. *(Organization of the Spinal Cord),* eds. J. C. Eccles and J. P. Schadé. Amsterdam: Elsevier

Romer, A. S. 1927. The development of the thigh musculature of the chick. *J. Morphol and Physiol.* 43:347–85

Romer, A. S. 1964. *The Vertebrate Body.* Philadelphia: Saunders

Rusoff, A., Easter, S. S. 1980. Order in the optic nerve of goldfish. *Science* 208: 311–12

Scalia, F., Fite, K. 1974. A retinotopic analysis of the central connections of the optic nerve in the frog. *J. Comp. Neurol.* 158:445–78

Schmidt, J. T. 1978. Retinal fibers alter tectal positional markers during the expansion of the half retinal projection in goldfish. *J. Comp. Neurol.* 177:279–300

Schmidt, J. T., Cicerone, C. M., Easter, S. S. 1978. Expansion of the half retinal projection to the tectum in goldfish: An electrophysiological and anatomical study. *J. Comp. Neurol.* 177:257–78

Scholes, J. H. 1979. Nerve fiber topography in the retinal projection to the tectum. *Nature (London)* 278:620–24

Scott, S. 1975. Persistence of foreign innervation on reinnervated goldfish extraocular muscles. *Science* 189:644–46

Scott, S. A. 1977. Maintained function of foreign and appropriate junctions on reinnervated goldfish extraocular muscles. *J. Physiol. (London)* 268:87–109

Scott, T. M., Lazar, S. 1976. An investigation into the hypothesis of shifting neuronal relationships during development. *J. Anat.* 121:485–96

Sharma, S. C. 1979. The development of ordering in the optic axons of *Rana pipiens. Neurosci. Abstr.* 5:179

Sharrard, W. J. W. 1955. The distribution of permanent paralysis in the lower limb in poliomyelitis. *J. Bone Jt. Surg.* 378:540–58

Sperry, R. W. 1945. The problem of central nervous reorganization after nerve regeneration and muscle transposition. *Q. Rev. Biol.* 20:311–69

Sperry, R. W. 1950. Myotypic specificity in teleost motoneurons. *J. Comp. Neurol.* 93:277–88

Sperry, R. W. 1950. Neuronal specificity. In *Genetic Neurology,* ed. P. Weiss, pp. 232–9. Chicago: Univ. of Chicago Press

Sperry, R. W. 1951. Regulative factors in the orderly growth of nerve circuits. *Growth Symp.* 10:63–87

Sperry, R. W. 1963. Chemoaffinity in the orderly growth of nerve fiber patterns and connections. *Proc. Natl. Acad. Sci. U.S.A.* 50:703–9

Sperry, R. W., Arora, H. L. 1965. Selectivity in regeneration of the oculomotor nerve in the cichlid fish *Astronotus ocellatus. J. Embryol. Exp. Morphol.* 14:307–17

Sperry, R. W., Deupree, N. 1956. Functional recovery following alterations in nerve-muscle connections of fishes. *J. Comp. Neurol.* 106:143–61

Stirling, R. V., Summerbell, D. 1979. The segmentation of axons from the segmental nerve roots to the chick wing. *Nature (London)* 278:640–43

Stirling, R. V., Summerbell, D. 1980. Evidence for non-selective motor innervation of rotated chick limbs. *J. Physiol. (London)* 300:7P

Straznicky, K. 1978. The acquisition of tectal positional specification in *Xenopus. Neurosci. Lett.* 9:177–84

Straznicky, K., Gaze, R. M. 1971. The growth of the retina in *Xenopus laevis:* An autoradiographic study. *J. Embryol. Exp. Morphol.* 26:67–79

Székely, G., Czéh, G. 1967. Localization of motoneurons in the limb moving spinal cord segments of *Ambystoma. Acta Physiol. Acad. Sci. Hung.* 32:3–18

Taylor, A. C. 1943. Development of the innervation pattern in the limb bud of the frog. *Anat. Rec.* 87:379–413

Udin, S. B. 1977. Rearrangements of the retino-tectal projection in *Rana pipiens* after unilateral caudal half-tectal ablation. *J. Comp. Neurol.* 173:561–82

Udin, S. B. 1978. Permanent disorganization of the regenerating optic tract in the frog. *Exp. Neurol.* 58:455–70

Weiss, P. 1936. Selectivity controlling the central-peripheral relations in the nervous system. *Biol. Rev.* 11:494–531

Weiss, P. 1950. An introduction to genetic neurology. In *Genetic Neurology,* ed. P. Weiss, pp. 1–39. Chicago: Univ. of Chicago Press

Whitelaw, V., Hollyday, M. 1980. Motoneuron target selectivity in chicks with deleted, duplicated, or reversed limb segments. *Neurosci. Abstr.* 6:647

Yip, J. W., Dennis, M. J. 1976. Suppression of transmission at foreign synapses in adult newt muscle involves reduction in quantal content. *Nature (London)* 260: 350–2

Yoon, M. 1971. Reorganization of the retino-tectal projection following surgical operations on the optic tectum in goldfish. *Exp. Neurol.* 33:395–411

Yoon, M. 1972. Transposition of the visual projection from the nasal hemiretina on to the foreign rostral zone of the optic tectum in goldfish. *Exp. Neurol.* 37:451–62

Yoon, M. 1976. Progress of topographic regulation of the visual projection in the halved optic tectum of the goldfish. *J. Physiol. (London)* 257:621–43

9 / *Determination of Central Retinal Connections*

S.C. SHARMA

HIGHLY STEREOTYPED NERVE PATTERNS exist in all parts of the adult nervous system. The question as to how such patterns develop has given rise to various hypotheses: chemotropism; mechanical guidance; metabolic attractions; temporal matching; random growth of axons, followed by selective retraction; and chemoaffinity. These hypotheses and the arguments for and against each of them have been summarized by Weiss (1955) and by Jacobson (1978). The "chemoaffinity" hypothesis has been extensively utilized to explain the formation of interneuronal connectivity (Sperry, 1941). This theory arises from Sperry's contention that hypotheses emphasizing mechanical factors and neural activity, which were popular at the time, could not account for the degree of selectivity demonstrated during regeneration of neuronal connections. Following experiments in which he showed that ganglion cell axons from rotated eyes in frogs grew back to their original terminal sites in the optic tectum, Sperry (1963) proposed the hypothesis of chemoaffinity, which states that "the establishment and maintenance of synaptic connections [are] regulated by highly specific cytochemical affinities that arise systematically among the different types of neurons involving self-differentiation, induction through terminal contacts and embryonic gradient effects." He further

I am indebted to Professor Viktor Hamburger for giving me the opportunity to work in his laboratory from 1968 to 1972. During this period I profited greatly from our innumerable discussions. The early part of the work reported here stemmed from Viktor's prodding, and the embryonic manipulations were learned directly from him. I consider myself fortunate indeed to have been guided and encouraged by a scientist of such distinction. It is with gratitude that these pages are written in his honor.

postulated that cells carry "individual identification tags" that allow them to be distinguished at the level of the single cell. Sperry (1945) also predicted the possibility of separate differentiation of the anteroposterior and dorsoventral axes of the retina. This axial system endows the individual ganglion cells with properties that enable their axons to terminate at specific sites on the corresponding optic tectum.

The manifestation of such neuronal properties has been studied extensively in the past three decades in the visual system of lower vertebrates. The early studies dealt with the behavioral demonstration of the embryonic determination of functional polarity of the eye (Székély, 1954, 1957). Székély (1954) rotated the eyefields of *Triturus vulgaris* at the medullary plate stage, and the resulting adult animals showed normal vision; however, when the eye was rotated at Harrison Stage 20 (closure of the medullary plate), reversed visual fields developed. By transplanting the right eye to the left orbit and rotating the anteroposterior and dorsoventral axes separately, Székély showed that the determination of the functional specificity occurred first in the anteroposterior axis, whereas the dorsoventral axis of the retina adapted to its new position. A separate dorsoventral axis determination was never shown. Székély (1957) further showed that after partial ablation of the eye primordium, the already determined parts reestablish their functional polarity by influencing the nondetermined part.

The independent determination of anteroposterior and dorsoventral axes in embryonic eyes of *Xenopus laevis* was shown by Jacobson (1967, 1968a). He rotated one eye cup 180 degrees at various stages of development prior to formation of the retinotectal connections. After metamorphosis, the projection from the rotated eye to the contralateral tectum was mapped electrophysiologically and compared with the projection of the normal eye to the tectum. The retinotectal projection map was found to be normal when the eye was rotated at embryonic Stage 28 or Stage 29 (Nieuwkoop and Faber, 1967), thereby suggesting that ganglion cells at that stage were unspecified. However, eye rotation at Stage 30 resulted in the inversion of the retinotectal projection map in the anteroposterior axis but not in the dorsoventral axis of the retina. After Stage 31, rotation of the eye resulted in complete inversion of the retinotectal projection map. Jacobson concluded that the anteroposterior axis of the retina was specified, or "polarized," at Stage 30 and was followed by the specification of the dorsoventral axis a few hours later, at Stage 31. At Stage 30, when the anteroposterior axis was determined, the ganglion cells were in the early neuroblast stage of development; their final central connections were determined about 20 hours before the initial growth of optic axons.

Jacobson (1968b) showed that precursors of the retinal ganglion cells ceased DNA synthesis and completed their final mitotic division shortly before the specification of the ganglion cells in *Xenopus* at Stage 29. He suggested that DNA synthesis and specification of the ganglion cells are mutually exclusive and that the cessation of DNA synthesis in the retina is correlated with the period of specification of the retina. He further speculated that specification involves the synthesis of macromolecules that uniquely label each ganglion cell according to its position in the retina.

A series of experiments by Hunt and Jacobson (1972a,b; 1973a,b) has given credence to these earlier observations concerning development and specification of positional-dependent information in the retinal ganglion cells of the *Xenopus* eye. They showed that when Stage 28 eyes were implanted in different orientations into the lateral body wall and allowed to develop past the critical stages and then transplanted back into the orbit, the final orientation of retinal axes was based on the orientation of the eye in the flank of the carrier embryo. In another study, they showed that the orientation of the retinal axes can be specified *in vitro* without any embryonic axial clues. Hunt and Jacobson (1973b) further showed that each separate region of the eye does not possess the retinal specification normally observed for that region and that the retinal axial specification during the critical period requires the structural integrity of the eye.

Since these remarkable observations are of central importance in understanding the principles for development of orientation of visuotopic maps and for the mechanics of formation of selective nerve connections, Dr. Joe Hollyfield and I tested the validity of these observations in a different amphibian species, *Rana pipiens* (Sharma and Hollyfield, 1974a). The results in *R. pipiens* showed that the eye is already specified at embryonic Stage 17 (early tail-bud stage) and that in this species there was no correlation with the cessation of DNA synthesis (which begins at Stage 18) and the time when the axes of the embryonic eye are specified. The important conclusion from the studies in *Rana pipiens* is that, regardless of whether the eye is rotated before or after the early tail-bud stage, the resultant visuotopic maps are always rotated through the same angle as is indicated by the position of the choroidal fissure.[1]

In our study, then, it became apparent that there was a definite correlation between the orientation of the eye in the orbit and its subsequent retinal projections. However, we were at first uncertain whether, after surgical manipulations, the eye anlage rerotates to a different position or whether the choroidal fissure develops in a new position.

Although a choroidal fissure is also present in *Xenopus laevis*, none of the earlier studies by Jacobson (1968a,b) or Hunt and Jacobson (1972a,b; 1973a,b) has commented on the relative position of this landmark in the eye in relation to the orientation of the retinotectal map. The consistency of our results in *R. pipiens* led us to reinvestigate in *Xenopus* the orientation of the retinotectal maps relative to the position of the choroidal fissure in eyes rotated at various stages of development, or reciprocally exchanged between the left and right orbits (Sharma and Hollyfield, 1980).

EYE OPERATIONS AND RETINOTECTAL MAPS IN XENOPUS

The types of eye operations that were performed on *Xenopus* embryos between Stage 24 and Stage 32 are shown in Figure 9-1. The eye rudiment was removed and then returned to the orbit with a 180° rotation *(Series a)*. The right eye anlage was exchanged with the left eye anlage, and vice versa, with zero-degree rotation *(Series b)*, or 180-degree rotation *(Series c)*. Electrophysiological mapping of the operated eyes in the postmetamorphic animals of *Series a* showed that whenever the eye appears

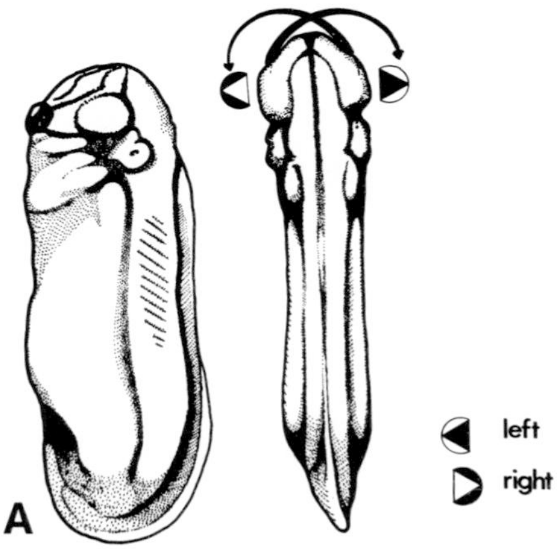

Figure 9.1 The diagrams in A show the *Xenopus* embryo at Stage 27. The drawing on the right shows eye exchanges from left to right and vice versa.
The diagram in B shows left-eye to right-eye transfer with 0° rotation. The lower diagram shows left-eye to right-eye transfer with 180° rotation. Letters inside the circle indicate the graft polarity. Letters outside the circle indicate polarity of the orbit: N, nasal; T, temporal; S, superior; I, inferior.

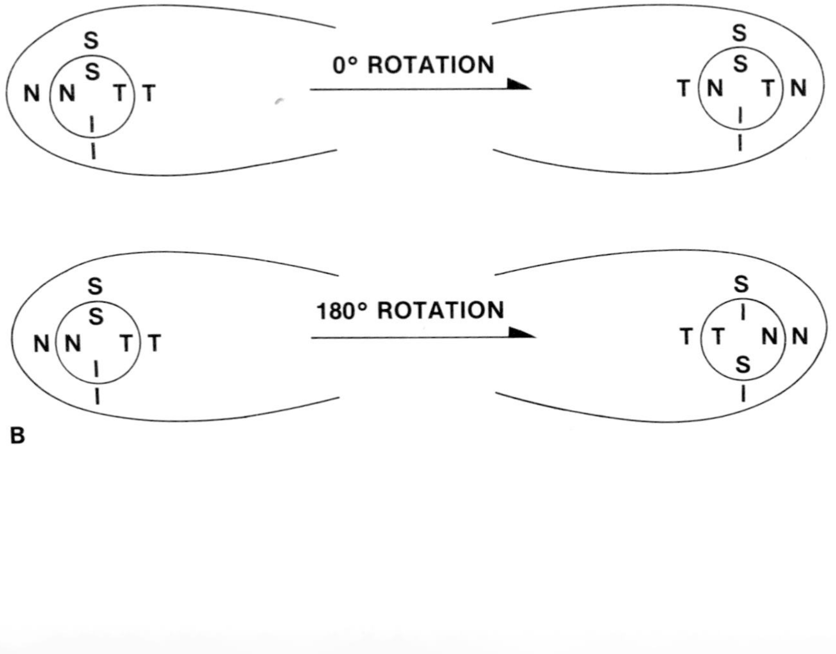

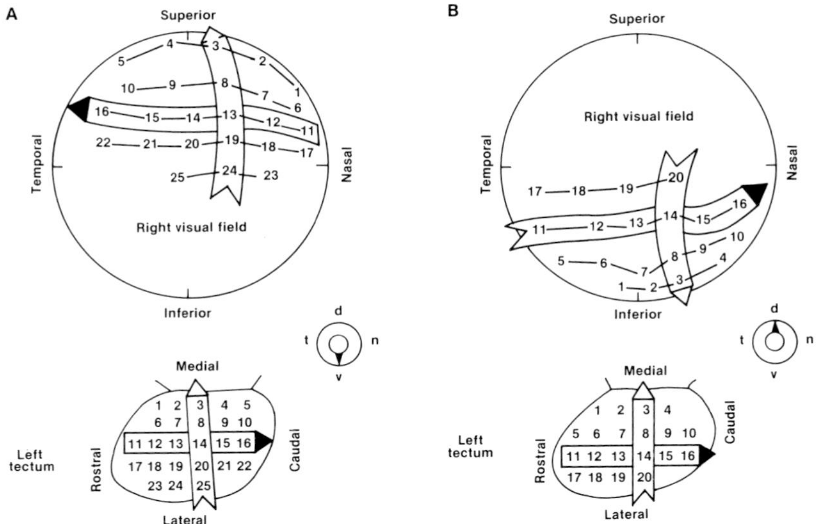

Figure 9.2 The visuotopic map from a normal right eye to the left optic tectum (A). Each number in the visual field shows the position of a visual stimulus that gave a localized response at the electrode position indicated by the same number on the tectum. The small circle on the right side shows the eye diagramatically with the choroid fissure pointing ventrally: d, dorsal; v, ventral; n, nasal; t, temporal. Similar conventions are used for the following diagrams. Right visual field projection map (B) to the left tectum in an animal whose eye was rotated 180° at embryonic Stage 24 (Series a). The map is rotated 180°. Fissure is at 180°

rotated anatomically to any degree (as measured by the appearance of the choroidal fissure in various positions), the resultant map is correspondingly rotated to a similar degree (Figure 9-2B). Furthermore, in the animals of *Series b* and *Series c,* with reciprocal eye exchanges, the retinotectal connections are expressed *according to the original axes of the eye.* For example, following eye exchange with 0° rotation, the choroidal fissure appears at its normal ventral position. In these cases, the nasotemporal axis of the eye will be reversed as compared with the body axis, but the dorsoventral axis will be aligned with the body axis; on the other hand, when the eye exchange is done with 180° rotation, the dorsoventral axis of the eye will be reversed, but the nasotemporal axis will be maintained in its normal relationship with respect to the nasotemporal axis of the host orbit (Fig. 9-1B). The visual map of the eye exchanges with zero-degree rotation shows a reversal of the nasotemporal axis but a normal dorsoventral axis (Fig. 9-3A). However, in the eye exchanges with 180-degree rotation, the choroid fissure is positioned at an altered

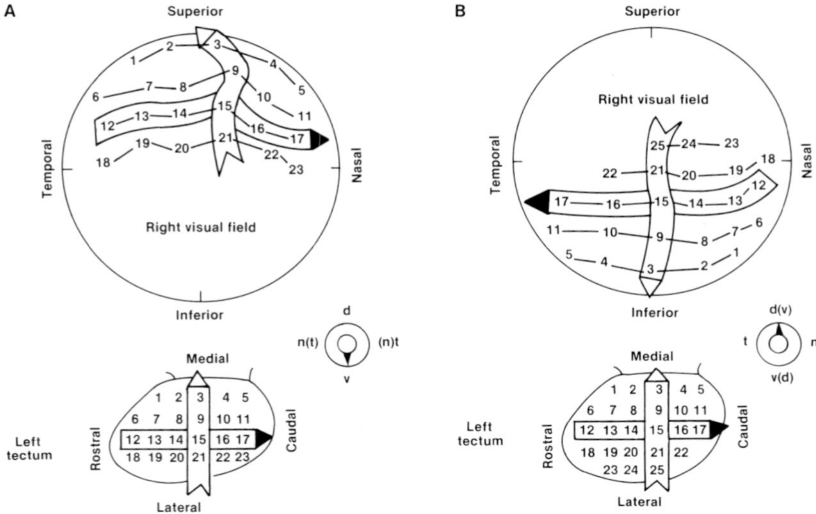

Figure 9.3 Right visuotopic map (A). In this case, an originally *left* eye, was transplanted into the *right* orbit at Stage 27 without rotation (Series b). The nasotemporal axis of the field is reversed, but the dorsoventral axis is normal (as compared with a normal right-eye map). In the eye diagram, the choroid fissure is ventral: (n) and (t) represent the original poles of the eye. The projection of the right onto the left tectum (B) in an animal in which the left eye had been transplanted into the right orbit at Stage 27 with a 180-degree rotation (Series c). The choroid fissure at the time of recording was dorsal: v and d indicate the original poles of the left eye. The dorsoventral axis in this map is rotated, but the nasotemporal axis is normal.

location, and the nasotemporal axis of the map is normal but the dorsoventral axis is rotated (Fig. 9-3B). Such projections are consistent whether the exchange is done before or after embryonic Stage 25. Furthermore, in the eye-exchange experiments, it appears that the eyes are never respecified by axial cues from the new orbit and that they consistently project to the contralateral optic tectum, which indicates the absence of any lateralizing specificity.

Recently, Gaze and his coworkers (1979) have provided strong evidence in *Xenopus* for the correspondence of the orientation of the visual maps with the orientation of the eye at the time of recording. The studies by Gaze and his coworkers have included eye operations in *Xenopus* as early as Stage 21/22. They similarly concluded that the eye anlage at Stage 21/22 is determined (or specified) in relation to the later orientation of the visuotopic maps and that repolarization of the eye rudiment following surgery *does not* occur.

A clue to the apparent conflict between the results obtained by us and by Gaze and his colleagues and those obtained by Jacobson (1968a,b) and by Hunt and Jacobson (1973a,b) seems to emerge from experiments where early eye rotation subsequently gave a "compound map." The characteristics of such maps is that one of the components of the map appears to be organized as if coming from a normally oriented retina, whereas the other component of the map is rotated (Fig. 9-4). Gaze and colleagues (1979) showed that such "compound maps" are of an anatomically compound origin and that the eyes frequently have two optic nerve heads although there is usually only a single optical system.

To demonstrate anatomically the compound nature of the eye, Gaze and his coworkers (1979) grafted a wild-type eye primordium (Stage 24/25) into the orbit of an albino host at a similar stage. In those cases in which a compound map was observed, an extensive albino component

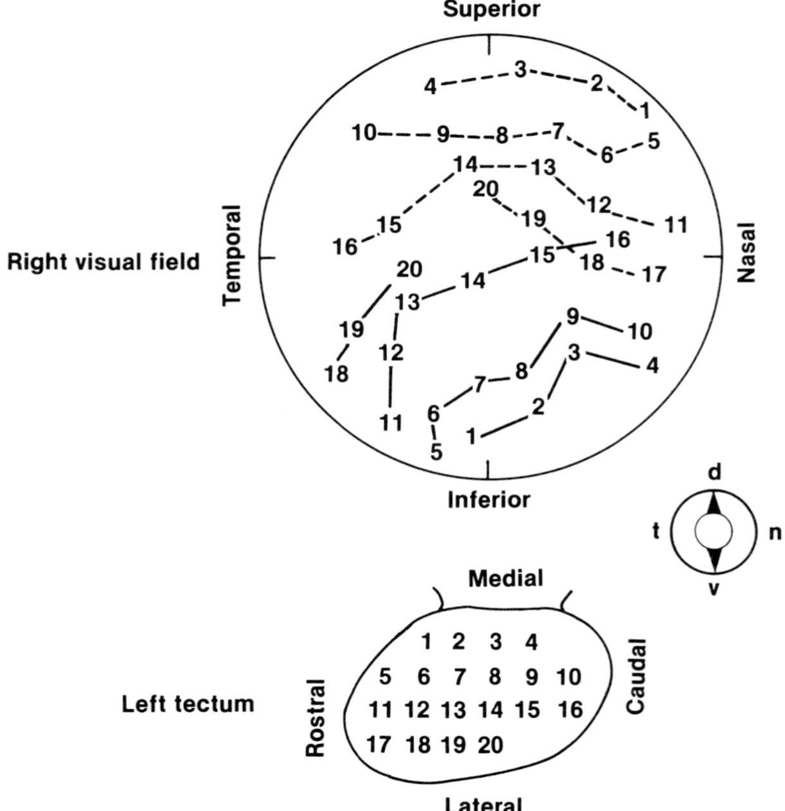

Figure 9.4 The projection of the right eye onto the left tectum in an eye which showed two fissures. The temporal half of this eye projected normally, whereas the nasal eye was rotated 180 degrees. While this eye displayed a duplicate map, optically it was a single eye with a single lens and cornea.

was seen in the eye. These apparently chimeric eyes produced a map in which the host's component of the eye showed a normal retinotectal map, whereas the donor eye, which had been rotated, gave a rotated map. The peculiarity in these cases is that the host-derived component of the eye constituted a very small area a few hours after the surgery; however, at the time of mapping, the host-derived eye constituted a fairly large portion of the eye. Gaze and his colleagues suggested the possibility that the operated eye is partially replaced by the newly emerging eye from the optic stalk cells, which are apparently left behind.

In a recent autoradiographic study, Holt (1980) reported that during optic cup formation in *Xenopus*, cells from the optic stalk region migrate into the ventral retina, an area where the future choroid fissure develops. It is conceivable that incomplete rotation of the eye anlage could leave the presumptive ventral retinal cells in the optic stalk. In this situation, a normal ventral retina would develop with a choroid fissure at its proper position. Since most retinotectal maps are recorded from the dorsal tectum (innervated by the ventral retina), a normal map in a presumably rotated eye would represent an unspecified stage (see Hunt and Jacobson, 1973a,b).

Together, these results supported the notion that patterning in the eye for the subsequent retinotectal map is determined from the early optic vesicle stage. Gaze and his colleagues further suggested the possibility that presumptive neural retina may have no specificity per se and that the presumptive pigment epithelium may determine the locus specificity of the retinal ganglion cells. It is interesting to note that during retinal regeneration in adult newts after various surgical manipulations (Levine, 1975), new retinal tissue is derived from the posterior retinal pigment epithelium and is apparently specified by the pigment epithelium.

Our failure to observe respecification of the rotated eye by the surrounding orbital tissues, and therefore the lack of any prior axial specification of the eye, was attributed earlier (Sharma and Hollyfield, 1974b) to sloppiness in the surgical manipulations. It was suggested that the rotated eyes in our experiments probably did not heal properly and hence the axial cues from the surrounding area could not override the axial information present within the eye. Since the total time from Stage 28 through Stage 31 (the critical period for axial specification according to Jacobson, 1968a) is approximately five hours, we extended our studies to Stage-24 embryos. In the light of our recent studies and of those by Gaze and his colleagues (1979), who have extended the eye rotations back to Stage 21/22, the postulated failure of the operated eyes to heal within a short time does not appear to be an appropriate explanation for our inability to find evidence for the axial specification of the eye.

The use of a low ionic strength operating medium has been implicated as leading to less viable tissue (Gaze et al., 1979). If, during embryonic manipulations, presumptive pigment epithelial tissue is left behind, cell death may occur in the tissues of the rotated eye, and this may lead to the proliferation of a new retina from the unrotated pigment epithelium; this new retina would then generate a normal map. It is possible that following the surgical maneuvers that were done in the earlier studies in which low ionic strength operating media were used (Hunt and Jacobson, 1973a,b; 1974), a new retina could have developed, and the findings might have been erroneously attributed to the respecification of the eye.

Hunt and Jacobson (1973b) and Hunt and Frank (1975) have provided evidence that one-half of an host eye may repolarize the other half of a compound eye when such eyes are made by fusing the right nasal and left temporal half retinas at Stage 32. However, in contrast to these studies, recent work by Straznicky and Gaze (1980) and by Gaze and Straznicky (1980) has shown that compound eyes, constructed surgically at Stage 32 by fusing various types of half eyes, give visuotopic maps that are oriented in accord with the orientation of the fragment constituting the compound eye. These authors have shown further that when the compound eyes are made in low ionic strength medium (thereby intentionally delaying the healing process) a mirror-image reduplication of the retinotectal projection occurs, a phenomenon that is consistent with the findings but not with the interpretation of Hunt and Jacobson (1973b). Gaze and Straznicky (1980) have concluded that mirror duplication may be due to the conditions under which the operation is performed, to responses to injury (in particular to the noxious influence of the operating medium), or to both.

Regarding the earlier behavioral studies of Stone (1960) in *Ambystoma punctatum*, is which it was claimed that separate axial specification of the eye occurs and that eye rotation before late tail-bud stage has no effect on the visual responses, it should be noted that eye rotation at later stages of development leads to confused visuomotor behavior. Moreover, Stone (1966) also described the appearance of a double choroidal fissure following early eye rotation. In the light of the more recent studies, it could be argued that the confused visuomotor behavior results from eyes with two fissures. The appearance of double choroidal fissures in experimentally perturbed eye primordia has been documented extensively in amphibians (Beckwith, 1927; Sato, 1933; Woerdeman, 1934).

Polarization of the avian retina has been studied in an extensive series of experiments by Goldberg (1976). Using morphological markers to delineate the horizontal and vertical polarity of the chick retina, Goldberg rotated optic vesicles prior to the appearance of the optic fiber layer and studied the subsequent changes in the appearance of optic fiber patterns. He has suggested that Stage 12 is the critical period in the chick embryo for the morphological polarization of the retina. Eye rotation before Stage 12 gave normal fiber patterns in the stratum opticum of the retina, whereas rotation after Stage 12 gave morphologically inverted retinas. In those cases in which the rotated eye grafts were small, the morphological markers suggested that the polarity of the retina had become fixed in the horizontal (i.e., nasotemporal) axis prior to the vertical (i.e., dorsoventral) axis. However, eye rotations of moderate size always led to a realignment of the eye, by its rerotation to its normal position; this is in contrast to the small eye grafts that underwent retinal repolarization without rerotation of the eye. Goldberg has emphasized the role of the equatorial zone of the graft in moderate and small eye grafts. In a few cases in which a small graft of chick eye was transplanted into the orbit of a quail embryo, subsequent histology showed that the pigment epithelium was of quail origin, whereas the neural retina was always of chick origin (Goldberg, personal communication). In these cases it was not possible that prospective pigment epithelium respecified the transplanted retina. In any event, in Goldberg's work only the morphological pattern of the retina has been studied. The subsequent connectional patterns of the eyes with the appropriate optic tecta has not been examined. On the other hand, in amphibians, it is the final connectional patterns of the retina that are correlated with the position of the choroidal fissure; in amphibians, the sharply defined pattern of optic nerve fibers and the circumferential fiber tracts, which are prominent in chick retinas, are not well developed.

THEORETICAL CONSIDERATIONS

A Cartesian coordinate system has been used to define axial polarity in the retina (Sperry, 1951; Jacobson, 1968a). In using this system, it has been assumed that the retinal ganglion cells acquire positional information within an orthogonal grid pattern that is based on the anteroposterior or nasotemporal and dorsoventral embryonic axes. The locus specificity of ganglion cells (Hunt and Jacobson, 1974) were thought to be determined by a chemospecific marker, whose character is based upon its position within the grid following embryonic axial specification. This model relied on the assumption of orthogonal chemical gradients present within the retina. More recently, MacDonald (1977) has proposed a polar coordinate model for the specification of the developing retina; according to this view, each ganglion cell position is defined both by its radial distance from the center of the retina and by its angular position along the circumference. The positional information of the entire ganglion cell population, would, accordingly, arise gradually during retinal development; as the cells become postmitotic, their radial distance from the center of the retina and their angular coordinate position would be established, and this would determine the angles of the fascicles that guide their axons to the optic nerve head. Each fascicle contains axons from an area in a common wedge-shaped sector of the retina, with the oldest axons being located near the optic nerve head and the youngest farther out toward the retinal periphery. Therefore, the first fascicle to be formed in the developing retina would assign angular coordinates for all the axons that develop later. The spatiotemporal model for the development of optic nerve fiber topography is, therefore, consistent with the polar coordinate positional information hypothesis.

Bodick and Levinthal (1980) have shown that in developing zebrafish a large optic bundle arises from cells in a pie-shaped sector of the retina and that cells from the opposite side of the choroidal fissure never merge into the same bundle. The separation, which leads to the formation of a crescent pattern by the fibers leaving the eye, is a consequence of the barrier presented by the choroidal fissure. These authors suggest that the opening in the crescent pattern in the developing retina is important since it leads to an inversion of the overall topography of the optic axons in the optic nerve, where the youngest fibers are on one side of the nerve and the oldest fibers are on the opposite side (see also Scholes, 1979). Bodick and Levinthal (1980) further suggest that the polar coordinates of the cell body position in the retina correspond to the rectangular coordinates in the optic nerve because of the inversion of overall topography of optic axons in the nerve. They propose that the ganglion cell axons follow pathways that are determined by the substrate along which they grow and that the locus specificities of the ganglion cells are determined by the time and place at which they initiate the growth of an axon. These rules, then, may perhaps determine both the locus specificity of the ganglion cells and the ordering of the optic axons within the optic nerve. However, whether similar rules apply to the mode of selective synapse formation on the target centers remains to be determined.

In conclusion, recent experiments on amphibian eye rudiments give no support for the view that specification of the retina for subsequent map formation involves polarization of the eye rudiment in Cartesian coordinates. Rather, specification of the eye is more likely to be controlled by polar coordinates. I have here outlined current concepts of eye specification. It is not clear exactly how the position-dependent properties of each ganglion cell are acquired. The role that pigment epithelium might play during this process remains, at present, inexplicably elusive.

NOTE

1. When the optic vesicle invaginates to form the optic cup, the optic stalk is located at the ventral aspect of the eye rudiment. The ventral surface of the optic stalk and cup invaginates to form a groove within which lie the hyaloid vessels. This groove finally closes and forms a continuous line from the ventral margin of the future pupil into the optic nerve. This line is the choroidal fissure.

REFERENCES

Beckwith, C. J. 1927. The effect of the extirpation of the lens rudiment on the development of the eye in *Amblystoma punctatum* with special reference to choroidal fissure. *J. Exp. Zool.* 49:217–59

Bodick, N., Levinthal, C. 1980. Growing optic nerve fibers follow their neighbours during embryogenesis. *Proc. Natl. Acad. Sci.* U.S.A. 77:4374–78

Gaze, R. M., Feldman, J. D., Cooke, J., Chung, S. H. 1979. The orientation of the visuotectal map in *Xenopus:* Developmental aspects. *J. Embryol. Exp. Morphol.* 53:39–66

Gaze, R. M., Straznicky, C. 1980. Stable programming for map orientation in disarranged embryonic eyes in *Xenopus. J. Embryol. Exp. Morphol.* 55:143–65

Goldberg, S. 1976. Polarization of the avian retina. Ocular transplantation studies. *J. Comp. Neurol.* 168:379–92

Holt, C. 1980. Cell movements in *Xenopus* eye development. *Nature* 287:850–52

Hunt, R. K., Frank, E. D. 1975. Neuronal locus specificity: Transrepolarisation of *Xenopus* embryonic retina after the time of axial specification. *Science* 189:563–65

Hunt, R. K., Jacobson, M. 1972a. Development and stability of positional information in *Xenopus* retinal ganglion cells. *Proc. Natl. Acad. Sci.* U.S.A. 69:780–83

Hunt, R. K., Jacobson, M. 1972b. Specification of positional information in retinal ganglion cells of *Xenopus:* Stability of the specified stage. *Proc. Natl. Acad. Sci.* U.S.A. 69:2860–64

Hunt, R. K., Jacobson, M. 1973a. Specification of positional information in retinal ganglion cells of *Xenopus:* Assays for analysis of the unspecified state. *Proc. Natl. Acad. Sci.* U.S.A. 70:507–11

Hunt, R. K., Jacobson, M. 1973b. Neuronal locus specificity: Altered pattern of spatial development in fused fragments of embryonic *Xenopus* eyes. *Science* 180:509–11

Hunt, R. K., Jacobson, M. 1974. Neuronal specificity revisited. *Curr. Top. Dev. Biol.* 8:203–59

Jacobson, M. 1967. Retinal ganglion cells: Specification of central connections in larval *Xenopus laevis. Science* 155:1106–8

Jacobson, M. 1968a. Development of neuronal specificity in retinal ganglion cells of *Xenopus. Dev. Biol.* 17:202–18

Jacobson, M. 1968b. Cessation of DNA synthesis in retinal ganglion cells correlated with the time of specification of their central connections. *Dev. Biol.* 17:219–32

Jacobson, M. 1978. *Developmental Neurobiology.* New York: Plenum

Levine, R. 1975. Regeneration of the retina in the adult newt, *Triturus cristatus,* following surgical division of the eye by a limbal incision. *J. Exp. Zool.* 192:363–80

MacDonald, N. 1977. A polar coordinate system for positional information in the vertebrate neural retina. *J. Theor. Biol.* 69:153–65

Nieuwkoop, P. D., Faber, J. 1967. *Normal Table of* Xenopus laevis (Daudin). Amsterdam: North-Holland

Sato, T. 1933. Über die Determination der fetal Aügenspolts bei *Triton taeniatus. Arch. Entwicklungomech. Org.* 128:342–77

Scholes, J. H. 1979. Nerve fiber topography in the retinal projection to the tectum. *Nature* (London) 278:620–24

Sharma, S. C., Hollyfield, J. G. 1974a. Specification of retinal central connections in *Rana pipiens* before the appearance of the first post-mitotic ganglion cells. *J. Comp. Neurol.* 155:395–408

Sharma, S. C., Hollyfield, J. G. 1974b. The retinotectal projection in *Xenopus laevis* following left-right exchanges of eye rudiment. (4th Annual Meeting of Society for Neuroscience) *Neurosci. Abstr.*, p. 421

Sharma, S. C., Hollyfield, J. G. 1980. Specification of retinotectal connections during development of the toad *Xenopus laevis. J. Embryol. Exp. Morphol.* 55:77–92

Sperry, R. W. 1941. The effect of crossing nerves to antagonistic muscles in the hind limb of the rat. *J. Comp. Neurol.* 75:1–19

Sperry, R. W. 1945. The problem of central nervous reorganization after nerve regeneration and muscle transposition. *Q. Rev. Biol.* 20:311–69

Sperry, R. W. 1951. Mechanisms of neural maturation. In *Handbook of Experimental Psychology,* ed. S. S. Stevens, pp. 236–80. New York: John Wiley

Sperry, R. W. 1963. Chemoaffinity in the orderly growth of nerve fiber patterns and connections. *Proc. Natl. Acad. Sci. U.S.A.* 50:703–10

Stone, L. S. 1960. Polarization of the retina and development of vision. *J. Exp. Zool.* 145:85–93

Stone, L. S. 1966. Development, polarization and regeneration of the ventral iris cleft (remnant of choroid fissure) and protactor lentis muscle in urodele eyes. *J. Exp. Zool.* 161:95–108

Straznicky, C., Gaze, R. M. 1980. Stable programming for map orientation in fused eye fragments in *Xenopus. J. Embryol. Exp. Morphol.* 55:123–42

Székély, G. 1954. Zür Ausbildung der lokalen funktionellen spezifitat der retina. *Acta Biol. Acad. Sci. Hung.* 5:157–67

Székély, G. 1957. Regulationstendenzen der Ausbildung der "functionellen spezifitat" der retinaanlag bie *Triturus vulgaris.* Wilhelm Roux' Arch. *Entwicklungsmech.* Org. 150:48–60

Weiss, P. 1955. Nervous system. In *Analysis of Development,* ed. B. H. Willier, P. Weiss, V. Hamburger, pp. 346–401. Philadelphia: W. B. Saunders

Woerdeman, M. W. 1934. On the development of the structure of the eye-lens in amphibians. *Proc. K. Ned. Akad. Wet. Ser. C* 27:324–28

10 / *Selective Formation of Synapses in the Peripheral Nervous System and the Chemoaffinity Hypothesis of Neural Specificity*

DALE PURVES

INTRODUCTION

No developmental problem presents a greater dilemma for neurobiologists than understanding the way that highly intricate yet stereotyped patterns of neural connections are established. Since the issue remains unsolved, it is to Viktor Hamburger's credit that he avoided this arena for more than 50 years to devote his efforts to other, more tractable problems. By the same token, those interested in neural specificity may see it as a favorable sign that Viktor has lately taken some experimental interest in this topic, and has contributed importantly to it by showing that supernumerary limbs become innervated in a systematic way by motoneuron pools that do not normally project to the limb (Hollyday, Hamburger, and Farris, 1977).

Since Viktor and I have argued about "neural specificity" on more than one occasion, I welcomed the opportunity to set down my point of view.

WHAT IS MEANT BY NEURAL SPECIFICITY?

The phrase "specificity of neural connections" refers to the process inferred to have created detailed patterns of connectivity observed in virtually every part of the nervous system. Although some treatments of specificity address themselves to a single putative mechanism, many regulatory influences must contribute to the mature pattern of synaptic connections. These include mechanisms of neuronal differentiation, proliferation, migration, and the outgrowth of axons to the vicinity of targets. A final step is the formation of synapses between appropriate partners. Thus, the proper starting point in any discussion of specificity is to state which aspect of this complex subject one intends to consider. In what follows, I shall limit the focus to synapse formation, and in particular to the evidence for *selective synapse formation.* By this phrase I mean the evidence that axons make a choice during synapse formation between different classes of neighboring target cells.

SELECTIVE SYNAPSE FORMATION IN THE CENTRAL NERVOUS SYSTEM (CNS)

The formation of selective synaptic connections in the CNS of developing vertebrates has been little explored because of the difficulty of studying neurons *in embryo;* the alternative of studying synapse formation during reinnervation of adult neurons is not possible because regeneration in the CNS of higher vertebrates is usually abortive. Since regeneration *does* occur in the CNS of lower vertebrates, the reinnervation of the optic tectum by retinal axons in amphibians and fish has for many years held the attention of workers in this field. The observation that has dominated thinking about specificity is that retinal axons grow back to the part of the tectum they originally innervated, even when the eye is rotated after cutting the optic nerve (Sperry, 1943; see also Attardi and Sperry, 1963). Although this result argues for some special relation between particular retinal axons and different tectal cells (see, however, Horder and Martin, 1979), I think it fair to say that the basis of accurate retinotectal reconnection remains quite unclear. In particular, it is not known whether specific reinnervation is a function of ordered affinities between presynaptic and postsynaptic elements, or whether axons are guided to their original tectal location where they then make synapses indiscriminately. The reason for this uncertain state of affairs is largely technical. The usual assay of specific regeneration in the lower vertebrate visual system is the arrangement of retinal projections onto the tectum: the tectum is explored with an extracellular electrode and the tectal points excited in response to stimulation of various retinal positions are noted. Although this technique seems straightforward, it does not tell the experimenter whether the observed retinotopic map reflects selective reinnervation of different *classes* of tectal cells, accurate regeneration of retinal axons to an appropriate *position* in the tectum, or some combination of these two possibilities. Thus the question of selective synapse formation remains moot in the retinotectal system.

SELECTIVE INNERVATION IN THE PERIPHERAL NERVOUS SYSTEM (PNS)

A prerequisite for demonstrating selective synapse formation is the identification of different classes of target cells based on some criterion other than their position within the target organ. It is relatively easy to make these identifications in the PNS because individual postsynaptic cells can be impaled with microelectrodes and their innervation characterized according to criteria that are unrelated to a cell's location.

As in the CNS, the formidable problems of working with embryonic neurons have largely precluded studies of selective synapse formation in developing peripheral targets. Thus, in the PNS as well, most information has come from studies of reinnervation after lesions in adult animals.

Selective Innervation of Muscle Cells

Several classes of experiments have explored the preference of skeletal muscle fibers for their original innervation. The first class of experiments asks whether the muscle as a whole has some special affinity for its native innervation. Such experiments have been done either by cutting a nerve containing the innervation to two or more different muscles composed of similar fiber types (Bernstein and Guth, 1963; Brushart and Mesulam, 1980) or by surgically rerouting a foreign motor nerve so that it competes with the native nerve during reinnervation (Weiss and Hoag, 1946; Gerding, Robbins, and Antosiak, 1977). Both approaches have, for mammalian muscles, given the same result: different muscles of similar fiber composition show very little preference for their original motor axons. In some lower vertebrates, the situation is quite different in that muscles *do* show a preference for their original innervation (Sperry and Arora, 1965); moreover, this phenomenon occurs at the level of the synapses since native axons can compete successfully with foreign ones, even after foreign synapses have become established. Thus Dennis and Yip (1978) and others have shown that in anurans the quantal content of the postsynaptic response to stimulation of an implanted foreign nerve gradually decreases when the native nerve returns. Examination of the remaining nerve terminals with a histochemical marker indicated that the foreign terminals are actually lost from the muscle surface. While it is attractive to suppose that this phenomenon betokens the ability of nerves to recognize different muscles as such, other explanations are not ruled out. For instance, since peripherally overextended motor axons in these animals give way to regenerating axons that are making relatively few synapses (Wigston, 1980), loss of the foreign nerve terminals could be on this basis.

A second class of experiments asks whether there is selective synapse formation based on functionally different fiber types within a muscle. In some amphibian muscles there are two distinct classes of muscle fibers: so-called fast fibers that contract rapidly in response to a conducted action potential, and slowly contracting fibers that lack an action potential mechanism and give only graded responses to nerve stimulation (Kuffler and Vaughan-Williams, 1953; Burke and Ginsborg, 1956). The axons innervating each type of fiber can be identified by their different conduction velocities: slowly conducting axons innervate slow muscle fibers while larger, rapidly conducting axons innervate the fast fibers (Kuffler and Vaughan-Williams, 1953). When such mixed muscles are reinnervated, both classes of muscle cell are initially contacted by the motor axons that originally innervated fast muscle fibers (Schmidt and Stefani, 1976). However, when slow motor axons, which take longer to regenerate, eventually reach the muscle, they apparently displace the synapses formed by fast axons on slow fibers so that the normal innervation of these two fiber types is ultimately restored (Elul, Miledi, and Stefani, 1970; Schmidt and Stefani, 1976; 1977). Again, other interpretations of these results are possible. For example, activity patterns are well-known to influence a number of mammalian muscle fiber properties, and although there is no evidence for such a transition, it is difficult to rule out a neurally induced change in muscle fiber type. A related uncertainty is whether fast and slow fibers are distinguished before they become innervated in development, or whether these phenotypes are a result of the innervation received by a single class of fiber. If the muscle fiber types normally differentiate in response to innervation, then their appropriate reinnervation in maturity would have little relevance to selective synapse formation in normal development.

In a third class of experiments, the ability of regenerating axons to distinguish between two muscles with *different* fiber compositions has been tested in amphibians (Hoh, 1971), birds (Feng, Wu, and Yang, 1965; Bennett and Pettigrew, 1976), and mammals (Hoh, 1975; see also Miledi and Stefani, 1969). In general, native nerves appear to show a preference for their original muscles in these experiments, presumably because of the different average fiber composition of the respective targets.

The muscle spindle offers a further opportunity to study selective innervation in muscle. Briefly, mammalian muscle spindles are composed of specialized muscle fibers (intrafusal fibers) that can be readily distinguished from ordinary ones (extrafusal fibers). (For a detailed review of spindle morphology, see Barker, 1974.) Intrafusal fibers receive motor innervation from a special class of motoneurons (γ-motoneurons), and are end organs for two classes of sensory axons called primary and secondary afferents. Functionally, intrafusal fibers, together with their motor and sensory innervation, convey two sorts of information about muscle to the CNS. When the muscle in which a spindle lies is stretched, the primary afferent discharges transiently to give information about the dynamic aspect of the length change, while both primary and secondary afferents fire in a sustained manner to give information about the final muscle length. Each of these responses is highly organized in the sense of being generated by different elements of the spindle, and each response is modulated by a different type of γ-motoneuron. Thus spindles not only receive innervation that is distinct from that of extrafusal fibers (although there is a class of motor axons—β-fibers—which is shared), but sensory and motor innervation within each spindle is specifically organized to serve two sensory functions. This intricate arrangement of spindle innervation provides a demanding test for regenerating axons, as errors in any aspect of sensory or motor reinnervation should disturb spindle function.

In fact, when spindles are studied at long intervals after interruption of a peripheral nerve, their responses are remarkably normal (Thulin, 1960; Bessou, Laporte, and Pages, 1965; Brown and Butler, 1976). On a gross level, stimulation of γ-motoneurons does not cause appreciable tension in reinnervated muscle, which suggests that γ-motoneurons do not reinnervate extrafusal fibers to any great degree. More critically, characteristic sensory and motor spindle responses are readily observed (Bessou et al., 1965; Brown and Butler, 1976). Thus when a reinnervated muscle is stretched, spindles generate typical dynamic and sustained responses; moreover, in the majority of cases, stimulating different γ-motor axons causes an appropriate modulation of the afferent discharge (Bessou et al., 1965; Brown and Butler, 1976). While this evidence of selective reinnervation cannot be considered firm in the absence of histological confirmation, the normal function of spindles after reinnervation provides further support for selective synapse formation between motor axons and functionally different sorts of target cells.

In summary, axons regenerating to muscle appear to show considerable selectivity at the level of synapse formation with muscle fibers of different types. In higher vertebrates, however, fibers of the *same* type in *different* muscles probably are not distinguished during reinnervation to any great degree, although such distinctions are apparently made in some lower vertebrates. Higher vertebrates presumably make similar distinctions during normal development, although this has not yet been demonstrated. Taken together, this information suggests that motor axons may be influenced by both functional and positional properties of the postsynaptic muscle cell during synapse formation. The evidence for a positional influence on synapse formation with muscle is so far limited to the reinnervation of skeletal muscle in some lower vertebrates, and, for the reasons given above, is somewhat uncertain.

Selective Innervation of Autonomic Ganglion Cells

Studies of the innervation (and reinnervation) of autonomic ganglion cells indicate that neurons, as well as muscle cells, are innervated selectively. Moreover, selective synapse formation in mammalian sympathetic ganglia offers a further example of the apparent importance of positional information in synaptogenesis. The following account summarizes what is known about the selective innervation of mammalian sympathetic ganglion cells.

The assessment of selective innervation in the sympathetic nervous system of mammals depends on two properties of sympathetic neurons: (1) sympathetic ganglion cells receive preganglionic innervation from a number of spinal cord segments of which the ventral roots can be stimulated separately; and (2) regeneration of preganglionic axons occurs quite readily, thus permitting studies of synapse formation in adult ganglia. The ganglion cells that have been analyzed in greatest detail are those of the superior cervical ganglion. (For a review, see Purves and Lichtman, 1978.) The postganglionic axons of this group of neurons innervate all of the sympathetically driven end organs of the head and neck (e.g., iris, hairs, blood vessels). If one simply watches end-organ responses in anesthetized animals while stimulating each of the spinal cord segments that contribute preganglionic axons to the superior cervical ganglion, it is apparent that activation of different spinal segments elicits different constellations of end-organ effects (Langley, 1892; Njå and Purves, 1977a). These differences appear to be based largely on the *position* of a given end organ, rather than its functional modality (Lichtman, Purves, and Yip, 1979). Thus stimulation of a particular spinal segment activates all the functional modalities in a limited region of the entire territory innervated by the ganglion; conversely, the responses of a single modality (e.g., piloerection) shift progressively in position as one stimulates successive spinal cord segments (Lichtman et al., 1979).

Intracellular recordings from ganglion cells made during stimulation of different spinal segments *in vitro* show the neuronal basis for these *in vivo* observations (Njå and Purves, 1977a). In the guinea-pig superior cervical ganglion, where most of this work has been carried out, each neuron is innervated by approximately 10 to 12 different axons that arise from an average of four of the eight different spinal cord segments that supply the ganglion as a whole. Almost invariably, the axons innervating a particular cell arise from a *contiguous* subset of the spinal cord segments that contribute innervation to the ganglion. A similar arrangement has been observed in the hamster (Lichtman and Purves, 1980), and in the rat (J. W. Lichtman, unpublished). Studies of ganglion cells that project to different regions of the postganglionic territory (Lichtman et al., 1979) confirm the observations of end-organ responses in anesthetized animals: the normal segmental innervation of ganglion cells relates the position of their postganglionic projection and the rostrocaudal position in the spinal cord of the preganglionic neurons that supply them (see also Rubin and Purves, 1980). The actual position of cells *within* a ganglion, however, seems irrelevant, since neurons innervated by different spinal segments are evenly distributed throughout the ganglion (Lichtman et al., 1979).

These studies of the normal sympathetic system show that ganglion cells are specifically innervated according to positional criteria, but they do not show that this pattern is the product of selective synapse formation. For example, ganglion cells might be innervated at random during development, and, by virtue of the innervation received, be induced to innervate targets in a particular region of the postganglionic territory. More direct evidence for the selective innervation of ganglion cells comes from studies of the reinnervation of adult ganglia. Within several months of cutting the preganglionic nerve trunk in the neck, axons regenerate and establish a pattern of ganglion cell innervation that is largely normal (Njå and Purves, 1977b; 1978; see also Langley, 1895; 1897). Thus ganglion cells are once more innervated by contiguous spinal cord segments whose activity produces postganglionic responses in appropriate peripheral locations. In addition, it seems likely that ganglion cells, like muscle cells, are reinnervated according to functional criteria (see Purves and Lichtman, 1978). For example, in frog sympathetic ganglia two distinct classes of ganglion cells (B and C cells) are correctly reinnervated (Feldman, 1979), as are choroidal and ciliary neurons in the chick ciliary ganglion (Landmesser and Pilar, 1970). This aspect of ganglionic innervation, however, has not been studied in any detail in mammals.

Support for the idea that mammalian ganglion cells are distinguished by positional (and perhaps functional) criteria during synaptogenesis is provided by studies of the segmental reinnervation of *different* ganglia in the sympathetic chain. In the fifth thoracic ganglion, for example, virtually all the cells are strongly innervated by midthoracic segments (Lichtman et al., 1980). This contrasts with the superior cervical ganglion in which some cells are strongly innervated by upper thoracic segments (C8–T3), while others are strongly innervated by midthoracic segments (T4–T7). The segmental innervation of the fifth thoracic ganglion is consistent with the observation that postganglionic projection is correlated with the rostrocaudal level of preganglionic neurons: because most fifth thoracic ganglion cells project to a limited peripheral territory via the fifth thoracic spinal nerve, a relatively narrow segmental innervation of this ganglion is expected (Lichtman et al., 1980). The importance of these findings in the present context is the suggestion that each sympathetic chain ganglion is composed of neurons with appreciably different selective properties.

The possibility that the neurons that make up two different ganglia can be distinguished by a given set of axons can be tested by transplanting the fifth thoracic ganglion to the neck of a host animal where it becomes reinnervated by the population of preganglionic axons that normally innervates the superior cervical ganglion (Purves, Thompson, and Yip, 1981; see also Purves and Thompson, 1979). By transplanting superior cervical ganglia as controls, one can compare the reinnervation of two different ganglia by the same preganglionic axons. The result of this experiment is that, on average, neurons in transplanted fifth thoracic ganglia are better reinnervated by axons arising from a more caudal set of spinal segments than are superior cervical ganglion cells. This finding supports the conclusion that sympathetic ganglion cells are endowed with some property, apparently permanent, that biases the innervation they receive. At present, this is the most direct evidence one has that mammalian neurons bear some sort of identifying label that is important during synapse formation.

CHEMOAFFINITY AS THE BASIS FOR NEURAL SPECIFICITY

The idea of a chemical affinity between appropriate presynaptic and postsynaptic classes was first put forward by J. N. Langley in 1895 to explain his studies of the reinnervation of sympathetic ganglia (Langley, 1895; 1897). Langley had no reason to extend his explanation to a general proposal about neural specificity, and little further attention was paid to his idea. It remained for R. W. Sperry in the 1940s to elevate chemoaffinity to the status of a widely discussed hypothesis (see Sperry, 1963). The chemoaffinity hypothesis states (1) that specific neural connections arise from matching chemical labels on the elements that are ultimately to be linked and (2) that such labels might be relevant to either the guidance of axons to their targets or to selective synapse formation (Sperry, 1963; see also Attardi and Sperry, 1963). The problem with this hypothesis is not so much the general idea (which still seems to me as viable as when it was first proposed) but rather how to relate the hypothesis to actual synaptogenic events.

The work on selective synapse formation in the peripheral nervous system that I have discussed may inform the general idea of chemoaffinity in several ways. Most important, perhaps, the evidence that muscle fibers and autonomic ganglion cells have some property that biases the innervation they receive offers the only direct support for the view that, amongst vertebrates, chemoaffinity might operate at the level of synapse formation. However, the nature of selective synapse formation in the peripheral nervous system raises a number of puzzling questions which suggest that patterns of synaptic connections cannot be due simply to matching labels on the presynaptic and postsynaptic elements.

First, a good deal of evidence indicates that trophic factors produced by target cells contribute to the control of both the formation and maintenance of synapses (see Purves, 1977; Purves and Lichtman, 1978). It seems likely that the protein nerve growth factor (NGF), discovered by Viktor and Rita Levi-Montalcini (Levi-Montalcini and Hamburger, 1953; Levi-Montalcini, Meyer, and Hamburger, 1954) is one such agent, that acts to regulate the sympathetic innervation of postganglionic targets. Such factors might actually be considered mediators of chemoaffinity in a broad sense, since NGF appears to promote innervation by sympathetic (and dorsal root) ganglion cells but not by other classes of neurons. Do such "chemoaffinities" provide a mechanism of synaptic modulation entirely apart from the kind of surface labels envisioned by Sperry and others?

Second, the *activity* of presynaptic and postsynaptic cells seems importantly related to the pattern of innervation ultimately established. Thus, the innervation of particular cells in the visual cortex of cats or monkeys is dramatically influenced by the function of the axons that contact them at an early stage of development (e.g., see Hubel, 1978). A similar influence of activity may explain some aspects of the pattern of synaptic connections in the peripheral nervous system as well (Purves and Lichtman, 1980). How is this functional influence related to the idea of chemoaffinity?

Third, the nature of selective synapse formation in the periphery suggests that postsynaptic cells are innervated according to both functional and positional criteria. Does this mean that two different labels encode these qualities? Or might one quality be expressed through a chemical label, and might the other be encoded in an entirely different fashion, for example, by feedback based on neural activity?

Finally, what is the relation of chemoaffinity at the level of synapse formation to the accurate outgrowth of axons to their targets? This is a particularly interesting question vis-à-vis sympathetic ganglia where a positional quality of the presynaptic and postsynaptic elements appears to be important in synaptogenesis. Since axons presumably make use of positional information in reaching their intended destination, might the same chemoaffinity label influence both an axon's direction of growth and its tendency to form synapses with particular cells?

Although the answers to these fundamental questions are a matter of speculation, it seems clear that chemoaffinity cannot provide a *complete* explanation of selective synapse formation, much less neural specificity.

CONCLUSION

Stent (1980) and others have commented on the importance of context in the success or failure of scientific ideas. Because the general idea of chemoaffinity has not matured into a more detailed statement of how neural specificity occurs, it has not, in this sense, succeeded. I think that this attractive idea has failed to mature because chemoaffinity (and neurospecificity in general) has been considered almost exclusively in the context of the retinotectal system of lower vertebrates. Another context in which to consider specificity is the peripheral nervous system. It is a richer context, in my view, because information about this part of the nervous system has been gained independently of particular arguments and preconceptions about the nature of neural specificity. Much of the richness of the peripheral context is a direct result of Viktor's work. The vital influence of peripheral targets on the neurons that innervate them, the importance of neuronal competition, and the discovery with Levi-Montalcini of nerve growth factor and its possible trophic role, are some of the major themes that Viktor has fostered over the years. If chemoaffinity is shown to affect peripheral innervation, the establishment of this influence will not represent an ultimate explanation, but will add another mechanism to those already known to be important in shaping the synaptic connections between neurons and their targets.

ACKNOWLEDGMENTS

I am grateful to J. W. Lichtman for many useful discussions, and to E. Rubin, R. I. Hume, D. J. Wigston, and D. A. Johnson for critical comments. My own work is supported by USPHS Grant NS 11699 and a grant from the Muscular Dystrophy Association.

REFERENCES

Attardi, D. G., Sperry, R. W. 1963. Preferential selection of central pathways by regenerating optic fibers. *Exp. Neurol.* 7:46–64

Barker, D. 1974. The morphology of muscle receptors. In *Muscle Receptors. Handbook of Sensory Physiology,* vol. III, ed. by C. C. Hunt, pp. 1–190. New York: Springer-Verlag

Bessou, P., Laporte, Y., Pages, B. 1965. Observations sur la reinnervation de fuseaux neuro-musculaires de chat. *C. R. Soc. Biol.* 160:408–11

Bennett, M. R., Pettigrew, A. G. 1976. The formation of neuromuscular synapses. *Cold Spring Harbor Symp. Quant. Biol.* 40:409–24

Bernstein, J. J., Guth, L. 1963. Non-selectivity in the establishment of neuromuscular connections following nerve regeneration in the rat. *Exp. Neurol.* 4:262–75

Brown, M. C., Butler, R. 1976. Regeneration of afferent and efferent fibres to muscle spindles after nerve injury in adult cats. *J. Physiol.* (London) 260:253–66

Brushart, T. M., Mesulam, M.-M. 1980. Alteration in connections between muscle and anterior horn motoneurons after peripheral nerve repair. *Science* 208:603–5

Burke, W., Ginsborg, B. L. 1956. The electrical properties of slow muscle fibre membrane. *J. Physiol.* (London) 132:587–98

Dennis, M. J., Yip, J. W. 1978. Formation and elimination of foreign synapses on adult salamander muscle. *J. Physiol.* (London) 274:299–310

Elul, R., Miledi, R., Stefani, E. 1970. Neural control of contracture in slow muscle fibres of the frog. *Acta Physiol. Lat. Am.* 20:194–226

Feldman, D. 1979. Specificity of reinnervation of frog sympathetic ganglia. *Neurosci. Abstr.* 5:625

Feng, T. P., Wu, W. Y., Yang, F. Y. 1965. Selective reinnervation of "slow" or "fast" muscle by its original motor supply during regeneration of mixed nerve. *Sci. Sin.* 14:1717–20

Gerding, R., Robbins, N., Antosiak, J. 1977. Efficiency of reinnervation of neonatal rat muscle by original and foreign nerves. *Dev. Biol.* 61:177–83

Hoh, J. F. Y. 1971. Selective reinnervation of fast-twitch and slow-graded muscle fibers in the toad. *Exp. Neurol.* 30:263–76

Hoh, J. F. Y. 1975. Selective and nonselective reinnervation of fast-twitch and slow-twitch rat skeletal muscle. *J. Physiol.* (London) 251:791–801

Hollyday, M., Hamburger, V., Farris, J. M. G. 1977. Localization of motor neuron pools supplying identified muscles in normal and supernumerary legs of chick embryo. *Proc. Natl. Acad. Sci. U.S.A.* 74:3582–6

Horder, T. J., Martin, K. A. C. 1979. Morphogenetics as an alternative to chemospecificity in the formation of nerve connections. *Symp. Soc. Exp. Biol.* 32:275–359

Hubel, D. 1978. The effects of deprivation on the visual cortex of cat and monkey. In *The Harvey Lectures,* Ser. 72, pp. 1–51. New York: Academic

Kuffler, S. W., Vaughan-Williams, E. M. 1953. Small nerve junction potentials. The distribution of small motor nerves to frog skeletal muscle and the membrane characteristics of the fibres they innervate. *J. Physiol.* (London 121:289–317

Landmesser, L., Pilar, G. 1970. Selective re-innervation of two cell populations in the adult pigeon ciliary ganglion. *J. Physiol.* (London) 211:203

Langley, J. N. 1892. On the origin from the spinal cord of the cervical and upper thoracic sympathetic fibres, with some observations on white and grey rami communicantes. *Phil. Trans.* (London) 183B:85–124

Langley, J. N. 1895. Note on regeneration of pre-ganglionic fibers of the sympathetic. *J. Physiol.* (London) 18:280–84

Langley, J. N. 1897. On the regeneration of pre-ganglionic and of post-ganglionic visceral nerve fibres. *J. Physiol.* (London) 22:215–30

Levi-Montalcini, R., Hamburger, V. 1953. A diffusible agent of mouse sarcoma, producing hyperplasia of sympathetic ganglia and hyperneurotization of viscera in the chick embryo. *J. Exp. Zool.* 123:233–88

Levi-Montalcini, R., Meyer, H., Hamburger, V. 1954. *In vitro* experiments on the effects of mouse sarcomas 180 and 37 on the spinal and sympathetic ganglia of chick embryos. *Cancer Res.* 14:49–57

Lichtman, J. W., Purves, D. 1980. The elimination of redundant preganglionic innervation to hamster sympathetic ganglion cells in early post-natal life. *J. Physiol.* (London) 301:213–28

Lichtman, J. W., Purves, D., Yip, J. W. 1979. On the purpose of selective innervation of guinea-pig superior cervical ganglion cells. *J. Physiol.* (London) 292:69–84

Lichtman, J. W., Purves, D., Yip, J. W. 1980. Innervation of sympathetic neurones in the guinea-pig sympathetic chain. *J. Physiol.* (London) 298:285–99

Miledi, R., Stefani, E. 1969. Non-selective reinnervation of slow and fast muscle fibres in the rat. *Nature* (London) 222:569–71

Njå, A., Purves, D. 1977a. Specific innervation of guinea-pig superior cervical ganglion cells by preganglionic fibres arising from different levels of the spinal cord. *J. Physiol.* (London) 264:565–83

Njå, A., Purves, D. 1977b. Re-innervation of guinea-pig superior cervical ganglion cells by preganglionic fibres arising from different levels of the spinal cord. *J. Physiol.* (London) 273:633–51

Njå, A., Purves, D. 1978. Specificity of initial synaptic contacts on guinea-pig superior cervical ganglion cells during regeneration of the cervical sympathetic trunk. *J. Physiol.* (London) 281:45–62

Purves, D. 1977. The formation and maintenance of neural connections. In *Function and Formation of Neural Systems,* ed. G. S. Stent, pp. 21–49. Berlin: Dahlem Conferenzen

Purves, D., Lichtman, J. W. 1978. Formation and maintenance of synaptic connections in autonomic ganglia. *Physiol. Rev.* 58:821–62

Purves, D., Lichtman, J. W. 1980. Elimination of synaptic connections in the developing mammalian nervous system. *Science* 210: 153–157

Purves, D., Thompson, W. 1979. The effects of post-ganglionic axotomy on selective synaptic connexions in the superior cervical ganglion of the guinea-pig. *J. Physiol.* (London) 297:95–110

Purves, D., Thompson, W., Yip, J. W. 1981. Re-innervation of ganglia transplanted to the neck from different levels of the guinea-pig sympathetic chain. *J. Physiol.* (London) (in press)

Rubin, E., Purves, D. 1980. Segmental organization of sympathetic preganglionic neurons in the mammalian spinal cord. *J. Comp. Neurol.* 192:163–74

Schmidt, H., Stefani, E. 1976. Re-innervation of twitch and slow muscle fibres of the frog after crushing the motor nerves. *J. Physiol.* (London) 258:99–123

Schmidt, H., Stefani, E. 1977. Action potentials in slow muscle fibres of the frog during regeneration of motor nerves. *J. Physiol.* (London) 270:507–17

Sperry, R. W. 1943. The effect of 180 degree rotation of the retinal field on visuomotor coordination. *J. Exp. Zool.* 92:263–79

Sperry, R. W. 1963. Chemoaffinity in the orderly growth of nerve fiber patterns and connections. *Proc. Natl. Acad. Sci. U.S.A.* 50:703–10

Sperry, R. W., Arora, H. L. 1965. Selectivity in regeneration of the oculomotor nerve in the cichlid fish, *Astronotus ocellatus. J. Embryol. Exp. Morphol.* 14:307–17

Stent, G. 1980. To the Stockholm station: Makers of the molecular-biological revolution. *Encounter* 54:79–85

Thulin, C. A. 1960. Electrophysiological studies of peripheral nerve regeneration with special reference to small diameter (γ) fibres. *Exp. Neurol.* 2:598–612

Weiss, P., Hoag, A. 1946. Competitive reinnervation of rat muscles by their own and foreign nerves. *J. Neurophysiol.* 9:413–18

Wigston, D. J. 1980. Suppression of sprouted synapses in *Axolotl* muscle by transplanted foreign nerves. *J. Physiol.* (London) 307:355–66

11 / Cellular Recognition During Neural Development

DAVID I. GOTTLIEB
LUIS GLASER

It now appears that the complicated nerve fiber circuits of the brain grow, assemble and organize themselves through the use of intricate chemical codes under genetic control. Early in development, the nerve cells, numbering in the billions, acquire and retain thereafter, individual identification tags, chemical in nature by which they can be recognized and distinguished from one another.

Roger W. Sperry, 1965

ROGER SPERRY'S CHEMOAFFINITY HYPOTHESIS, elegantly summarized in the preceding quotation (Sperry, 1965), is one of the central ideas of contemporary neurobiology. When Sperry first proposed the chemoaffinity hypothesis, the experimental evidence for it consisted entirely of behavioral and anatomical studies. The proposed identification tags were necessary to explain the way axons grow during the regeneration of altered nervous pathways; the hypothesis gained widespread support be-

Reproduced, with permission, from the *Annual Review of Neuroscience,* 1980, 3:303–18.

Authors' note: At the time Max Cowan approached us about a contribution to the present volume, we had just completed a review for the *Annual Review of Neuroscience.* The field under consideration, namely, studies on the molecular basis of cell recognition in the nervous system, is hardly noted for its rapid pace, and so a new review of this area seemed out of the question. So, too, did the possibility of not joining in this effort to congratulate Viktor Hamburger on the occasion of his 80th birthday, since we owe him two distinct debts. Scientifically, he is one of the true founders of the discipline in which we labor. Much that is conceptually clear and enduring in neuroembryology can be traced to his experiments and writings. As a colleague and friend, he has almost defined the standard for scholarship, fairness, and integrity. It is a privilege to dedicate our review to Viktor Hamburger.

cause it was extremely difficult to imagine other reasonable explanations for the anatomical and behavioral results.

Experimental support for the chemoaffinity hypothesis could theoretically be of two distinct kinds. In the first, the nervous system is altered and the assembly of nerve circuits is observed and compared to that in normal animals. If chemical identification tags offer a parsimonious explanation for the observed patterns of growth and connections, then the theory gains support. More direct support must come from experiments of a second kind, which study actual molecular content, in order to test the proposition that certain macromolecules mediate the formation of specific connections. The detailed structure and localization of the molecules must be given and perturbation of the molecules shown to lead to perturbations in the formation of pathways.

Research on neuronal specificity is now at an exciting juncture. A large body of experiments of the first kind described above have greatly strengthened and extended the experimental support for chemoaffinity as the basis for the normal assembly of nerve circuits. Simultaneously, a number of research groups have made impressive gains in studying the molecular basis of *in vitro* models for cell recognition in the nervous system. In light of these recent developments, we address ourselves to three tasks in our review: (1) we cite a number of experiments on neurospecificity done since Sperry's formulation of the chemoaffinity hypothesis, which add strong and novel support to that view; (2) we review recent work dealing with the molecular basis of embryonal cell adhesion and recognition in developing nervous tissues; (3) we enumerate some of the technical obstacles that must be overcome if molecular studies of synaptic recognition are to proceed and review hopeful developments in this area.

FURTHER EVIDENCE IN FAVOR OF CHEMOAFFINITY

Sperry's pioneering studies spawned a large field of research which has been treated in a number of comprehensive reviews (Gaze, 1970; Hunt and Jacobson, 1974; Jacobson, 1978; Cowan, 1978). Of the many experiments, we will cite a few that provide particularly strong support for the view first enunciated by Sperry. Attempts to characterize synaptic recognition at a molecular level are likely to be frustrated in the short-term by technical constraints, but by marshalling the strongest evidence for chemoaffinity, we hope to underscore the incentive for studying various model systems of cell-cell recognition.

Sperry's original data suffered from two major limitations. The first was that all of his critical experiments were performed on animals undergoing regeneration of nervous pathways (Sperry, 1943, 1944, 1945; Attardi and Sperry, 1960, 1961). Although the extension of these results to original embryonic outgrowth seemed natural, direct evidence that the foundation of embryonic pathways and their regeneration follow the same principles was lacking. Second, in many cases the patterns of pathway regeneration was inferred from behavioral results rather than from direct anatomical analysis. In some cases anatomy was done, but only to the level of following major fiber bundles with stains for normal material (Attardi and Sperry, 1961). The studies of Jacobson and his colleagues on the developing frog visual system (reviewed in Hunt and Jacobson, 1974) have shown that primary embryonic outgrowths do indeed show the same principles first discovered for regenerating pathways. Studies by Cowan and his colleagues on the visual system of the chick have made a number of important contributions to this issue. In one of these studies (Crossland et al., 1974) parts of the early developing retina were surgically removed. The remaining intact part of the retina projected to the tectum. Several crucial observations were made:

1. The remaining parts of the retina projected to the appropriate parts of the tectum.
2. Fibers growing from partial retinas crossed over large portions of the uninnervated tectum, ignoring countless opportunities to form synapses with sites normally innervated by excised portions of the retina.
3. The actual terminals of the axons from partial retinae were studied and shown to be indistinguishable from normal retinal terminals. These facts find their most plausible explanation in the chemoaffinity hypothesis.

In its simplest form, the chemoaffinity hypothesis also predicts that cells of the tectum should also be specified. To determine if this is so, experiments have been performed in which portions of the adult tectum in frogs and fishes have been excised and repositioned. In many cases, there was a clearcut reorientation of the retinal projection as would be predicted from chemoaffinity (Yoon, 1975; Jacobson and Levine, 1975a,b). Finally, Chung and Cooke (1975) succeeded in rotating the optic tectum of the developing frog before the stage at which innervation by the optic nerve occurs. In cases in which a portion of the diencephalon was included with the rotated piece of tectum, inverted retinal maps were formed. Thus some positional information must reside in the central nervous system.

Recent studies show that the innervation of the embryonic limb seems to proceed by similar mechanisms as those discussed for the visual system. When motoneuron axons first innervate the developing limb, they immediately go to the appropriate developing muscle mass rather than search for it by trial and error (Landmesser and Morris, 1975).

It has long been clear that each class of axon in the central nervous system chooses to synapse with only a very limited repertoire of cell types. Until recently this ability could be explained either by chemoaffinity or by elaborate timing mechanisms of axon and dendrite outgrowth.

However, thalamocortical afferents in the reeler mouse find their appropriate target cells in spite of the gross rearrangement of cortical cellular laminae (Caviness and Rakic, 1978). The thalamocortical afferents follow curved trajectories to their target cells; the suggestion that the axons are following chemical cues is strong, and it is difficult to see how a timing mechanism could generate this pattern. Similar conclusions are drawn from the study of regenerating nerve pathways in invertebrates. In the case of the leech, certain central axons regenerate and reinnervate their original targets, while ignoring countless inappropriate cells (Baylor and Nicholls, 1971, Jansen and Nicholls, 1972). In the house cricket, cercal afferents whose development has been experimentally delayed establish connections with appropriate central neurons (Edwards and Palka, 1971; Palka and Edwards, 1974). It is difficult to reconcile these observations with a mechanism that assigns a central role to timing of axon outgrowth in the specification of nerve circuits.

Although we have emphasized axonal connections, there are many other patterns of cell-cell association in the nervous system that suggest specific affinities between complementary cell surfaces. The dramatic, characteristic lamination of neurons in many parts of the central nervous system and the specific association between glia and neurons come readily to mind. The fact that some of these patterns can be reconstructed from dissociated cells *in vitro* is most easily explained by the chemoaffinity theory (Garber and Moscona, 1972a,b; Delong and Sidman, 1970).

STUDIES OF CELL ADHESION

As discussed in the previous section, cell recognition in the nervous system is presumed to be an important component for the three-dimensional arrangement of cells in any given region of the nervous system as well as for synaptogenesis. These two situations probably represent different degrees of complexity in that in the former, one need only postulate a very limited number of cell surface components responsible for cell recognition, whereas in the latter case, a large number of cell surface components or combination of components have to be invoked in order to account for specific patterns of synapse formation. As detailed above, the specificity for these interactions is determined at least initially by complementary surface molecules present on the interacting cells, a model first presented by Sperry (1963) and elaborated into specific chemical models by other investigators (e.g., Barondes, 1970; Barbera, Marchase, and Roth, 1973, Barbera, 1975). This hypothesis is by no means proven, but would seem most reasonable in light of our knowledge of cell surfaces. It appears less likely that cellular organization is totally the consequence of gradients of soluble molecules, although these may be important in specific instances (Crick, 1970; Wolpert, 1971)

Recently it has been suggested that reinnervation of muscle at the original synaptic site may be determined by the basal lamina rather than by the surface of the muscle cell (Sanes, Marshall, and McMahan, 1978). While regeneration of a neuromuscular junction may be a special situation not directly applicable to normal neuronal development, the possible contribution of the extracellular matrix to the organization of cells is an important topic for histotypic differentiation, but one outside the scope of this review.

The Sperry model fundamentally assumes not only that the presence of complementary molecules on the surface of interacting cells governs cellular adhesion but also that during development, equilibrium is reached such that within a cell population those cells will bind to each other that have either the largest number of complementary ligands, or the highest affinity, or both. Implied in such a model is functional reversibility of adhesion. The high specificity required by such a model implies that at least one of the specific cell surface ligands involved in cell adhesion is a protein.

The study of the molecules involved in cellular recognition, either in the nervous system or elsewhere, is complicated by the fact that standard biochemical methods apply, to a large extent, only to large numbers of cells or to components extracted from such cells. Currently this means that within the nervous system, cellular recognition can only be studied in very heterogenous collections of cells. This contrasts strikingly with developmental studies that suggest precise specification of cell interactions and allow distinctions to be made between similar cells in adjacent regions (e.g., the neuroretina as discussed in the previous section).

An ideal system for the study of cell recognition would require the availability of large populations of homogenous cells, appropriately differentiated, wherein specific cell recognition can be recognized by the presence of either physiological or morphological alterations in the interacting cells. With very few exceptions, this ideal situation is at present unattainable. Therefore a large amount of work in this area has been devoted to the study of two types of model systems utilizing (1) heterogenous cell populations of known anatomical origin and (2) homogenous populations of cells in culture, whose anatomical origin is not always known. An additional complication arises in these studies from the fact that alterations of the cell surface brought about by the procedures involved in cell dissociation and so on cannot always be adequately controlled.

Any given cell may interact with a variety of other cells; different cell surface ligands are likely to be involved in these interactions. For example, a given cell specifically interacts with adjacent cells to form a particular region or layer in the nervous system; this interaction presumably involves ligands different from the interaction of this same cell with axons derived from other regions of the nervous system.

It is convenient to discuss the model systems that have been studied based on the methodology used in the investigation:

Studies of Cell Recognition Using Intact Cells

These methods all seek to discover whether a particular cell can adhere to another cell of either homologous or heterologous type. The measurement that is usually but not always made (e.g., see Beug and Gerisch, 1972) is one of rate of adhesion, which is assumed to reflect the number of sites and the types of adhesive sites present on the two cell surfaces. A number of methods have been devised for such studies, most of which immobilize a large number of cells of a given type and ask whether cells of other types can adhere to these immobilized cells (Roth, 1968; Roth, McGuire, and Roseman, 1971; Grady and McGuire, 1976a,b; McGuire and Burdick, 1976; Walther, Ohman, and Roseman, 1973; Gottlieb and Glaser, 1975). These methods have been reviewed recently and are not discussed in detail (Frazier and Glaser, 1979). The adhesion of one cell to another under these conditions appears to be a complex phenomenon that can be seen as a reversible step followed by functionally irreversible events (Umbreit and Roseman, 1975). Some of the existing data suggest that the first reversible step represents the binding of a small number of ligands to each other, whereas the irreversible step may represent the binding of a large number of ligands from one cell to the other (Moya, Silbert, and Glaser, 1979). While the minimum number of molecules required to obtain specific cell adhesion is one or two different molecules, the complexity of the process suggests that a larger number of molecules will be involved.

One of the favorite systems for the study of cell adhesion has been the embryonal nervous system, and in particular the retinotectal system. This is due in part to the Sperry hypothesis for the establishment of retinotectal connectivity, but also because pioneering studies by Moscona and coworkers showed that embryonic chick neural retina can readily be dissociated into single cells which, under appropriate conditions, will reaggregate and undergo histiotypic differentiation (Moscona, 1965; Fujisawa, 1973; Moscona and Hausman, 1977). Such aggregates will also show spatial segregation, not only between cells from different organs (e.g., liver and retina) but also between cells from different regions of the nervous system (Garber and Moscona, 1972a,b). This classical demonstration of cell specificity does not readily provide a tool for the identification of the molecules involved in this sorting out process because of the complexity of the systems involved and the relatively long time period required for segregation.

The available evidence suggests that neuronal cells can be easily distinguished from nonneuronal cells in adhesion assays (e.g., liver or heart) but the distinction between different regions of the embryonal nervous system is less apparent. A discussion of these differences should start with a discussion of the studies of adhesive specificity in the neuroretina of the chick embryo.

Neural retina cells, obtained by gentle dissociation with trypsin, will rapidly adhere to each other to form large aggregates. Because of the general interest in retinotectal connectivity, Roth and coworkers examined the adhesion of such cells to tectal halves (Barbera et al., 1973; Barbera, 1975, Marchase, Barbera, and Roth, 1975; Marchase, 1977). They found that retinal cells prepared from the ventral half of the neural retina adhered preferentially (faster) to dorsal tectal halves as compared to ventral tectal halves; the converse was true for cells obtained from the dorsal half of the neural retina. It was therefore tempting to conclude that this asymmetry is related to retinotectal connectivity in that axons from the dorsal half of the neuroretina connect to the ventral half of the tectum, and conversely axons from the ventral half of the neuroretina connect to the dorsal half of the tectum. (Similar results have been obtained using monolayers of tectal cells rather than tectal halves [Gottlieb and Arington 1979].) Implicit in this interpretation is the notion that at least at early stages of development all neural retina cells carry the same cell surface markers.

In related studies, Gottlieb and coworkers (1976) found an adhesive gradient within the neural retina such that cells from the extreme dorsal area of the neural retina adhere preferentially (faster) to a monolayer of cells from the extreme ventral area of the neural retina and then adhere progressively less well (slower) to cells from increasingly dorsal areas of the retina. The converse is true of cells from the extreme ventral area of the neural retina. Neuraminic acid residues on glycoproteins may be involved in this preferential adhesion (Cafferata, Panosian, and Bordley, 1979).

A more extensive series of studies of adhesive preference of dissociated cells obtained from the nervous system has led to the following conclusion (Gottlieb and Arington, 1979): Homotypic cell adhesion is not necessarily the preferred mode of cell adhesion, and within any defined anatomical region there occur significant differences in cellular adhesion rate (which may represent qualitative or quantitative changes in the adhesive component). These differences occur along one of the major axes of the embryo, and they reflect the presence of an adhesive gradient which may have considerable developmental significance in that it may provide embryonal cells with information regarding the orientation of the embryo. Within any given area of the nervous system, adjacent cells do not necessarily show the highest adhesive preference to each other. Finally, adhesive preference is shown by cells that normally are not synaptically related.

The chemical basis for the adhesive specificity shown in these tests is not understood, and it is unlikely to be understood entirely unless some attempt can be made to isolate the cell surface components responsible for cell adhesion. A preliminary attempt has been made to identify such molecules on the basis of their susceptibility to hydrolytic enzymes (Marchase, 1977). These observations have suggested that adhesion between retinal cells and tectal halves is mediated in part by a carbohydrate-binding protein specific for *N*—acetyl—D—galactosamine (GalNAc) residues and a GalNAc-containing molecule, possibly the ganglioside GM-2. These two molecules are assumed to be distributed assymetrically in the retina and tectum with highest concentration of the binding protein on the ventral half of each structure. It should be noted that the adhesive characteristic of these cells cannot be explained entirely by these ligands because treatment with appropriate hydrolytic enzymes prevents selective cell adhesion but allows adhesion to take place at the same rate for dorsal and ventral cells.

The data obtained by these methods are clearly of interest in that they define differences in cell surface adhesive components and reveal the presence of a gradient, or gradients, of those components that are likely to be of importance for cell orientation during development. The methods also have serious limitations in that mixtures of cell types are used, and there is some uncertainty as to whether the adhesive components revealed by these *in vitro* assays are biologically functional. The amount of chemical information obtained in this type of study is limited, although, as discussed below, specific antibodies may prove to be extremely useful in identifying adhesive molecules.

Studies Using Plasma Membranes

Alternative approaches to the study of cell recognition require the development of systems in which the binding of a cell surface component to cells is measured; such systems would allow purification and identification of these surface components.

In several systems it has been possible to demonstrate that plasma membrane fractions retain the ability to bind to cells and do so with specificity, that is, they bind very much faster to one cell type than to another (Merrell and Glaser, 1973; Gottlieb, Merrell, and Glaser, 1974, Obrink, Kuhlenschmidt, and Roseman, 1977a; Obrink, Warmergard, and Pertoft, 1977b; Santala, Gottlieb, and Littman, 1977). In addition to binding assays, membranes have also been tested as agents that prevent cell aggregation,[1] presumably by competing for cell adhesive sites (Merrell and Glaser 1973, Gottlieb et al., 1974), or in the case of liver, promote aggregation by crosslinking cells (Obrink, Kuhlenschmidt, and Roseman, 1977a). In the nervous system these membranes have indicated the presence of temporal changes in cell surface properties of retina and tectum and have demonstrated an interesting unidirectional reactivity between retina and tectum in that retinal membranes inhibit aggregation of both retinal and tectal cells, but tectal membranes appear to only inhibit tectal cell aggregation (Gottlieb et al., 1974).

Until recently only modest success was obtained in isolating the membrane components responsible for specific adhesion in these systems (Merrell, Gottlieb, and Glaser, 1975a). However, recently a small molecular weight protein that is at least partly responsible for the inhibition by retinal cell membranes of cell-to-cell adhesion has been isolated as a pure component and partially characterized (Jakoi and Marchase, 1979).

Specific adhesion can also be demonstrated with permanent cell lines derived from the nervous system and with plasma membranes derived from these cells (Santala et al., 1977). The origin within the nervous system of these established cell lines is not known and therefore the relevance of these observations to adhesive components in the nervous stystem is not always apparent. Work with such cell lines has, however, led to several conclusions that are of general applicability. These conclusions could have been obtained only with these cell lines because they are homogenous cell populations.

Any given cell contains a number of different molecules on its surface which can participate in specific cell adhesion. Molecules may be present on a cell that allow it to adhere to heterologous cells but that are not involved in homologous cell adhesion (Santala et al., 1977; Stallcup, 1977). Cell adhesion can be influenced by trophic factors (Merrell el al., 1975b) and the nature of the adhesive factors present on the cell surface can be altered by culture conditions (Santala and Glaser, 1977)—most notably cell density and cell-to-cell contact.[2] When extrapolated to the nervous system, these observations suggest that cells' adhesive properties will alter during development and may alter as a consequence of contact with either heterologous or homologous cells. This last point is of particular interest in that normal development must be necessity involve both adhesion between cells and the breaking of such connections. "De-adhesion" may therefore arise as a consequence of changes in cell surface compositions induced by trophic factors or contacts with other cells (see also Changeux and Danchin, 1976).

Very little information is available regarding the reversal of adhesion. Observations in this regard have been made in *Chlamydomonas* (Snell and Roseman, 1977) but this system may not be an appropriate model for the nervous system. In culture what appears to be "synapse" formation between retina cells and muscle has been noted, but this is reversed with time. The survival of appropriate cells, or their ability to synthesize transmitter, is not well enough understood under these culture conditions to be certain that this really represents a reversal of adhesion between cells for which synapse formation is inappropriate (Ruffolo et al., 1978). The suggestion that under the same culture conditions stable synapses are formed between cells that normally synapse *in vivo* has not yet been reported in full.

Aggregation Promoting Factors

Factors that promote cell aggregation can generally be considered in two categories. The first are lectins or antibodies, which are multivalent ligands that can aggregate cells. A discussion of such ligands is outside the scope of this review, except to point out that multivalent carbohydrate binding proteins have been found in vertebrate cells (see Frazier and Glaser, 1979), including cells in the nervous system; although in principle these could be involved in cell-cell adhesion, evidence in this regard (except for liver) (Obrink et al 1977a, Schnaar et al., 1978) is lacking.

Proteins that specifically promote the formation of large aggregates of neuronal calls and that show regional specificity within the nervous system have been described through extensive work in the laboratory of A. A. Moscona (Hausman and Moscona, 1975, 1976; Moscona and Hausman, 1977). One of these proteins has been obtained from neural retina in pure form; it stimulates the formation of large retinal cell aggregates, after 24-hr incubation, while the same cells incubated in the absence of the protein show only the formation of small aggregates. The precise function of this protein is not known, although it has been shown to be located on the surface of the retinal cells (Hausman and Moscona, 1979) and can be isolated from a surface membrane fraction. Whether this and other factors represent adhesive molecules or trophic agents, they provide a striking molecular correlate of anatomical differences between major areas of the nervous system. Taurine, a small molecular-weight amino acid, functions as an aggregation-promoting factor for liver (Sankaran et al., 1977).

Use of Antibodies to Study Cell Recognition

Antibodies have generally been prepared against determinants in different regions of the nervous system with the expectation of generating antibodies specific for a unique cell type. In a few examples antibodies have been obtained that interfere with cell adhesion (Rutishauser et al., 1976, Brackenbury et al., 1977; Thiery et al., 1977; McClay, Gooding, and Fransen, 1977; Rutishauser, Gall, and Edelman, 1978a; Rutishauser et al., 1978b). The most extensively studied of these is an antibody against a cell surface molecule designated as CAM (cell adhesion molecule). CAM is widely distributed in the nervous system and anti-CAM Fab′ fragments prevent cell adhesion in a variety of assays. Anti-CAM Fab′ also prevents the formation of neurite bundles in cultures of dorsal root ganglia. CAM appears to be particularly enriched in neurite membranes.

This antibody defines a unique molecule (mol wt 140,000) present in the surface of neuronal cells and is a component of the adhesive mechanism. Since this molecule does not show regional specificity, it most likely represents a common component of the cell surface adhesive mechanism.

A related approach is the use of lectins, or antibodies to yeast or bacterial carbohydrates, to block cellular interactions. Interesting data in these systems have been obtained in studies of cerebellar development in normal and mutant mice. Whether such antibodies are specific enough to identify unique cell surface ligands remains to be established, as it seems likely that such reagents will react with a variety of cell surface molecules, although only a few of these will be responsible for the observed developmental change (Trinkner, 1978).

Biological Consequences of Cell Adhesion

A totally different approach would measure adhesive events that result in functional consequences to cells. This approach has the unique advantage that the biological meaning of the adhesive molecules is never in doubt and the biological assay provides an amplification of the binding event. This approach has only been used in a limited number of systems, and we will discuss only one of these in detail.

In 1975, Wood and Bunge (see also Wood, 1976) showed that rat dorsal root ganglia could be freed of fibroblasts and that such ganglia showed extensive developments of neurites and proliferation of Schwann cells. Removal of the ganglia from this culture left a bed of Schwann cells that ceased to divide but remained viable and could be induced to resume division if a new ganglion (free of Schwann cells) was added to the culture. The mitogenic effect of neurons on Schwann cells appears to require contact between the Schwann cells and neurites.

It has been possible to prepare a neurite membrane fraction that is mitogenic to Schwann cells (Salzer, Glaser, and Bunge, 1977; Bunge et al., 1979). The data obtained so far indicate that the mitogenic signal resides in the surface of the neurites (it is inactivated by mild trypsin treatment of intact cells and neurites) and is absent from the neuritic cytosol. The neurite membrane component appears to be highly specific since it is absent from a variety of other plasma membrane fractions obtained from cultured neuronal and nonneuronal cell lines. Mitogenic activity for Schwann cells prepared by different procedures have been noted for cholera toxin as well as components present in pituitary extracts and other cells (Raff et al., 1978a, Hanson and Partlow, 1978). The precise relation of such cells to those isolated according to the protocol of Wood (1976) is not yet clear. It should also be understood that the presence of mitogenic factors for Schwann cells in pituitary extracts is in no way incompatible with the presence of a specific membrane bound mitogenic factor on the surface of neurites.

Although this systems suffers from the disadvantage that the relevant cells can only be obtained in small quantities and are thus difficult objects for biochemical investigation, the system is ideal from other points of view—it uses defined cells and it measures a physiological consequence of cell adhesion. The fact that the biological effect can be reproduced at the membrane level strongly suggests that further purification and characterization of the membrane components should be feasible. A similar approach in a nonneuronal system, which has also made considerable progress, is the study of contact inhibition of growth of Swiss 3T3 cells (Whittenberger and Glaser, 1977; Whittenberger et al., 1978, Whittenberger, Raben, and Glaser, 1979; Bunge et al., 1979) wherein it has been possible to extract the membrane components responsible for growth control, but these have not yet been obtained in purified form.

FUTURE APPROACHES TO CELLULAR RECOGNITION

In the preceding sections, we have reviewed some of the successful attempts to probe the molecular basis of cell-cell recognition in the early developing nervous system. These studies have been successful insofar as they have given evidence for the existence of cell surface molecules involved in cell surface recognition. Nevertheless, the studies have important limitations. In many cases, heterogeneous cellular systems are being used and it is unclear on which particular cells the recognition molecules in question actually reside. In other cases, activities have been identified on whole cells or derivatives from them, but they have not been purified to homogeneity. Two important technical problems tend to limit further progress in this area. The first is the difficulty of obtaining purified populations of cells from the developing nervous system. Recent advances in the culture of both normal and transformed cells from the nervous system make it appear that this limitation will gradually be reduced. But even if we assume the availability of reasonable quantities of pure neuronal cells, there is a further problem that limits progress in this area, and that is the extreme complexity of the repertoire of cell surface proteins on any vertebrate cell. For this reason, progress in the study of cell recognition will be intimately related to progress in being able to characterize cell surfaces at the molecular level. Fortunately, there have been several recent advances in this regard that are pertinent to the present discussion. The ability to discriminate proteins on two-dimensional polyacrylamide gel electrophoresis systems has enabled investigators to definitively separate complex mixtures of proteins, such as we would expect to find in growing axons and dendrites. This technique has been applied to developing systems in general (e.g., Levinson et al., 1978) and to neurons in particular (Stone, Wilson, and Hall, 1978), with great success. The problem of mapping cell surface macromolecules in particular has been addressed by separating proteins by SDS-PAGE electrophoresis and staining glycoproteins with radioactive lectin preparations. Although first applied to nonneuronal cells (Burridge, 1976), this method has now been extended to the developing nervous system (Mintz and Glaser, 1978). It has been clearly shown that in the chick retina five glycoproteins are developmentally regulated. Three glycoproteins can be shown to be present in later stages of development although they are absent in earlier stages, and in the case of two other proteins, the converse is true. These are present during early development and disappear somewhat later. Since these studies are performed with very small masses of tissue, they can be extended to a number of developmental situations in nervous tissues.

Immunological methods also could potentially be of great value in the study of cell-cell recognition in the nervous system. These methods have already provided probes that clearly distinguish between different cell types in the nervous system. For example, Brockes and coworkers (1977) have shown that a particular antigen called RAN 1 is present on Schwann cells but not on fibroblasts; conversely, the antigen, theta, is present on fibroblasts and neurons but not on Schwann cells. Using immunological techniques, Raff and his colleagues (Raff et al., 1978b) have also shown that the glycolipid galactocerebroside is a rather specific marker for developing oligodendrocytes. The potential of immunological techniques has been greatly heightened by the introduction of the hybridoma method for obtaining antibodies. The unique advantage of this methodology is that purified antibodies can be obtained from antigens which have themselves not been purified. This method has already proven to be very useful in obtaining antibodies against specific subpopulations of lymphocytes (Williams, Galfre, and Milstein, 1977), and its extension into the nervous system should provide very refined maps of the cells surface of various types of central nervous system cells.

ACKNOWLEDGMENTS

Work in the authors' laboratory has been supported by Grants GM18405 and NSF PCM 77-15972 to L.G. and NS 12867 to D.I.G.

NOTES

1. The limitation to these inhibition assays have been discussed in detail previously (Merrell, Gottlieb, and Glaser, 1976).
2. Nerve growth factor has been shown to alter the adhesion of PC-12 pheochromocytomic cells not only to each other but also to substratum under culture conditions (Schubert and Whitlock, 1977). The importance of substratum for tissue culture cells in general has recently been reviewed and is outside the scope of this article (Grinnell, 1978; Grinnel and Hays, 1978). It is a subject of particular interest in the study of the nervous system, because successful culture of cells (as well as some cell separation methods) depends on the presence of a suitable surface on which cells can grow.

REFERENCES

Attardi, G., Sperry, R. W. 1960. Central routes taken by regenerating optic fibers. *Physiologist* 3:12

Attardi, G., Sperry, R. W. 1961. Preferential selection of central pathways by regenerating optic fibers. *Exp. Neurol.* 4:262–75

Attardi, G., Sperry, R. W. 1963. Preferential selection of central pathways be regenerating optic fibers. *Exp. Neurol.* 7:46–64

Barbera, A. J. 1975. Adhesive recognition between developing retinal cells and the optic tecta of the chick embryo. *Dev. Biol.* 46:167–91

Barbera, A. J., Marchase, R. B., Roth, S. 1973. Adhesive recognition and retinotectal specificity. *Proc. Natl. Acad. Sci. U.S.A.* 70:2482–86

Barondes, S. H. 1970. Brain glycomacromolecules and interneuronal recognition. In *Neurosciences Second Study Program,* ed. F. O. Schmitt, pp. 747–60. New York: Rockefeller

Baylor, D. A., Nicholls, J. G. 1971. Patterns of regeneration between individual nerve cells in the central nervous system of the leech. *Nature* (London) 232:268–70

Beug, H., Gerisch, G. 1972. A micromethod for routine measurement of cell agglutination and dissociation. *J. Immunol. Methods* 2:49–57

Brackenbury, R., Thiery, J. P., Rutishauser, U., Edelman, G. M. 1977. Adhesion among neural cells of the chick embryo. I. An immunological assay for molecules involved in cell-cell binging. *J. Biol. Chem.* 252:6835–40

Brockes, J. P., Fields, K. L., Raff, M. C. 1977. A surface antigenic marker for rat Schwann cells. *Nature* (London) 266:364–66

Bunge, R., Glaser, L., Lieberman, M., Raben, D., Salzer, J., Whittenberger, B., Woolsey, T. 1979. Growth control by cell to cell contact. *J. Supramol. Struct.*

Burridge, K. 1976. Changes in cellular glycoproteins after transformation: Identification of specific glycoproteins and antigens in sodium dodecyl sulfate gels. *Proc. Natl. Acad. Sci. U.S.A.* 73:4457–61

Cafferata, R., Panosian, J., Bordley, G. 1979. Developmental and biochemical studies of adhesive specificity among embryonic retinal cells. *Dev. Biol.* 69:108–17

Caviness, V. S. Jr., Rakic, P. 1978. Mechanisms of cortical development: A view from mutations in mice. *Ann. Rev. Neurosci.* 1:297–326

Changeux, J. P., Danchin, A. 1976. Selective stabilization of developing synapses as a mechanism for the specification of neuronal networks. *Nature* (London) 264: 705–12

Chung, S. H., Cooke, J. 1975. Polarity of structure and of ordered nerve connections in the developing amphibian brain. *Nature* (London) 258:126–32

Cowan, W. M. 1978. Aspects of neural development. In *International Review of Physiology Neurophysiology III*, ed. R. Porter, 17:149–89. Baltimore: Univ. Park Press

Crick, F. 1970. Diffusion in embryogenesis. *Nature* (London) 225:420–22

Crossland, W. J., Cowan, W. M., Rogers, L. A., Kelly, J. P. 1974. The specification of the retino-tectal projection in the chick. *J. Comp. Neurol.* 155:127–64

DeLong, G. R., Sidman, R. L. 1970. Alignment defect of reaggregating cells in cultures of developing brain of reeler mutant mice. *Dev. Biol.* 22:584–600

Edwards, J. S., Palka, J. 1971. Neural regeneration: Delayed formation of central contacts by insect sensory cells. *Science* 172:591–94

Frank, E., Jansen, J. K. S., Rinvik, E. 1975. A multisomatic axon in the central nervous system of the leech. *J. Comp. Neurol.* 159:1–14

Frazier, W., Glaser, L. 1979. Surface components and cell recognition. *Ann. Rev. Biochem.* 48:491–523

Fujisawa, H. 1973. The process of reconstruction of histological architecture from dissociated retina cells. *Wilhelm Roux' Arch. Entwicklungsmech. Org.* 171:312–30

Garber, B. B., Moscona, A. A. 1972a. Reconstruction of brain tissue from cell suspensions. I. Aggregation patterns of cells dissociated from different regions of the developing brain. *Dev. Biol.* 27:217–34

Garber, B. B., Moscona, A. A. 1972b. Reconstruction of brain tissue from cell suspensions. II. Specific enhancement of aggregation of embryonic cerebral cells by medium from homologous cell cultures. *Dev. Biol.* 27:235–43

Gaze, R. M. 1970. *The Formation of Nerve Connections*, pp. 1–128. London & New York: Academic

Gottlieb, D. I., Arington C. 1979. Patterns of adhesive specificity in the developing central nervous system of the chick. *Dev. Biol.* 71:260–73

Gottlieb, D. I., Glaser, L. 1975. A novel assay of neuronal cell adhesion. *Biochem. Biophys. Res. Commun.* 63:815–21

Gottlieb, D. I., Merrell, R., Glaser, L. 1974. Temporal changes in embryonal cell surface recognition. *Proc. Natl. Acad. Sci. U.S.A.* 71:1800–2

Gottlieb, D. I., Rock, K., Glaser, L. 1976. A gradient of adhesive specificity in developing avian retina. *Proc. Natl. Acad. Sci. U.S.A.* 73:410–14

Grady, S. R., McGuire, E. J. 1976a. Intercellular adhesive selectivity. III. Species selectivity of embryonic liver intercellular adhesion. *J. Cell Biol.* 71:96–106

Grady, S. R., McGuire, E. J. 1976b. Tissue selectivity of the initial phases of cell adhesion. *J. Cell Biol.* 70:346a

Grinnell, F. 1978. Cellular adhesiveness and extracellular substrata. *Int. Rev. of Cytol.* 53:65–144

Grinnell, F., Hays, D. G. 1978. Cell adhesion and spreading factor. *Exp. Cell Res.* 115:221–29

Hanson, G. R., Partlow, L. M. 1978. Stimulation of non-neuronal cell proliferation in vitro by mitogenic factors present in highly purified sympathetic neurons. *Brain Res.* 159:195–210

Hausman, R. E., Moscona, A. A. 1975. Purification and characterization of a retina specific aggregation factor. *Proc. Natl. Acad. Sci. U.S.A.* 72:916–20

Hausman, R. E., Moscona, A. A. 1976. Isolation of retina specific aggregating factor from membranes of embryonic neural retina. *Proc. Natl. Acad. Sci. U.S.A.* 73:3594–96

Hausman, R. E., Moscona, A. A. 1979. Immunological detection of retina cognin on the surface of embryonic cells. *Exp. Cell Res.* 119:191–207

Hunt, R. K., Jacobson, M. 1974. Neuronal specificity revisited. *Curr. Top. Dev. Biol.* 8:203–59

Jacobson, M. 1978. *Developmental Neurobiology*, pp. 345–433. New York: Plenum

Jacobson, M., Levine, R. L. 1975a. Plasticity in the adult frog brain: Filling the visual scotoma after excision or translocation of parts of the optic tectum. *Brain Res.* 88:339–45

Jacobson, M., Levine, R. L. 1975b. Stability of implanted duplicate tectal positional markers serving as targets for optic axons in adult frogs. *Brain Res.* 92:468–71

Jakoi, E. R., Marchase, R. B. 1979. Ligatin from embryonic chick neuroretina. *J. Cell Biol.* 80:642–50

Jansen, J. K. S., Nicholls, J. G. 1972. Regeneration and changes in synaptic connections between individual nerve cells in the central nervous system of the leech. *Proc. Natl. Acad. Sci. U.S.A.* 69:636–39

Landmesser, L., Morris, D. G. 1975. The development of functional innervation in the hind limb of the chick embryo. *J. Physiol.* (London) 249:301–26

Levinson, J., Goodfellow, P., Vadeboncoeur, M., McDevitt, H. 1978. Identification of stage-specific polypeptides synthesized during murine preimplantation development. *Proc. Natl. Acad. Sci. U.S.A.* 75:3332–36

Marchase, R. B. 1977. Biochemical studies of retino-tectal specificity. *J. Cell Biol.* 75:237–57

Marchase, R. B., Barbera, A. J., Roth, S. 1975. A molecular approach to retina-tectal specificity. *Ciba Symp.* 29:315–27

McClay, D. R., Gooding, L. R., Fransen, M. E. 1977. A requirement for trypsin sensitive cell surface components for cell-cell interaction of embryonic neural retina cells. *J. Cell Biol.* 75:56–66

McGuire, E. J., Burdick, C. L. 1976. Intercellular adhesive selectivity. I. An improved assay for the measurement of embryonic chick intercellular adhesion (liver and other tissues). *J. Cell Biol.* 68:80–89

Merrell, R., Glaser, L. 1973. Specific recognition of plasma membranes by embryonic cells. *Proc. Natl. Acad. Sci. U.S.A.* 70:2794–98

Merrell, R., Gottlieb, D. I. Glaser, L. 1975a. Embryonal cell surface recognition, extraction of an active plasma membrane component. *J. Biol. Chem.* 250:5655–59

Merrell, R., Gottlieb, D. I., Glaser, L. 1976. Membranes as a tool for the study of cell surface recognition. In *Neuronal Recognition*, ed. S. H. Barondes, pp. 249–73. New York: Plenum

Merrell, R., Pulliam, M. W., Randono, L., Boyd, L. F., Bradshaw, R. A., Glaser, L. 1975b. Temporal changes in tectal cell surface specificity induced by nerve growth factor. *Proc. Natl. Acad. Sci. U.S.A.* 72:4270–74

Mintz, G., Glaser, L. 1978. Specific glycoprotein changes during development of the chick neural retina. *J. Cell Biol.* 79:132–37

Moscona, A. A. 1965. Recombination of dissociated cells and the development of cell aggregates. In *Cells and Tissues in Culture*, ed. E. N. Willmar, pp. 489–529. New York: Academic

Moscona, A. A., Hausman, R. E. 1977.

Biological and biochemical studies on embryonic cell recognition. In *Cell and Tissue Interactions,* ed. J. W. Lash, M. M. Burger, pp. 173–86. New York: Raven

Moya, F., Silbert, D. F., Glaser, L. 1979. The relation of temperature and lipid composition to cell adhesion. *Biochim. Biophys. Acta* 550:485–99

Obrink, B., Kuhlenschmidt, M. S., Roseman, S. 1977a. Adhesive specificity of juvenile rat and chicken liver cells and membranes. *Proc. Natl. Acad. Sci. U.S.A.* 74:1077–81

Obrink, B., Warmergard, B., Pertoft, H. 1977b. Specific binding of rat liver plasma membranes by rat liver cells. *Biochem. Biophys. Res. Commun.* 77:665–70

Palka, J., Edwards, J. S. 1974. The cerci and abdominal giant fibers of the house cricket. *Acheta domesticus.* II. Regeneration and effects of chronic deprivation. *Proc. R. Soc. London Ser. B* 185:105–21

Raff, M. C., Abney, E., Brockes, J. P., Smith, A. H. 1978a. Schwann cell growth factors. *Cell* 15:813–22

Raff, M. C., Mirsky, R., Fields, K. L., Lisak, R. P., Dorfman, S. H., Silberberg, D. H., Gregson, N. A., Leibowitz, S., Kennedy, M. C. 1978b. Galactocerebroside is a specific cell surface antigenic marker for oligodendrocytes in culture. *Nature* 274:813–16

Roth, S. 1968. Studies on intercellular adhesive specificity. *Dev. Biol.* 18:602–31

Roth, S., McGuire, E. J., Roseman, S. 1971. An assay for intercellular adhesive specificity. *J. Cell Biol.* 51:525–35

Ruffolo, R. R. Jr., Eisenbarth, S. S., Thompson, J. M., Nirenberg, M. 1978. Synapse turnover: A mechanism for acquiring synaptic specificity. *Proc. Natl. Acad. Sci. U.S.A.* 75:2281–85

Rutishauser, U., Gall, W. E., Edelman, G. M. 1978a. Adhesion among neural cells of the chick embryo. IV. Role of the cell surface molecule CAM in the formation of neurite bundles in cultures of spinal ganglia. *J. Cell Biol.* 79:382–93

Rutishauser, U., Thiery, J. P., Brackenbury, R., Sela, B. H., Edelman, G. M 1976. Mechanism of adhesion among cells from neural tissue of the chick embryo. *Proc. Natl. Acad. Sci. U.S.A.* 73:577–81

Rutishauser, U., Thiery, J. P., Brackenbury, R., Edelman, G. M. 1978b. Adhesion among neural cells of the chick embryo. III. Relationship of the surface molecule CAM to cell adhesion and the development of histotypic patterns. *J. Cell Biol.* 79:371–81

Salzer, J., Glaser, L., Bunge, R. P. 1977. Stimulation of Schwann cell proliferation by a neurite membrane fraction. *J. Cell Biol.* 75:75

Sanes, J. R., Marshall, L. M., McMahan, U. J. 1978. Reinervation of muscle fiber basal lamina after removal of myofibers. *J. Cell Biol.* 78:176–98

Sankaran, L., Proffitt, R. T., Petersen, J. R., Pogell, B. M. 1977. Specific factors influencing histotypic aggregation of chick embryo hepatocytes. *Proc. Natl. Acad. Sci. U.S.A.* 74:4486–90

Santala, R., Glaser, L. 1977. The effect of cell density on the expression of cell adhesion properties in a cloned rat astrocytoma (C-6). *Biochem. Biophys. Res. Comm.* 79:285–91

Santala, R., Gottlieb, D. I., Littman, D., Glaser, L. 1977. Selective cell adhesion of neuronal cell lines. *J. Biol. Chem.* 252:7625–34

Schnaar, R. L., Weigel, P. H., Kuhlenschmidt, M. S., Lee, Y. C., Roseman, S. 1978. Adhesion of chicken hepatocytes to polyacrylamide gels derivatized with *N*-acetylglucosamine. *J. Biol. Chem.* 253:7940–51

Schubert, D., Whitlock, C. 1977. Alteration of cellular adhesion by nerve growth factor. *Proc. Natl. Acad. Sci. U.S.A.* 74:4055–58

Snell, W. J., Roseman, S. 1977. A quantitative assay for the adhesion and deadhesion of *Chlamydomonas reinhardii. Fed. Proc.* 36:811.

Sperry, R. W. 1943. Visuomotor coordination in the newt *(Triturus viridescens)* after regeneration of the optic nerve. *J. Comp. Neurol.* 79:33–55

Sperry, R. W. 1944. Optic nerve regeneration with return of vision in anurans. *J. Neurophysiol.* 7:57–69

Sperry, R. W. 1945. Restoration of vision after crossing of optic nerves and after contralateral transplantation of eye. *J. Neurophysiol.* 8:15–28

Sperry, R. W. 1963. Chemoaffinity in the orderly growth of nerve fiber patterns and connections. *Proc. Natl. Acad. Sci. U.S.A.* 50:703–10

Sperry, R. W. 1965. In *Organogenesis,* ed. R. L. DeHaan, H. Ursprung, p. 170. New York: Holt, Rinehart, Winston

Stallcup, W. B. 1977. Specificity of adhesion between cloned neural cell lines. *Brain Res.* 126:475–84

Stone, G. C., Wilson, P. L., Hall, M. E. 1978. Two-dimensional gel electrophoresis of proteins in rapid axoplasmic transport. *Brain Res.* 144:287–302

Thiery, J. P., Brackenbury, R., Rutishauser, U., Edelman, G. M. 1977. Adhesion among neural cells of the chick embryo. II. Purification and characterization of a cell adhesion molecule from neural retina. *J. Biol. Chem.* 252:6841–45

Trinkner, E. 1978. Postnatal cerebellar cells of staggerer express immature components on their surface. *Nature* (London) 277:566–67

Umbreit, J., Roseman, S. 1975. A requirement for reversible binding between aggregating embryonic cells before stable adhesion. *J. Biol. Chem.* 250:9360–68

Van Essen, D., Jansen, J. K. S. 1976. Repair of specific neuronal pathways in the leech. *Cold Spring Harbor Symp. Quant. Biol.* 40:495–502

Walther, B. T., Ohman, R., Roseman, S. 1973. A quantitative assay for intercellular cell adhesion. *Proc. Natl. Acad. Sci. U.S.A.* 70:1569–73

Whittenberger, B., Glaser, L. 1977. Inhibition of DNA synthesis in cultures of 3T3 cells by isolated surface membranes. *Proc. Natl. Acad. Sci. U.S.A.* 74:2251–55

Whittenberger, B., Raben, D., Glaser, L. 1979. Regulation of the cell cycle of 3T3 cells in culture by a surface membrane enriched cell fraction. *J. Supramol. Struct.* 10

Whittenberger, B., Raben, D., Lieberman, M. A., Glaser, L. 1978. Inhibition of growth of 3T3 cells by extract of surface membranes. *Proc. Natl. Acad. Sci. U.S.A.* 75:5457–61

Williams, A. F., Galfre, G., Milstein, C. 1977. Analysis of cell surfaces by xenogeneic myeloma-hybrid antibodies: Differentiation antigens of rat lymphocytes. *Cell* 12:663–73

Wolpert, L. 1971. Positional information and pattern formation. *Curr. Top. Dev. Biol.* 3:183–225

Wood, P. M. 1976. Separation of functional Schwann cells and neurons from normal peripheral nerve tissue. *Brain Res.* 115:361–75

Wood, P. M., Bunge, R. P. 1975. Evidence that sensory axons are mitogenic for Schwann cells. *Nature* (London) 256:662–64

Yoon, M. G. 1975. Topographic polarity of the optic tectum studied by reimplantation of the tectal tissue in adult goldfish. *Cold Spring Harbor Symp. Quant. Biol.* 40:503–19

12 / *Programs of Early Neuronal Development*

NICHOLAS C. SPITZER
JANET E. LAMBORGHINI

INTRODUCTION

The road from embryological determination to behavior proceeds via the differentiation of cellular properties. The development of the nervous system can be subdivided into a number of events, occurring for the most part serially, although often partly in parallel. This has been well summarized by Cowan (1979). Following induction, proliferation, migration, and selective aggregation, cytodifferentiation of neurons occurs, both with respect to their membrane and to their cytoplasmic properties, and their characteristic form or geometry is elaborated. These cells establish a specific set of afferent and efferent connections, and there is selective death of some of the neurons initially generated. Finally, there is validation of some synaptic connections while other connections are eliminated.

The normal developmental timetables or patterns of expression of certain phenotypes have been determined *in vivo* for frog spinal neurons, grasshopper ganglion neurons, moth peripheral sensory neurons, and less extensively for other neuronal populations. Neurite outgrowth, the onset of excitability, and the development of neurotransmitter synthetic capability are among the phenotypes that have been examined. From this work it emerges that there is no fixed stereotypic sequence of developmental events that characterizes the differentiation of neurons across the phylogenetic scale, although there may be some common elements.

Our studies of amphibian neural development suggest that certain phases of early cytodifferentiation proceed independent of cellular interactions. Evidence of this independence of cytodifferentiation comes from experiments in which neurons dissociated in culture in minimal medium differentiate as they do *in vivo* with respect to the sequence and timing of development. Autonomous development can be said to occur when phenotypes continue to be expressed in the absence of further instructions. It should be noted, however, that the independent character of differentiation of nerve cells does not preclude modulation of phenotypes by cellular interactions. Both neurotransmitter synthesis and neurotransmitter receptor expression can be altered in cultured neurons by changes in environment, as shown by studies of other systems (Patterson, 1978; Baccaglini and Cooper, 1979). It has been repeatedly demonstrated that later stages of differentiation require cellular interactions for further neuronal maturation or maintenance.

One of the central themes running through Viktor Hamburger's many scientific contributions is the exploration and definition of the interactions between different elements in developing nervous systems. There can be no question, as he has noted, that "for the experimental embryologist, independent development is much less interesting than dependency" (Hamburger, 1980). It appears that there is a free-running program for cellular development in amphibian neurons, one that proceeds in the absence of cellular interactions. This independence simplifies the system and makes it potentially easier to analyze. However, there remains the possibility that there is interdependency among the events within that program. This article will examine some recent studies that have spelled out the order of appearance of differentiated phenotypes of particular neurons and those that have begun to investigate the interrelationships of their expression.

EXPERIMENTAL EMBRYOLOGICAL FOUNDATIONS

The descriptive phase of any scientific discipline precedes the analytic one, and a substantial debt is owed to the pioneers who first characterized the stages of development of different embryos. The development of the chick was described by Hamburger and Hamilton (1951), that of *Xenopus* by Nieuwkoop and Faber (1957), that of *Ambystoma* by Harrison (1969) and that of *Rana pipiens* by Shumway (1940) and by Taylor and Kollros (1946). In many respects the more recent studies have involved simply increasing the degree of detail of the characterization. On the one hand, investigations have focused on a small portion of the organism, such as a nucleus of the brain, a single population of cells, or an identified neuron. On the other hand, investigators have used new techniques of intracellular recording and tracer injection, autoradiography, electron microscopy, immunohistochemistry, and microchemical analysis. The description of stages of cellular rather than organismal development is a necessary prerequisite for the description of the controls of development at the molecular level. In fact, these cellular studies have only recently yielded sufficient detail to permit analytical approaches that could lead to an understanding of the interrelation of events in the neuronal program.

EXPERIMENTAL STRATEGY

It would be well to outline the approaches taken in the studies to be described and to make explicit the rationale for each.

1. In studying the cytodifferentiation of neurons, there seems to be a clear value in examining the expression of enzymatic activities associated with soluble cytoplasmic proteins, or the expression of biophysical properties associated with integral membrane proteins, in preference to examination of a more complex differentiation like neurite outgrowth. In this way one is more likely to be looking at single rather than multiple gene expression. The control of this simpler differentiation may be easier to understand at the molecular level.

2. There is substantial value in looking at identified nerve cells, or at populations of such cells, rather than at an entire nucleus or whole brain preparation. In this way, the differentiation of one cell, or set of cells, is not masked or confused with that of others. For technical reasons, especially for ease of impalement with microelectrodes or chemical analysis, there is a distinct preference for relatively large cells.

3. It is valuable to compare the differentiation of different neurons, both within the same animal and from animal to animal across the phyla. Is the temporal order in the cytodifferentiation of specific neuronal properties constant? Are there common rules—that the expression of one phenotype must precede the expression of others? Comparative studies will yield the answers.

4. It is of great value to examine the differentiation of neurons in an *in vitro* system after studying development *in vivo.* If development in explant or dissociated cell culture proceeds as it does *in vivo,* the stage is set for analytic perturbation experiments. It is possible to manipulate the environment more easily, more completely, and in a more defined way when the cells or tissues are growing outside the animal, as Levi-Montalcini, Meyer, and Hamburger (1954) were among the first to show in their studies of nerve growth factor (NGF). Furthermore, one can use reversible metabolic inhibitors for brief periods to block the expression of differentiated phenotypes. One can be confident that the drugs are reaching the cells in the desired dosages, and the difficulties of side effects on the organism as a whole may be avoided.

In sum, these considerations have governed the choice of preparations and experiments that have been most fruitful in expanding our knowledge of the patterns of neuronal cytodifferentiation.

AMPHIBIAN SPINAL NEURONS

The Rohon-Beard neurons located in the spinal cords of amphibia, fish, and elasmobranchs have been useful for the study of vertebrate neuronal cytodifferentiation. They are likely to be primary sensory neurons. DuShane (1938) observed that removal of dorsal neural tube anlage resulted in a later sensory behavioral deficit in the region of the operation, consistent with the histological absence of the cells. More recently Roberts (1974, 1977) has recorded impulse activity from the dorsal spinal cord in response to tactile stimulation of the flank of the tadpole. Rohon-Beard neurons probably play a role similar to that of the dorsal root ganglion cells, which develop later.

We have focused our attention on the Rohon-Beard cells of *Xenopus laevis,* the African clawed frog. Their large size (25 μm), superficial position (on the dorsal aspect of the spinal cord), and the large number of cells (about 200 per embryo) over a broad range of developmental stages have been essential to our progress (Fig. 12-1). From the anatomist's point of view, these parameters have made the identification of the cells an explicit process, facilitating the determination of birthdate and the study of morphology; from the physiologist's point of view, they have made intracellular recordings feasible for the examination of membrane properties. It seems likely that from the biochemist's point of view, microtechnology will permit analysis of the neurotransmitter synthetic capacity of these cells.

Rohon-Beard neurons have been birthdated by conventional autoradiography. Series of embryos were injected with tritiated thymidine at successive stages of development and examined at a later time to determine the stage at which these cells no longer incorporate the label. Perhaps the most striking finding was that these neurons complete their final round of DNA synthesis at a very early stage of development (Lamborghini, 1980). The cells appear unlabeled when the tracer is injected during gastrulation, before the appearance of the neural plate and long before closure of the neural tube (Fig. 12-2). This early birthday is not a unique feature of Rohon-Beard neurons, since some primary motor neurons, the Mauthner neurons, trigeminal ganglion cells, and several other populations also originate during gastrulation. These are the first neurons to originate in the development of this central nervous system. The Rohon-Beard neurons arise remarkably synchronously, with 90 percent being generated during a six-hour period. Given the cell cycle times

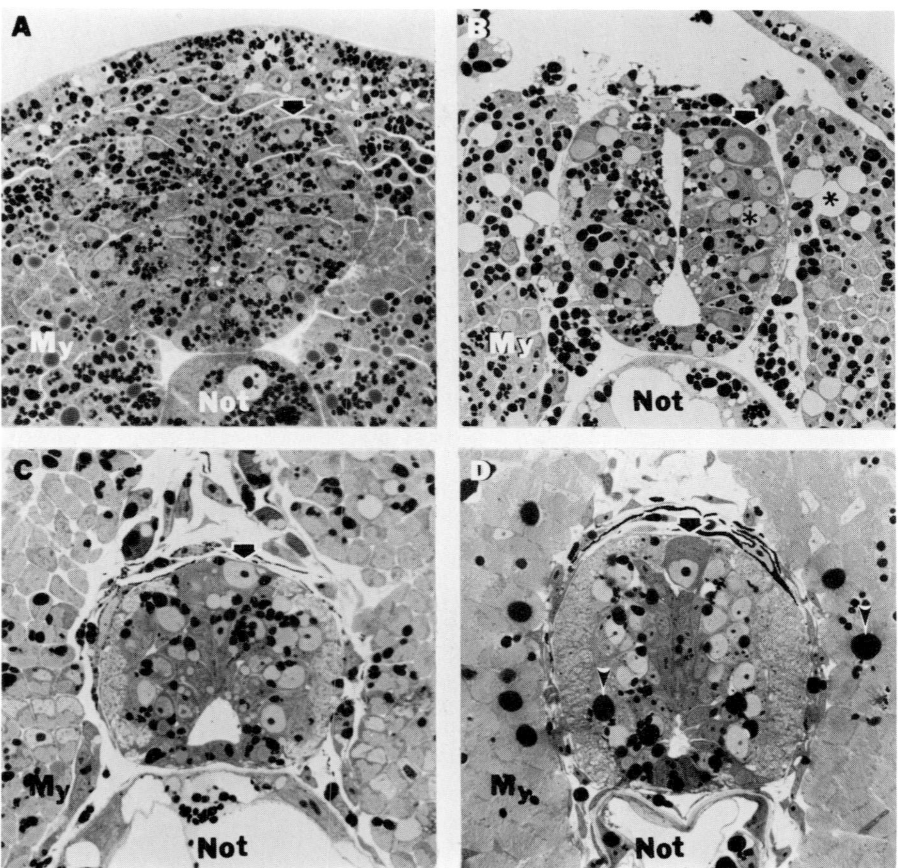

Figure 12.1 The size and position of Rohon-Beard cells (large arrows) at each of four stages. Light micrographs of transverse sections through the midtrunk level of *Xenopus laevis* spinal cords. A, tail-bud stage, 1 day; B, mid tail-bud stage 1½ days; C, early larval stage, 2 days; D, larval stage, 3 days. Magnification ×320. From Lamborghini et al. (1979), with permission.

from other amphibian studies (Gurdon, 1967; Flickinger, Freedman, and Stambrook, 1967), the final mitosis could be complete within another one and one-half hours.

The early birthdate of these neurons is quite close to the time of primary induction. It is clear that approximately half the number of Rohon-Beard neurons finish DNA synthesis before exposure of the presumptive neurectoderm to the archenteron roof is completed (Maleyvar and Lowery, 1973; Lamborghini, 1980). One possibility is that the cells first induced by the mesoderm rapidly signal other cells in the anterior neurec-

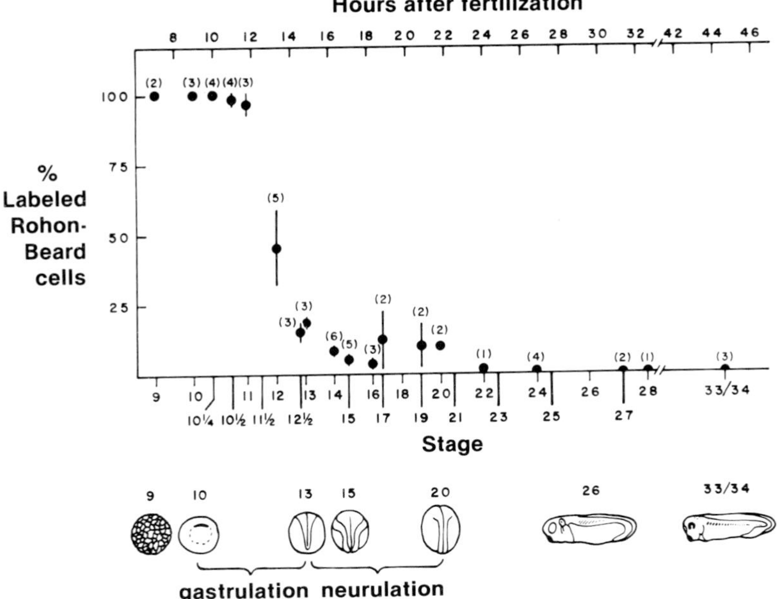

Figure 12.2 The kinetics of birth of Rohon-Beard cells. The percentage of labeled cells present in the spinal cord of larval (3-day) animals is plotted against the stage at which the injection of ^{3}H-thymidine was administered. The data are presented as means and standard errors of the means; (n) = number of animals scored for that stage. The morphology of the embryo at several stages is diagramatically represented. From Lamborghini (1980), with permission.

toderm. Another possibility is that induction and the determination of other cell properties occurs after the final round of DNA synthesis. The neuronal birthday has been suggested as the time at which precursor cells make decisions about the fate of their progeny (Jacobson, 1978). However, some neurons acquire characteristic differentiated phenotypes prior to their birthdate (Cohen, 1974; Rothman, Gershon, and Holtzer, 1978); others may become determined later.

It has been known for some time that Rohon-Beard cells have axons that run in the spinal cord and dendrites that project to skin and muscle (Hughes, 1957). The projection of axons to the anterior end of the spinal cord has been confirmed electrophysiologically (Spitzer, 1976a), and the ultrastructure of free nerve endings in the periphery, presumably those of Rohon-Beard cells, have been described (Roberts, 1977). Silver stains reveal that the peripheral processes have grown out into the myotomes by the time of closure of the neural tube (Muntz, 1964). However, the precise timing of initial neurite outgrowth and the manner in which different processes are elaborated remains to be worked out.

The development of cell membrane properties has proceeded by the visualization of cells with interference contrast optics and impalement with intracellular microelectrodes. The cells can be recognized in dissected preparations at an early neurula stage, just prior to the closure of the neural tube, in spite of yolk and lipid inclusions that render the tissue quite opaque to transillumination.

The egg and blastula cells are incapable of generating action potentials (Palmer and Slack, 1970; Slack and Warner, 1975), and in the early neurula, Rohon-Beard cells also show only passive membrane properties to brief, injected current pulses of either polarity. However, these cells often produce action potentials when examined at the time of closure of the neural tube (Baccaglini and Spitzer, 1977). These are overshooting events arising in the cell body and are hundreds of milliseconds in duration. The inward current is carried chiefly by Ca^{2+}, as indicated by their ionic dependence, pharmacologic responses, the magnitude of the conductance increase, their activation potential, and their kinetics. At intermediate stages of development, from early tail bud to early larva, these neurons produce action potentials that consist of a spike followed by a plateau and are tens of milliseconds in duration. The spike depends on an influx of Na^{+}, whereas the plateau is due to a Ca^{2+} current. At larval stages the cells make brief action potentials about a millisecond in duration in which the inward current is carried largely by Na^{+} in an environment of normal physiological saline (Fig. 12-3). These changes are continuous, and Rohon-Beard neurons gradually lose the Ca^{2+} channels from their cell bodies. This developmental change in the capability of the cells is rapid, and is nearly complete in three to four days. The ionic dependence of the inward current of the action potential shifts from Ca^{2+}, to Ca^{2+}

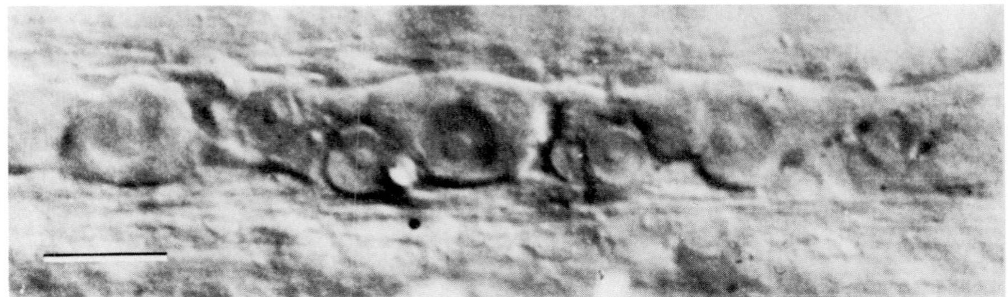

Figure 12.3 Rohon-Beard cells on the dorsal surface of the spinal cord in a dissected preparation, viewed with Nomarski interference contrast optics; six-day-old tadpole. The nuclei and prominent nucleoli of three cells are seen in the midline. Other smaller neurons can be discerned, adjacent to Rohon-Beard cells, and deeper in the cord. The processes of cells course along the surface, on either side of the cord. Calibration is 20 μm. From Spitzer (1976a), with permission.

and Na^+, to Na^+ alone. Calculations indicate that a single Ca^{2+} action potential could raise the internal Ca^{2+} concentration more than a hundredfold. Since the overshoot of the action potential does not decrease during repetitive firing, this Ca^{2+} must be rapidly bound. Studies of the ultrastructure of these perikarya suggest that this is the case, as the mitochondria contain numerous dense granules that are absent when the ionic basis of the impulse changes to Na^+ (Lamborghini, Revenaugh, and Spitzer, 1979). In contrast, there appears to be no change in the ionic dependence of the outward current of the impulse, carried by K^+ from the time that Ca^{2+} action potentials are first recorded (Baccaglini and Spitzer, 1977; Spitzer, 1979).

An important question that arises here is the extent to which the neurons normally fire impulses at different stages of development, when not driven by the instrumentation of the investigator. Bursts of "spontaneous" impulses are occasionally recorded following a penetration (Spitzer, unpublished results), but it is difficult to exclude the possibility of damage during the impalement. The answer is at present unknown, but will influence consideration of the role of Ca^{2+} action potentials during development. For example, the long duration of these events may reverse the sign of the membrane potential for a significant period, promoting insertion of other membrane proteins, or the Ca^{2+} ions themselves may function as an intracellular messenger.

Many embryonic cells are electrically coupled to one another at early stages of development, and subsequently become uncoupled. This has recently been observed also for Rohon-Beard neurons. Polarization of one cell by a small amount causes a voltage change in nearby cells; the coupling does not show rectification. However, even prior to the appearance of impulse capability, these cells show a different form of electrical excitability at the earliest stages at which recordings have been made. When one cell is sufficiently polarized by a current pulse of long duration, it becomes relatively uncoupled from other cells; the coupling coefficient falls to about 10 percent of its initial value (Spitzer, 1980). The uncoupling has a latency of a second or more, that decreases with increasing depolarization or hyperpolarization. The cells usually become recoupled at the termination of the injected current pulse. Voltage-dependent uncoupling has been previously observed between pairs of amphibian blastomeres dissociated and artificially pushed together (Spray, Harris, and Bennett, 1979). Other cells in the spinal cords of the same embryos exhibit coupling that is not voltage dependent. Voltage-dependent uncoupling persists at the time of appearance of Ca^{2+} action potentials in the Rohon-Beard neurons. However, these impulses are generally not large and long enough to uncouple the cells, and a Ca^{2+} action potential in one cell often elicits one in an adjacent cell to which it is coupled. This voltage dependent feature of electrical coupling remains until the cells become permanently uncoupled, at about the time the Na^+ component of the action potential appears. Electrophysiological evidence indicates that this coupling occurs via the processes rather than via the cell bodies of the neurons; it is presumably mediated by gap junctions.

The function of voltage dependent uncoupling remains to be determined. Since the Ca^{2+} action potentials do not uncouple the cells electrically, the intracellular channels still permit the passage of current, probably carried by K^+. One possibility is that the cells are in fact uncoupled by these impulses for the movement of larger molecules between cell interiors.

Neurons characteristically possess specific neurotransmitter receptors, and it has been of interest to examine the chemosensitivity of Rohon-Beard neurons. The cells are initially insensitive to a variety of agents, but become sensitive to γ-aminobutyric acid (GABA) around the time of appearance of the Na^+ component of the action potential. The cells are depolarized by superfusion or iontophoretic application of this transmitter, which produces conductance increases that are blocked by curare and picrotoxin. The ionic basis of this response does not appear to change during development. The responses that are occasionally recorded from cells with Ca^{2+} action potentials generally involve no conductance increase; they are probably the result of electrotonic spread of depolarization resulting from activation of receptors on other electrically distant cells that are at a slightly more advanced stage of development. The cells are unaffected by concentrations of one milli-molar (mM) of acetylcholine, norepinephrine, serotonin, dopamine, glycine, or glutamate (Spitzer, 1976b). It would be of substantial interest to know if chemical synapses are made on these cells. Ultrastructural examination of the cell bodies over a wide range of stages has not revealed any (Lamborghini et al., 1979); if present, they are likely to occur out on the processes.

The phenomenon of neuronal death during normal development is widespread, a phenomenon that Viktor Hamburger has done much to elucidate (Cowan, 1973; Hamburger, 1980). Rohon-Beard neurons are perhaps exceptional in that instead of the death of roughly one-half to two-thirds of the population, as normally observed for other cells, all of them disappear (Nieuwkoop and Faber, 1956; Hughes, 1957). It is clear that they actually die rather than undergo some metaplasia, to reappear as another cell type, since cells containing tritiated thymidine label in their nuclei disappear by late stages of development (Lamborghini, unpublished results). The kinetics of their disappearance are shown in Figure 12-4A. Prior to their death the cells stain intensely for alkaline phosphatase, a lysosomal enzyme (Fig. 12-4B), and ultrastructural examination reveals a proliferation of lysosomes in the cytoplasm (Lamborghini, unpublished results). The process of elimination appears quite rapid, as it is difficult to catch a cell in the midst of autolysis. After the cells are

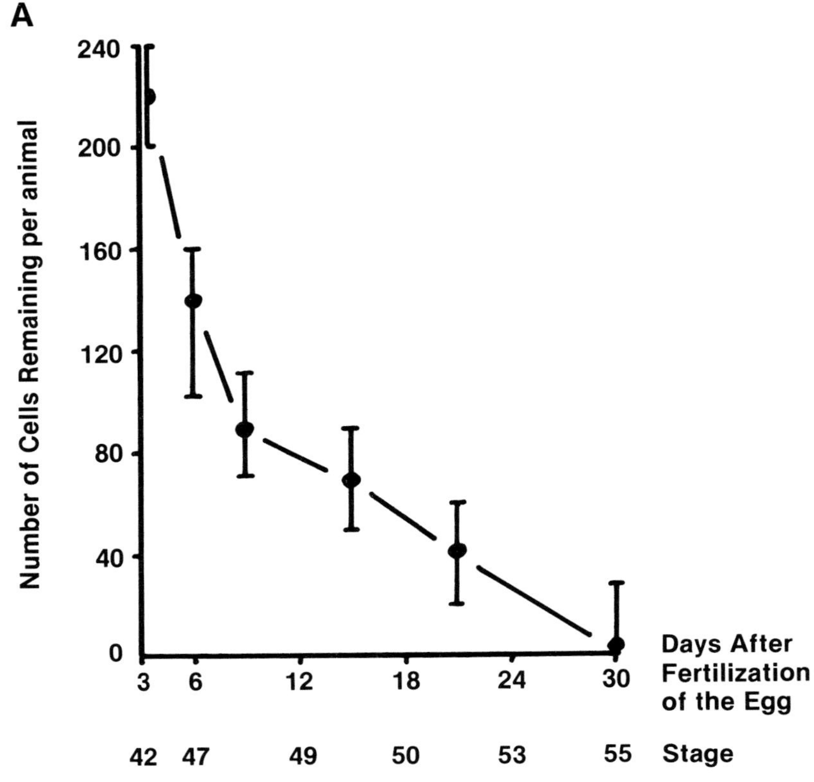

Figure 12.4 The kinetics of Rohon-Beard cell death. The number of cells remaining per animal is plotted as a function of age; the mean and range are indicated (A). The data are a composite of counts from sibling sets maturing at different rates, which may account in part for the variability observed. Whole mount of the spinal cord of a 7-day-old embryo (B), stained for acid phosphatase; 8 Rohon-Beard neurons are densely stained (arrows). Calibration is 100 μm. From Lamborghini (unpublished results).

gone, craters can be seen for a while on the dorsal surface of the spinal cord in both dissected living preparations and in sections of fixed tissue. The trigger for the death of these cells is unknown, but unlike the expression of the other earlier phenotypes, it seems likely to involve cellular interactions. Reduction of the periphery leads to premature death (Bacher, 1973). It may be that competition with the developing dorsal root ganglion cells normally provides the appropriate cues, leading to the death of Rohon-Beard cells.

The differentiated phenotypes described so far can be arranged in the temporal sequence of their expression, creating a developmental timetable for Rohon-Beard neurons *in vivo* (Fig. 12-5). Such a timetable provides a concise summary of a body of work and places the differentiation of a population of cells in the general context of the development of the organism. Furthermore, timetables for different neurons facilitate comparisons as we shall see later.

One attractive feature of frog spinal neurons as a preparation for the developmental neurobiologist is the ease with which they can be grown in dissociated cell culture (Harrison, 1910). The techniques for

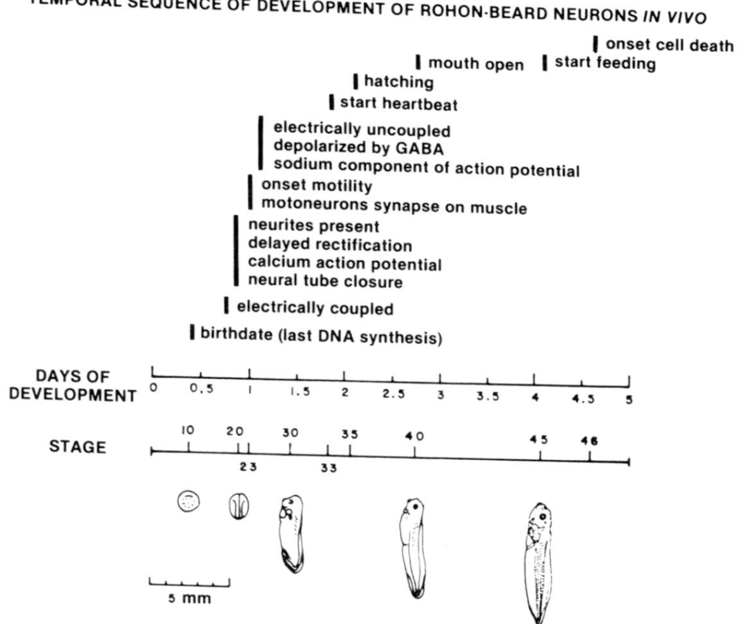

Figure 12.5 The developmental timetable for Rohon-Beard neurons of the *Xenopus* embryo. The time of onset, or first appearance, of a phenotype in these neurons is indicated by a vertical bar. The morphology of the embryo at different stages is indicated below. From Spitzer and Lamborghini (unpublished results).

preparation of the cultures are exceptionally simple, probably in large part because the holoblastic development of the embryo insures that each cell carries its own source of nutrients. Cells can be grown for periods up to five days in minimal, and wholly defined media, on uncoated tissue culture plastic (Jones and Elsdale, 1964; Spitzer and Lamborghini, 1976). Cultures prepared from single embryos at the mid neural plate stage yield a low density of neurons (10–100/dish); some of them are likely to be Rohon-Beard cells because they have the early birthdate and are depolarized by GABA. The attractive element of these cultures is that the cells appear to develop normally in several aspects and differentiate much as they do *in vivo.* This result has been obtained with respect to the development of the action potential mechanism and the development of chemosensitivity. It is now possible to study neuronal properties that are difficult to study directly in the animal and to manipulate the environment in a more controlled manner.

The observation that the neurons differentiate in cultures of cells dissociated at the mid neural plate stage indicates that the determination of these cells has occurred despite disruption of normal cellular relationships. One alternative is that determination occurred prior to dissociation. The other alternative, that determination is occurring *in vitro* as the result of signals transmitted over relatively long distances through the medium, seems unlikely, in view of the extensive evidence that indicates the impor-

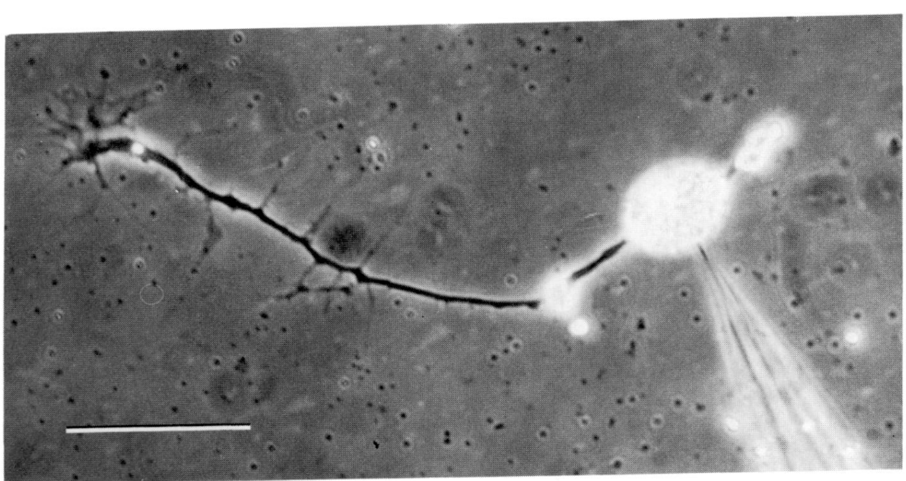

Figure 12.6 A neuron from a dissociated *Xenopus* neural plate; the cell was grown in culture for 16 hours and viewed with phase contrast optics. Note the phase dark process and growth cone; phase bright yolk granules obscure the nucleus in the cell body, which is indicated by the microelectrode. Calibration is 50 μm. From Spitzer (1979), with permission.

tance of short-range interactions for primary induction in amphibians. However, modulation of transmitter synthesis in cultures of mammalian sympathetic neurons has been shown to be mediated by signals carried in the medium (Patterson, 1978).

The cultures have also permitted study of the development of the excitability of neurites, because they are clearly visible on the bottom of the dish (Fig. 12-6). Electrical stimulation of the processes while recording from the cell body gives information about the capabilities of the neurites to generate action potentials; the event that is recorded in the cell body can be distinguished from the somatic action potential by its smaller amplitude (Willard, 1980). Because the cells are plated out before they have extended processes, it is possible to examine neurites as they are growing out for the first time. The neurites appear to be electrically excitable from the start, and like the cell bodies, they generate Ca^{2+} action potentials first and Na^{+} action potentials later. Because the shift in ionic dependence occurs in the neurites before it occurs in the cell body, it raises the possibility that electrical excitability first appears in the neurites.

RAT CEREBELLAR PURKINJE CELLS

How does the pattern of development of the neurons considered above compare with those of other vertebrate nerve cells? The Purkinje cells of the rat cerebellum have their birthdates between E14 (embryonic day 14) and E17, and they then migrate from the ventricular germinal zone into the mantle of the cerebellar plate. The axon and an apical dendrite are present at birth (E21), but the cell bodies thin out to form a monolayer only within two to three days postnatally. The elaboration of the Purkinje cell dendrites is delayed until after birth and then occurs over the extended period in which the granule cell layer arises. The tips of the dendrites reach the pia at about three weeks postnatally (Jacobson, 1978). When examined shortly after birth, these cells generate Na^{+} action potentials in their cell bodies; Ca^{2+} spikes are recorded in the dendrites at the time they begin to extend their characteristic arbor (5 to 6 days postnatally) (Llinas and Sugimori, 1978). The Ca^{2+} action potentials are a feature of adult Purkinje cell dendrites; it is not known if there is any early change in the ionic dependence of the action potentials in the cell bodies.

At birth the spontaneous firing activity of Purkinje cells can already be modulated by the iontophoretic application of norepinephrine, serotonin, GABA, and glutamate, which indicates that chemosensitivity to these putative neurotransmitters has developed prior to the extension of most dendritic branches (Woodward et al., 1971). The precise times of onset of some of these different forms of excitability remain to be established. Synapses are not formed on the Purkinje cells until postnatal day 3, at which time they can be driven by the climbing fibers; a few synapses can be distinguished morphologically at this time, but they become more readily observable over the next few days. Activation by the parallel fibers of the granule cells begins at the same time, or shortly thereafter. Inhibitory inputs from basket and stellate cells appear later, at about 11 days postnatally (Shimono, Nosaka, and Sasaki, 1976). The early-formed synapses by the climbing fibers may play a role in triggering the differentiation of the dendritic tree (Kornguth and Scott, 1972), but they clearly arrive after most of the differentiation of the neuronal membrane is complete.

SUPERIOR CERVICAL GANGLION NEURONS

The development of superior cervical ganglion neurons, like that of other sympathetic ganglion cells and parasympathetic neurons as well, is first indicated by the migration of these cells out of the neural crest. The pathway of migration and the site of aggregation appear to be instrumental in determining that these particular autonomic neurons will become noradrenergic (LeDouarin et al., 1975), probably as the result of interaction with the mesenchyme (Norr, 1973). The expression of neurotransmitter synthesis occurs very early in the differentiation of these cells. In the chick some sympathetic neurons have already begun to synthesize catecholamines and can take up norepinephrine while they are still dividing (Cohen, 1974; Rothman et al., 1978). The development of neurotransmitter synthetic capability has been studied in detail in the rat *in vivo* (Cochard, Goldstein, and Black, 1978; 1979). After migration is complete and the cells have largely aggregated, tyrosine hydroxylase, dopamine β-hydroxylase, and catecholamines all appear synchronously, although aggregation may not be necessary since isolated ectopic cells also show this form of differentiation. Although the cells have very short processes at this time, axonal outgrowth and target innervation do not occur for several days (deChamplain et al., 1970; Coughlin et al., 1978). Furthermore, only a fraction of 1 percent of the synaptic input to these cells is present as late as the time of birth, and these synapses appear immature (Black, Hendry, and Iverson, 1971; Smolen and Raisman, 1980). Of special interest is the finding that this early biochemical maturation does not require NGF in the mouse superior cervical ganglion (Coughlin, Boyer, and Black, 1977; Coughlin et al., 1978); similarly, chick sympathetic and sensory ganglia do not respond to NGF during an early period in their development (Levi-Montalcini and Hamburger, 1951). In contrast, postnatal development requires target innervation, preganglionic synaptic input, and the presence of NGF (reviewed by Black, 1978). Thus there may be a period of relatively autonomous early development for these neurons also.

GRASSHOPPER GANGLIONIC NEURONS

The embryonic central nervous system of the grasshopper, *Schistocerca nitens,* has been found to be an excellent preparation for the study of neuronal cytodifferentiation. The dorsal unpaired median (DUM) cells are interneurons located on the dorsal surface of the thoracic and abdominal ganglia, near the midline. They constitute a population of from 80 to 100 cells per ganglion, many of them large and accessible. Many, and perhaps all, of these cells can be identified as unique neurons, constant in form and function, from one embryo to the next. The meroblastic character of cleavage results in a transparent embryo developing in close apposition to the yolk mass in the egg; separation of the two permits clear visualization of the developing nervous system. This feature allows description of cell lineage by serial examination of dissected preparations of increasing age. Within a ganglion, cells can be visualized as they arise by asymmetric mitotic divisions of a single identified dorsal precursor cell (the DUM neuroblast) and lie on the dorsal surface of the ganglion; these ganglion mother cells undergo a symmetrical division to produce cells that differentiate into neurons. The neuroblasts are 30 μm in diameter, and their progeny are 10 μm initially and enlarge rapidly—some exceed 40 μm during embryogenesis. Finally, the mature properties of these neurons are known from studies of adults. All of the cells have a single neurite that bifurcates into bilaterally symmetrical processes, and all stain with neutral red, indicating that they contain monoamines. The largest (and possibly all) of the cells contain octopamine, and these large cells have somata that generate overshooting action potentials. Individual neurons have axons that project out characteristic roots: DUMETi, the DUM neuron to the extensor tibiae muscle, has been best studied. The development of these neurons has been followed from E5, when the precursor cells have already differentiated from the epidermis, up to hatching (E20) (Goodman and Spitzer, 1979; 1980a; Goodman et al., 1979).

At early embryonic stages all the cell types examined are electrically coupled to one another. The large population of ventral neuroblasts are coupled to each other and to the single dorsal neuroblast; coupling is present between neuroblasts to the left and right of the midline, and anteriorly and posteriorly from one ganglion to the next. Neuroblasts are also coupled to epithelial cells near them and even to cells out in the limb bud. The DUM neuroblast begins dividing on E6, and its progeny are initially electrically coupled to it. Furthermore, dye coupling is extensive at these early stages, as shown by the intracellular spread of Lucifer Yellow. Dye moves from the DUM neuroblast to its progeny on E7, and between neuroblasts when injected as late as E8 of development. However, dye does not spread from neuroblasts out to the limb bud.

Development entails selective uncoupling. At E8 the oldest progeny of the DUM neuroblast become dye uncoupled from their parent, although these cells are still electrically coupled. At E10 the DUM neuroblast becomes electrically uncoupled from cells other than its progeny, and on E11, the oldest progeny finally become electrically uncoupled. The mature DUM neurons are not electrically coupled to one another. Ventral neuroblasts become electrically uncoupled from limb buds on E10, but they remain coupled to each other until they degenerate by E15.

The identities of three of the oldest progeny of the DUM neuroblast have been examined. One of these turned out to be DUMETi (the cell so well studied in the adult nervous system) on the basis of the criteria outlined above. Investigation of the development of this cell was pursued in greatest detail, since it proved possible to recognize it easily in living preparations.

The outgrowth of the processes of DUMETi was followed from E7, when it begins with the extension of a single median neurite, by examination of preparations in which the cell body was injected with Lucifer Yellow. This process bifurcates on E8, and the distal tips of the axons reach the edges of the ganglion by E9. By the following day the axons have started to grow out into the peripheral nerves, and by E12 they have reached the extensor tibiae muscle, which they rapidly cover. There are multiple branches, terminating in growth cones, that are extended in inappropriate directions during this time; these subsequently disappear. After reaching its target, there is an enlargement of the cell body and the central arborization of processes.

The oldest neurons first become sensitive to neurotransmitters on E8, while still electrically coupled. Although they show no response to bath application of 1 mM acetylcholine or GABA on E7, they are depolarized and hyperpolarized, respectively, by the superfusion or the iontophoretic application of these agents the following day. The processes of the cells yield responses over their entire length, and it appears that sensitivity is distributed over the whole surface of the cell at the time, or soon after, it first appears. The cells seem to become sensitive to both transmitters at the same time, and they may be sensitive to other agents as well. The response to acetylcholine involves a conductance increase to Na^+ that is reduced by curare, hexamethonium, and decamethonium; a variety of α-toxins have no effect. The response to GABA involves a conductance increase to Cl^- that is blocked by picrotoxin. The ionic dependence, the reversal potential, and the pharmacology of these responses appear to be constant during development, since they are similar on E8, E13, and E18.

The oldest developing DUM neurons first become electrically excitable on E11, when voltage dependent outward K^+ currents (delayed rectification) can first be detected. Voltage dependent inward currents are often detected later the same day in the form of small Na^+ action potentials arising out in the axons. A slightly larger action potential arising in the median neurite is recorded shortly afterwards, and on E12 overshooting action potentials that arise in the cell body can be elicited for the first time. These action potentials depend on an influx of Na^+ and Ca^{2+}, although either ion alone is sufficient to sustain the impulse. There is a shift in the ionic dependence of these somatic action potentials during further development, since both ionic currents are necessary to sustain the action potential in adult animals. However, there is no evidence of a chiefly Ca^{2+}-dependent action potential at early stages.

Since the cells are often electrically coupled as they are becoming electrically excitable, it is reassuring to find that the cells are truly inexcitable initially, when mechanically uncoupled from their neighbors or exposed to veratridine: coupling does not appear to mask excitability. The cells become gradually uncoupled during the period that they become electrically excitable, and there is some variability in the timing of the process. Pairs of cells can generate the same kinds of action potentials in two different embryos and yet be coupled in one instance and not in another (Goodman and Spitzer, 1980b; 1980c).

The population of DUM neurons demonstrates a spectrum of electrical excitability. The firstborn cells, with long peripheral axons, come to generate action potentials in their cell bodies and in their processes. Other cells generate spikes only in their axons. Many of the progeny of the single DUM neuroblast do not generate action potentials at all, when examined in normal saline around the time of hatching and in the adult. These cells are local intraganglionic neurons (Goodman, Pearson, and Spitzer, 1980).

The neurotransmitter of the adult DUM neurons is octopamine, and this compound is first detected by radioenzymatic assays of the embryonic thoracic ganglia on E13, after the axons of the oldest DUM neurons have reached their target. No octopamine was detected before this time, even when ganglia were pooled to increase the volume of material. Three of the oldest DUM neurons begin to stain with neutral red on E14, which is consistent with the accumulation of the monoamine. The octopamine appears to be located in the DUM neurons, and the sequential appearance of neutral red staining may indicate the sequential appearance of octopamine in different cells (Goodman et al., 1979). Spontaneous synaptic potentials are recorded as early as E15 from unidentified inputs. The DUM neuroblast disappears the following day. The order of these developmental events is summarized in a developmental timetable (Fig. 12-7). The extent to which these different phenotypes develop independently remains to be determined.

The development of another set of neurons on the dorsal surface of embryonic grasshopper ganglia reveals some intriguing similarities to and differences from the development of the DUM neurons. Unlike most of the neurons in the ganglia that are descendents of the major class of precursors, the neuroblasts, these neurons are derived from a small set of seven cells, the midline precursor (MP) cells (Bate and Grunewald, 1980). The MP cells undergo a single symmetric mitotic division to yield two cells that differentiate into neurons. These cells are the first, or among the first, to lay down the fiber pathways of the central nervous system; the axons of the later progeny of the neuroblasts often follow these pioneering paths.

The two MP3 cells have been most extensively studied because of their accessibility to direct observation and intracellular recording. In the metathoracic ganglion these cells first extend processes late on E5. However, on E8 one of the two cells extends a second neurite that becomes further elaborated and assumes a characteristic H-shaped arborization by E12. The original process of this cell disappears between E11

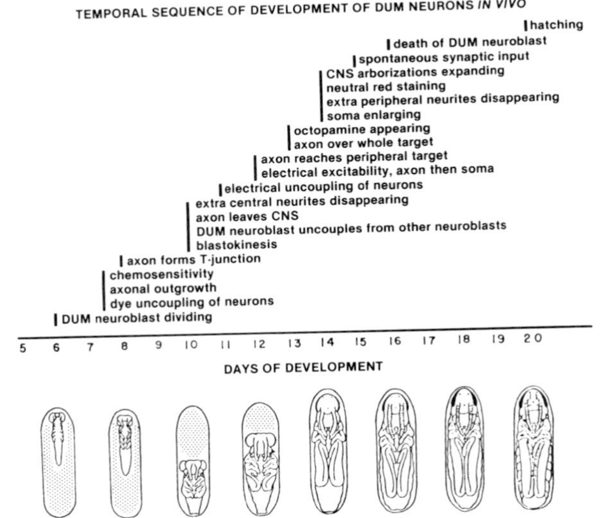

Figure 12.7 The developmental timetable for the oldest DUM neurons of the grasshopper embryo. The time of onset, or first appearance, of a phenotype in these neurons is indicated by a vertical bar. The morphology of the embryo, viewed from the ventral surface in the egg case, is illustrated below as traced from camera lucida projections. From Goodman and Spitzer (1979), with permission.

and E12. Either one of the two can undergo this transformation, but the one being transformed morphologically undergoes a physiological transformation as well—becoming sensitive to acetylcholine and GABA, electrically excitable with overshooting action potentials, and electrically uncoupled from its neighbors, the DUM neurons. Furthermore, the time course of this differentiation parallels that of the DUM neurons, being delayed by about a day. Its sibling retains its original process and develops only small action potentials that arise out in this process (Goodman, Bate, and Spitzer, 1981).

Both the DUM neurons and some of the MP3 progeny develop similar phenotypes. Both share the same environment, but are the immediate descendents of different precursor cells. These observations suggest that environment may play an important role in determining the kind of differentiation the cells express, and in fact the transforming H cell appears to be contacted by axons of the oldest DUM neurons at E8, when the cells are electrically coupled. This suggestion is supported by the finding that the fate of the two MP3 progeny varies from one segment to the next and that their morphology, physiology and survival, depend on their segmental position (Bate et al., submitted).

MOTH PERIPHERAL SENSORY NEURONS

The antennae of the moth, *Manduca sexta,* develop from clumps of epidermal cells, the imaginal disks, during metamorphosis. They each contain a large population of primary sensory neurons (2.5×10^5), that have been useful for correlated anatomical, biochemical, and physiological studies of neuronal cytodifferentiation.

The flagellum of the antenna is divided into about 80 annuli that contain small sensory organs, the sensilla. There are four recognizably distinct types of sensilla; each consists of cuticular hairs, and several cells including two or more neurons. These structures develop during the pupal stage as the larva is being transformed into an adult; this period lasts about 18 days. The regions in which the sensilla will arise can be recognized before overt differentiation occurs, and the differentiation of the annuli is synchronous along the length of the antenna (Sanes and Hildebrand, 1976a).

The birthdate of these neurons in male moths was determined by thymidine autoradiography. The neurons are all born during a period between the end of Day 1 and the middle of Day 3 of pupation. In sensilla that contain two neurons, both are born at the same time, and the non-neuronal cells originate simultaneously, all probably derived from a single precursor cell. The cells arise synchronously along the length of the antenna, no proximal-distal gradient is present. There appears to be no neuronal cell death in this developmental process because the number of neurons generated is the same as the number of cells found in the adult.

The neurons of the different sensilla differentiate in the same way and at the same time, but the long trichoid sensilla lie in an orderly array and have been most extensively studied. The two neurons of each sensillum, with somata about 8 μm in diameter, extend axons as early as Day 3, and by Day 4 the nerve branches beneath the epidermis are nearly their final size. These axons grow down the lumen of the antenna toward the brain, along preexisting pupal nerves from a group of several hundred neurons near the antennal tip. Axonal growth is essentially complete by Day 10. The axons are gradually ensheathed by glia. The dendrites grow out soon after the axons (as early as Day 4) and have characteristic ciliary specializations. The trichogen cells of the sensilla begin to grow out of the epidermis on Day 5 and Day 6, creating the sensory hair, and encircle the outgrowing dendrites. A glial cell lies between the trichogen cell and the neurons, which it has begun to ensheath. Between Day 7 and Day 12 the sensory hair elongates, and the trichogen cell begins to secrete cuticle. Following this secretion it is largely retracted from the hair between Day 13 and Day 15, leaving the two dendrites extending nearly to the tip of the 400 μm long hair. Neuronal cell bodies migrate from the basal to the apical margin of the epidermis between Day 10 and Day 14 (Sanes and Hildebrand, 1976b).

The antennal sensory neurons probably use acetylcholine as their neurotransmitter. Mature antennae contain it, have the appropriate enzymatic machinery for its synthesis and degradation, and do not manufacture a variety of other transmitter candidates. This compound first appears on Day 4 of pupal development, prior to the time at which axons reach their synaptic targets in the brain. Acetylcholine levels rise rapidly to a steady state by Day 13. Choline acetyltransferase activity first appears between Day 3 and Day 4 and rises at a comparable rate, accounting for the accumulation of transmitter. Acetylcholinesterase activity begins to rise at Day 4 and increases until Day 16 (Sanes and Hildebrand, 1976c).

The development of electrical activity in the antennal nerves has been assayed by examining the electroantennogram (EAG) responses to various odorants at different stages. The EAG is believed to be the summed extracellular record of receptor potentials. Odorants that elicit substantial responses from adult antennae (e.g., trans-2-hexenal) yield no response in male animals on Day 14 or younger. Responses are first recorded on Day 15, and they increase in amplitude with further development; this is about the time that the neurons have attained morphological and biochemical maturity. Sensitivity to a female pheromone begins several days later. The response to mechanical stimulation with a puff of

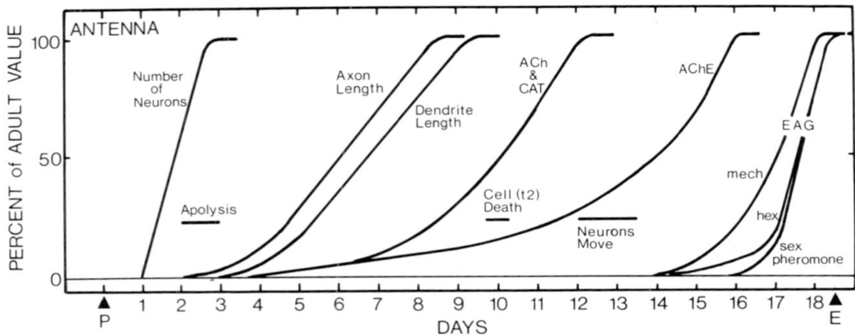

Figure 12.8 The developmental timetable for the antennal sensory neurons of the moth, *Manduca sexta*. P, pupation; E, eclosion. From Hildebrand (1980), with permission.

clear air appears on Day 14, prior to the response to either odorant (Schweitzer, Sanes, and Hildebrand, 1976). The order of appearance of these phenotypes is summarized in a developmental timetable (Fig. 12-8; Hildebrand, 1980).

Most, and perhaps all, of the differentiated phenotypes of the antennal sensory neurons develop in the absence of at least one important class of cellular interactions—the synaptic connections of their axons with the brain. Surgical removal of the brain during the first few days of pupation does not compromise the survival and development of the animal, and the parameters of morphological, biochemical, and electrophysiological development described above appear largely unchanged when examined on Day 15 (Sanes, Hildebrand, and Prescott, 1976). Thus the period of early differentiation may proceed independently in these neurons.

EXPERIMENTAL PERTURBATIONS OF CYTODIFFERENTIATION

In recent years a number of drugs that act specifically on different phases of intermediary metabolism have been used to define the metabolic requirements for differentiation of non-neural tissues (Rutter et al., 1968). The application of these drugs to neurons developing in culture has already begun to provide similar information about the role and timing of ribonucleic acid (RNA) and of protein synthesis necessary for the expression of different phenotypes. A temporal separation of the effect of inhibitors on cellular differentiation has been demonstrated by Bloom and Black (1979), who have studied the maturation of embryonic mouse superior cervical ganglion *in vitro*, which parallels development *in vivo*. At an early stage the cells do not require NGF, and intrinsic developmental mechanisms could be examined without interference by this growth factor. The addition of actinomycin D, or puromycin, to block RNA or protein synthesis, respectively, had no effect on neurite elaboration and elongation, although either drug was sufficient to block the normal development of tyrosine hydroxylase activity. These results indicate that the transcriptional and translational control of neurite outgrowth is exercised in advance of that of neurotransmitter synthetic enzymes. Process extension after a certain stage thus involves assembly of preformed units, independent of RNA and of protein synthesis. In cultures of dissociated *Drosophila* neuroblasts and myoblasts, a similar independence of neurite outgrowth from RNA synthesis was seen after an initial phase of sensitivity, and further elongation then occurred in the presence of the drug (Donady, Seecof, and Dewhurst, 1975). However, the blockade of RNA synthesis in cultures in which the neurons had passed the stage of sensitivity with respect to axon formation was still capable of inhibiting choline actyltransferase activity in these neurons (Seecof, 1977).

Studies of this sort will provide information about the timing of the macromolecular syntheses necessary for normal differentiation. In addition, they make it possible to ask if there is any causal connection between different phenotypes in a developmental timetable. The use of reversible inhibitors for brief periods may permit the pharmacological excision of a single phenotype, normally expressed at a particular time. We can then hope to find out whether or not the phenotypes that are normally expressed later, do, in fact, appear. Such experiments should tell us whether the neuronal phenotypes differentiate in an interdependent or autonomous fashion.

CONCLUSIONS

For specific neuronal populations, there are distinct and repeatable sequences of development that can be summarized by developmental timetables. Comparison of these timetables reveals that the order of appearance of different neuronal phenotypes is not the same from one set of neurons to the next. However, there are a number of similarities: outgrowth of processes is an early event; cells are electrically coupled while other membrane properties are differentiating; cell processes become electrically excitable before their cell bodies; a change often occurs in the ionic dependence of the action potential, although this change varies in degree; and the ionic basis of the responses to neurotransmitters remain constant. There are also some clear differences: neurotransmitter synthesis begins before, close to, or a relatively long time after the final mitosis (in superior cervical ganglion cells, antennal sensory neurons and

DUM neurons, respectively); processes can be electrically inexcitable or excitable as they first grow out (DUM neurons and amphibian spinal neurons, respectively). One feature of development is that many phenotypes are constant in the form in which they are initially expressed, while others are not; some processes are retracted as others grow out, and the ionic dependence of the action potential often changes.

It seems likely that the cellular mechanisms for neuronal development are similar in vertebrates and invertebrates. There are differences in the sequence of developmental events among the invertebrates (e.g., in the time of onset of transmitter synthesis), and there may be differences among the vertebrates as well; at present, the differences between invertebrates and vertebrates do not seem significant.

One important characteristic of early neuronal development, which emerges from the studies cited here, is its apparent autonomy from some or all of the normal histotypic intercellular contacts. The preparations discussed provide evidence for varying degrees of independence. Both the differentiation of membrane properties (e.g., in the Rohon-Beard neurons) and of cytoplasmic properties (e.g., in the superior cervical ganglion cells) may occur autonomously. It seems evident, however, that the specific phenotypes expressed during the period of autonomous development are different for different sets of neurons.

In attempting to understand the rules and mechanisms of neuronal cytodifferentiation, it is helpful to investigate the causality that leads to a particular sequence of developmental events. Does the expression of one phenotype (or the protein or the RNA synthesis associated with it) permit or promote the subsequent expression of others? Causality cannot be inferred from the temporal order of events, except in the negative sense that later events cannot be the cause of earlier ones. Thus, action potentials can play no role in axonal outgrowth and synaptogenesis when they are not yet present. (However, it should be noted that similar results have been obtained in a number of cases in which action potentials were present, but blocked—see review by Harris, 1980.) In general, it has been difficult to find agents that specifically and exclusively block the function of particular phenotypes in order to examine the consequences this blocking may have for subsequent development. In the future, monoclonal antibodies to specific functional proteins may prove to be very useful in this regard.

Experiments with reversible metabolic inhibitors will provide information about the timing of RNA and of protein syntheses necessary for the expression of specific cellular properties. The brief application of these drugs during a critical period, to prevent expression of one phenotype, should allow investigation of the effect of the absence of one character on the appearance of others.

ACKNOWLEDGMENT

The authors' work is supported by grants from the Office of Naval Research and the National Institutes of Health.

REFERENCES

Baccaglini, P., Cooper, E. 1979. Effect of environment on ACh sensitivity of developing nodose neurons *in vitro*. *Neurosci. Abstr.* 9:152

Baccaglini, P. I. , Spitzer, N. C. 1977. Developmental changes in the inward current of the action potential of Rohon-Beard neurones. *J. Physiol.* (London) 271:93–117

Bacher, B. E. 1973. The peripheral dependence of Rohon-Beard cells. *J. Exp. Zool.* 185:209–16

Bate, M., Goodman, C. S., Spitzer, N. C. 1981. Embryonic development of identified neurons: Segment-specific differences in the H cell homologues. J. Neurosci. *1*:103–106

Black, I. B. 1978. Regulation of autonomic development. *Ann. Rev. Neurosci.* 1:183–214

Black, I. B., Hendry, I. A., Iversen, L.L. 1971. Transsynaptic regulation of growth and development of adrenergic neurones in a mouse sympathetic ganglion. *Brain Res.* 34:229–40

Bloom, E. M., Black, I. B. 1979. Metabolic requirements for differentiation of embryonic sympathetic ganglia cultured in the absence of exogenous nerve growth factor. *Dev Biol.* 68:568–78

Cochard, P., Goldstein, M., Black, I. B. 1978. Ontogenetic appearance and disappearance of tyrosine hydroxylase and catecholamines in the rat embryo. *Proc. Natl. Acad. Sci.* U.S.A. 75:2986–90

Cochard, P., Goldstein, M., Black, I. B. 1979. Initial development of the noradrenergic phenotype in autonomic neuroblasts of the rat embryo *in vivo*. *Dev. Biol.* 71:100–114

Cohen, A. 1974. DNA synthesis and cell division in differentiating avian adrenergic neuroblasts. In *Wenner-Gren Center International Symposium Series, Vol. 22, Dynamics of Degeneration and Growth in Neurons,* ed. K. Fuxe, L. Olson, Y. Zotterman, pp. 359–70. Oxford: Pergamon

Coughlin, M. D., Boyer, D. M., Black, I. B. 1977. Embryologic development of a mouse sympathhetic ganglion *in vivo* and *in vitro*. *Proc. Natl. Acad. Sci.* U.S.A. 74:3438–42

Coughlin, M. D., Dibner, M. D., Boyer, D. M., Black, I. B. 1978. Factors regulating development of an embryonic mouse sympathetic ganglion. *Dev. Biol.* 66:513–28

Cowan, W. M. 1973. Neuronal death as a regulative mechanism in the control of cell number in the nervous system. In *Development and Aging,* ed. M. Rockstein, pp. 19–41. New York: Academic

Cowan, W. M. 1979. Selection and control in neurogenesis. In *The Neurosciences Fourth Study Program,* ed. F. O. Schmitt, pp. 59–79, Cambridge: MIT

deChamplain, J., Malmfors, T., Olson, L., Sachs, Ch. 1970. Ontogenesis of peripheral adrenergic neurons in the rat: Pre- and postnatal observations. *Acta Physiol. Scand.* 80:276–88

Donady, J. J., Seecof, R. L., Dewhurst, S. 1975. Actinomycin D-sensitive periods in the differentiation of *Drosophila* neurons and muscle cells *in vitro. Differentiation* 4:9–14

DuShane, G. P. 1938. Neural fold derivatives in the amphibia: Pigment cells, spinal ganglia and Rohon-Beard cells. *J. Exp. Zool.* 78:485–503

Flickinger, R. A., Freedman, M. L., Stambrook, P. J. 1967. Generation times and DNA replication patterns of cells of developing frog embryos. *Dev. Biol.* 16:457–73

Goodman, C. S., Bate, M., Spitzer, N. C. 1981. Embryonic development of identified neurons: Origin and transformation of the H cell. *J. Neurosci.* 1:94–102

Goodman, C. S., O'Shea, M., McCaman, R., Spitzer, N. C. 1979. Embryonic development of identified neurons: Temporal pattern of morphological and biochemical differentiation. *Science* 204: 1219–22

Goodman, C. S., Pearson, K. G., Spitzer, N. C. 1980. Electrical excitability: A spectrum of properties in the progeny of a single embryonic neuroblast. *Proc. Natl. Acad. Sci.* U.S.A. 77:1676–80

Goodman, C. S., Spitzer, N. C. 1979. Embryonic development of identified neurones: Differentiation from neuroblast to neurone. *Nature* (London) 280:208–14

Goodman, C. S., Spitzer, N. C. 1980a. Embryonic development of neurotransmitter receptors in grasshoppers. In *Receptors for Neurotransmitters, Hormones and Pheromones in Insects,* ed. D. B. Satelle et al., pp. 195–207. New York: Elsevier/North Holland

Goodman, C. S., Spitzer, N. C. 1980b. The mature electrical properties of identified neurons in grasshopper embryos. *J. Physiol.* (London), In press

Goodman, C. S., Spitzer, N. C. 1980c. The development of electrical properties of identified neurons in grasshopper embryos. *J. Physiol.* (London), In press

Gurdon, J. B. 1967. African clawed frogs. In *Methods in Developmental Biology,* ed. F. H. Wilt, N. K. Wessells, pp. 75–84. New York: T.Y. Crowell

Hamburger, V. 1980. Trophic interactions in neurogenesis: A personal historical account. *Ann. Rev. Neurosci.* 3:269–78

Hamburger, V., Hamilton, H. 1951. A series of normal stages in the development of the chick embryo. *J. Morphol.* 88:49–92

Harris, W. A. 1980. Neural activity and development. *Ann. Rev. Physiol.* 43:689–710

Harrison, R. G. 1910. The outgrowth of the nerve fiber as a mode of protoplasmic movement. *J. Exp. Zool.* 9:787–846

Harrison, R. G. 1969. In *Organization and Development of the Embryo,* ed. S. Wilens, pp. 44–66, New Haven: Yale Univ. Press

Hildebrand, J. G. 1980. Development of putative acetylcholine receptors in normal and deafferented antennal lobes during metamorphosis of *Manduca Sexta.* In *Receptors for Neurotransmitters, Hormones and Pheromones in Insects,* ed. D. B. Satelle et al., pp. 209–20. New York: Elsevier/North Holland

Hughes, A. 1957. The development of the primary sensory neurons in *Xenopus laevis (Daudin). J. Anat.* 91:323–38

Jacobson, M. 1978. *Developmental Neurobiology.* New York: Holt

Jones, K. W., Elsdale, T. R. 1963. The culture of small aggregates of amphibian embryonic cells *in vitro. J. Embryol. Exp. Morphol.* 11:135–54

Kornguth, S. E., Scott, G. 1972. The role of climbing fibers in the formation of Purkinje cell dendrites. *J. Comp. Neurol.* 146:61–82

Lamborghini, J.E. 1980. Rohon-Beard cells and other large neurons in

Xenopus embryos originate during gastrulation. *J. Comp. Neurol.* 189:323–33

Lamborghini, J. E., Revenaugh, M., Spitzer, N. C. 1979. Ultrastructural development of Rohon-Beard neurons: Loss of intramitochondrial granules parallels loss of calcium action potentials. *J. Comp. Neurol.* 183:741–52

LeDouarin, N. M., Renaud, D., Teillet, M. A., LeDouarin, G. H. 1975. Cholinergic differentiation of presumptive adrenergic neuroblasts in interspecific chimeras after heterotopic transplantations. *Proc. Natl. Acad. Sci.* U.S.A. 72:728–32

Levi-Montalcini, R., Hamburger, V. 1951. Selective growth-stimulating effects of mouse sarcoma on the sensory and sympathetic nervous system of the chick embryo. *J. Exp. Zool.* 116:321–62

Levi-Montalcini, R., Meyer, H., Hamburger, V. 1954. *In vitro* experiments on the effects of mouse sarcoma 180 and 37 on the spinal and sympathetic ganglia of the chick embryo. *Cancer Res.* 14:49–57

Llinas, R., Sugimori, M. 1978. Dendritic calcium spiking in mammalian Purkinje cells: *In vitro* study of its function and development. *Neurosci. Abstr.* 8:66

Maleyvar, R. P., Lowery, R. 1973. The pattern of mitosis and DNA synthesis in the presumptive neuroectoderm of *Xenopus laevis (Daudin).* In *The Cell Cycle in Development and Differentiation,* eds. M. Balls, F. S. Billet, pp. 240–55. Cambridge: Cambridge Univ. Press

Muntz, L. 1964. Neuromuscular foundations of behaviour in embryonic and larval stages of the anuran, *Xenopus laevis.* PhD thesis, University of Bristol, Bristol.

Nieuwkoop, P. D., Faber, J. 1956. *Normal Table of Xenopus Laevis (Daudin).* Amsterdam: North Holland, 252 pp.

Norr, S. C. 1973. *In vitro* analysis of sympathetic neuron differentiation with chick neural crest cells. *Dev. Biol.* 34:16–38

Palmer, J. F., Slack, C. 1970. Some bioelectric parameters of early *Xenopus* embryos. *J. Embryol. Exp. Morphol.* 24:535–54

Patterson, P. H. 1978. Environmental determination of autonomic neurotransmitter functions. *Ann. Rev. Neurosci.* 1:1–17

Roberts, A., Hayes, B. P. 1977. The anatomy and function of "free" nerve endings in an amphibian skin sensory system. *Proc. R. Soc. London Ser. B* 196:415–29

Roberts, A., Smyth, D. 1974. The development of a dual touch sensory system in embryos of the amphibian *Xenopus laevis. J. Comp. Physiol.* 88:31–42

Rothman, T. P., Gershon, M. D., Holtzer, H. 1978. The relationship of cell division to the acquisition of adrenergic characteristics by developing sympathetic ganglion cell precursors. *Dev. Biol.* 65: 322–41

Rutter, W. J., Clark, W. R., Kemp, J. D., Bradshaw, W. S., Sanders, T. G., Ball, W. D. 1968. Multiphasic regulation in cytodifferentiation. In *Epithelial-Mesenchymal Interactions,* eds. R. Fleischmajer, R. E. Billingham, pp. 114–131. Baltimore: Williams & Wilkins

Sanes, J. R., Hildebrand, J. G. 1976a. Structure and development of antennae in a moth, *Manduca sexta. Dev. Biol.* 51:282–99

Sanes, J. R., Hildebrand, J. G. 1976b. Origin and morphogenesis of sensory neurons in an insect antenna. *Dev. Biol.* 51:300–19

Sanes, J. R., Hildebrand, J. G. 1976c. Acetylcholine and its metabolic enzymes in developing antennae of the moth, *Manduca sexta. Dev. Biol.* 52:105–20

Sanes, J. R., Hildebrand, J. G., Prescott,

D. J. 1976. Differentiation of insect sensory neurons in the absence of their normal synaptic targets. *Dev. Biol.* 52:121–27

Schweitzer, E. S., Sanes, J. R., Hildebrand, J. G. 1976. Ontogeny of electroantennogram responses in the moth, *Manduca sexta*. *J. Insect Physiol.* 22:955–60

Seecof, R. L. 1977. A genetic approach to the study of neurogenesis and myogenesis. *Am. Zool.* 17:577–84

Shimono, T., Nosaka, S., Sasaki, K. 1976. Electrophysiological study on the postnatal development of neuronal mechanisms in the rat cerebellar cortex. *Brain Res.* 108:279–94

Shumway, W. 1940. Stages in the normal development of *Rana pipiens*. *Anat. Rec.* 78:139

Slack, C., Warner, A. E. 1975. Properties of surface and junctional membranes of embryonic cells isolated from blastula stages of *Xenopus laevis*. *J. Physiol.* (London) 248:97–120

Smolen, A., Raisman, G. 1980. Synapse formation in the rat superior cervical ganglion during normal development and after neonatal deafferentation. *Brain Res.* 181:315–23

Spitzer, N. C. 1976a. The ionic basis of the resting potential and a slow depolarizing response in Rohon-Beard neurones of *Xenopus* tadpoles. *J. Physiol.* (London) 255:105–35

Spitzer, N. C. 1976b. Chemosensitivity of embryonic amphibian neurons *in vivo* and *in vitro*. *Neurosci. Abstr.* 6:204

Spitzer, N. C. 1979. Ion channels in development. *Ann. Rev. Neurosci.* 2:363–97

Spitzer, N. C. 1980. Electrical uncoupling of vertebrate spinal cord neurons during development. *Neurosci. Abstr.* 10:287

Spitzer, N. C., Lamborghini, J. E. 1976. The development of the action potential mechanism of amphibian neurons isolated in culture. *Proc. Natl. Acad. Sci.* U.S.A. 73:1641–45

Spray, D. C., Harris, A. L., Bennett, M. V. L. 1979. Voltage dependence of junctional conductance in early amphibian embryos. *Science* 204:432–34

Taylor, A. C., Kollros, J. R. 1946. Stages in the normal development of *Rana pipiens* larvae. *Anat. Rec.* 94:7

Willard, A. L. 1980. Electrical excitability of outgrowing neurites of embryonic neurones in cultures of dissociated neural plate of *Xenopus laevis*. *J. Physiol.* (London) 301:115–28

Woodward, D. J., Hoffer, B. J., Siggins, G. R., Bloom, F. E. 1971. The ontogenetic development of synaptic junctions, synaptic activation and responsiveness to neurotransmitter substances in rat cerebellar Purkinje cells. *Brain Res.* 34:73–97

13 / *Strength and Weakness of the Genetic Approach to the Development of the Nervous System*

GUNTHER S. STENT

Wie an dem Tag, der dich der Welt verliehen,
Die Sonne stand zum Grusse der Planeten
Bist alsobald und fort und fort gediehen
Nach dem Gesetz wonach du angetreten.

Goethe

THE GENETIC APPROACH

Of all the contributors to this volume, I am probably the last one to have come to know Viktor Hamburger personally. When I visited St. Louis in the spring of 1978, I spent an unforgettable afternoon at his house, whose ambience awakened in me long-buried souvenirs of the lifestyle that I had known in my childhood in Berlin before the Second World War. On leaving Viktor that evening, I came to regret deeply having met him so late in my scientific *Werdegang.* A cruel turn of fate had conspired against me for all these years, so that despite our having many mutual friends, our paths had never crossed. I have little doubt that I would otherwise have taken up much earlier the study of developmental neurobiology on which I only lately embarked. Of course, I cannot

With the exception of the introduction, this chapter appeared earlier, in essentially the same form, in the *Annual Review of Neuroscience* (1981, 4:163–193) and is reproduced with the permission of Annual Reviews Inc., Palo Alto, California.

complain about the mentor that *did* shape my career for more than 30 years, namely Max Delbrück, who was a physics student at the University of Berlin when Viktor was a young postdoctoral fellow there at the Kaiser Wilhelm Institute (and I was getting ready to enter nursery school). In 1948, as a fresh Ph.D. in physical chemistry from the University of Illinois, I joined Delbrück at Cal Tech, to enlist in the project of lifting the mystery of the gene. But within a few months of my arrival in Pasadena, Delbrück's interest in the gene problem seemed to wane, to be replaced by a fascination with the nervous system, especially with its sensory aspects. So, to speed his entry into what for him was a new field, Delbrück organized a weekly seminar at which his students and postdoctoral fellows had to give lectures on sensory neurophysiology. Since most of my fellow disciples, such as Renato Delbecco and Seymour Benzer, knew at least *something* about the subject, they wisely picked from Delbrück's list all the papers on vision; for me, whose ignorance was total, they kindly left the coverage of von Békésy's collected works on hearing. Largely on the basis of that initial traumatic encounter with the neurosciences, I resolved to stick with bacterial viruses and to seek my fortune in the field that eventually would style itself "molecular biology," rather than to follow Delbrück in his new venture dedicated to lifting the mystery of the brain.

But when in the mid 1960s the mystery of the gene *had* been lifted, the genetic code cracked, and the mechanism of protein synthesis elucidated, many protomolecular biologists, including myself, were ready to disengage themselves from their now seemingly jejune field. By contrast, neurobiology appeared to us as an appealing new area of research, whose progress we might well speed by means of our genetic expertise. And, thanks to this shift in interest by the oldtimers, research projects began to be formulated that attempted to wed the disciplines of genetics and developmental neurobiology. In particular, the idea gained currency that the deep biological problem posed by the metazoan nervous system, namely, how its cellular components and their precise interconnection arise during ontogeny, could (and even should) be approached by focusing on genes. For instance, Benzer, the person more responsible than any other for the transformation of the classical abstract concept of the gene into its modern molecular-genetic version, outlined such a project in his Lasker Award Lecture entitled "From Gene to Behavior" (Benzer, 1971). Similarly, Sydney Brenner, another leading pioneer of molecular biology, set forth the merits of genetic methodology for the solution of neurological problems in an essay entitled "The Genetics of Behavior" (Brenner, 1973). Since then, the genetic approach to neurobiology has gained many adherents and given rise to a substantial literature, including numerous reviews (Bentley, 1976; Pak and Pinto, 1976; Dräger, 1977; Ward, 1977; Hall and Greenspan, 1979; Kankel and Ferrus, 1979; Mullen and Herrup, 1979; Palka, 1979; Quinnn and Gould, 1979; Macagno, 1980; Anderson, Edwards, and Palka, 1980). The purpose of this essay is not to add one more account of recent advances in neurogenetics to that already extensive corpus of reviews; it is rather to attempt a critical appraisal of the genetic approach.

Some Basic Concepts

To make that appraisal it is convenient to consider two related, yet distinct conceptual aspects of the genetic approach to developmental neurobiology: the *ideological* and the *instrumental* aspect. The ideological aspect of the genetic approach confronts us with the basic belief that the structure and function of the nervous system, and hence the behavior of an animal, is specified by its genes, which are held to "contain the information for the circuit diagram" (Benzer, 1971). Admittedly, it cannot be the case that the genes really contain enough information that, if we could only dicipher it, would allow us to draw a neuron-by-neuron schematic of the nervous system (Brindley, 1969). But, all the same, the circuit might somehow be *implicit* in the genetic information, since the kind of purely quantitative information-theoretical arguments advanced by Horridge (1968) against the notion of genetic determination of the nervous system are clearly invalid (Stent, 1978). And so the ideological aspect presents the discovery of how genes play their determinative role in the genesis of the nervous system and its behavioral output as the ultimate goal of developmental neurobiology. As Hall and Greenspan (1979) express this ideology: "The single gene approach to behavioral genetics and neurogenetics starts out with the knowledge that genes blatantly specify the assembly of the nervous system and the components that underlie the function of the cells in that system. Our 'only' questions, then, revolve around trying to find out *how* specific genes control neurobiological phenomena." Indeed, this discovery is not merely a distant goal but one that has already been brought within grasp, since "work with single gene mutations in a variety of organisms is yielding insights on how genes build nerve cells and specify the neural circuits which underlie behaviour" (Quinn and Gould, 1979).

By contrast, as seen from its instrumental aspect, the genetic approach does not necessarily entail a belief in genetic specification of the nervous system or present as its goal the discovery of how specific genes control neurobiological phenomena. Instead, from the instrumental aspect, the genetic approach appears as the study of differences in neurologic phenotype between animals of various genotypes, without any particular interest (other than methodologic) in the concept of genetic specificiation. For instance, Mullen and Herrup (1979) find that "the rationale for using genetic mutations involving the nervous system is two-sided. On one side, they are exciting and unique means for lesioning the system *at the level of cell interactions* [emphasis added] and to help expose the ground rules of the developmental process. On the other side, they are models for inherited diseases that unfortunately are as common in human beings as they are in mice." Similarly, Pak and Pinto (1976) see the genetic approach merely as "one of many techniques that could be applied to the nervous system and that . . . alone is unlikely to solve any of the major problems in neurobiology. One of the major contributions that the mutant approach could make may well be to provide convenient experimental preparations to which other techniques can be applied. Thus the ultimate success of the genetic approach is likely to depend on the further development and application of suitable other techniques."

In order to assess the strength and weakness of the genetic approach from both of its aspects, we will consider a by no means exhaustive, but, we hope, a representative, sample collection of studies in which genetic techniques have been brought to bear on problems connected with the development of the nervous system. These studies make use of two distinct, indeed diametrically opposite, procedures. Under the first, less-common procedure, differences in neurologic phenotype are noted among the individuals of a genetically *homogeneous* population. Here the aim is to ascertain the degree of the precision with which the developmental process gives rise to the structures of the nervous system. Under the second, more-common procedure, differences in neurologic phenotype are noted among the individuals of a genetically *heterogeneous* population. Here the aim is to establish a causal link between an observed difference in neurologic phenotype and known differences in genotype, in the hope that this link can be accounted for in terms of developmental mechanisms. We first consider some results obtained by study of genetically homogeneous populations.

PHENOTYPIC VARIANCE

The variance of phenotypes among individuals of a single species has long been appreciated. One part of such phenotypic variance is attributable to intraspecific variations in genotype, or genetic polymorphisms, that are of importance for evolutionary biology. But another part of phenotypic variance simply reflects what Waddington (1957) called "developmental noise." The study of this noise—its sources and the physiological processes that reduce its consequences—seems to be an underdeveloped area of contemporary developmental biology, even though an understanding of developmental noise bids fair to lead to significant insights into developmental mechanisms. The nervous system is particularly suitable for the study of developmental noise, in that here the existence of subtle phenotypic variance can be more readily assessed, and can have more pronounced functional consequences, than in many other organs. To ascertain the degree which developmental noise actually occurs in the formation of some part of the nervous system, it is necessary to examine a set of isogenic animals (i.e., all sharing exactly the same genotype). Any variance found under these conditions can then be attributed to noise rather than to genetic polymorphisms. (It should be noted that in the context of phenotypic variance the concept of "genotype" has to refer to the genetic constitution of the zygote, or the germ line, rather than of the somatic tissues. This distinction is important because during development a variety of processes, such as chromosome loss or somatic recombination, may give rise to differences in genetic constitution between somatic cells. But insofar as random somatic generation diversity does give rise to phenotypic variance, it would merely represent one source of developmental noise.) The aspects of specifically identifiable neuronal structures whose phenotypic variance may be examined from individual to individual include the number of neurons of which the structure is composed, the relative position of these neurons within the structure, the morphology of their axonal and dendritic processes, and the pattern of the synaptic connections.

The nervous system of presumptively isogenic individuals of the nematode *Caenorhabditis* has been examined from these aspects of phenotypic variance (Ward et al., 1975; Ware et al., 1975; Albertson and Thomson, 1976; White et al., 1976). As far as cell number is concerned, the *Caenorhabditis* nervous system was found to be highly invariant, consisting of exactly 258 identifiable neurons. This constant neuron number is attributable to an invariant pattern of cell cleavage and cell death during development. However, the location of the cell bodies of these neurons is subject to some significant variation, owing to a random element in the migration of neuroblast precursor cells (Sulston, 1976; Sulston and Horvitz, 1977). Furthermore, the morphology and synaptic connection pattern of the nematode neurons manifest a considerable degree of phenotypic variance. Although the major geometrical features of the processes of each identifiable neuron are quite constant, the fine details of their branching pattern are quite variable. Similarly, although the overall topology of the neuronal network is quite constant, the synaptic sites possessed by individual identified neurons are quite variable with regard to their number and location.

Another detailed study of phenotypic variance was carried out on the visual system of isogenic individuals in parthogenetic clones of the crustacean *Daphnia* (Macagno, Lopresti, and Levinthal, 1973). The *Daphnia* compound eye consists of 22 facets, or ommatidia. As is generally true for arthropod compound eyes, every ommatidium contains eight primary photoreceptor cells, each of which occupies a characteristic position within the ommatidium. The axons of the eight receptor cells project as a bundle to an underlying layer of 110 secondary visual neurons, the optic lamina. There the eight photoreceptor axons of each bundle make synaptic contact with five contiguous laminar interneurons that form an optic cartridge. The axons of the second order laminar neurons in turn project to and make synaptic contacts with an underlying layer of about 320 third-order visual neurons, the medulla. Examination of more than 100 individuals of an isogenic *Daphnia* clone revealed only one that did not have exactly 22 bundles of eight photoreceptor axons; that specimen had one nine-axon bundle (Macagno, 1980). No variance was observed in the position of the photoreceptors within the ommatidial array, and in 100 optic cartridges examined, the cell body of only one second-order neuron was seen to be in an atypical location (Levinthal, Macagno, and Levinthal, 1975). Analysis of the fine structural details of the synaptic contacts between the eight photoreceptor axons and the five second-order laminar neurons showed that there exist "strong" and "weak" connections. Strong connections consist of 30 or more synaptic contacts between a photoreceptor axon terminal and a laminar neuron, whereas weak connections consist of no more than five such synaptic contacts. The pairs of first-order and second-order neurons linked by strong connections were found to be the same from specimen to specimen, although the number of actual synaptic contacts between any given strongly linked pair varied from about 30 to 100. By contrast, the identity of the neuronal pairs found to be linked via weak connections showed a high degree of variance.

A low variance in the projection pattern of compound-eye photoreceptor axons to the lamina was established also in nonisogenic specimens of the fly *Calliphora* (Horridge and Meinertzhagen, 1970). (In order to establish some upper limit to the developmental noise in the formation of some neuronal structure, the population of animals studied need not be isogenic since true isogenicity would only serve to reduce phenotypic variance.) Here two of the eight photoreceptors of each ommatidium project their axons directly to the optic medulla without forming synaptic contacts in the lamina. Of the remaining six photoreceptors, each projects its axon to a different optic cartridge in the lamina. At that cartridge, the axon meets the axon terminals of five other photoreceptors, each from a different neighboring ommatidium. This optical system is constructed so that all six photoreceptors whose axons project to the same cartridge receive light from the same point in the visual field. Examination of the projection pattern of about 650 photoreceptors, originating in 120 different *Calliphora* ommatidia failed to turn up a single case of an axon entering the wrong optic cartridge, that is, a cartridge in which an axon would have met axons reporting incident light from a different point in the visual field.

The number and morphology of identifiable second-order neurons contacted by axons of ocellar (rather than ocular) photoreceptors have been examined for their phenotypic variance in isogenic individuals of parthenogenetic clones of the locust *Schistocerca* (Goodman, 1974, 1976, 1977, 1978). Of these ocellar interneurons more than 70 can be individually identified from animal to animal. Study of 11 parthogenetic locust clones showed that one particular interneuron was present in duplicate in about one-half to one-third of the individuals belonging to three such clones. Among the individuals of the remaining eight clones, duplications of that interneuron were rarely seen. No individuals *lacking* that particular interneuron occurred in any of these parthogenetic clones, although in some nonisogenic populations of *Schistocerca* that interneuron is not infrequently absent. A survey of the location of the cell bodies of the identified interneurons showed that among the individuals of one clone the relative position of a particular interneuron was more variable than was the case in the other clones, and the major branching pattern of the axon of a third identified interneuron showed significant abnormalities in the majority of the individuals of two of the clones. The major branching pattern of the other interneurons was, however, quite invariant in these two clones. As for the fine branching pattern of the axons, it was found to be highly variable for all the identified ocellar interneurons. That variance was equally great for interclone and intraclone comparisons. These differences in phenotypic variance between individual locust clones indicate that the genotype can influence the degree of developmental noise to which a particular neuron or neuronal structure is subject.

A large phenotypic variance was found to obtain in the properties of higher order visual interneurons of the locusts *Schistocerca* and *Locusta* (Pearson and Goodman, 1979). Although the locusts examined were merely highly inbred, rather than authentically isogenic, it seems most likely that the variance observed is attributable largely to developmental noise rather than to genetic polymorphism. Even the major branches of the interneuronal axons were found to be so variable from individual to individual that no "normal" pattern could be described. And this morphological variability of the interneurons was matched by a corresponding variability in its functional synaptic connections to a set of motoneurons innervating leg and wing muscles. That is to say, functional connections were present in some individuals that were absent in many others.

Thus these few studies of phenotypic variance show that if the nervous system and its components are examined at progressively finer levels, some point is eventually reached at which the effects of developmental noise become manifest. In some cases that point is reached already at the relatively gross level of cell number, whereas in other cases it is reached only at the level of fine branching of neuronal processes and individual synaptic sites. One can account for this fact by resorting to the facile, all-purpose evolutionary explanation that in the development of any part of the nervous system just that amount of developmental noise is tolerated which is still compatible with adaptive function. The nematode, whose nervous system of only 258 cells generates a behavior of complexity not far behind that shown by animals with nervous systems consisting of hundreds or thousands of times more neurons, can probably ill afford any departure from the normal cell number. Similarly, it can be argued that for the compound eye to provide its high-resolution image of the visual surround for effective guidance of visual behavior, the connections between primary ommatidial photoreceptors and optic cartridges must be of high percision. By contrast, since the ocellus appears to respond mainly to changes in overall light intensity (Chappell and Dowling, 1972) and provides little spatial resolution of the visual surround (Wilson, 1978), it can be imagined that here less precision and more developmental noise can be tolerated in the establishment of its central neuronal connections. Among the mechanisms that appear to be available for reducing the effect of developmental noise to acceptable levels is an initial overproduction of neurons and synaptic connections followed by functional testing and elimination of the unwanted excess (Stent, 1978). But resort to any mechanism for the reduction of developmental noise represents, in the parlance of sociobiology, an investment on the part of the animal, which it would be wise to avoid unless it were absolutely necessary.

NEUROLOGIC MUTANTS

We now consider results obtained from the study of genetically *heterogeneous* populations, and in particular from the comparison of normal animals and conspecific genetic mutants exhibiting abnormal neurologic phenotypes. As one of many possible examples illustrating this approach, let us consider the role of sensory afference in the development of the central nervous system of crickets such as *Acheta* or *Teleogryllus.* From the rear of the abdomen of the cricket, there projects a pair of sensory appendages, the cerci, that carry several thousand chemosensory and mechanosensory organs. One type of organ consists of a filiform hair that is set into directionally selective vibration by sound or wind stimulation. The hairs are innervated by the endings of sensory neurons whose axons project via the cercal nerve to the abdominal ganglion, where they form synaptic contacts, particularly with the ipsilateral medial giant interneuron (MGI) of the ventral nerve cord. Two sound-sensitive and wind-insensitive cricket mutants were isolated by Bentley (1973, 1975, 1977). In one mutant the filiform hairs are partially absent and in the other they are mising altogether, which is the reason why exposure of the cerci of these mutants to the normally adequate mechanical stimulus evokes no response. Despite their lack of functional hairs, the mutant cerci carry normal-looking sensory cells. Moreover, intracellular recordings taken from the MGI of the mutants showed a normal postsynaptic response following direct electrical stimulation of the hairless sensory cells. The hairless sensory cells are therefore electrophysiologically functional, albeit inactive for lack of natural stimuli. Upon neuroanatomical examination of these mutants Bentley found, however, that the dendrites of their MGIs are stunted and significantly thinner than normal. What developmental inferences can be made from these findings? First, it follows that the development of functional mechanosensory cells on the cerci does not require the postembryonic presence of functional hair cells. Second, development of functional central synaptic contacts between the sensory axon terminals and the MGI does not require a normal electrical activity pattern on the part of the presynaptic sensory cells. And third, normal structural development of the MGI dendrites does require a normal afferent synaptic input pattern from their presynaptic sensory cells.

Some good will is actually necessary for the acceptance of the third, and most important, conclusion regarding the morphogenetic role of sensory afference; since *all* tissues, and not just the cerci, of the mutant crickets are of mutant genotype, it is formally possible that the abnormal development of the MGI represents a *direct* morphogenetic effect of the gene mutation on interneuron development. In that case, the inferred indirect effect of the absence of functional cercal hairs mediated via the lack of normal activity of the cercal sensory cells and their consequent lack of normal afferent synaptic input to the developing MGI might not exist. However, independent, nongenetic experiments make that possibility unlikely. For a very similar stunted anatomy of MGI dendrites is found in genetically normal crickets that were raised throughout their postembryonic development while their cerci had been surgically removed (Murphey et al., 1975). Thus, sensory deprivation achieved here by surgical means causes a similar morphogenetic effect on the central nervous system as that observed under analogous deprivation achieved by genetic means. It is fortunate that Bentley troubled to ascertain the state of the cercal hairs of his sound-insensitive and wind-insensitive cricket mutants instead of proceeding directly to an examination of their neuroanatomy. Otherwise, some believers in the genetic approach might have concluded from the abnormal MGI morphology that the mutation is located in a gene that specifies the structure of the cricket central nervous system. That the abnormal neurologic phenotype engendered by a particular gene mutation is the result of a cascade of pleiotropic effects set in train by a (usually unknown) primary effect of the mutation is a feature inhering in nearly all the neurogenetic examples to be considered in the following sections.

GENETIC MOSAICS

In view of the difficulty of making straightforward developmental interpretations of the phenotypes of mutant animals whose tissues are *all* of mutant genotype, it is fortunate that there exists a genetic method that can be called on to ascertain whether the neurologic abnormality of structure or function A is really a direct result of the mutation or merely a secondary, cascading result of the abnormal phenotype of B. This method consists of the production of *genetic mosaics,* that is, of animals in which some cells are of mutant genotype and others are of normal genotype. The mosaic method is especially useful in situations where the kind of nongenetic, surgical intervention cannot be carried out which, as in the case of the cricket mutants, gives support to the interpretation of an influence of the activity of the presynaptic sensory cell on the morphology of the postsynaptic interneuron. It is not accidental that the use of genetic mosaics for developmental neurobiology has thus far been limited mainly to *Drosophila* (for reviews, see Hall et al., 1976; Palka, 1979) and to the mouse (for a review, see Mullen and Herrup, 1979), because they are some of the few animals for which the diversity of mutants with altered neurologic phenotype and the very high level genetic technology required for the experimental production of mosaics is at present available.

Drosophila

In *Drosophila,* two different methods for the production of genetic mosaics can be used. One of these consists of the experimental generation of gynanders, or individuals whose tissues form a mosaic of male and female cells. For this purpose a female zygote of XX genotype is produced in which one of the pair of X sex chromosomes is ring-shaped rather than, as is normally the case, rectilinear. There is a high probability that the ring-X chromosome is lost during one of the first few mitotic nuclear divisions in the developing early embryo, giving rise to a daughter nucleus that carries only a single X chromosome, or is of XO genotype. (Although in *Drosophila* the normal male genotype is XY, zygotes of XO karyotype also develop the male phenotype.) In this way there arises in the female embryo a clone of male cells that eventually gives rise to a patch of contiguous male tissue in the postembryonic animal. The position of the boundary between male and female tissues in the gynandromorph mosaic is highly variable, because in *Drosophila* embryogenesis the first nuclear division in the fertilized egg is oriented at random with respect to the future body axes. In order to identify the patch of male tissue, the rectilinear X chromosome usually carries a mutation whose expression is recessive to the dominant allele carried by the ring-X chromosome subject to elimination. Thus only the male cells of XO karyotype show the recessive mutant phenotype.

The second method for generating mosaic flies relies on the induction of somatic recombination of homologous chromosomes during nuclear divisions at stages of early embryogenesis. For this purpose, a heterozygote is produced which carries a recessive mutation on one chromosome and the dominant allele on the homolog of that chromosome. Exposure to low doses of x-rays of the heterozygote nuclei at a stage of the mitotic cycle when their chromosomes have already replicated (i.e., are present as sister chromatids) induces crossing-over between a pair of homologous nonsister chromatids. Such somatic crossing-over can result in a diploid homozygote nucleus in which both homologous chromosomes carry the recessive mutant locus. In this way there arises a clone of homozygotic cells that eventually produces a patch of contiguous tissue in which the recessive mutant phenotype is expressed, while the remainder of the body shows the phenotype of the dominant allele. As in the case of the gynanders, the position of the boundary between the two types of tissue in somatic recombinant mosaic flies is highly variable. However, in contrast to the gynander method, the somatic recombinant method permits experimental control over the size of the mosaic patch, in that the patch will be generally smaller the later the developmental stage at which the embryo is x-irradiated. Moreover, whereas the gynander method is limited to studying the developmental effects of mutant loci on the X-chromosome, the somatic recombinant method is applicable also to gene mutations carried on the three other (autosomal) *Drosophila* chromosomes.

Fate Mapping At an early developmental stage, the *Drosophila* embryo consists of a single, superficial layer of 5,000 to 10,000 cells, the blastoderm. As was first realized by Sturtevant (1929), the distribution of the boundary between male and female tissues in postembryonic *Drosophila* gynanders should provide an indication of the spatial distribution of the precursor cells of various body parts over the blastoderm. For, so Sturtevant reasoned, the more frequently the male-female tissue boundary happens to fall between any two given body parts in a population of gynanders, the more distant should the precursor cells of these body parts, or landmarks, be on the blastoderm. By observing the pairwise frequency of the male-female difference between an ensemble of such landmarks, a two-dimensional map can be constructed that represents the relative disposition of the landmark precursor cells over the blastoderm. In order to apply such blastoderm fate mapping to landmarks of the *Drosophila* nervous system, a recessive mutation leading to the loss of the histochemically identifiable enzyme acid phosphatase in neurons of male genotype was introduced into the X-chromosome (Kankel and Hall, 1976). By these means, it is possible to show that the precursor cells of the central nervous system lie in the ventral area of the blastoderm, while the precursors of the supraesophageal, subesophageal, and optic parts of the head ganglion and the thoracic and abdominal parts of the thoracicoabdominal ganglion are disposed in an anteroposterior sequence. By searching for the smallest patch of male or female tissue that can be observed within a single mosaic ganglion, it was also possible to estimate the number of blastoderm cells that constitute the precursor ensemble of individual ganglia. Since no mosaic patches smaller than one-third to one-tenth of the total ganglionic mass were seen, it can be concluded that only a few blastoderm cells (from 3 to 10) are the precursors of each major ganglion on the right or left side of the body.

Fate mapping studies also showed that the precursor cells of the Drosophila compound eye are situated in the anterior part of the blastoderm, but at a level more dorsal than that of the precursors of the ganglia of the central nervous system. How do these ocular precursor cells eventually give rise to the highly regular array of ommatidia? Early studies on the development of the arthropod compound eye had led to the conclusion that the eight primary photoreceptor cells of each ommatidium form a clone, that is, they are the decendants of a single founder cell (Bernard, 1937). Both radiological and genetic experiments later seemed to provide support for this view of the development of the 700 to 800 ommatidia of the *Drosophila* compound eye (Becker, 1957). However, more recent findings with genetic mosaics of *Drosophila* produced by the somatic recombination method argue against this view (Hofbauer and Campos-Ortega, 1976; Ready, Hanson, and Benzer, 1976). In these studies heterozygous fly embryos carrying a recessive white (i.e., pigmentless) eye color mutation and its dominant red (i.e., normally pigmented) allele were x-irradiated at an early developmental stage to induce somatic recombination. Among the resulting adult flies, there were many with a mosaic eye, having one patch of ommatidia with normally pigmented photoreceptors and another patch of ommatidia with pigmentless photoreceptors. Contrary to the expectations of the view of the ommatidium as a single cell clone, the border between pigmented and pigmentless patches was found to be formed by ommatidia that were themselves of mosaic character. That is to say, of the eight photoreceptors within a single ommatidium, some were pigmented and others not. Hence they could not all be the descendants of a single founder cell. Thus the morphogenetic selection of eight photoreceptor cells for formation of a given ommatidium is not strictly determined by their developmental line of descent. It is nevertheless still possible that the *Drosophila* photoreceptors do arise as eight-cell clones but that the members of a single clone are not constrained to participate in the formation of the same ommatidium (Campos-Ortega and Hofbauer, 1977).

Organization of the Optic Lamina How, in the development of the arthropod visual system, is the low phenotypic variance achieved in the connection of photoreceptor axons with the second order visual neurons of the optic cartridges? A partial answer to this question is that the ommatidia of the compound eye play an organizational role in the development of the underlying optic lamina. One of the most convincing demonstrations of this developmental interaction was provided by the use of *Drosophila* mosaics (Meyerowitz and Kankel, 1978). For this purpose, two recessive mutations, *rough* eye and *glass* eye, and one dominant mutation, *glued* eye, were studied. The phenotype attributable to any one of these three mutations consists of severe abnormalities in the structure of both the ommatidia and the optic cartridges. In order to ascertain whether the abnormal phenotype of one of these components of the arthropod visual system is merely a secondary consequence of the abnormal phenotype of the other, mosaic flies were produced by both the gynander and the somatic recombination methods. In these mosaics either the ommatidia or the optic lamina were of normal genotype while the other component was of mutant genotype. (Here the genotype of the photoreceptor cells was identified by means of an eye-color mutation linked genetically to the eye structure mutation; the genotype of the optic lamina neurons was identified by means of a linked mutation leading to loss of acid phosphatase.) The result of this experiment was that in all cases in which a patch of the eye was of mutant genotype, both the ommatidia of that patch and the underlying optic cartridges were of abnormal (i.e., structurally disrupted) phenotype. But in the genotypically normal remainder of the mosaic eye, both the ommatidia and their correspondent optic cartridges were of normal phenotype, even when the underlying optic lamina neurons were of mutant genotype. Thus it can be inferred that the effect of the three eye-structure mutations studied here is more proximal to the development of the ommatidia than it is to that of the optic lamina. More importantly, it can be concluded that the normal development of the optic cartridges depends on an interaction of the neurons of the optic lamina with a structurally normal array of ommatidia and, presumably, on a normal projection pattern of photoreceptor axons.

Behavioral Mosaics As was shown by Hotta and Benzer (1972, 1976), the genetic mosaic method can be used also to pinpoint the neural structures, or "foci" that are in control of behavioral routines. One such routine that was examined in *Drosophila* gynanders is the wing vibration associated with male courting behavior. These studies showed that mosaic flies with male heads and female bodies extend their wings toward females, as if in courtship. But the actual wing vibration is produced only if, in addition to the male head, there is some male tissue present also in the thoracic body segments. The neurologic interpretation of these findings was in some doubt, however, because the sexual genotype of head and thorax had been ascertained by examining the male-female boundary of the external cuticle of mosaic flies rather than of the underlying neural tissues. Subsequently, courting behavior was studied also in gynanders whose neurons were scored for their male or female genotype by the acid phosphatase method. In this way it was shown that for there to be any vestige of male courting behavior, such as extension of the wing towards the female, it is necessary that the dorsoposterior part of the gynander head ganglion is of male genotype (Hall 1979). But for wing vibration to occur, some male tissue must also be present in the thoracic ganglion (von Schilcher and Hall, 1977). Thus, it can be concluded that one part of the male-specific neural circuits required for this behavioral routine (presumably the "command" component) is located in the head ganglion and another part (presumably the "motor-pattern generator") is located in the thoracic ganglion of the ventral nerve cord. Within the context of developmental neurobiology, these findings imply that the formation of a particular neural network can depend on the local genotype of the neurons that compose it rather than on bodywide (e.g., hormonal) influences.

Homeotic Mutants There exists a class of mutants of *Drosophila,* termed homeotic, in which one part of the body of the fly is transformed into another (Lewis, 1963; Ouweneel, 1976). One example of such a mutant is *Antennapedia,* in which the distal part of the antenna is transformed into a mesothoracic (or middle) leg. Another is *bithorax-postbithorax (bx pbx),* a double mutant that shows a transformation of the metathorax (or hind thorax) into a second meso thorax (or middle thorax) and is thus endowed with a second pair of wings in place of the pair of halteres, balancing organs that aid the fly in navigation.

Since their abnormally positioned, or ectopic, structures are innervated, homeotic mutant flies are highly promising objects for the study of the mechanisms that govern the formation of neural connections. For instance, application of a sugar solution to the homeotic leg of an *Antennapedia* mutant evokes the proboscis-extension reflex evoked also by stimulation of normally positioned legs, but not by stimulation of normal antennae (Deak, 1976; Stocker, 1977). This response indicates that despite its ectopic attachment to the head, the homeotic leg carries chemoreceptors that make functional connections in the central nervous system capable of eliciting a normal motor output. Anatomical study of the sensory fibers projecting centrally from the ectopic leg show that they enter the head ganglion via the antennal nerve and ramify in the head ganglion in a manner similar to the normal antennal sensory fibers (Stocker et al., 1976). The sensory fibers from the ectopic leg do not appear to project rearward to the thoracicoabdominal ganglion that is the ordinary destination of the sensory fibers from the normally positioned mesothoracic leg. The sensory fibers from the normally positioned leg do not in turn project frontward to the head ganglion. In view of this neuroanatomical hiatus, it seems a considerable challenge to try to explain how such a functional pathway from sensory receptors to motoneurons responsible for the proboscis extension reflex manages to be formed in the case of the ectopic *Antennapedia* leg.

The anatomy of wing sensory fibers of *bx pbx* mutants has also been studied in considerable detail (Ghysen, 1978; Palka, Lawrence, and Hart, 1979). As for the normal (mesothoracic) wings, they carry three anatomically different types of mechanosensory structures: bristles, large campaniform sensilla (or s.c.), and small s.c., all of which project their axons into the nerve cord via a mesothoracic nerve. The bristles send fine sensory axons into the ventral region of mesothoracic segment of the thoracicoabdominal ganglion. The large s.c. send thick axons that ramify in the ventral region of all three thoracic segments of that ganglion. And the small s.c. send axons into a dorsal tract of the nerve cord, where they bifurcate and project both frontward to the head ganglion and rearward to the metathoracic segment of the thoracicoabdominal ganglion. By contrast, the normal (metathoracic) halteres carry only small s.c., which send axons via the (posteriorly situated) haltere nerve into the dorsal nerve tract. There they project frontward to the head ganglion, and ramify in both metathoracic and mesothoracic segments along the way. No axonal projections are manifest from normal halteres to the ventral region of the nerve cord, destination of the sensory axons of wing bristles and large s.c.

The homeotic (metathoracic) wing of the *bx pbx* mutant resembles the normal wing in carrying bristles as well as large s.c. and small s.c. The axons of these homeotic sensory structures all enter the central nervous system at the site normally occupied by the haltere nerve. Once in the central nervous system, the central projections of the axons of the sensory structures on the homeotic wings form a pattern that would be expected of posterior serial homologs of the corresponding sensory structures on the normal wings. Thus the bristles on the homeotic wing send their fine axons to the ventral region of their own metathoracic segment, where they form a pattern similar to that formed in the mesothoracic segment by the bristle axons from the normal wing. The thick axons of the large s.c. on the homeotic wing ramify in the ventral region of all three thoracic segments of the thoracicoabdominal ganglion, where they overlap with the axon endings of the large s.c. on the normal wing. And the axons from the small s.c. on the homeotic wings enter the dorsal tract of the nerve cord, ramify within the metathoracic segment, and project frontward to the head ganglion, just as do the axons of the small s.c. on the normal wings. In this last case, of course, the axons of the small s.c. on the homeotic wings can also be said to project centrally as if they still belonged to halteres. Indeed, the pattern of ramification of the small s.c. axons on the homeotic wing in the metathoracic segment is indistinguishable from that of small s.c. on halteres in normal flies.

How is one to account for these findings with *bx pbx* mutants? As far as the large s.c. on the homeotic wing are concerned, one could imagine that their sensory axons are destined to seek the same central target neurons as the homologous axons coming from the large s.c. on the normal wing, even though the former enter the nerve cord on an abnormal (posterior) level. The same explanation can be given for the small s.c. on the homeotic wing, although in this case an even simpler explanation is available, namely, that the small s.c. on the homeotic wing are destined for the same central targets as the small s.c. of the absent haltere. But a more complicated explanation is required for the bristles on the homeotic wing, whose axons end in the metathoracic segment of the thoracicoabdominal ganglion not normally reached by axons from any bristle cells. One likely explanation is that in the metathoracic segment there normally arise neurons that are appropriate targets for wing bristle axons but are not ordinarily reached by them because bristle cell axons of the normal wing are somehow destined to remain in the segment at which they entered the nerve cord. But there is also an alternative explanation that could be considered for the observed projection pattern of the bristle cells on the homeotic wing: contrary to the general finding that the proximal phenotypic effect of homeotic mutations is confined to tissues derived from imaginal disks (of which the nerve cord is not one [Ghysen, 1978]), it would still be formally possible that, owing to their own mutant genotype, the neurons of the metathoracic segment of the nerve cord take on the quality of mesothoracic neurons. In that case, the central projection pattern of the homeotic wing sensory cells would reveal little about the mechanisms underlying the normal case. In order to test for this possibility, Palka and coworkers (1979), produced mosaic flies with patches of homeotic bithorax wings embedded within an otherwise normal haltere, and with a central nervous system of nonhomeotic genotype. Neuroanatomical analysis of the sensory projections from such homeotic wing patches to the genotypically normal metathoracic segment of the thoracicoabdominal ganglion showed the presence of fine axons in the ventral region, characteristic of the sensory fibers from homeotic wing bristle cells. It follows, therefore, that in the case of *bx-pbx* mutant flies it is not the mutant genotype of the central nervous system that is responsible for directing bristle cell fibers to a metathoracic region, which is ordinarily never reached by such axons.

Mouse

In the early 1960s a technique became available by means of which mice (and by now, some other mammals) can be produced whose tissues are mosaics of cells of two different genotypes (Tarkowski, 1961; Mintz, 1962, 1965). Under this technique an early embryo, or morula, consisting of only a few cells is removed from the oviduct of a pregnant female mouse and dissociated into its constituent cells by enzyme treatment. These cells are mixed with a second set of dissociated cells similarly obtained from another pregnant female mouse carrying a morula of a different genotype. Upon incubation for some hours, the mixed cells reassociate, continue to cleave and form a blastocyst. This blastocyst is then reimplanted into the uterus of a female host animal. There the artificial hybrid blastocyst continues normal embryonic development and is eventually brought to term as a morphologically normal mouse, some of whose cells are derived from one pair of parents and others from another pair of parents. The mosaic character of such tetraparental mice is readily manifest if the two parental pairs differed from each other genetically in some external characteristic, such as coat color. In mosaic mice, the actual proportion of the total cell population descended from either of the two mixed morulas varies within wide limits; some mosaic specimens are 50–50 mixtures of the two cell genotypes, in others one genotype constitutes as little as 1 percent of the total, and yet others show no evidence of mosaicism at all. This variability is the reflection of a developmental indeterminacy, regarding which few morula cells contribute any descendants at all to the postnatal mouse and the developmental fate of the progeny of any morula cell that does make a contribution. This method can thus generate a large variety of mosaic patterns. For use of such mosaic mice in the context of developmental neurobiology, cells of two morulas are mixed, of which one is of normal genotype and the other is of a mutant genotype that gives rise to some particular neurologic abnormality. In addition, the two morulas are made to differ at some other genetic locus so as to produce a histologically detectable but neurologically neutral difference between the nerve cells derived from either source. Two neutral cell-marker loci favorable for this purpose are the structural genes for β-glucoronidase (Condamine, Custer, and Mintz, 1971; Feder, 1976; Mullen, 1977a,b) and for β-galactosidase (Dewey, Gervais, and Mintz, 1976), at which mutations affect the intensity of staining produced by an enzyme-specific histologic reaction.

Retinal Degeneration Mutants The first neurological use of mosaic mice was made in the study of a retinal degeneration (or *rd*) mutation that leads to the postnatal degeneration of the retinal photoreceptor cells (Mintz and Sanyal, 1970; Wegmann, La Vail, and Sidman, 1971). Here the mixing of morulas of a normal and a *rd* genotype resulted in mice whose retinae consisted of a mosaic of structurally normal patches and patches in which the photoreceptors had degenerated. The neutral cell marker showed that the pigment epithelium underlying the photoreceptors was also of genotypically mosaic character. But the phenotype of the photoreceptors in any retinal patch was found to be unrelated to the genotype of the underlying pigment epithelium. Thus it can be inferred that the degenerative effect of the *rd* mutation is expressed directly within individual photoreceptor cells, rather than being the result either of a retinawide influence of the presence of the mutant gene in some other tissue or of a local influence of its presence in the contiguous pigment epithelium. This finding is to be contrasted with a seemingly similar photoreceptor degeneration phenotype in the rat. In that case retinal disorganization was found to be a secondary effect of the presence of the mutant genotype in the underlying pigment epithelium (La Vail and Mullen, 1976; Mullen and La Vail, 1976). Another instance of an indirect, or secondary, effect was uncovered upon study of a muscular dystrophy (or *dy* mutation). Here a mosaic mouse was found with a muscle of predominantly dystrophic phenotype, whose muscle fibers, according to the neutral cell marker, were of predominantly normal genotype. Conversely, in another mosaic mouse, a muscle of predominantly normal phenotype was found whose muscle fibers were, according to the neutral cell marker, of predominantly *dy* mutant genotype. Thus dystrophy appears to be attributable to some disturbance of a normal muscle function by action of the *dy* mutation at a distant, extramuscular site (Peterson, 1974).

Cerebellar Mutants Thus far, mosaic mice have found their most extensive neurological application in the study of the dozen or so mutations that are known to perturb the development of the cerebellum and give rise to easily recognizable deficits in locomotory behavior (Sidman, Green, and Appel, 1965). One such case examined by this method is the Purkinje cell degeneration *(pcd)* mutation that causes postnatal degeneration of all the Purkinje cells in the cerebellar cortex. Here it was found that in mosaic mice of mixed *pcd* and normal genotype some Purkinje cells degenerate and some survive. The neutral cell marker indicated that all the surviving Purkinje cells are of normal genotype and that all the mutant Purkinje cells degenerate (Mullen, 1977a). Hence it appears that the cerebellar effect of the *pcd* mutation resembles the retinal effect of the *rd* mutation in that its degenerative action is expressed within individual cells of mutant genotype. It is to be noted, however, that the effect of the *pcd* mutation is not restricted to the cerebellar Purkinje cells, since it engenders also the degeneration of photoreceptor, mitral, and sperm cells (Mullen and La Vail, 1975; Mullen, Eicher, and Sidman, 1976).

Reeler. The effects of the two other well-known, cerebellar mutations, *reeler (rl)* and *staggerer (sg),* have also been studied within the mosaic mouse context. The *reeler* mutation, rather than leading to degeneration of the Purkinje cells, prevents their migration during development of the embryonic cerebellum. This migration normally proceeds outward from the deep ventricular zone of origin of the Purkinje cells to the cerebellar cortex, where their cell bodies eventually form a single layer below the molecular layer formed by the parallel fibers and above the granule cell layer. But in *reeler* mutants, the vast majority of Purkinje cells fails to migrate outward and remains below the granule cell layer, which is itself composed of far fewer than the normal number of granule cells (Hamburgh, 1963; Caviness, 1977; Mariani et al. 1977).

Upon production of *reeler*/normal mosaic mice, specimens were found with a cerebellum whose gross morphology was intermediate between the fully normal and the fully *reeler* phenotype: some Purkinje cells had migrated to their appropriate positions in the cerebellar cortex whereas other Purkinje cells had remained in the underlying deep zone. As shown by the neutral cell marker, many of the normally positioned cells were of the mutant *reeler* genotype and many of the abnormally positioned cells were of the normal genotype (Mullen and Herrup, 1979). This finding shows that its failure to migrate outward is the result of an indirect, secondary effect of the *rl* mutation on the Purkinje cell, with the primary effect being exerted at another, extrinsic site. In thus exerting its effect indirectly, or transcellularly, the *reeler* mutation seems to resemble the *dy* muscular dystrophy mutation. However, the interspersion of phenotypically normal and abnormal Purkinje cells makes it appear that each of these cells is positioned individually during histogenesis of the cerebellar cortex and that the signals mediating the transcellular effect of the *rl* mutation aborting normal migration reach the target cells not in the form of a homogenous field, such as would be produced by diffusion of a chemical substance, but as a fine-grained mosaic. Two possible explanations of such a fine-grained mosaic of transcellular signals are that the migration of each Purkinje cell is individually stimulated by early synaptic contact with the axon terminals of another type of neuron (Caviness and Sidman, 1972) or is individually guided by contact with a particular glial cell (Caviness and Rakic, 1978). Here the eventual position of a Purkinje cell would depend not on its own *reeler* genotype but on that of the presumptive presynaptic stimulatory neuron or guiding glial cell with which it happens to come into contact (Mullen, 1977b).

As is the case for the *pcd* mutation, the effect of the *rl* mutation is not confined to the cerebellar Purkinje cells. In *reeler* mice the structure of the cerebral cortex is also highly abnormal in that here too neurons fail to migrate during development. Just as do the Purkinje cells, the neurons of the cerebral cortex arise in the deep ventricular zone and later migrate outward. The nature of the migration in the cerebral cortex is such that the first neurons to arise remain as the deepest layer, with neurons arising at progressively later developmental stages migrating outward past the neurons that have arisen before them. The normal cerebral cortex is thus composed of a set of cell layers of which the most superficial layer contains the last—and the deepest layer the first—neurons to have arisen. Because this migration is aborted in the *reeler* mutant, the structure of its cerebral cortex is inside out: the most superficial layer contains the first neurons to have arisen, and the deepest layer contains the last (Caviness and Sidman, 1973). But despite this spatiotemporal inversion of the normal structure of the cortical cell layers, the afferent nerve fibers of the optic radiation manage to project to cell layers of appropriate age in the visual cortex of reeler mutants (Caviness, 1977). Moreover, neurophysiological study of the visual neurons of the *reeler* cortex showed that their respective field structure and their retinotopic organization are quite normal (Dräger, 1977). No studies have as yet been reported of the visual cortex of *reeler*/normal mosaics, but the present findings with *reeler* mutants nevertheless allow at least one significant conclusion regarding cortical development: the eventual identity of cortical neurons with respect to both morphology and function is more closely related to their "birthdates" than to their positions along the radial (inside-outside) axis. And as a corollary, it can be concluded that the formation of neuronal synaptic contacts within the cortex is not so much governed by cell position along the radial axis as it is by cell identity.

Staggerer. The *staggerer* mutation leads to a poorly developed cerebellum nearly devoid of the normal granule cell layer. Although during embryonic development of *staggerer* mice the granule cells arise more or less normally on the outer surface of the cerebellum, the granule cells subsequently degenerate during their inward postnatal migration across the molecular and Purkinje cell layers and fail to reach the deeper zone at which the granule cell layer is ordinarily present. The morphology of the Purkinje cells of the *staggerer* cerebellum is also abnormal: these Purkinje cells do not form their characteristic single-cell layer; and they have very sparsely developed dendritic trees whose much fewer dendrites lack the spines at which the granule cell axons, or parallel fibers, normally make synaptic contacts (Landis and Sidman, 1978; Landis and Reese, 1977; Sotelo and Changeux, 1974; Yoon, 1977). Upon production of *staggerer*/normal mosaic mice, individuals were found that, as in the case of *reeler*/normal mosaics, had a cerebellum whose gross morphology was intermediate between fully normal and fully mutant phenotypes (Mullen and Herrup, 1979). In these mosaics, there was present a nearly normal and well-developed granule cell layer, but the Purkinje cells formed a patchwork of cells, of which some had abnormal positions and sparse and spineless dendrites and others had normal positions and dendritic trees. The status of the neutral cell marker indicated that the Purkinje cells with abnormal positions and sparse dendrites were of *staggerer* mutant genotype and that cells with normal positions and normal dendrites were of normal, nonmutant genotype. Thus in the directness of its effect on Purkinje cell development, the *sg*, (or *staggerer*) mutation resembles the *pcd* (or Purkinje cell degeneration) mutation and differs from the indirectly acting *rl* (or *reeler*) mutation. Because of the small size of the granule cells, it has not been possible thus far to establish their genotype by means of a neutral cell marker. Thus it has not been ascertained whether the normally appearing granule cell layer of the *staggerer* mosaic mice comprises cells of both normal and *staggerer* mutant genotype. On other grounds (Landis and Sidman, 1978; Messer and Smith, 1977), it seems probable however, that the death of granule cells in *staggerer* mutant mice is a secondary consequence of the primary effect of the *sg* mutation on Purkinje cell development. In other words, the postnatal survival and the normal inward migration of granule cells appear to require their interaction with Purkinje cells having normal positions and dendrite structure, a requirement that would explain also the granule cell deficiency of the *reeler* cerebellum. Hence one would expect that the granule cell layer of *staggerer* mosaic mice actually contains cells of normal phenotype that nevertheless carry the *staggerer* genotype.

THE SIAMESE CAT

Finally, as our last example, we consider a familiar mammalian mutant phenotype, the Siamese cat (for a review, see Guillery, Casagrande, and Uberdorfer, 1974). The visual pathway of mammals is generally so arranged that the visual cortex receives visual input from the nasal part of the contralateral retina and from the temporal part of the ipsilateral retina. To produce this projection pattern, the optic nerve axons of ganglion cells originating in the nasal part of the retina cross over, or decussate, at the optic chiasm to the contralateral lateral geniculate nucleus (LGN), whereas the axons originating in the temporal part of the retina do not cross over at the optic chiasm and proceed to the ipsilateral LGN. In the ordinary domestic cat, as in other carnivores and in primates, the retinal line of demarcation for decussation of ganglion cell axons is just midway between the nasal and temporal edges. This is a necessary correlate of the fact that these species have binocular vision over nearly the whole of their visual field, since here left nasal and right temporal half retinas both receive visual input from the left half of the field, whereas right nasal and left temporal half retinas both receive visual input from the right half of the field.

The feline LGN consists of three layers, of which the outer two layers receive the axons projecting from the contralateral (nasal) retina and the middle layer receives the axons from the ipsilateral (temporal) retina. The entire projection occurs in a retinotopic order: the retinal ganglion cells located near the retinal midline project to the medial edge of each LGN layer; ganglion cells located at progressively more nasal sites in the (contralateral) retina project to progressively more lateral sites in the outer two LGN layers; and ganglion cells located at progressively more temporal sites in the (ipsilateral) retina project to progressively more lateral sites in the middle LGN layer. The axons of the higher order LGN neurons of these layers in turn travel to the ipsilateral visual cortex via the optic radiation, where they form a *binocular* retinotopic projection: axons reporting visual input from corresponding parts of the contralateral visual field received through either eye project to common sites on the visual cortex, in a characteristic nasotemporal order.

As was noted by Guillery (1969), this normal projection pattern is significantly perturbed in the visual pathway of Siamese cats. As far as the projection from retina to LGN is concerned, the line of demarcation for decussation of retinal ganglion cell axons is displaced from the dorsoventral retinal midline towards the temporal edge of the retina: some retinal ganglion cell axons from the temporal half of the retina project to the contralateral instead of ipsilateral LGN. There their endings occupy those positions of the middle layer which, in normal cats, are occupied by the ganglion cell axons from corresponding locations in the ipsilateral retina (Guillery and Kaas, 1971; Shatz 1977a; Kaas and Guillery, 1973). And as for the projection from LGN to visual cortex, that projection is reorganized in Siamese cats in one of two quite distinct modes. Under one of these modes the axons from the contralaterally (i.e., abnormally) innervated part of the LGN middle layer pass to the cortical area normally innervated by them, but in an order that is the reverse of normal. The rest of the projection from the LGN is compressed in a normal retinotopic order onto the remainder of the visual cortex. The result of this reorganization is the reconstitution of a normal nasotemporal sequence of visual field representation, but with two domains of monocular rather than binocular visual input (Hubel and Wiesel, 1971; Shatz, 1977b). Under the other reorganization mode, the anatomical pattern of the projection from LGN to visual cortex is apparently normal, but the whole of the input from the LGN middle layer is functionally suppressed. The result of this other reorganization mode is that these Siamese cats can see only the nasal part of their visual field (Kaas and Guillery, 1973).

These findings suggest that in the development of the mammalian visual system, innervation of the LGN layers is guided mainly by the position of origin of retinal ganglion cell axons. That is to say, neither the ipsilateral or contralateral origin of a retinal axon nor the visual coherence of the eventual projection appears to play a determinative role in the establishment of the normal retinotopic order on the LGN. By contrast, the innervation of the visual cortex cannot be guided mainly by the position of origin of the LGN cell axons but must be governed by a process that demands visual coherence of axons making functional connections at the same cortical site.

Although the genome of the Siamese cat certainly differs from that of ordinary cats in more than one genetic locus, it has been possible to identify—by induction rather than by conventional genetic crosses—the mutant gene whose presence is responsible for the abnormal decussation mode of retinal ganglion cell axons at the optic chiasm. It is the gene that encodes the structure of the enzyme tyrosinase that catalyzes a reaction step in the biosynthesis of the dark pigment melanin. In the Siamese cat this gene carries a mutation that renders the mutant protein unable to carry out its catalytic function at 37°C and thus prevents melanin synthesis at body temperature (Searle, 1968). It is this mutation that is responsible for the characteristic Siamese coat color, namely, the lightly pigmented body fur framed by black hair patches on the tips of the ears, the paws and the snout. Despite the lack of any demonstration that in crosses of Siamese to normal cats neurologic phenotype invariably segregates with coat color, there can be little doubt that it really is the mutant tyrosinase gene (and the lack of melanin formation at 37°C) which is responsible for the abnormal retinogenicular projection pattern. For it turns out that the impaired projection of axons from the temporal retina to the ipsilateral LGN is a general property of albino (i.e., melanin-deficient) mutants of mammalian species, such as tigers, mice, rats, ferrets, and mink. Some of these mutants owe their lack of dark pigment to a mutation in a gene that encodes an enzyme other than tyrosinase in the pathway of melanin synthesis (Guillery, 1974). But what is the possible connection between the formation of melanin and the ingrowth of optic nerve axons to the right or left LGN? And why does the absence of pigment produce their aberrant decussation, particularly in view of the fact that the retinal ganglion cells themselves do not normally contain noticeable amounts of melanin? The answers to these questions are not yet available, but the most likely developmental link between optic nerve axon projection pattern and melanin is provided by the layer of melanin-containing cells that form the pigment epithelium which underlies the retina and shields the photoreceptors from stray light. Thus the absence of pigment from that epithelium would engender some retinal disturbance during the outgrowth of the optic nerve axons that causes some of them to end up on the wrong side of the brain. And as we already saw in the case of the photoreceptor degeneration mutant of the rat, an abnormal genotype of the pigment epithelium can have a disruptive influence on normal development of the retina.

CONCLUSION

The Instrumental Aspect

The sample collection of experimental findings presented in the foregoing indicates that the genetic approach to developmental neurobiology offers considerable strength, even though, as Pak and Pinto (1976) have remarked, "no major breakthrough has yet been achieved as a result of this approach. Nevertheless the studies that have been carried out to date show sufficient promise to be encouraging." Instead of major breakthroughs, the results of neurogenetic studies can be seen to have provided further support to (or placed on a more secure basis) notions generally held by developmental neurobiologists. Thus the detailed neuroanatomical comparison of members of isogenic clones has brought into clearer focus the quantitative aspects of developmental noise, of whose existence few embryologists could have had any doubt. Similarly, the general proposition that presynaptic sensory neurons can play a morphogenetic or organizational role in the development of postsynaptic sensory interneurons, as demonstrated by the cases of the sound-insensitive and wind-insensitive cricket mutants, the *Drosophila* eye mutants, and the visual cortex of the Siamese cat, lies squarely in the conceptual mainstream of developmental neurobiology. And this is also the case for the converse proposition, which is applicable to the development of motor systems: as demonstrated by the case of the granule cells in the *reeler* and *staggerer* mouse cerebellum, survival of the presynaptic neuron is contingent on contact with an appropriate target cell. Finally, the conclusions devolving from *Drosophila* homeotic mutants (i.e., that axons of peripheral sensory neurons can make appropriate central terminations despite an abnormal point of entry into the central nervous system) and those devolving from the innervation pattern of the LGN of the Siamese cat (i.e., that at their cerebral terminus sensory cell axons are recognized according to their relative positions of origin) had also been previously reached on other grounds (Frank et al., 1977). Study of the Purkinje cells in the *reeler* mutants, however, revealed a previously unknown feature of histogenetic cell migration, namely, that migrating cells can be positioned individually by contact with a fine-grained, possibly cellular, mosaic rather than a homogeneous morphogenetic field. It is difficult to see how this insight could have been reached by any method other than genetic mosaics. In any case, the neurologic mutants have been of great value in unravelling the underlying mechanisms in the particular cases to which they pertain and in providing working material for the developmental neurobiologist. Some of the most fascinating instances among this working material, but possibly also the most difficult for future analysis, are the male/female behavioral mosaics of *Drosophila.* As for fate mapping (i.e., tracing the origin of the nervous system to specific sites of the blastula), this idea was first conceived and realized experimentally by direct observation of leech embryos by Whitman more than a century ago (1878). Nevertheless, the extension of genetic methodology to fate mapping in embryos of species such as *Drosphila,* whose complexity precludes direct observational methods, constitutes an important technical advance for developmental neurobiology. All these remarks pertain, of course, to the instrumental aspect of the genetic approach, to ignore which, according to Pak and Pinto (1976) "may well be to neglect one of the potentially most powerful tools in the study of the nervous system."

One limitation of the instrumental aspect of the genetic approach to neural development should be noted. Thus far at least, its contributions have furthered the understanding mainly of systems, such as the arthropod compound eye or the mammalian visual system and cerebellum, for which prior neuroanatomical, neurophysiological, and behavioral investigations had already provided a fairly high level of understanding. The reason for this is that whereas for *any* approach to the nervous system, be it anatomical, physiological, or behavioral, the investigator must bring some preunderstanding to his material, this requirement is much more demanding in the case of the genetic approach. As Kankel and Ferrus (1979) have pointed out "the devising of screening procedures and the subsequent analyses of isolated mutations imply that one already knows at least some of the relevant questions to ask. As always, the latter is the most difficult of problems and is one for which genetics alone provides no unique answer."

The Ideological Aspect

Whereas the main strength of the genetic approach lies in the instrumental aspect, its main weakness derives from the ideological aspect. For the viewpoint that the structure and function of the nervous system of an animal is specified by its genes provides too narrow a context for actually understanding developmental processes and thus sets a goal for the genetic approach that is unlikely to be reached. Here "too narrow" does not mean that a belief in genetic specification of the nervous system necessarily implies a lack of awareness that in development an interaction occurs between genes and environment, a fact of which all practitioners of the genetic approach are certainly aware. Rather "too narrow" means that the role of the genes, which, thanks to the achievements of molecular biology, we now know to be the specification of the primary structure of protein molecules, is at too many removes from the processes that actually "build nerve cells and specify neural circuits which underlie behavior" to provide an appropriate conceptual framework for posing the developmental questions that need to be answered. In this regard, the ideological aspect of the genetic approach resembles the quantum mechanical approach to genetics (Jordan, 1938) that had some vogue in the 1930s and 1940s. Since, so it was then thought, the structure and function of genes is obviously determined by the atoms of which they are composed, the goal of genetic research should be to give an account of heredity in terms of interatomic forces that govern the formation of chemical bonds. Eventually the problem of the mechanism of heredity was solved at the macromolecular rather than atomic level, although atomic interactions such as the hydrogen bond did turn out to have a crucial part in the story. Similarly, the insights into developmental mechanisms thus far available suggest that the solution to the problem of development lies at a cellular and intercellular rather than genetic level, although genes will undoubtedly figure in some crucial part, but only a part, of that solution.

What Is a Program? Those who speak of a genetic specification of the nervous system, and hence of behavior, rarely spell out what it actually is that they have in mind. On information-theoretical grounds, the genes obviously cannot embody a neuron-by-neuron circuit diagram; and even if they did, the existence of an agency that reads the diagram in carrying out the assembly of the component parts of a neuronal Heathkit would still transcend our comprehension. So a seemingly more reasonable view of the nature of the specification of behavior would be that the genes embody, not a circuit diagram, but a *program* for the development of the nervous system (Brenner, 1973, 1974). But this view is rooted in a semantic confusion about the concept of "program." Once that confusion is cleared up, it becomes evident that development from egg to adult is unlikely to be a programmatic phenomenon. Development belongs to that large class of regular phenomena which share the property that a particular set of antecedents generally leads, via a more or less invariant sequence of intermediate steps, to a particular set of consequents. However, of the large class of regular phenomena, programmatic phenomena form only a small subset, almost all the members of which are associated with human activity. For membership of a phenomenon in the subset of programmatic phenomena, it is a necessary condition that, in addition to the phenomenon itself, there exists a second thing, the "program," whose structure is isomorphic with (i.e., can be brought into one-to-one correspondence with) the phenomenon. For instance, the onstage events associated with a performance of "Hamlet," a regular phenomenon, are programmatic since there exists Shakespeare's text with which the actions of the performers are isomorphic. But the no-less-regular offstage events, such as the actions of house staff and audience, are mainly nonprogrammatic, since their regularity is merely the automatic consequence of the contextual situation of the performance. One of the very few regular phenomena independent of human activity that can be said to have a programmatic component is the formation of proteins. Here the assembly of amino acids into a polypeptide chain of a particular primary structure is programmatic because there exists a stretch of DNA polynucleotide chain—the gene—whose nucleotide base sequence is isomorphic with the sequence of events that unfolds at the ribosomal assembly site. However, the subsequent folding of the completed polypeptide chain into its specific tertiary structure lacks programmatic character, since the three-dimensional conformation of the molecule is the automatic consequence of its contextual situation and has no isomorphic correspondent in the DNA.

When we extend these considerations to the regular phenomenon of development, we see that its programmatic aspect is confined mainly to the assembly of polypeptide chains (and of various species of RNA). But as for the overall phenomenon, it is most unlikely—and no credible hypothesis has as yet been advanced how this *could* be the case—that the sequence of its events is isomorphic with the structures of any second thing, especially not with the structure of the genome. The fact that mutation of a gene leads to an altered neurologic phenotype shows that genes are part of the causal antecedents of the adult organism, but such a mutation does not in any way indicate that the mutant gene is part of a program for development of the nervous system.

But are not polemics about the meaning of words such as "program" just a waste of time for those who want to get on with the job of finding out how the nervous system develops? As J. H. Woodger (1952) showed in his Tarner Lectures "Biology and Language," published shortly before Watson and Crick's discovery of the DNA double helix and Benzer's reform of the gene concept, semantic confusion about its fundamental terms, such as "gene," "genotype," "phenotype," and "determination" had become the bane of classical genetics. It would be well to avoid reconstituting that confusion in the context of developmental biology and to remember Woodger's advice (p. 6) that "an understanding of the pitfalls to which a too naive use of language exposes us is as necessary as some understanding of the artifacts which accompany the use of microscopical techniques."

Development as History The notion of genetic specification of the nervous system is defective not only at the conceptual level but also represents a misinterpretation of the knowledge already available from developmental studies, including those that have resorted to the genetic approach. As Székely (1979) has pointed out, we know enough about its mode of establishment already to make it most unlikely that neuronal circuitry is, in fact, prespecified; rather, all indications (including those provided by the study of phenotypic variance reviewed here) point to stochastic processes as underlying the apparent regularity of neural development. That is to say, development of the nervous system, from fertilized egg to mature brain, is not a programmatic but a historical phenomenon under which one thing simply leads to another. To illustrate the difference between programmatic specification and stochastic history as alternative accounts of regular phenomena, we may consider the establishment of ecological communities upon colonization of islands (Simberloff, 1974) or the growth of secondary forests (Whittaker, 1970). Both of these examples are regular phenomena in the sense that a more or less predictable ecological structure arises via a stereotyped pattern of intermediate steps, a pattern in which the relative abundances of various types of flora and fauna follow a well-defined sequence. The regularity of these phenomena is obviously not the consequence of an ecological program encoded in the genome of the participating taxa. Rather it arises via a historical cascade of complex stochastic interactions between various biota (in which genes play an important role, of course) and the world as it is.

Although, compared to all the other cases of neurologic mutants considered here, our present knowledge of the genotypic differences between Siamese and normal cats is very crude, and the production of Siamese/normal mosaic cats has yet to be achieved, this case illuminates better than any other the strength of the instrumental and the weakness of the ideological aspect of the genetic approach to the development of the nervous system. Here mutation to loss of function of a single gene participating in a known metabolic pathway leads to a series of obviously interrelated changes in the structure and function of the central nervous system (in addition to the change in coat color). From the instrumental aspect, the existence of the Siamese cat (and of other mammalian albino mutants) is likely to be of enormous help in the elucidation of the mechanisms underlying the development of the mammalian visual pathway. First, study of the role of the presence of melanin in the retinal pigment epithelium in the pattern of optic fiber decussation at the chiasm should throw light on the mechanism that normally determines whether a retinal ganglion cell axon projects to the contralateral or ipsilateral LGN. Second, study of the mode of (inappropriate) innervation of the middle layer of the LGN from the contralateral retina should provide valuable information on how the relative retinal position of origin of ganglion cell axons is normally recognized upon arrival at the LGN. And third, study of the rearranged cortical projection pattern of LGN axons to the visual cortex should reveal insights into the hitherto quite mysterious processes by which coherence of binocular visual input plays its role in the establishment of cortical innervation by afferent neurons. But as for the ideological aspect, the case of the Siamese cat reveals the conceptual poverty of the notion that genes specify the neural circuits that underlie behavior. All that the gene mutated in the Siamese cat can usefully be said to "specify" is the amino acid sequence of an enzyme that takes part in melanin synthesis, presumably in retinal epithelial cells. That gene is evidently one of the causal antecedents in the developmental history of the mammalian visual system, in that the absence of the polypeptide chain it specifies sets off a cascade of dysfunctional, albeit specific aberrations, which eventually lead to a specific reorganization of the feline brain. But despite these specific cerebral aberrations there is, in this paradigmatic case, no trace of genetic specification of neural circuitry.

Summary Appraisal

By way of a summary appraisal, it can be said that the genetic approach is certainly of great practical and technical significance. First, within the context of neurophysiology, the availability of neurologic mutants can be of great assistance for the analysis of nerve cell functions and networks. For instance, an abnormal behavior and a concomitant abnormal neural structure of a mutant animal can obviously provide insights into how the normal circuitry generates normal behavior. Second, within the context of developmental neurobiology, the perturbation of developmental processes by mutant genes can help us to recognize the functional relations that create the normal pathways that lead to the genesis of the adult nervous system. Finally, within the context of psychology and medicine, it is of the utmost importance to understand the hereditary contribution to differences in behavior. For instance, if it could be shown that schizophrenia is attributable to a particular mutant gene, then the value of this knowledge would be in no way diminished by the realization that this gene does not "specify" brain development. But from the conceptual point of view the examples considered show that the focus on genetic specification is not likely to be a very fruitful approach to explaining the development of the nervous system. Rather, they indicate that the goal of developmental neurobiology ought *not* to be phrased as the understanding of how genes build nerve cells and specify the neural circuits which underlie behavior, but as the discovery of epigenetic functional relations, or algorithms. The horizon of that epigenetic approach to developmental neurobiology extends far beyond the genes and must encompass the universe of nonprogrammatic, contextually governed intracellular and intercellular interactions that underlie the historical phenomenon of metazoan ontogeny.

ACKNOWLEDGMENTS

I thank John Palka and David Weisblat for helpful criticisms of the manuscript. My research has been supported by NIH grant NS 12818 and NSF grant BN577-19181.

REFERENCES

Albertson, D. G., Thomson, J. N. The pharynx of *Caenorhabditis elegans. Phil. Trans. R. Soc. London Ser. B* 275:299–325

Anderson, H., Edwards, J. S., Palka, J. 1980. Developmental neurobiology of invertebrates. *Ann. Rev. Neurosci.* 3:97–139

Becker, H. J. 1957. Über Röntgenmosaikflecken und Defektmutationen am Auge von Drosophila melanogaster und die Entwicklungsphysiologie des Auges. *Z. Indukt. Abstamm. Vererbungsl.* 88:333–73

Bentley, D. 1973. Postembryonic development of insect motor systems. In *Developmental Neurobiology of Arthropods,* ed. D. Young, pp. 147–77. Cambridge: Cambridge Univ. Press

Bentley, D. 1975. Single gene cricket mutations: Effects on behavior, sensilla, sensory neurons and identified interneurons. *Science* 187:760–64

Bentley, D. 1976. Genetic analysis of the nervous system. In *Simpler Networks and Behavior,* ed. J. C. Fentress, pp. 126–39. Sunderland, Mass: Sinauer Associates

Bentley, D. 1977. Development of insect nervous systems. In *Identified Neurons and Behaviour of Arthropods,* ed. G. Hoyle, pp. 461–81. New York: Plenum

Benzer, S. 1971. From gene to behavior. *J. Am. Med. Assoc.* 218:1015–22

Bernard, F. 1937. Recherches sur la morphogénèse des yeux composés d'arthropodes. *Bull. Biol. Fr. Belg. (Suppl.)* 23:1–162

Brenner, S. 1973. The genetics of behaviour. *Brit. Med. Bull.* 29: 269–71

Brenner, S. 1974. The genetics of *Caenorhabditis elegans. Genetics* 77:71–94

Brindley, G. S. 1969. Nerve net models of plausible size that perform many simple learning tasks. *Proc. R. Soc. London Ser. B* 174:173–91

Campos-Ortega, J. A., Hofbauer, A. 1977. Cell clones and pattern formation in the lineage of photoreceptor cells in the compound eye of *Drosophila. Wilhelm Roux's Arch. Dev. Biol.* 181:227–45

Caviness, V. S. 1977. Reeler mouse mutant: A genetic experiment in developing mammalian cortex. In *Society for Neuroscience Symposia* Vol II, ed. W. M. Cowan, J. A. Ferendelli, pp. 27–46. Bethesda, Md: Society for Neuroscience

Caviness, V. S., Jr., Rakic, P. 1978. Mechanisms of cortical development: A view from mutations in mice. *Ann. Rev. Neurosci.* 2:297–326

Caviness, V. S., Jr., Sidman, R. L. 1972. Olfactory studies of the forebrain in the reeler mutant mouse. *J. Comp. Neurol.* 145:85–104

Caviness, V. S., Jr., Sidman, R. L. 1973. Time of origin of corresponding cell classes in the cerebral cortex of normal and reeler mutant mice: An autoradiographical analysis. *J. Comp. Neurol.* 148:141–51

Chappell, R. L., Dowling, J. E. 1972.

Neural organization of the median ocellus of the dragonfly. I. Intracellular electrical activity. *J. Gen. Physiol.* 60:121–47

Condamine, H., Custer, R. P., Mintz, B. 1971. Pure-strain and genetically mosaic liver tumors histochemically identified with the β-glucoroindase marker in allophenic mice. *Proc. Natl. Acad. Sci.* U.S.A. 68:2032–36

Deak, I. I. 1976. Demonstration of sensory neurons in the ectopic cuticle of *spineless-aristapedia,* a homeotic mutant of Drosophila. *Nature* (London) 260:252–54

Dewey, M. J., Gervais, A. G., Mintz, B. 1976. Brain and ganglion development from two genotypic classes of cells in allophenic mice. *Dev. Biol.* 50:68–81

Dräger, V. C. 1977. Abnormal neural development in mammals. In *Function and Formation of Neural Systems,* ed. G. S. Stent, pp. 111–38. Berlin: Dahlem Konferenzen

Feder, N. 1976. Solitary cells and enzyme exchange in tetraparental mice. *Nature* (London) 263:67–69

Frank, E., Fillenz, M., Jansen, J. K. G., Kandel, E. R., Kuffler, S. W., Landmesser, L. T., Lømo, T., McMahan, U. J., Nicholls, J. G., Parnas, I., Patterson, P. H., Purves, D. Formation and maintenance of neural connections. See Dräger, 1977, pp. 225–52

Ghysen, A. 1978. Sensory neurones recognize defined pathways in *Drosophila* central nervous system. *Nature* (London) 274:869–72

Goodman, C. S. 1974. Anatomy of locust ocellar interneurons: Constancy and variability. *J. Comp. Physiol.* 95:185–201

Goodman, C. S. 1976. Constancy and uniqueness in a large population of small interneurons. *Science* 193:502–4

Goodman, C. S. 1977. Neuron duplications and deletions in locust clones and clutches. *Science* 197:1384–86

Goodman, C. S. 1978. Isogenic grasshoppers: Genetic variability in the morphology of identified neurons. *J. Comp. Neurol.* 182:681–705

Guillery, R. W. 1969. An abnormal retinogeniculate projection in Siamese cats. *Brain Res.* 14:739–41

Guillery, R. W. 1974. Visual pathways in albinos. *Sci. Am.* 230:44–54

Guillery, R. W., Casagrande, V. A., Uberdorfer, M. D. 1974. Congenitally abnormal vision in Siamese cats. *Nature* (London) 252:195–99

Guillery, R. W., Kaas, J. H. 1971. A study of normal and congenitally abnormal retinogeniculate projections in cats. *J. Comp. Neurol.* 143:73–100

Hall, J. C. 1979. Control of male reproductive behavior by the central nervous system of Drosophila: Dissection of a courtship pathway by genetic mosaics. *Genetics* 92:437–57

Hall, J. C., Gelbart, W. M., Kankel, D. R. 1976. Mosaic systems. In *The Genetics and Biology of Drosophila,* ed. M. Ashburner, E. Novitski, pp. 265–314. New York: Academic

Hall, J. C., Greenspan, R. J. 1979. Genetic analysis of Drosophila neurobiology. *Ann. Rev. Genet.* 13:127–95

Hamburgh, M. 1963. Analysis of the postnatal developmental effects of "reeler," a neurological mutation in mice. A study in developmental genetics. *Dev. Biol.* 8:165–85

Hofbauer, A., Campos-Ortega, J. A. 1976. Cell clones and pattern formation: genetic eye mosaics in *Drosophila melanogaster. Wilhelm Roux's Arch. Dev. Biol.* 179:275–89

Horridge, G. A. 1968. *Interneurons,* p. 321. San Francisco: W. H. Freeman

Horridge, G. A., Meinertzhagen, I. A. 1970. The accuracy of the patterns of connexions of the first- and second-order neurones of the visual system of Calliphora. *Proc. R. Soc. London Ser. B* 175:69–82

Hotta, Y., Benzer, S. 1972. Mapping of behavior in *Drosophila* mosaics. *Nature* (London) 240:527–35

Hotta, Y., Benzer, S. 1976. Courtship in *Drosophila* mosaics: Sex-specific foci of sequential action patterns. *Proc. Natl. Acad. Sci.* U.S.A. 73:4154–58

Hubel, D. H., Wiesel, T. N. 1971. Aberrant visual projections in the Siamese cat. *J. Physiol.* (London) 218:33–62

Jordan, P. 1938. Zur Frage einer spezifischen Anziehung zwischen Genmolekülen. *Phys. Z.* 39:711–14

Kaas, J. H., Guillery, R. W. 1973. The transfer of abnormal visual field representations from dorsal lateral geniculate nucleus to visual cortex in Siamese cats. *Brain Res.* 59:61–95

Kankel, D. R., Ferrus, A. 1979. Genetic analyses of problems in the neurobiology of *Drosophila.* In *Neurogenetics: Genetic Approaches to the Nervous System,* ed. X. Breakefield, pp. 27–66. New York: Elsevier-North Holland

Kankel, D. R., Hall, J. C. 1976. Fate mapping of nervous system and other internal tissues in genetic mosaics of *Drosophila melanogaster. Dev. Biol.* 48:1–24

Landis, D. M. D., Reese, T. S. 1977. Structure of the Purkinje cell membrane in staggerer and weaver mutant mice. *J. Comp. Neurol.* 171:247–60

Landis, D. M. D., Sidman, R. L. 1978. Electron microscopic analysis of postnatal histogenesis in the cerebellar cortex of staggerer mutant mice. *J. Comp. Neurol.* 179:831–64

LaVail, M. M., Mullen, R. J. 1976. Role of pigment epithelium in inherited retinal degeneration analyzed with experimental mouse chimeras. *Exp. Eye Res.* 23:227–45

Levinthal, F., Macagno, E. R., Levinthal, C. 1975. Anatomy and development of identified cells in isogenic organisms. *Cold Spring Harbor Symp. Quant. Biol.* 40:321–31

Lewis, E. B. 1963. Genes and developmental pathways. *Am. Zool.* 3:33–56

Macagno, E. R. 1980. Genetic approaches to invertebrate neurogenesis. *Curr. Top. Dev. Biol.* 15(Part I):319–345

Macagno, E. R., Lopresti, V., Levinthal, C. 1973. Structure and development of neuronal connections in isogenic organisms: Variations and similarities in the optic system of *Daphnia magna. Proc. Natl. Acad. Sci.* U.S.A. 70:57–61

Mariani, J., Crepel, F., Mikoshiba, K., Changeux, J.-P., Sotelo, C. 1977. Anatomical, physiological and biochemical studies of the cerebellum from reeler mutant mouse. *Phil. Trans. R. Soc. London Ser. B* 281:1–28

Messer, A., Smith, D. M. 1977. In vitro behavior of granule cells from staggerer and weaver mutants of mice. *Brain Res.* 130:13–23

Meyerowitz, E. M., Kankel, D. R. 1978. A genetic analysis of visual system development in *Drosophila melanogaster. Dev. Biol.* 62:63–93

Mintz, B. 1962. Formation of genotypically mosaic mouse embryos. *Ann. Zool.* 2:432

Mintz, B. 1965. Genetic mosaicism in adult mice of quadriparental lineage. *Science* 148:1232–33

Mintz, B., Sanyal, S. 1970. Clonal origin of the mouse visual retina mapped from genetically mosaic eyes. *Genetics (Suppl.)* 64:43–44

Mullen, R. J. 1977a. Site of *pcd* gene action and Purkinje cell mosaicism in cerebella of chimeric mice. *Nature* (London) 270:245–47

Mullen, R. J. 1977b. Genetic dissection of the CNS with mutant-normal mouse and rat chimeras. In *Society for Neuroscience Symposia,* vol II, ed. W. M. Cowan, J. A. Ferendelli, pp. 47–65. Bethesda, Md: Society for Neuroscience

Mullen, R. J., Eicher, E. M., Sidman, R. L. 1976. Purkinje cell degeneration, a new neurological mutation in the mouse. *Proc. Natl. Acad. Sci.* U.S.A. 73:208–12

Mullen, R. J., Herrup, K. 1979. Chimeric analysis of mouse cerebellar mutants. In *Neurogenetics: Genetic Approaches to the Nervous System,* ed. X. Breakefield, pp. 173–96. New York: Elsevier-North Holland

Mullen, R. J., LaVail, M. M. 1975. Two new types of retinal degeneration in cerebellar mutant mice. *Nature* (London) 258:528–30

Mullen, R. J., LaVail, M. M. 1976. Inherited retinal distrophy: Primary defect in pigment epithelium determined with experimental rat chimeras. *Science* 192:799–801

Murphey, R. K., Mendenhall, B., Palka, J., Edwards, J. S. 1975. Deafferentation slows the growth of specific dendrites of identified giant interneurons. *J. Comp. Neurol.* 159:407–18

Ouweneel, W. J. 1976. Developmental genetics of homeosis. *Adv. Genet.* 18:179–248

Pak, W. L., Pinto, L. H. 1976. Genetic approach to the study of the nervous system. *Ann. Rev. Biophys. Bioeng.* 5:397–448

Palka, J. 1979. Mutants and mosaics: Tools in insect developmental neurobiology. *Soc. Neurosci. Symp.* 4:209–27

Palka, J., Lawrence, P. A., Hart, H. S. 1979. Neural projection patterns from homeotic tissue of *Drosophila* studied in *bithorax* mutants and mosaics. *Dev. Biol.* 69:549–75

Pearson, K. G., Goodman, C. S. 1979. Correlation of variability in structure with variability in synaptic connection of an identified interneuron in locusts. *J. Comp. Neurol.* 184:141–65

Peterson, A. C. 1974. Chimera mouse study shows absence of disease in genetically dystrophic muscle. *Nature* (London) 248:561–64

Quinn, W. G., Gould, J. L. 1979. Nerves and genes. *Nature* (London) 278:19–23

Ready, D. F., Hanson, T. E., Benzer, S. 1976. Development of the *Drosophila* retina, a neurocrystalline lattice. *Dev. Biol.* 53:217–40

Searle, A. G. 1968. *Comparative Genetics of Coat Color in Mammals,* pp. 146–47. New York: Academic

Shatz, C. J. 1977a. A comparison of visual pathways in Boston and Midwestern Siamese cats. *J. Comp. Neurol.* 171:205–28

Shatz, C. J. 1977b. Anatomy of interhemispheric connections in the visual system of Boston Siamese and ordinary cats. *J. Comp. Neurol.* 173:497–518

Sidman, R. L., Green, M. C., Appel, S. H. 1965. *Catalog of the Neurological Mutants of the Mouse.* Cambridge: Harvard Univ. Press

Simberloff, D. S. 1974. Equilibrium theory of island biography and ecology. *Ann. Rev. Ecol. Syst.* 5:161–82

Sotelo, C., Changeux, J. -P. 1974. Transsynaptic degeneration "en cascade" in the cerebellar cortex of staggerer mutant mice. *Brain Res.* 67:519–26

Stent, G. S. 1978. *Paradoxes of Progress,* pp. 169–89. San Francisco: W. H. Freeman

Stocker, R. F. 1977. Gustatory stimulation of a homeotic mutant appendage, *Antennapedia,* in *Drosophila melanogaster. J. Comp. Physiol.* 115:351–61

Stocker, R. F., Edwards, J. S., Palka, J., Schubiger, G. 1976. Projections of sensory neurons from a homeotic mutant appendage, *Antennapedia,* in *Drosophila melanogaster. Dev. Biol.* 52:210–20

Sturtevant, A. H. 1929. The claret mutant type of *Drosophila simulans:* A study of chromosome elimination and cell lineage. *Z. Wiss. Zool.* 135:325–56

Sulston, J. E. 1976. Post-embryonic development in the ventral cord of *Caenorhabditis elegans. Phil. Trans. R. Soc. London Ser. B* 275:287–97

Sulston, J. E., Horvitz, H. R. 1977. Post-embryonic cell lineages of the nematode, *Caenorhabditis elegans. Dev. Biol.* 56:110–56

Székely, G. 1979. Order and plasticity in the nervous system. *Trends Neurosci.* (Oct):245–48

Tarkowski, A. K. 1961. Mouse chimera developed from fused eggs. *Nature* (London) 190:857–60

von Schilcher, F., Hall, J. C. 1979. Neural topography of courtship song in sex mosaics of *Drosophila melanogaster. J. Comp. Physiol.* 128:85–95

Waddington, C. H. 1957. *The Strategy of the Genes.* London: Allen & Unwin

Ward, S. 1977. Invertebrate neurogenetics. *Ann. Rev. Genet.* 11:415–50

Ward, S., Thomson, H., White, J. G., Brenner, S. 1975. Electron microscopical reconstruction of the anterior sensory anatomy of the nematode *Caenorhabditis elegans. J. Comp. Neurol.* 160:313–38

Ware, R. W., Clark, D., Crossland, K., Russell, R. L. 1975. The nerve ring of the nematode *Caenorhabditis elegans.* Sensory input and motor output. *J. Comp. Neurol.* 162:71–110

Wegmann, T. G., LaVail, M. M., Sidman, R. L. 1971. Patchy retinal degeneration in tetraparental mice. *Nature* (London) 230:333–34

White, J. G., Southgate, E., Thomson, J. N., Brenner, S. 1976. The structure of the ventral nerve cord of *Caenorhabditis elegans. Phil. Trans. R. Soc. London Ser. B.* 275:327–48

Whitman, C. O. 1878. The embryology of *Clepsine. Q. J. Microsc. Sci.* (N.S.) 18:215–315

Whittaker, R. H. 1970. *Communities and Ecosystems.* New York: Macmillan. 158 pp.

Wilson, M. 1978. The functional organisation of locust ocelli. *J. Comp. Physiol.* 24:297–316

Woodger, J. H. 1952. *Biology and Language.* Cambridge: Univ. Press. 364 pp.

Yoon, C. H. 1977. Fine structure of the cerebellum of "staggerer-reeler," a double mutant of mice affected by staggerer and reeler condition. II. Purkinje cell anomalies. *J. Neuropathol. Exp. Neurol.* 36:427–39

14 / *Cues and Constraints in Schwann Cell Development*

RICHARD P. BUNGE
MARY B. BUNGE

PROLOGUE

When those of us who study development of the vertebrate embryo look up periodically from the corners of our particular interest to view the whole embryo evolving, we cannot escape a sense of wonderment at the harmony of events that create the final vertebrate form. We sense that Viktor Hamburger's life is constantly enriched by this sense of wonder for he has often said that the embryo has always been his greatest teacher. Taking the embryo rather than fallible human authority as teacher is certainly one way to avoid being disappointed by the flaws of human error. To paraphrase Bertrand Russell, the embryo cannot err, for unlike professors, it makes no postulates. The embryo can, of course, develop abnormally, but in doing so it is often doubly instructive rather than disgraced. Thus, in taking the embryo as his teacher, Viktor Hamburger could be assured of a source of wonder that may surprise, but would never disappoint. Undoubtedly this *Anschauung* is one of the sources of Viktor's unceasing spirit of optimism. This optimism seems rooted in the belief that if we continue to study, we can understand, and as we understand, we will come to respect the processes of life and thus become more willing to protect and cherish its diversities.

Just as the embryo has unceasingly inspired Viktor Hamburger, so he in turn has continually inspired those of us fortunate enough to come into contact with him. His example has taught us to spend the greater part of our time in experimenting and watching cells interact, and less time in making "models." We remember his impatience at being asked to generalize without reviewing experimental data. He has seen too many generalities demolished by one new experimental fact to fall into that trap. As a teacher, he has thus stressed the review of decisive experiments, never the recitation of dogma.

Viktor Hamburger has often experimentally modified embryonic development and studied the result; our approach has been to take tissues out of the embryo, place them in tissue culture dishes, and observe the result. This allows direct visualization of the interacting cells and control of the immediate cellular milieu. The question is often raised, of course, whether the tissue-culture approach may lead to error, for cells in culture dishes may fail to mimic their behavior *in vivo.* Indeed they may, but the cells make no error for they know no hypothesis; they can only labor to express their capabilities in the new environment. Whereas we may err in interpreting what their behavior means, their actions, we believe, do more to instruct than to deceive. We will attempt to support this view as we consider below the earlier accrual of our knowledge of Schwann cell biology as well as the recent advances in our understanding of Schwann cell behavior, which derive largely from tissue-culture experimentation.

For the reasons listed above, and many others, it is a distinct pleasure to dedicate this essay to Viktor Hamburger. We have not only learned from him what understanding he has gained from his lifelong and affectionate affair with the nervous system of the embryo, we have also learned from him how to learn.

AN INTRODUCTION WITH HISTORICAL NOTES

Peripheral nerve trunks appear during the somite stage of the vertebrate embryo as nerve fibers grow into body tissues from the developing neural tube and from peripheral sensory and autonomic ganglia. Early in their development these nerve fibers are populated by Schwann cells derived from the neural crest (Harrison, 1924). During this stage a large number of Schwann cells will be generated to provide for the cytoplasmic ensheathment of the small unmyelinated nerve fibers and myelin ensheathment of the larger axons (Webster, 1975). The domain of Schwann cells extends proximally to the point at which the glia limitans provides a border between the central and peripheral nervous systems (CNS, PNS); distally Schwann cells are excluded from certain peripheral tissues such as epidermis. Subsequent to this developmental period, the number of Schwann cells along the nerve trunk is stabilized, but the resident Schwann cells will respond with proliferation to a variety of injuries (Asbury, 1975). Their participation in nerve regrowth after injury has been extensively discussed by Ramón y Cajal (1928) (for a brief recent review, see R. Bunge, 1980).

Selected aspects of the cellular mechanisms involved in the proliferation, differentiation, and functional expression of Schwann cells during development (and their response to certain types of injury) will be the subject of this essay. We begin by presenting the historical context of those particular aspects of Schwann cell biology that will receive our greatest attention. We then review recent experiments utilizing tissue culture and nerve transplantation techniques that have substantially altered our concepts of the ways in which Schwann cells achieve their mature functional state. Throughout, we emphasize the cues that are known to signal Schwann cell proliferation, axonal ensheathment and myelination, and the conditions that appear to provide constraints regulating where Schwann cells are able to function within the body tissues.

Schwann's initial study of the peripheral myelin-related cell (Schwann, 1839) was undertaken in part to add weight to his contention that cells were the building blocks of all body tissues and responsible for the formation of substances as unique as the fatty material (myelin) in peripheral nerve. His observations clearly related the cell that would come to bear his name to the myelin sheath surrounding the nerve fiber but, as we will develop below, it was to be more than a century before the precise relationship between the Schwann cell and the myelin sheath would be elucidated. In the interim, the abundance of Schwann cells that become associated with nerve fibers prior to the formation of myelin and the close contact established between these cells and the nerve fibers were to provide the scenario for a long battle regarding the origin of the nerve fiber. The thin and poorly stained developing nerve fiber was in such intimate consort with Schwann cell cytoplasm that the formation of the axon from the linearly disposed Schwann cells (rather than by extension of the nerve cell) seemed entirely reasonable. But the description of the axonal growth cone by Ramón y Cajal (1952) and the direct observation of axonal extension in tissue culture by Harrison (1910) were to clearly demolish the catenary hypothesis of axon formation. Little concern was subsequently expressed that this basic conceptual change left the Schwann cell related to the unmyelinated nerve fiber with no known function in peripheral nerve development.

The general presumption appeared to be that these cells, formed in such substantial numbers early in development, were, in fact, awaiting future responsibilities. The role of the Schwann cell in the processes of regeneration after nerve injury was recognized early in this century and was the object of considerable study, particularly by Ramón y Cajal. The proliferation of Schwann cells in response to injury, their faithful wait in the amputated nerve stump, and their apparent ability to guide the regenerating nerve fiber and nourish it, led Ramón y Cajal to write an almost romantic description of Schwann cell function in peripheral nerve regeneration (Ramón y Cajal, 1928).

We can now clarify certain of these aspects of Schwann cell development and function; others remain unclear. There is evidence, which we review below, that the quite sudden generation of a large number of Schwann cells early in peripheral nerve development can be ascribed to the presence of a mitogenic signal on the surface of the growing axon. The differences between this mechanism for the control of Schwann cell numbers and the mechanism operative following peripheral nerve injury are also considered in our review.

Regarding the question of the function of the Schwann cells related to unmyelinated nerve fibers, some definitive—and some rather speculative—comment can now be made. Certain peripheral nerve cells, if isolated in tissue culture, can be maintained in a functional state without the benefit of Schwann cells; this fact establishes that peripheral nerve cell function is not directly dependent upon Schwann cell association. But the culture media for these preparations must contain certain trophic agents for neuronal survival (e.g., nerve growth factor for adrenergic autonomic neurons). These trophic agents (or some substitute) can be provided as secretory products of Schwann cells *in vivo,* as we will discuss below.

What can now be stated with some certainty is that Schwann cells related to nerve cells secrete a number of products. As we indicate below, these products include several types of collagen and almost certainly other substances as well. These products provide for the deposition of a basal lamina over the surface of the axon-related Schwann cell as well as a small amount of associated fibrillar collagen (M. Bunge et al., 1980). The direct observation of the formation of basal lamina on cultured axon-related Schwann cells (in the absence of fibroblasts) clarifies a long-debated aspect of Schwann cell function in the production of peripheral nerve components. The bulk of these components—the striking array of typical collagen fibrils of the endoneurium, the cellular sheath of the perineurium, and the substantial collagenous layers of the epineurium—appear to result from the activity of fibroblasts recruited from the surrounding mesenchyme.

This Schwann cell attribute—the overcoat of basal lamina material—had considerable influence on earlier concepts of Schwann cell function. It seems likely that this component was visualized, along with the plasma membrane of the Schwann cell, as an external covering of Schwann cell cytoplasm which earlier histologists termed the membrane (or sheath) of Schwann. Ramón y Cajal (1928) described this membrane as follows: "The most external layer of the nerve fiber is composed of a thin transparent sheath, situated outside Schwann's cell to which it is closely applied and without any apparent structure. This is a so-called sheath of Schwann."

His knowledge that this membrane was not found as a part of the myelin sheath of the CNS led him to a mistaken conclusion regarding the relationship between the Schwann cell and the peripheral myelin sheath. He reasoned as follows (Ramón y Cajal, 1928): The "myelin and its accompaniments [are] . . . a dependence of the axon, and as such, an apparatus foreign to the constitution of Schwann's cells. Such an opinion . . . is strongly supported by the fact that in the medullated fibres of the centres [the CNS] there are no Schwann cells . . . [p. 46] . . . it is well known that the fibres of the centres . . . lack a membrane of Schwann. . ." [p. 55].

We see here, as well as elsewhere in Ramón y Cajal's writing, that he was very much influenced by his view that because there were no Schwann cells within the white matter of the CNS, then the myelin sheath must be a product of the axon. We cannot fault him for his failure to recognize that the oligodendrocyte was the CNS counterpart of the cell of Schwann, for it was Del Rio-Hortega (1921) who was to define and describe the interfascicular oligodendrocyte as the myelin-related cell of the CNS. In many of Del Rio-Hortega's pictures it can be seen that, during early stages of development, the relationship of certain oligodendrocytes to the forming myelin sheath bears a striking similarity to that between the Schwann cell and the underlying myelin sheath. But what are we to make of Ramón y Cajal's statement that the nerve fibers of the CNS lack a membrane of Schwann? This observation was correct, for within the CNS the axon-myelin unit is not surrounded by a demonstrable connective tissue component, and adjacent myelin sheaths are in direct contiguity with one another without the interruption of basic lamellar spacing. A substantial portion of what follows bears on this basic difference between the CNS and PNS, for there appears to be a linkage between Schwann cell secretion (as evidenced by basal lamina production) and the ability of the Schwann cell to effect axonal ensheathment, whether this be the simple engulfment of unmyelinated nerve fibers or the more complicated formation of myelin sheaths.

We will develop the concept that the presence of a secretory phase in Schwann cell development, without evidence for a parallel phase in oligodendrocyte development, may represent one of the very fundamental differences between the activities of these two cell types. This aspect of Schwann cell development may explain the proclivity of Schwann cells to ensheathe all axons in their vicinity by enclosing them within troughs of cytoplasm. This ensheathment phase in Schwann cell development appears to be a necessary prelude to the subsequent myelination that occurs in relation to the larger peripheral nerve fibers. (The absence of ensheathment of unmyelinated fibers within the CNS may be related to the absence of a similar phase in the development of the oligodendrocyte.)

But we have digressed from our discussion of the historical landmarks in the accumulation of knowledge regarding Schwann cell development following the early decades of this century. The remainder of the story is well-known. The electron microscope allowed the observation that myelin was a layered structure formed during development by the addition of lamella after lamella of membrane. It was recognized that these lamellae of myelin were directly continuous with the plasma membrane of the Schwann cell, and it was then observed that, in fact, myelin was deposited by the spiral deposition of plasma membrane around the developing nerve fiber.

Despite intense investigation of the developmental biology of the cell of Schwann during the last several decades, many fundamental questions remain unanswered. Among them is the question whether the spiral deposition of surface membrane to form myelin is accomplished by the circumnavigation of the Schwann cell, or by the advancement of the inner Schwann cell cytoplasmic lip. It is also true to say that, despite the fact that we now know that the Schwann cell contributes the myelin sheath, we have little specific knowledge regarding other types of support the Schwann cell provides for the nerve fiber. We have instead a series of experiments indicating that the Schwann cell influences a nerve fiber not only during development but also during the steady mature state and during regeneration. But we do not know any of the specific chemical compounds that may be exchanged between these two cell types in their mutual support of one another. We will touch on the latter subject briefly in the last part of this essay.

CONTROL OF SCHWANN CELL NUMBER

Observations on Isolated Schwann Cells

Several methods are now available for the preparation of pure populations of Schwann cells in tissue culture (see, among others, Wood, 1976; Brockes, Fields, and Raff, 1979). These preparations allow direct observations on agents and conditions that influence Schwann cell proliferation. Studies by Raff and Brockes and their colleagues have utilized Schwann cells prepared from neonatal peripheral nerve and have concentrated on the identification of agents that act directly on isolated Schwann cells; here, in this section we briefly review their work. The work of Wood and Salzer and their colleagues, which has primarily explored the influence of axons on the control of Schwann cell numbers, is discussed in the following section.

Brockes and his coworkers (1979) have prepared Schwann cells by subjecting neonatal rat sciatic nerve to trypsin digestion and then culturing the mixture of Schwann cells and fibroblasts thus obtained. Because the fibroblasts proliferate more actively in these mixed cultures (in the presence of serum), they are more susceptible to treatment with antimitotic agents than are the slowly proliferating Schwann cells. A few fibroblasts remain after this treatment; these are removed by exposure to an antibody to the rat fibroblast surface component, thy-1. The cells that remain form an essentially pure population of Schwann cells. The thy-1 surface protein also serves as an identifying marker to distinguish fibroblasts from Schwann cells. The antigens Ran-1 and S-100 serve to mark the Schwann cell population. The purified Schwann cell population can be further expanded by the addition of a factor derived from pituitary gland.

In systematic studies of Schwann cells prepared by this method, these workers have observed two distinct pathways for stimulating division (Raff et al., 1978). Treatments that elevate the intracellular level of cyclic AMP (the most effective agent being cholera toxin) stimulate Schwann cell DNA synthesis. Other factors (present in extracts of bovine brain and pituitary gland) stimulate Schwann cell division by mechanisms independent of changes in intracellular cyclic AMP levels. An agent similar to that found in pituitary gland was also found in brain but not in several nonneural tissues. This agent was inactivated by proteolytic digestion and by boiling; it also required one or more factors in fetal calf serum in order to stimulate Schwann cell division. As has been confirmed by Salzer and R. Bunge (1980), Raff and colleagues (1978) observed the stimulation of Schwann cell division to be a specific process not influenced by a variety of growth factors, mitogens, or hormones known to stimulate other cell types. Whether or not the pituitary factor that Brockes and his coworkers (1979) have isolated plays a role in Schwann cell proliferation, either during development or in disease, is not yet known. The discovery of a readily available agent to expand Schwann cell populations in culture, however, provides promising new opportunities for the study of Schwann cell properties. It may also assist in preparing large populations of Schwann cells for use in promoting remyelination in demyelinated areas of the CNS.

Influences on Schwann Cell Number During Development

Alternative methods for preparing Schwann cell populations in tissue culture have provided data that appear to bear more directly on Schwann cell development *in vivo.* Wood (1976) has described methods for culturing dorsal root ganglia to promote neurite outgrowth with associated Schwann cell proliferation in the absence of fibroblasts. With subsequent ganglion removal (and consequent axonal degeneration), the outgrowth is composed entirely of Schwann cells. Subsequent to the demise of the axons, Schwann cell proliferation ceases, and a "quiescent" Schwann cell population is obtained (Figs. 14-1 and 14-2).

By direct observation of the axon-Schwann cell interaction in this type of tissue culture preparation, it has been possible to demonstrate that the number of Schwann cells is controlled, at least in part, by a mitogenic signal present on the neuronal surface. By utilizing both chick and rodent tissues, several laboratories have now reported this result. In the initial experiments, Schwann cells were prepared in pure populations (Wood and R. Bunge, 1975). When axons from a nearby sensory ganglion grew into the region of this Schwann cell population and established contact with the quiescent Schwann cells, a substantial proliferation of Schwann cells (as measured by thymidine incorporation) was observed. In adjacent regions of the culture dish (where Schwann cells were in residence without axonal contact) no increase in incorporation of thymidine by Schwann cells was observed. In similar experiments conducted

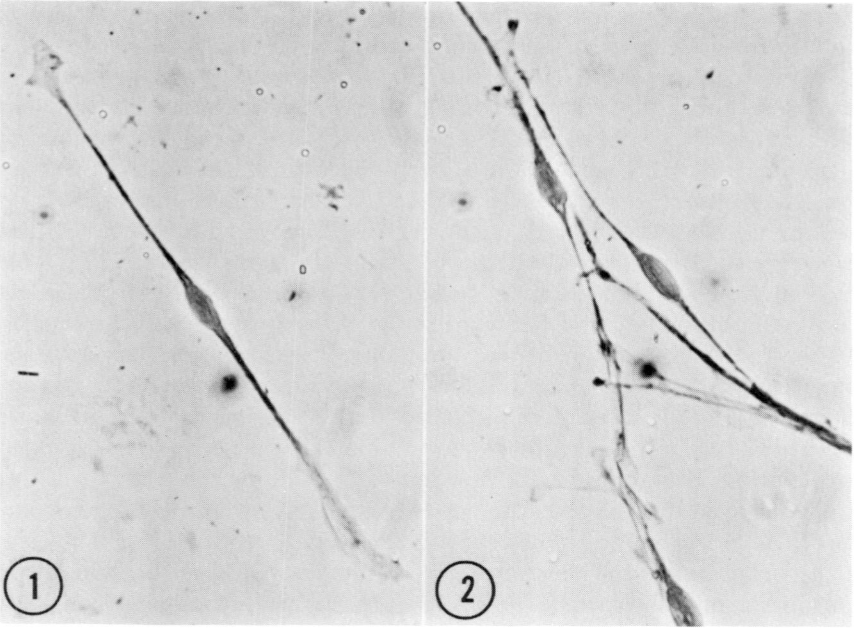

Figures 14.1 and 14.2 Examples of Schwann cells in cultures from which accompanying neurons and fibroblasts have been removed. These highly elongated cells are enlarged in the area of the nucleus and exhibit motile cytoplasmic expansions at each end. ×600.

with autonomic tissues taken from chick embryos, McCarthy and Partlow (1976) observed that the number of Schwann cells generated in a tissue culture preparation is in direct proportion to the number of neurons growing in the same culture dish. From both of these types of experiments, it has become evident that there is a direct influence of the neuron and its axon on Schwann cell populations.

In a recent series of papers Salzer and his colleagues have explored in detail the ability of the axon, and particularly the axolemmal membrane, to influence Schwann cell proliferation (Salzer et al., 1980a, 1980b). In their initial experiments they established that fractions prepared by the homogenization of growing axons, when supplied in tissue culture medium, stimulated thymidine incorporation into mitotically quiescent populations of cultured Schwann cells. It was thus possible to demonstrate that not only the living axon, but preparations containing axolemma, were instrumental in stimulating the proliferation of Schwann cells. Addition of membrane preparations from other cell types was not mitogenic. It was also reported that if the axons used to prepare particulate material were subjected to trypsin treatment prior to harvest, the efficacy of the axonal particulate fraction was substantially reduced. This observation led these authors to propose that there exists on the surface of the axonal plasma membrane a signal that, when delivered to the Schwann cell, is able to stimulate Schwann cell proliferation.

Salzer and coworkers (1980a) have also presented a number of experiments designed to test whether the signal delivered from axon to Schwann cell requires direct contact between the two cell types. When Schwann cells and axons were grown on opposite sides of a permeable collagen layer, they observed no axonal influence on the proliferative activity of Schwann cells. McCarthy and Partlow (1976) observed a similar phenomenon when they discovered that neurons in the same culture dish—but not in contact with Schwann cells—had no influence on the proliferative activity of the cocultured Schwann cells. *In vivo* observations support the idea that the total number of Schwann cells generated in a particular nerve trunk is proportional to the number of axons present. Aguayo and colleagues (1976b) have demonstrated that if the number of axons in the cervical sympathetic trunk of the rodent is decreased by experimental manipulation, then the number of Schwann cells populating this nerve trunk is likewise diminished. This concept does not explain how the expansion of the Schwann cell population is eventually stopped; for it is clear that as the number of Schwann cells is increased and ensheathment of individual axons is attained, the proliferation of Schwann cells slows and eventually reaches a very low level (Asbury, 1975).

On the basis of the experiments reviewed below, we now speculate that the mitogenic signal passed from axon to Schwann cell may be modified by the induction of secretory activity in the Schwann cell; this secretory activity may yield a product that, when released from the Schwann cell, masks the mitogenic signal. (In the next section we discuss the evidence that Schwann cells in one stage of their development initiate secretion.) Because the total repertoire of Schwann cell secretory products has not yet been determined, it is entirely possible that among these products there is one that binds to the axolemma. This would explain the paradox that as the axon and Schwann cell first come into contact, a potent mitogenic signal is delivered from axon to Schwann cell; but, as Schwann cells come into greater contact with axonal surfaces during ensheathment, either the signal, or the efficacy of the signal, diminishes. We will discuss in detail below our recent observation that Schwann cells grown with neurons in a defined medium are at the same time defective in secretory activity and deficient in providing axonal ensheathment. When we studied Schwann cell proliferation in these cultures, we observed that substantial mitosis continued for several weeks in culture and that it did not undergo the expected decline seen when ensheathment of nerve fibers progresses normally (R. Bunge and Wartels, unpublished data). In Figures 14-3 and 14-4 we illustrate the ongoing Schwann cell proliferation that occurs in this defined medium as well as the cessation of Schwann cell proliferation that occurs when the axonal mitogenic signal is removed by the excision of the neuronal population from the culture dish.

Schwann Cell Proliferation in Response to Injury

The tissue culture observations discussed above provide a reasonable explanation for the control of Schwann cell number during initial development, but they are inadequate to explain the Schwann cell proliferation observed during certain types of neuronal injury. It had been reported repeatedly that subsequent to nerve severance or crush the distal part of the nerve trunk, devoid of axons but rich in resident Schwann cells, exhibits a period of Schwann cell proliferation beginning about four days after nerve injury and continuing for a period of several weeks thereafter (reviewed by Asbury, 1975). This proliferation occurs even though regrowing nerve fibers are precluded from entering the amputated distal stump. It had been recognized for some time that this Schwann cell proliferation in response to axonal severance occurred primarily in nerves that contain substantial numbers of myelinated nerve fibers and was much less marked (essentially absent) in nerves that contain primarily unmyelinated fibers (for recent *in vivo* experiments, see Romine, Bray, and Aguayo, 1976).

Recent tissue culture observations offer a direct explanation for the differences observed in these two types of nerves (Salzer and R. Bunge, 1980). In tissue cultures containing explants of sensory ganglia that had

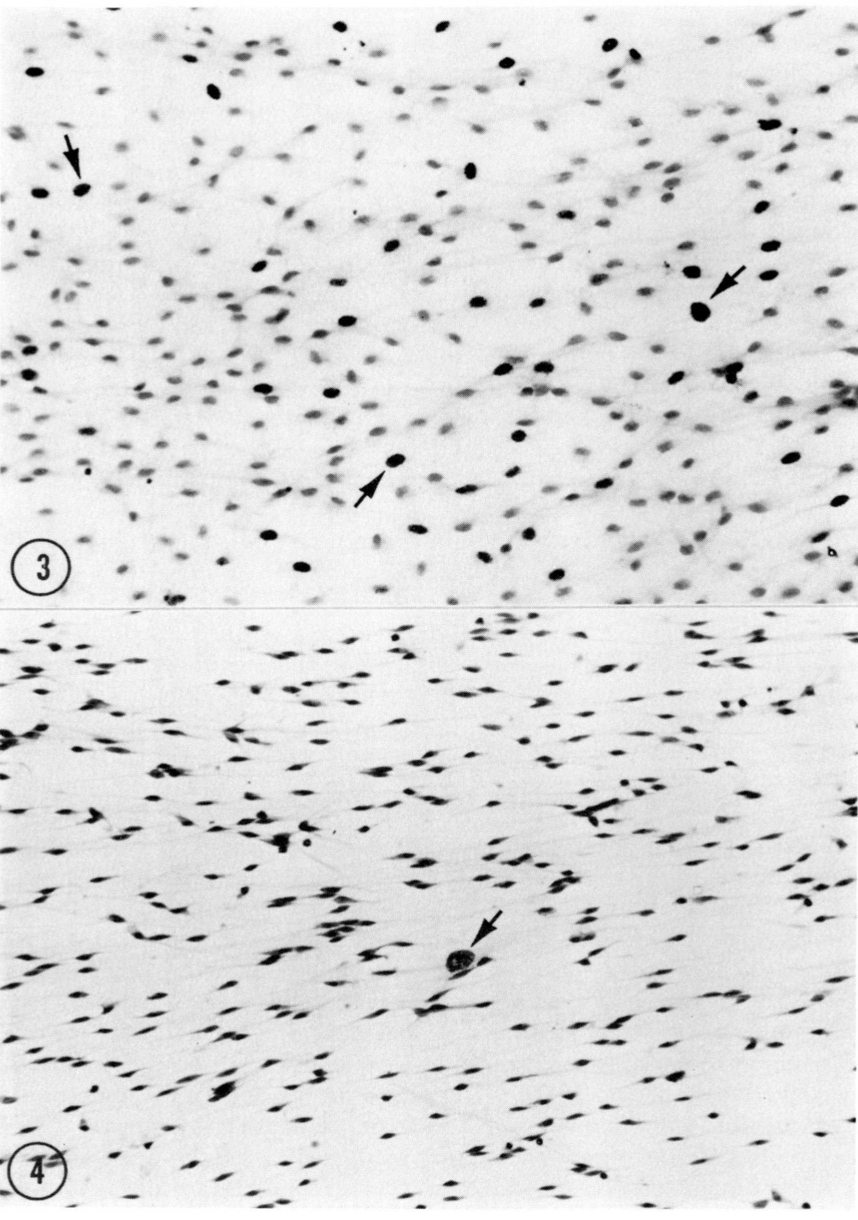

Figures 14.3 and 14.4 Autoradiograms of outgrowth regions in rat autonomic ganglion explant cultures given ^{3}H-thymidine 48 hours before fixation. Figure 14.3 illustrates the high number of labeled Schwann cell nuclei (some of which are indicated by arrows) in a neuron-Schwann cell preparation. Figure 14.4 is a sibling culture from which the neurons had been removed five days before ^{3}H-thymidine administration. The only labeled nucleus (arrow) in this field belongs to one of the few remaining fibroblasts in the culture. These autoradiograms demonstrate the dependence of Schwann cell proliferation upon the presence of neurites. Defined (N2) medium; five weeks *in vitro;* ×200.

generated a neuritic outgrowth containing both myelinated and unmyelinated nerve fibers, it was observed, subsequent to axonal removal, that (1) proliferative activity ceased within several days for those Schwann cells that were related to the outgrowing unmyelinated nerve fibers, and (2) proliferation was initiated about four days after explant removal specifically among those Schwann cells that had been related to myelin segments. This tissue culture experiment allowed direct demonstration of the incorporation of tritiated thymidine into Schwann cells in the process of digesting remnants of the myelin sheaths to which they had formerly been related.

These experiments also provided some evidence that unmyelinating Schwann cells may be particularly receptive to mitogenic stimulation during the period in which axonal degeneration is occurring within the Schwann cell (Salzer and R. Bunge, 1980). Schwann cells were directly damaged by mechanical injury at the same time that neurons were removed from the culture dish. It was then observed that, in contrast to the observations cited above, the concomitant direct injury to the unmyelinating Schwann cell and the digestion of debris from axons it formerly ensheathed caused a substantial proportion of these Schwann cells to incorporate tritiated thymidine. This finding may explain the observation that unmyelinated nerve fascicles show little inclination for Schwann cell proliferation in the region of their distal stumps but that Schwann cell proliferation is frequently observed in the region of the crush itself (Romine et al., 1976). Taken together with the observations on the reaction of myelinated nerve fibers to injury, these results suggest that the state of the Schwann cell (whether directly injured) and the mechanisms employed in disposing of axonal and myelin debris interact to determine if cell proliferation is initiated after nerve injury.

CONTROL OF SCHWANN CELL FUNCTIONAL EXPRESSION

Control of Schwann Cell Secretion and Ensheathment

In the peripheral tissues of the body, the axon and Schwann cell exist within the connective tissue framework of the peripheral nerve trunk. The connective tissue content of peripheral nerve includes the largely collagenous components of the endoneurium. These longitudinally oriented collagen fibrils are interspersed among occasional fibroblasts within the endoneurial space. Surrounding the nerve fibers, Schwann cells, and endoneurial connective tissue, there is a cellular sleeve that provides a barrier between the endoneurial space and the tissue spaces of the nerve environment. This sleeve, termed the perineurium, is composed of a series of flattened cells that are joined to one another by tight junctions, which together provide the basis for the substantial barrier to the passage of large molecules into the endoneurium. A series of nerve fascicles, each defined by perineurium, is held together by the looser but still substantial connective tissue of the epineurium.

There has been a longstanding debate regarding the sources of a number of the endoneurial extracellular matrix components. The collagen may derive from Schwann cell secretory activity; this has been frequently suggested because of the plethora of Schwann cells in this portion of the nerve. Alternatively, the far fewer fibroblasts that occupy this space may provide, by continued activity, collagen fibrils both for their immediate environment and for areas at some distance from their perinuclear regions. The source of the basal lamina that surrounds each axon-Schwann cell unit (and, in fact, aids in identifying the Schwann cell) has also been questioned. The availability of new types of tissue culture preparations has afforded some new answers to these questions.

The critical tissue culture preparation for the study of the contribution of Schwann cells to endoneurial connective tissue components contains neurons and their processes and Schwann cells without the usually attendant fibroblasts. The presence of nerve cells is critical; for we have observed that without axonal contact the formation of basal lamina in relation to the Schwann cell does not occur. This observation derives from studies of *in vivo* Schwann cells that were in a migratory phase without substantial axonal contact (Billings-Gagliardi, Webster, and O'Connell, 1974) and from studies of *in vitro* Schwann cells from which axons had been removed (Williams, Wood, and M. Bunge, 1976; M. Bunge, Williams, and Wood, 1979).

The neuron-Schwann cell culture preparations have been studied by both biochemical and electron microscopic techniques. It has been observed that in these preparations (after full maturation in culture), the Schwann cell surface is covered, as it is *in vivo*, by basal lamina, and that external to the basal lamina, there are sparse bundles of relatively thin collagenous fibrillar elements (Fig. 14-5) (M. Bunge et al., 1980). Whereas the basal lamina appears relatively continuous, it does not appear quite so substantive as it does when fibroblasts are added to the culture system. The basal lamina formed in neuron-Schwann cell cultures is only partially susceptible to enzymatic digestion with collagenase but is completely removed by treatment with the proteolytic enzyme, trypsin. It is reformed after trypsin treatment only if the Schwann cell is left in contact with an axon (Figs. 14-6 and 14-7) (M. Bunge et al., 1979). The thin collagenous fibrils that are seen in relation to the basal lamina in the neuron-Schwann cell cultures are trypsin-resistant but entirely digestible with collagenase, and they exhibit faint repeating cross-striations when sectioned longitudinally (M. Bunge et al., 1980). When the collagenous material that is generated in neuron-Schwann cell cultures is analyzed by standard biochemical techniques, it appears that collagens of Type 1 and Type 3, AB collagens (now commonly called Type 5), and in addition possibly Type 4 collagen are formed (M. Bunge et al., 1980). Their morphological distribution, and the types of collagens present in the Schwann cell basal lamina and thin extracellular fibrils, have not yet been determined. The larger-diameter collagen fibrils that occupy peripheral nerve endoneurial spaces are not observed in neuron-Schwann cell cultures, and there is no perineurial or epineurial ensheathment.

If, however, fibroblasts are added to neuron-Schwann cell cultures, then one observes that more collagen fibrils of larger diameter are formed (M. Bunge, Jeffrey, and Wood, 1977) and, in addition, perineurium encloses numerous axon-Schwann cell bundles, as in peripheral nerve *in*

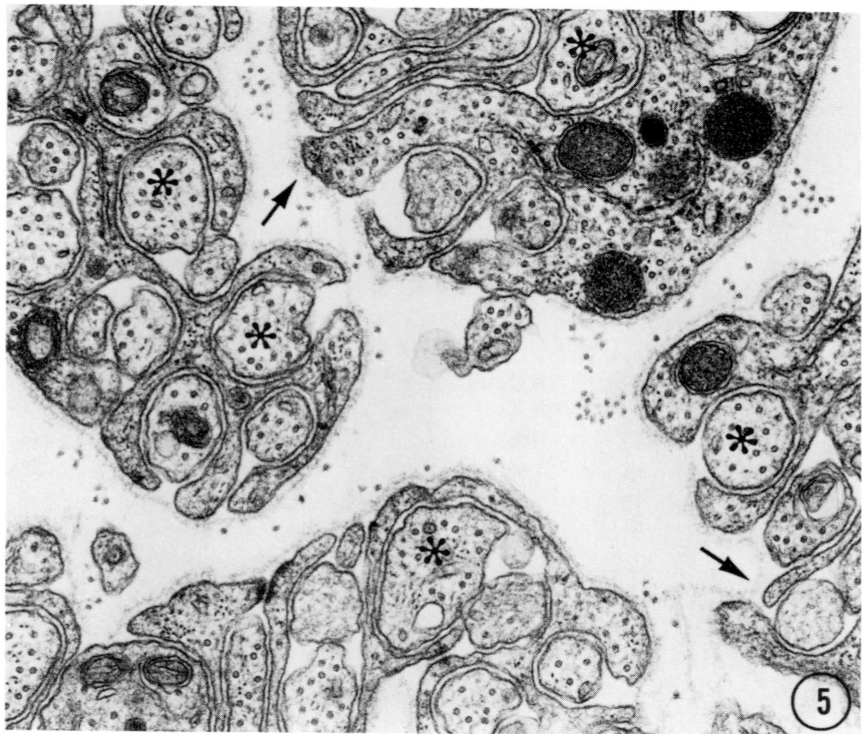

Figure 14.5 Electron micrograph of a neuron-Schwann cell culture. Axons, some of which are designated by asterisks, are ensheathed with Schwann cell cytoplasm. The Schwann cell exterior is coated with basal lamina, shown to best advantage at the arrows. Small bundles of transversely sectioned, thin collagen fibrils may be seen in the extracellular space. This and subsequent figures illustrate outgrowth regions of rat dorsal root ganglion cultures. Four weeks *in vitro;* ×39,000.

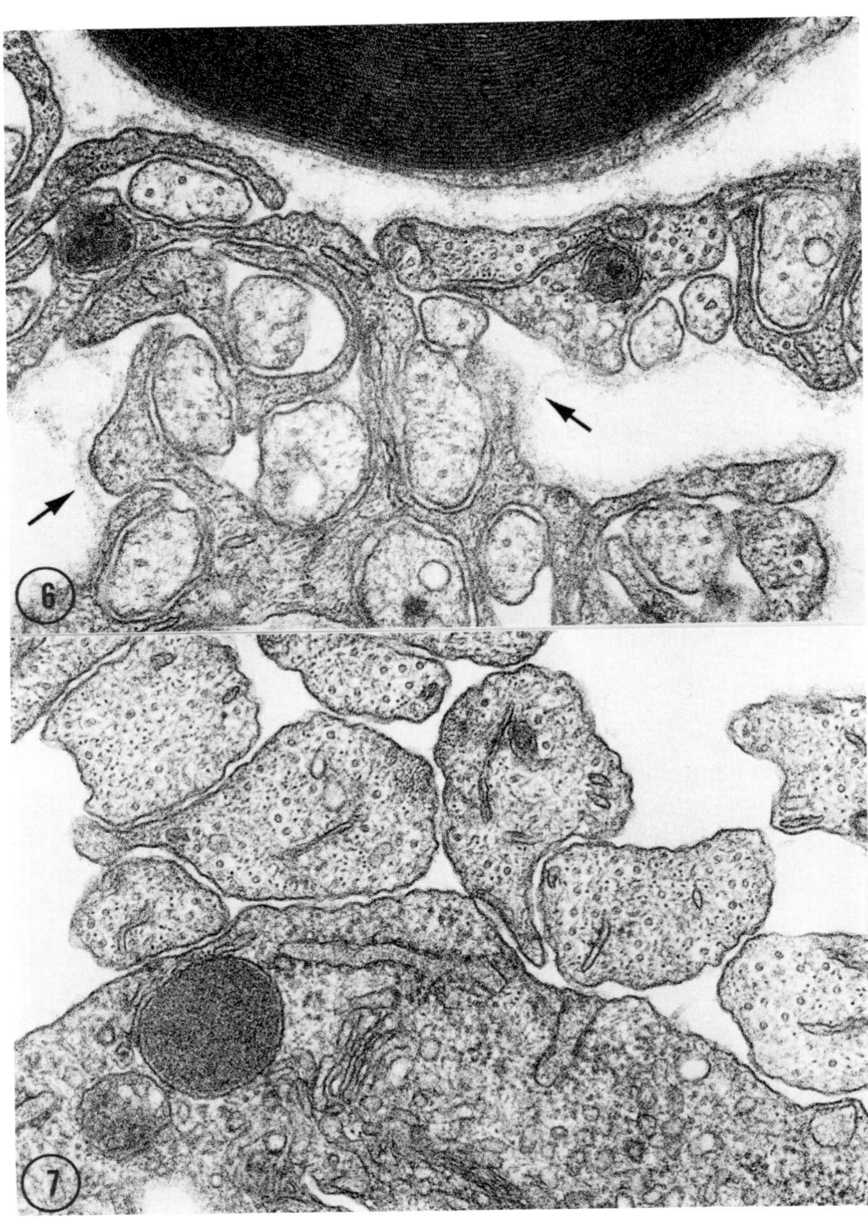

Figures 14.6 and 14.7 Illustration of an experiment designed to test if the presence of nerve cells is required for the generation of basal lamina on Schwann cells. A neuron-Schwann cell culture (Figure 14.6) and a culture preparation of Schwann cells alone (Figure 14.7) were subjected to trypsin treatment to remove basal lamina. One month later, basal lamina (arrows) is seen to coat the exterior of Schwann cells related to nerve fibers in contrast to Schwann cells bereft of neuronal contact. (Part of a myelin sheath appears at the top of Figure 14.6.) ×48,000.

vivo. These observations suggest that the perineurium is normally of fibroblast origin and that the typically larger endoneurial collagen fibrils are also of fibroblast origin. Alternatively, the Schwann cell and fibroblast may collaborate in the formation of the bulk of endoneurial extracellular matrix components. For example, a secretory product of the fibroblast may lead to heightened production of endoneurial collagen by the neuron-related Schwann cell or increased extracellular assembly of Schwann cell-derived collagen.

These tissue culture experiments appear to establish the secretory capacity of neuron-related Schwann cells, at least for some of the extracellular matrix components. What has not been systematically analyzed is whether a substantial number of additional compounds are secreted by Schwann cells in contact with axons. In our most recent experiments we have had an opportunity to study the functional capacity of Schwann cells in a situation that appears to interfere with normal Schwann cell secretion. These observations suggest that Schwann cell secretion may be involved in more than the provision of basal lamina and some collagenous components to the endoneurium; in these preparations the failure of Schwann cell secretion appears to be related to an inability of the Schwann cell to ensheathe an axon normally or to myelinate an axon.

These experiments have utilized a defined medium, N2, that was formulated by Bottenstein and Sato (1979) to maintain proliferation of neuroblastoma cells in culture. When used in standard peripheral nerve tissue cultures prepared from embryonic rats, this medium permits a substantial amount of neurite outgrowth from the neuronal explant, and it fosters continuing Schwann cell proliferation, which results in a heavy population of the neurite fields by Schwann cells (Moya, M. Bunge, and R. Bunge, 1980). At this point in development, however, there is a startling arrest in Schwann cell differentiation and the cultures do not progress to the expected steps of axonal ensheathment with subsequent myelination of larger axons, even after weeks in culture (Moya et al., 1980; R. Bunge et al., 1980b). On electron microscopic observation it is observed that Schwann cells are related to axonal bundles but that they have failed to enclose individual axons within troughs of their cytoplasm as occurs normally (Fig. 14-8). Larger axons are seen to be in contact with, or only partially ensheathed by, Schwann cells; and despite many weeks in culture, they do not acquire myelin. Also, basal lamina and extracellular fibrillar material do not appear (Fig. 14–8). This failure of Schwann cell differentiation is rectified within several days if the culture is fed defined medium to which embryo extract and serum have been added. Under these conditions axons are ensheathed by Schwann cells, basal lamina is formed, and the first signs of myelination occur in relation to the larger axons (Moya et al., 1980; R. Bunge et al., 1980b). The presence of engorged cisterns of rough endoplasmic reticulum in Schwann cells cultured in

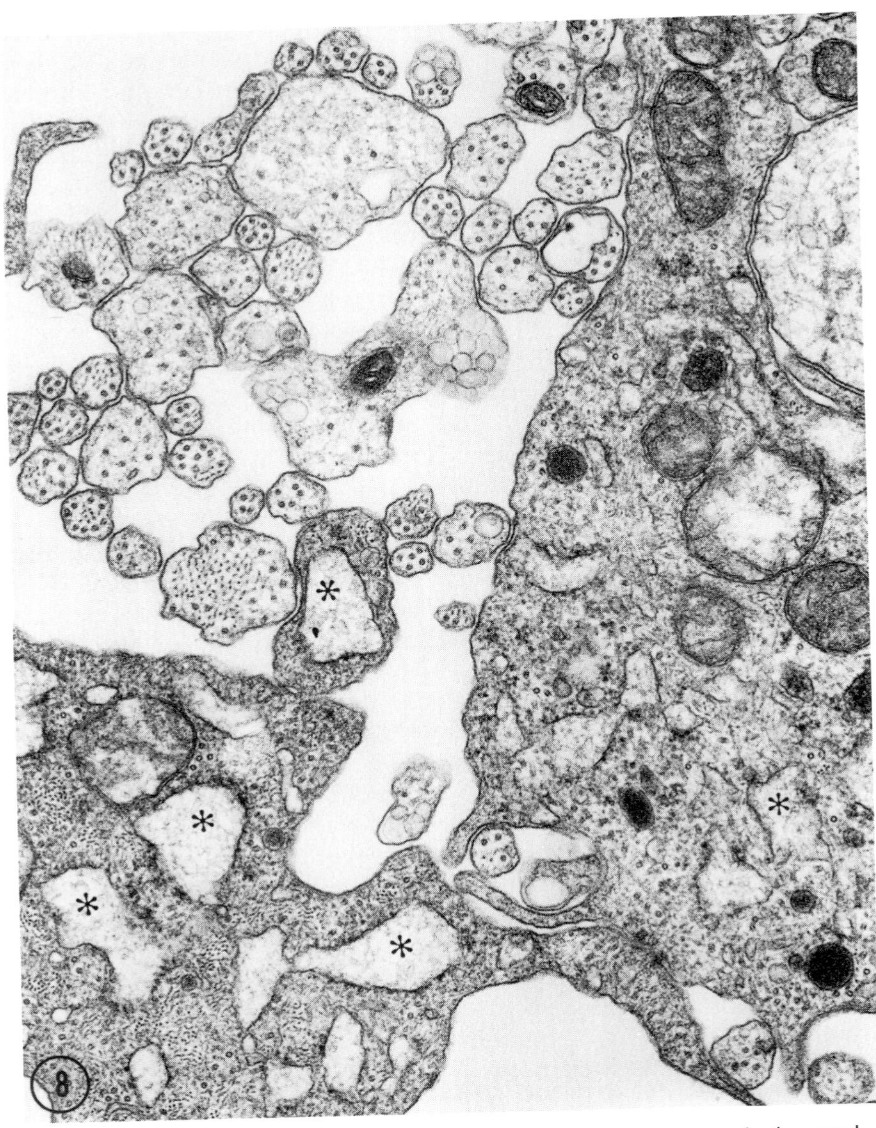

Figure 14.8 This electron micrograph demonstrates aspects of abnormal Schwann cell function in a neuron-Schwann cell culture maintained in defined (N2) medium for four weeks prior to fixation. Axons, in a fascicle at the upper left, are unensheathed, and basal lamina and extracellular fibrils are absent. Also notable is the presence of distended cisterns (*) of rough endoplasmic reticulum in both Schwann cells illustrated. ×33,000.

N2 medium also points to diminished export of Schwann cell secretory products.

As we noted in the section above, Schwann cell proliferation in the arrested state (i.e., in N2 medium that has not been supplemented with serum and embryo extract) is protracted well beyond the time when a slowing of proliferation usually occurs in control cultures, and there is some evidence of Schwann cell loss accompanying the ongoing proliferation (R. Bunge and Wartels, unpublished observations). When these cultures are reversed to medium containing embryo extract and serum, the proliferative activity of the Schwann cells is reduced in parallel with the development of normal ensheathment and basal lamina formation. This would appear to support the contention, presented above, that the secretory activity of Schwann cells may be instrumental in masking the axolemmal mitogen responsible for at least a substantial degree of Schwann cell proliferation during normal peripheral nerve development.

We have also noted that this apparent requirement of Schwann cells for a component in embryo extract or serum for full differentiation is in contrast to the demonstrable nutritional requirements for the production of myelin by oligodendrocytes. Several reports now indicate that CNS cultures which are grown in entirely defined medium may progress to the state of myelination without the addition of serum or other undefined components (Honegger, Lenoir, and Favrod, 1979; Kleinschmidt and R. Bunge, 1980). These new observations raise the question whether there may be fundamental differences in the mechanisms by which Schwann cells and oligodendrocytes ensheathe axons. The inability of the Schwann cell to progress in its differentiative course if it is prevented from passing through what appears to be a secretory phase would suggest that products of Schwann cell secretion may be instrumental in implementing the cytoplasmic enclosure of the axon by the Schwann cell. This intimacy of ensheathment is the hallmark of unmyelinated fibers and a prelude to myelination for the larger fibers of the PNS; there is no comparable ensheathment of CNS-unmyelinated axons by oligodendrocytes.

This view regarding the linkage between Schwann cell secretion (as manifested by basal lamina production) and expression of Schwann cell function (in its capacity to ensheathe and myelinate) is also supported by recent observations on Schwann cell malfunction in the dystrophic *(dy)* mouse. It has been realized since the seminal observations of Bradley and Jenkison (1973) that there is a very unusual failure of Schwann cell function in the nerve roots (particularly at the lumbar and cervical levels) of the *dy* mouse. The relationship between this nerve root lesion and the abnormalities expressed in muscle dysfunction is not yet clear and is not an appropriate subject for discussion here. An important aspect of the observations on *dy* nerve is the unusual behavior of some Schwann cells. This behavior is particularly manifest in the nerve root lesion but is also expressed in the more distal portions of the PNS. Most recently, Jaros and Bradley (1979) have reported that in the most severely affected of these *dy* mouse strains (ReJ 129) it is possible to identify substantial morphological abnormalities of Schwann cell basal lamina throughout the PNS. Concomitant with this abnormality there are substantial alterations in Schwann cell configuration including abnormalities at the node of Ranvier, distorted and/or shortened myelin internodes, a tendency for Schwann cell processes to wander from contact with one axon to contact with adjacent axons, and even the oligodendrocytelike configuration of a Schwann cell relating to more than one myelinated axon. They also comment on the interrelationship between the failure of basal lamina production and these abnormalities of Schwann cell differentiation.

It should be noted, however, that within portions of the nerve root (as opposed to the nerve trunk), the abnormality of Schwann cell function is more severe (for review, see Okada, R. Bunge, and M. Bunge, 1980). Here the Schwann cell fails to encircle (and thus ensheathe) axons and provide a basal lamina. Within these regions there is also an absence of collagen fibrils from the interaxonal spaces. The reason for the more substantial failure of the Schwann cell in the nerve root as opposed to the more distal nerve regions is not yet clear. In our studies of the expression of the *dy* nerve lesion in culture (Okada et al., 1980; R. Bunge et al. 1980a), we suggested that the cause may be related to the usual paucity of fibroblasts within the endoneurial spaces of the nerve root as compared with the more distal peripheral nerve. We shall return to this subject when we discuss the control of Schwann cell position in the next section of this chapter. What can be said here regarding the relationship of Schwann cell ensheathment and secretion is (1) that in regions of the nerve root lesion where Schwann cell ensheathment fails entirely, there is also little Schwann cell basal lamina production, and (2) that the cells present among and adjacent to these unensheathed nerve fibers have only marginal similarities to Schwann cells. This is a reminder of the fact that without the telltale production of basal lamina or axon ensheathment, identification of the cell of Schwann is difficult.

It is possible to experimentally disrupt Schwann cell synthesis of collagenous components in tissue culture by the use of analogs of proline, such as cis-hydroxyproline (as in Uitto et al., 1978). In recent experiments Copio and M. Bunge (1980) have observed that, in a dose-related manner, this agent interferes with basal lamina and thin collagenous fibril production in long-term cultures of Schwann cells and neurons. They also found, again depending upon the dose of cis-hydroxyproline, deficient Schwann cell ensheathment of unmyelinated fibers and abnormalities in myelin ensheathment; at the highest dose employed, myelin formation was markedly inhibited.

Thus, in the ongoing development of the relationship between Schwann cell and axon, first there appears to be a proliferative signal passed from the axonal surface to the Schwann cell. As the Schwann cells divide and increase their contact with axons, there appears to be a second signal which leads to the production of some materials that can be recognized by direct observation in the electron microscope. Newer observations on the behavior of Schwann cells under tissue culture conditions that appear to prevent this secretion would suggest that the presence of Schwann cell secretion is, in fact, linked to the progression of Schwann cell differentiation through the steps of axonal ensheathment. Without axonal ensheathment, myelination also fails. These observations also suggest that there is in the development of the Schwann cell an important stage in which the cell must become secretory in order to stabilize its relationship with the axon. In addition, full differentiation of the Schwann cell requires (as we discuss in a later part of this chapter) another step, namely, contact of the Schwann cell not only with an axon but also with a component of the extracellular matrix.

Axonal Control of Myelin Formation by Schwann Cells

In the last several years, reports of new types of investigation, which confirm earlier less rigorous experiments, indicate that the specificity for myelin formation is expressed not by the Schwann cell but by the axon. The question at issue is whether the axon controls myelin formation by the Schwann cell, or whether a certain subpopulation of Schwann cells is programmed to express myelin formation in relation to axons. In experiments by Weinberg and Spencer (1975), a nerve containing predominantly myelinated nerve fibers was anastomosed with a nerve that normally contains unmyelinated nerve fibers. It was observed that as the axons of the myelinated nerve grew into the stump of the previously unmyelinated nerve, they apparently recruited the resident Schwann cells for ensheathment and signaled these Schwann cells to produce myelin. In experiments designed to approach this same question, Aguayo, Charron, and Bray (1976a) labeled the nuclei of Schwann cells within an unmyelinated nerve trunk and, after transplantation into a gap within a largely myelinated nerve, established that labeled nuclei were involved in myelin formation around the axons that grew into the region of the implant. In the light of what is written above, it should be noted that there is as yet no evidence for axonal specificity in the signals that control the proliferation of Schwann cells and the induction of secretory activity of Schwann cells.

CONTROLS DETERMINING THE LOCATION OF FUNCTIONAL SCHWANN CELLS

Schwann cells provide the types of ensheathment described in the foregoing sections in essentially all portions of the PNS, whether this be sensory, somatic motor, or visceral. There are, however, important exceptions. It can be noted that where nerve fibers penetrate the basal lamina of the epidermis and pass among the epidermal cells, they are not followed in their intraepidermal course by ensheathing Schwann cells. It should also be noted that there are substantial portions of the enteric nerve plexus, particularly the plexuses of Auerbach and Meissner, in which the supporting cells do not have the characteristics of Schwann cells but instead show many characteristics of astrocytes. This configuration has been noted particularly by Gabella (1971) in relation to the regions containing nerve cell bodies in this portion of the autonomic nervous system. Schwann cells are, of course, also excluded from the CNS at the points of contact between the dorsal and ventral nerve roots and the spinal cord. Here the glia limitans is interposed between the typical connective tissue components and Schwann cells of the nerve root and the parenchyma of the CNS. The glia limitans is a series of astrocytic expansions along the surface of the cord which are bounded externally by a basal lamina that is continuous with that of the adjacent Schwann cells.

The studies of Blakemore and others (for review, see Blakemore, 1978) indicate that Schwann cells may, under certain circumstances, enter the CNS and function there as ensheathing cells for CNS axons. The conditions under which CNS invasion by Schwann cells may occur are not yet completely defined. Blakemore has emphasized the fact that Schwann cell invasion of the CNS appears to require a breakdown in the glia limitans of the spinal cord. In this situation Schwann cells may invade CNS parenchyma and provide myelin sheaths of the distinctively peripheral type in close proximity to myelin-related oligodendrocytes. It has been noted, particularly by Gilmore and Duncan (1968), that after demyelinating episodes in the region of the dorsal columns, Schwann cell invasion occurs by way of the dorsal root entry zone and that the first remyelination is observed in this area. It has also been recently observed by Blakemore (1977) that Schwann cells, introduced into a region of CNS demyelination by implanting a small fragment of peripheral nerve, remyelinate CNS axons but largely in regions adjacent to the implant. In somewhat similar experiments, Duncan and coworkers (1979) described transplantation of pure populations of Schwann cells into regions of demyelination in the dorsal columns of adult mouse spinal cord. Their observations indicated that, as in the experiments of Blakemore (1977), Schwann cells remyelinate demyelinated CNS axons primarily in the immediate vicinity of the implanted cells. Although the domain of Schwann cell function is limited to the region directly adjacent to the operative site, substantial portions of the demyelinated cord are remyelinated by oligodendrocytes. These observations raise the basic question of the controls that are operative in determining sites in which Schwann cells are able to function normally.

We again bring forward tissue culture observations that we believe have a bearing on this question. The observations derive from cultures of normal peripheral ganglia in which nerve fibers are populated by Schwann cells that are not also in contact with the substratum of the tissue culture dish (R. Bunge and M. Bunge, 1978). The substratum in cultures of peripheral nervous tissues is usually reconstituted rat-tail collagen. In certain cultures, particularly those grown in the absence of fibroblasts, axons course from the explant region to an area of the substratum some distance away. In the interval between the explant and the point of attachment, these axons are suspended in tissue culture medium and remain in this configuration for many weeks in culture. On these "guy-roping" fascicles, small aggregates of Schwann cells are frequently seen, but the relationship between the Schwann cells and axons in these regions is highly abnormal (Figs. 14-9 and 14-10). The Schwann cells, in contact with only the more peripheral axons of the suspended fascicles, exhibit a sparse coating of material which resembles basal lamina. The Schwann cells do not ensheathe the smaller axons and do not provide myelin sheaths for the larger axons.

As in the experiments employing defined medium as described above, Schwann cell differentiation is arrested at a very early stage. One cannot fail to note the similarity between the disposition of Schwann cells in relation to axons in these suspended nerve fascicles from normal ganglia in tissue culture and the faulty ensheathment by Schwann cells within nerve roots of the *dy* mouse, as has been discussed above. In comparing these Schwann cells to those observed in cultures carried in defined medium, however, there is a very clear difference. Schwann cells suspended on nerve roots exhibit a coating of extracellular matrix material which is morphologically quite similar to basal lamina. These cells, then, are not failing completely to secrete material as are Schwann cells grown in defined medium; yet their relationship to the axon does not progress to ensheathment and myelination. They appear only to lack contact with the rat-tail collagen substratum (or substances bound to the substratum).

The disposition of Schwann cells in relation to suspended axons can be corrected within hours if they are forced into contact with the reconstituted collagen substratum (R. Bunge and M. Bunge, 1978). This is accomplished by placing a clear plastic strip coated with rat tail collagen over the suspended fascicles (Fig. 14-9). In the same culture dish, suspended fascicles not subjected to this corrective treatment retain abnormal clus-

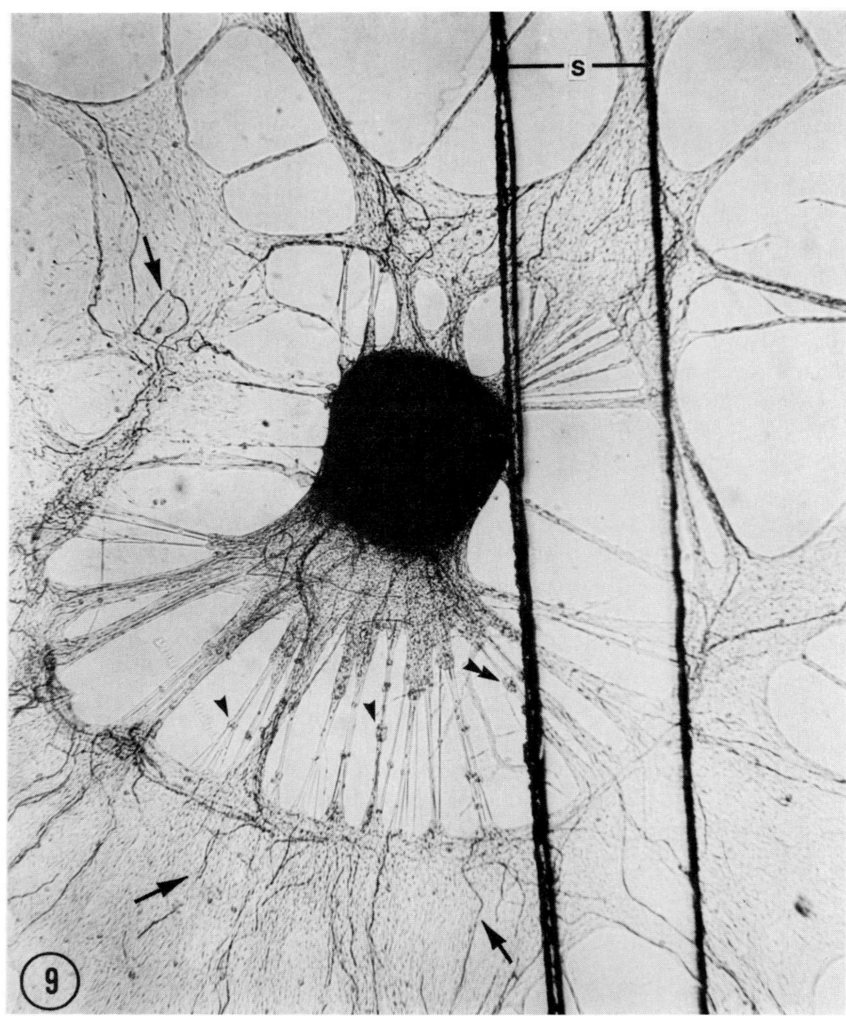

Figure 14.9 Neuron-Schwann cell culture grown to maturity (60 days) in the usual medium (containing serum and embryo extract). Peripheral outgrowth in contact with the collagen substratum is normal in Schwann cell configuration and ensheathment characteristics, including the formation of numerous myelin sheaths (arrows). Closer to the explant, however, where some fascicles do not rest in the substratum, Schwann cells are found in atypical clusters (arrowheads). When a plastic strip (S) coated with collagen is placed on fascicles similar to those in the region of the arrowheads, within a day Schwann cells begin to migrate from the clusters and elongate in relation to the axons. This micrograph, taken ten days after addition of the strip, shows that the highly abnormal fascicles have achieved a normal appearance, including the presence of myelin sheaths (seen at higher magnification). That this correction is directly related to the strip can be seen in the area of the double arrowhead; fascicle areas just outside the strip remain abnormal in contrast to those regions of the same fascicles underneath the strip. ×45.

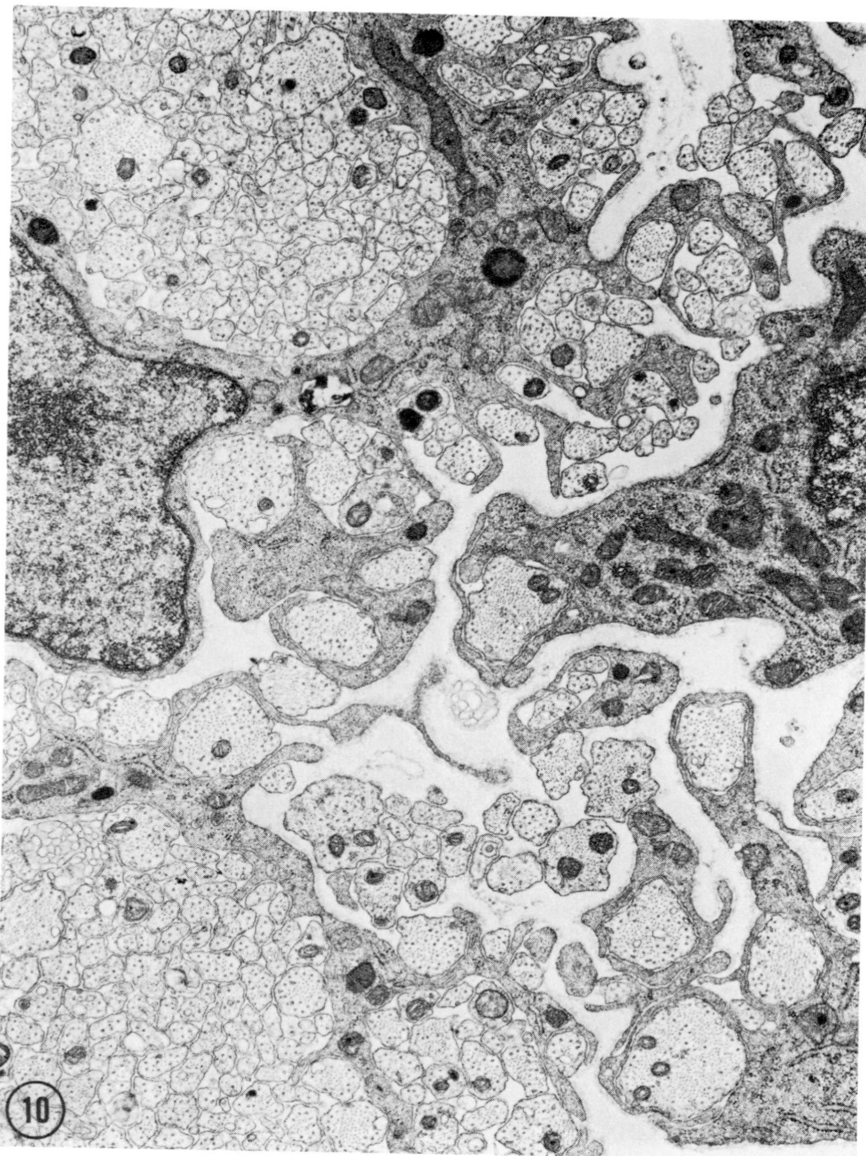

Figure 14.10 Electron micrograph of an area of one of the Schwann cell clusters on a suspended fascicle, as illustrated in Figure 14.9. Near the periphery of the cluster (upper right hand corner) a few of the larger diameter axons are related to Schwann cells but most neurites, seen here in two large bundles, have not become ensheathed despite the long culture period (60 days). A patchy lamina has been formed on the Schwann cell surface. ×15,000.

ters of Schwann cells despite the provision of a complete tissue culture medium (with serum and embryo extract) and despite the collagen substratum being just a few microns away. Electron microscopic observations confirm that contact with reconstituted collagen (or material bound to it) rapidly corrects the Schwann cell abnormality, that within six days unmyelinated nerve fibers are fully ensheathed by Schwann cells exhibiting a basal lamina covering, and that some larger nerve fibers then begin to be myelinated.

These observations have led us to suggest that during the development of Schwann cells, there is a requirement for an inductive signal, which is provided in our tissue culture dish by contact with a collagen surface. In this case, because the Schwann cell appears to secrete some basal lamina material, we assume that the substance required for contact is one that the Schwann cell itself does not make. This assumption leads us to the suggestion that Schwann cells may only be operative in regions where fibroblasts are able to provide a substance (collagen?) that provides the signal which is adequate for the induction of full Schwann cell function.

This explanation could explain the failure of Schwann cells to follow nerve fibers into the epidermis of the skin for, as the nerve fiber traverses the basal lamina underlying the epidermis, it enters a region that does not contain extracellular matrix materials but is occupied by closely packed cells of the epidermis. It may also explain the failure of Schwann cells to occupy certain regions of the neuronal aggregates of the enteric plexus if, in fact, in these regions the extracellular matrix components differ significantly from those in other regions of the body and are not capable of providing the signal for this newly proposed inductive step.

Whether this explanation pertains to those instances in which Schwann cells enter the CNS is not yet clear. Schwann cells may be operative within the CNS only in regions where there has been some fibrosis resulting from the invasion of fibroblasts not normally present. Certainly in the case of transplantation of nerve trunks into the CNS, connective tissue and fibroblasts are implanted with the nerve. These may provide the necessary inductive signal for full Schwann cell development. In the case of implantation of pure Schwann cell populations by Duncan and colleagues (1979), it should be noted that Schwann cells were placed into the CNS region along with the collagen substratum on which they had originally grown. It was observed that myelination of CNS fibers by the implanted Schwann cells occurred primarily in the region surrounding the collagenous component of the implant. The operative procedure may also have induced some fibrosis in the region of the implant.

These observations raise additional questions. Is the necessity for Schwann cell contact with a component in addition to the axon limited to only one point in embryonic development? Is the Schwann cell subsequently capable of being functionally operative without this sustained "third element" contact? Our own tissue culture observations suggest this possibility, for we have observed that fully developed nerve fascicles with typical Schwann cell ensheathment retain normal ensheathment if they subsequently become detached from the tissue culture substratum. This observation suggests that the "third element" requirement for Schwann cell differentiation is needed only during a certain developmental period. The question whether this contact is also a requirement for differentiation after Schwann cell proliferation in response to injury in the adult animal is not known.

POSTULATED TROPHIC ASPECTS OF SCHWANN CELL FUNCTION

We have mentioned above Ramón y Cajal's statements regarding the functions of the cell of Schwann in promoting peripheral nerve regeneration. Toward the end of his monograph on degeneration and regeneration in the nervous system (1928), he provided a description of the kind of factor that he believed the Schwann cell must release to explain its action on promoting axonal regrowth. He described it as follows:

> The nutritive and tutorial functions of the cells of Schwann had already been recognized by the embryologists, many of whom believed that the successive thickening of young fibres and the development of the medullary sheath were the principal functions of these cells . . . (p. 391) There is only a step between the recognition of these cells as a nutritive placenta and the admission in them of secretions capable of stimulating and orienting the sprouts that are wandering in the scar . . . It is difficult, in the present state of knowledge, to imagine what is the nature of the stimulating substance. As a tentative hypothesis we have supposed that the substance contained in the sheaths of Schwann of the peripheral stump should be conceived . . . as a ferment or catalytic agent which stimulates the assimilation of the axonic protoplasm and which does not become used up while acting on the nervous protoplasm. (p. 392)

Recent experimentation provides several lines of evidence that support this view, but as yet there is very little specific information regarding the nature of the substances that may be exchanged between Schwann cells and axons. By using modern tissue culture techniques, it is possible to demonstrate that certain types of cells are able to provide direct trophic support for neurons in coculture (Barde et al., 1978; Ebendal, Jordell-Kylberg, and Soderstrom, 1978; Lindsay, 1979). Up to this time many of these experiments have been done with mixed supporting cells of central origin; they have not utilized pure Schwann cell populations. In their earlier work, Varon and colleagues (1974) clearly demonstrated that nonneuronal cells derived from peripheral nerve will support the survival of nerve growth factor-dependent neurons in a coculture that lacks nerve growth factor in the culture medium. Because these experiments utilized mixed supporting cell populations, they did not clarify whether fibroblasts or Schwann cells were instrumental in supporting the nerve cells in culture. On several occasions, we have undertaken similar experiments with pure populations of Schwann cells prepared by the methods discussed above. Although very preliminary results have appeared (R. Bunge, 1980), we have not generally been able to obtain adequate control of quantification, for it is difficult to provide a predetermined number of Schwann cells for a given number of neurons. However, the results that have been obtained thus far certainly support the idea that the Schwann cell is able to preserve the vitality of a number of neuronal types in coculture.

Direct evidence that material may be exchanged between Schwann cells and axons is available from studies of lower vertebrate and invertebrate tissues. In initial experiments conducted by Singer and Salpeter (1966), indirect evidence was provided that material is transferred from Schwann cells to axons in peripheral nerve. Subsequent experiments were undertaken with giant axons of invertebrates in which it was possible to directly sample materials appearing within the axon after labeling compounds manufactured by adjacent Schwann cells. These experiments, conducted by Lasek, Gainer, and Barker (1977) and Gainer, Tasaki, and Lasek (1977), have clearly established that a number of compounds synthesized within the Schwann cell appear in adjacent axons and thus provided direct evidence that material is exchanged between Schwann cells and axons in invertebrates.

There is also more indirect evidence available from studies on vertebrate peripheral nerve. It has been observed that the diameter of an axon is systematically decreased in a number of experimental demyelinating neuropathies in which Schwann cells are abnormal (for discussion, see Aguayo, Bray, and Perkins, 1979). These workers have observed that correction of the reduction in axonal girth occurs after remyelination of demyelinated axons or provision of normal Schwann cells for ensheathment of abnormal axons in mutant mice. The demonstration that the diameter of an axon is systematically reduced in regions in which normal Schwann cell ensheathment does not occur is offered as another example of trophic support that Schwann cells provide for axons.

The availability of a number of methods for preparing pure Schwann cell populations suggests that the kinds of experiments necessary to define agents that may pass between Schwann cell and axon during development, that maintain the mature state, and that help regeneration should be amenable to experimentation. We have discussed above the evidence that the vertebrate Schwann cell is secretory and are hopeful that the definition of compounds that Schwann cells are capable of synthesizing will help to clarify the question concerning the nature of Schwann cell support for the neuron.

SUMMARY

We present in Figure 14-11 a diagrammatic summary of the steps in Schwann cell development proposed on the basis of evidence reviewed in this essay. We have not discussed the step, here labeled protodifferentiation, in which a sheath cell (Schwann cell) population appears among the developing neuroblasts (for discussion, see Tennyson, 1970; Noden, 1978). The number of Schwann cells related to nerve cells and axons is expanded rapidly by proliferation engendered (at least in part) by a mitogenic signal present on the nerve cell surface. Next, we propose that Schwann cell secretion of a number of products(*) is a requisite step in the development of full Schwann cell function. One of these products may mask the axolemmal mitogen. Some of the evidence for axon-related Schwann cell secretion is the appearance, at a later stage, of basal lamina (xx) on the Schwann cell surface and thin collagen fibrils in association with the basal lamina. Tissue culture observations suggest that during this developmental period the Schwann cell must have contact with certain, as yet incompletely defined, components of extracellular matrix in order to complete axonal engulfment and thus to provide for the type of axonal ensheathment typical of unmyelinated nerve fibers. This en-

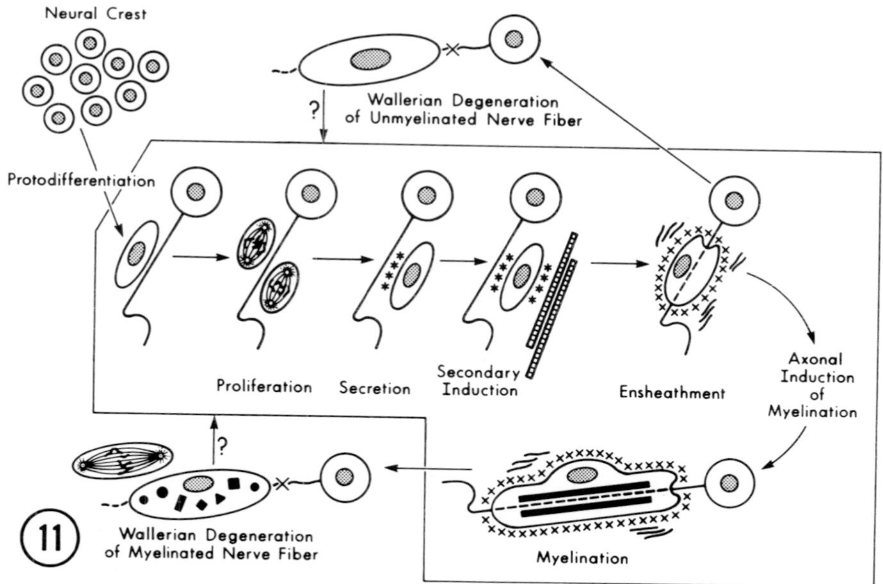

Figure 14.11 Diagrammatic representation of postulated stages in Schwann cell development. Refer to "Summary" for explanation.

sheathment includes the deposition of a complete basal lamina over the surface of the axon-related Schwann cell. Larger axons engender a further differentiative step in the Schwann cell resulting in the formation of a myelin sheath. These normal steps in Schwann cell development are contained within the box in Figure 14-11.

When fully differentiated Schwann cells are removed from axonal contact, their response varies depending upon whether they had been related to unmyelinated or to myelinated axons; those related to myelinated fibers undergo a substantial period of proliferation whereas those related to unmyelinated fibers do not. The functional capabilities of these cells to remyelinate axons is established, but it is not certain which of the diagrammed developmental steps (if any) they must again "experience" in order to express this functional capacity. The secretory capacity of the Schwann cell has been emphasized above; among these secretory products may also be agents instrumental in the trophic support of neurons.

ACKNOWLEDGMENTS

It is a great pleasure to acknowledge the laboratory assistance and collaboration of Ann Williams, Lisa Wartels and Robert Smith, Vincent Argiro, David Copio, and Donn Kleinschmidt; and of Drs. Michael Cochran, Carson Cornbrooks, John Jeffrey, Luis Glaser, Fernando Moya, Eiko Okada, James Salzer, Jouni Uitto, and Patrick Wood. We also wish to thank Susan Mantia for secretarial aid and Marc Davis and Robert Freund for photographic assistance. Our laboratory work reviewed here has been supported by United States Public Health Service Grant NS 09923 and National Multiple Sclerosis Grant RG 1118.

REFERENCES

Aguayo, A. J., Bray, G. M., Perkins, S. C. 1979. Axon-Schwann cell relationships in neuropathies of mutant mice. *Ann. N.Y. Acad. Sci.* 317:512–31

Aguayo, A. J., Charron, L., Bray, G. M. 1976a. Potential of Schwann cells from unmyelinated nerves to produce myelin: A quantitative ultrastructural and radiographic study. *J. Neurocytol.* 5:565–73

Aguayo, A., Peyronnard, J., Terry, L., Romine, J., Bray, G. M. 1976b. Neonatal neuronal loss in rat superior cervical ganglia: Retrograde effects on developing preganglionic axons and Schwann cells. *J. Neurocytol.* 5:137–55

Asbury, A. K. 1975. The biology of Schwann cells. In *Peripheral Neuropathy,* eds. P. J. Dyck, P. K. Thomas, E. H. Lambert, pp. 201–12. Philadelphia; W.B. Saunders

Barde, Y. A., Lindsay, R. M., Monard, D., Thoenen, H. 1978. New factor released by cultured glioma cells supporting survival and growth of sensory neurones. *Nature* (London) 274:818

Billings-Gagliardi, S., Webster, H. deF., O'Connell, M. F. 1974. In vivo and electron microscopic observations of Schwann cells in developing tadpole nerve fibers. *Am. J. Anat.* 141:375–92

Blakemore, W. F. 1977. Remyelination of CNS axons by Schwann cells transplanted from the sciatic nerve. *Nature* (London) 266:68–69

Blakemore, W. F. 1978. Observations on remyelination in the rabbit spinal cord following demyelination induced by lysolecithin. *Neuropathol. Appl. Neurobiol.* 4:47–59

Bottenstein, J. E., Sato, G. H. 1979. Growth of a rat neuroblastoma cell line in serum-free supplemented medium. *Proc. Natl. Acad. Sci.* U.S.A. 76:514–17

Bradley, W. G., Jenkison, M. 1973. Abnormalities of peripheral nerve in murine muscular dystrophy. *J. Neurol. Sci.* 18:227–47

Brockes, J., Fields, K., Raff, M. 1979. Studies on cultured rat Schwann cells. I. Establishment of purified populations from cultures of peripheral nerve. *Brain Res.* 165:105–18

Bunge, M., Jeffrey, J., Wood, P. 1977. Different contributions of Schwann cells and fibroblasts to the collagenous components of peripheral nerve. *J. Cell Biol.* 75:161a

Bunge, M. B., Williams, A. K., Wood, P. M. 1979. Further evidence that neurons are required for the formation of basal lamina around Schwann cells. *J. Cell Biol.* 83:130a

Bunge, M. B., Williams, A. K., Wood, P. M., Uitto, J., Jeffrey, J. J. 1980. Comparison of nerve cell and nerve cell plus Schwann cell cultures, with particular emphasis on basal lamina and collagen formation. *J. Cell Biol.* 84:184–202

Bunge, R. P. 1980. Some observations on the role of the Schwann cell in peripheral nerve regeneration. In *Nerve Repair and Regeneration: Its Clinical and Experimental Basis*, eds. D. Jewett, H. McCarroll, pp. 58–67. St. Louis: Mosby

Bunge, R. P., Bunge, M. B. 1978. Evidence that contact with connective tissue matrix is required for normal interaction between Schwann cells and nerve fibers. *J. Cell Biol.* 78:943–50

Bunge, R., Bunge, M., Okada, E., Cornbrooks, C. 1980a. Abnormalities expressed in cultures prepared from peripheral nerve tissues of trembler and dystrophic mice. In *Neurological Mutants Affecting Myelination*, ed. N. Baumann, pp. 433–46. New York: Elsevier North-Holland

Bunge, R., Moya, F., Bunge, M. 1980b. Observations on the role of Schwann cell secretion in Schwann cell-axon interactions. In *Neurosecretion and Brain Peptides: Implications for Brain Function and Neurological Disease*, eds. J. B. Martin, S. Reichlin, K. L. Bick, pp. 229–42. New York: Raven

Copio, D. S., Bunge, M. B. 1980. Use of a proline analog to disrupt collagen synthesis prevents normal Schwann cell differentiation. *J. Cell Biol.* 87:114a

Del Rio-Hortega, P. 1921. Estudios sobre la neuroglia. La glia de escasas radiaciones (oligodendroglia). *Bol. R. Soc. Esp. Hist. Nat.* 21:63–92

Duncan, I. D., Aguayo, A. J., Bunge, R. P., Wood, P. 1979. Transplantation of rat Schwann cells cultured in vitro into demyelinated areas of the mouse spinal cord. *Neurosci. Abstr.* 5:510a

Ebendal, T., Jordell-Kylberg, A., Soderstrom, S. 1978. Stimulation by tissue explants on nerve fibre outgrowth in culture. *Zoon* 6:235–43

Gabella, G. 1971. Glial cells in the myenteric plexus. *Z. Naturforsch.* 26b:244–45

Gainer, H., Tasaki, I., Lasek, R. J. 1977. Evidence for the glia-neuron protein transfer hypothesis from intracellular perfusion studies of squid giant axons. *J. Cell Biol.* 74:524–30

Gilmore, S. A., Duncan, D. 1968. On the presence of peripheral-like nervous and connective tissue within irradiated spinal cord. *Anat. Rec.* 160:675–90

Harrison, R. G. 1910. The outgrowth of the nerve fiber as a mode of protoplasmic movement. *J. Exp. Zool.* 9:787–846

Harrison, R. 1924. Neuroblast versus sheath cell in the development of peripheral nerves. *J. Comp. Neurol.* 37:123–205

Honegger, P., Lenoir, D., Favrod, P. 1979. Growth and differentiation of aggregating fetal brain cells in a serum-free defined medium. *Nature* (London) 282:305–8

Jaros, E., Bradley, W. G. 1979. Aytpical axon-Schwann cell relationships in the common peroneal nerve of the dystrophic mouse: an ultrastructural study. *Neuropathol. Appl. Neurobiol.* 5:33–147

Kleinschmidt, D. C., Bunge, R. P. 1980. Myelination in cultures of embryonic rat spinal cord grown in a serum-free defined medium. *J. Cell Biol.* 87:66a

Lasek, R. J., Gainer, H., Barker, J. L. 1977. Cell-to-cell transfer of glial proteins to the squid giant axon. The glia-neuron protein transfer hypothesis. *J. Cell Biol.* 74:501–23

Lindsay, R. M. 1979. Adult rat brain astrocytes support survival of both NGF-dependent and NGF-insensitive neurones. *Nature* (London) 282:80–84

McCarthy, K. D., Partlow, L. M. 1976. Neuronal stimulation of ^{3}H-thymidine incorporation by primary cultures of highly purified non-neuronal cells. *Brain Res.* 114:415–26

Moya, F., Bunge, M. B., Bunge, R. P. 1980. Schwann cells proliferate but fail to differentiate in defined medium. *Proc. Natl. Acad. Sci.* U.S.A. 77:6902–6

Noden, D. 1978. Interactions directing the migration and cytodifferentiation of avian neural crest cells. In *The Specificity of Embryological Interactions,* ed. D. Garrod. London: Chapman and Hall pp. 3–49

Okada, E., Bunge, R. P., Bunge, M. B. 1980. Abnormalities expressed in long term cultures of dorsal root ganglia from the dystrophic mouse. *Brain Res.* 194:455–70

Raff, M., Abney, E., Brockes, J., Hornby-Smith, A. 1978. Schwann cell growth factors. *Cell* 15:813–22

Ramón y Cajal, S. 1928. *Degeneration and Regeneration in the Nervous System,* vols. 1 and 2. New York: Hafner

Ramón y Cajal, S. 1952. *Histologie du système nerveux de l'homme and des vertébrés,* vol. 1. Tr. L. Azoulay. Madrid: Instituto Ramón y Cajal

Romine, J., Bray, G., Aguayo, A. 1976. Multiplication of unmyelinated Schwann cells after crush injury. *Arch. Neurol.* (Chicago) 33:49–54

Salzer, J. L., Bunge, R. P. 1980. Studies of Schwann cell proliferation. I. An analysis in tissue culture of proliferation during development, Wallerian degeneration, and direct injury. *J. Cell Biol.* 84:739–52

Salzer, J. L. Bunge, R. P., Glaser, L. 1980a. Studies of Schwann cell proliferation. III. Evidence for the surface localization of the neurite mitogen. *J. Cell Biol.* 84:767–78

Salzer, J. L., Williams, A. K., Glaser, L., Bunge, R. P. 1980b. Studies of Schwann cell proliferation. II. Characterization of the stimulation and specificity of the response to a neurite membrane fraction. *J. Cell Biol.* 84:753–66

Schwann, T. 1839. *Mikroskopische Untersuchungen über die Uebereinstimmung in der Struktur und dem Wachstum der Tiere und Pflanzen.* Berlin: Sander

Singer, M., Salpeter, M. M. 1966. The transport of [^{3}H]-L-histidine through the Schwann and myelin sheaths into the axon. *J. Morphol.* 120:281–316

Tennyson, V. M. 1970. The fine structure of the developing nervous system. In *Developmental Neurobiology,* ed. W. A. Himwich, pp. 47–116. Springfield, Ill.: Thomas

Uitto, U. J., Uitto, J., Kao, W. W.-Y., Prockop, D. J. 1978. Procollagen polypeptides containing cis-4-hydroxy-2-proline are over glycosylated and secreted as nonhelical pro-γ-chains. *Arch. Biochem. Biophys.* 185: 214–21

Varon, S., Raiborn, C. W. Jr., Burnham, P. A. 1974. Comparative effects of nerve growth factor and ganglionic non-neuronal cells on purified mouse ganglionic neurons in culture. *J. Neurobiol.* 5:355–71

Webster, H. deF. 1975. Development of peripheral myelinated and unmyelinated nerve fibers. In *Peripheral Neuropathy,* eds. P. J. Dyck, P. K. Thomas, E. H. Lambert, pp. 37–61. Philadelphia: W. B. Saunders

Weinberg, H., Spencer, P. 1975. Studies on the control of myelinogenesis. I. Myelination of regenerating axons after entry into a foreign unmyelinated nerve. *J. Neurocytol.* 4:395–418

Williams, A. K., Wood, P. M., Bunge, M. B. 1976. Evidence that the presence of Schwann cell basal lamina depends upon interaction with neurons. *J. Cell Biol.* 70:138a

Wood, P. M. 1976. Separation of functional Schwann cells and neurons from normal peripheral nerve tissue. *Brain Res.* 115:361–75

Wood, P. M., Bunge, R. P. 1975. Evidence that sensory axons are mitogenic for Schwann cells. *Nature* (London) 256:662–64.

15 / *Development of Connectivity in the Cerebal Cortex*

E.G. JONES

INTRODUCTION

> *"It seems very pretty," she said . . . "but it's* rather *hard to understand! . . . somehow it seems to fill my head with ideas—only I don't exactly know what they are!"*
>
> Lewis Carroll

Anyone attempting to unravel the developmental history of connection-formation in the mammalian cerebral cortex might well agree with these words of Alice when confronted by the Jabberwocky poem. Yet, later in *Through the Looking Glass,* Carroll explains that many of the expressions that baffle Alice are but portmanteau combinations into which two or more simpler meanings are packed. Perhaps this can serve to reaffirm our intuitive belief that the organizational complexity of the cerebral cortex depends upon the same fundamental principles that govern the growth and development of simpler neural systems. These fundamental principles of neural development were established by a sometimes tenuous but fortunately unbroken chain of investigators extending back to His and Ramón y Cajal. They form the foundations of the science of developmental neurobiology, foundations that were laid by Viktor Hamburger, his teachers, and contemporaries. Now that the science is at its strongest and now that advances in experimental techniques have provided us with more adequate access to the developing mammal, it is to be hoped that the way in which these principles of neuronal development manifest themselves in the cerebral cortex will become apparent to us.

Because of its complicated intrinsic organization and connectional relationships with so many other parts of the central nervous system, the mammalian neocortex has challenged the ingenuity of many generations of neuroscientists. The structural and functional complexity of this part of the brain must offer a similar challenge to the developmental neurobiologist. Apart from establishing cellular layers and architectonic fields and sorting out its intrinsic circuitry, over a brief time span most cortical areas are receiving several classes of ingrowing afferent fibers and are themselves growing axons to no less than nine or ten extrinsic sites: ipsilateral cortex, contralateral cortex, striatum, thalamus, tectum, pons, reticular formation, medullary relay nuclei, spinal cord. Furthermore, during this seemingly rapid course of events, the very great orderliness of cortical organization is created so that the somata of cortical cells sending axons to different sites come to occupy different layers; topographic order and laminar and columnar patterns of distribution are set up in the afferent and efferent fiber systems; and in certain instances such as the thalamus and the opposite cortex, a precise reciprocity of input and output connections is established (Jones, 1981). Is this orderliness inherent in the developing cortex, or is it created out of apparent chaos? And to what extent is it under the influence of environmental, hormonal, nutritional, and other factors?

At the moment, we can say little about the nature of the signals that cause intracortical neurons to mature and that cause axons headed to or away from the cortex to grow towards their targets and to establish their definitive patterns of connections. But we are in a position to provide a fairly comprehensive, descriptive account of the temporal sequence of events leading to the formation of the mature pattern of organization. In what follows, I shall try to outline the general course of events by reference to our studies on the development of afferent and efferent connections in the sensorimotor cortex of the rat and cat.

GROWTH OF CALLOSAL AND THALAMIC AXONS INTO THE SOMATIC SENSORY CORTEX

> *"The first thing I've got to do," said Alice to herself, as she wandered about the wood, "is to grow to my right size . . . and the second thing is to find my way into that lovely garden."*

The major experimental preparation we have used for studying callosal and thalamic afferent fiber distribution in maturing animals is the "flattened" rat brain first introduced by Dr. Carol Welt (Fig. 15-1; Welker, 1976). When the cerebral hemisphere is gently compressed between glass slides during postfixation, it is possible subsequently to cut sections in such a plane that they pass more or less continuously through single cortical layers. In this way, Welt was able to show that the internal granular layer of the somatic sensory cortex is not homogeneous but consists of large aggregations of granule cells separated by less granular zones. Cells in individual granular aggregations respond to stimuli applied to a particular part of the body surface, while the intervening and surrounding agranular zones are nonresponsive—at least in the anesthetized preparation. Subsidiary clusters of cells within the granular aggregations form the basis of the finer representation of the periphery. The best known of these are the "barrels" (Welker and Woolsey, 1974) found in the aggregation to which the trigeminal system projects. Each of these represents a sinus hair on the face, head, or lower jaw.

Our attention was attracted to the potential developmental significance of this maplike facsimile of the rat's body surface when we discovered that, in the adult rat, fibers of the corpus callosum emanating from the opposite cortex, together with the parent cells of those projecting reciprocally back, were distributed only to the agranular zones (Wise and Jones, 1976) (Fig. 15-2). Conversely, axons entering the cortex from

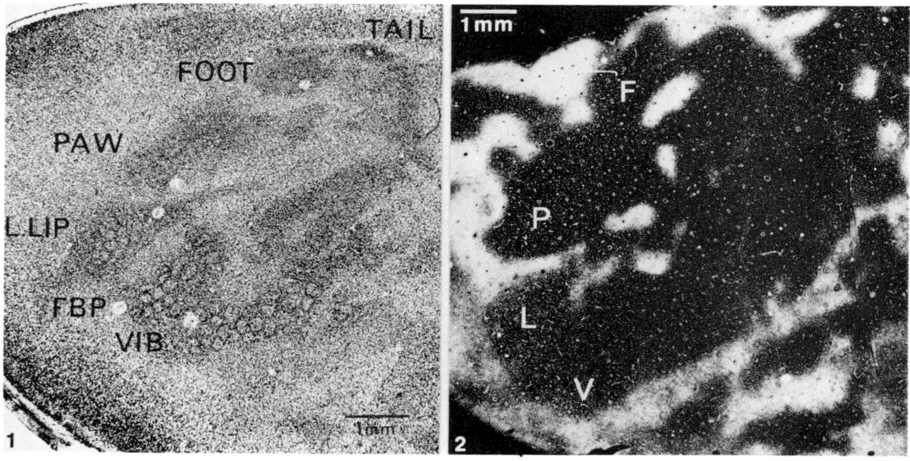

Figure 15.1 Section from a "flattened" preparation of the rat somatic sensory cortex, reproduced by kind permission from Dr. Carol Welt (Welker, 1976). This shows the interrupted granule cell aggregations, each of which represents a body part (L. Lip: lower lip; FBP: fatty buccal pad; VIB: mystacial vibrissae). Nissl stain.

Figure 15.2 Darkfield photomicrograph of a similar "flattened" preparation (Wise and Jones, 1976), showing distribution of degenerating callosal fibers (white) around and between the granular aggregations following section of the corpus callosum. Silver stain for degenerating axons. F: foot; P: paw; L: lip; V: vibrissae.

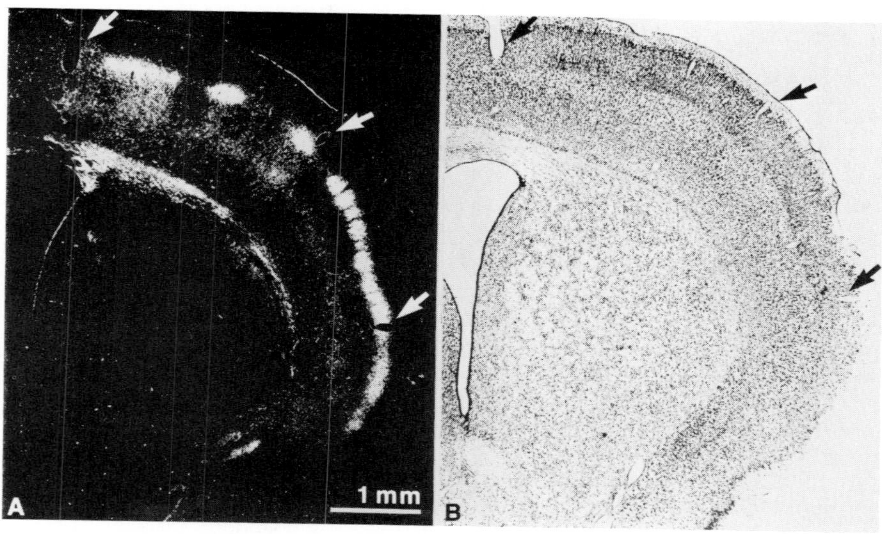

Figure 15.3 Alternate frontal sections at the same magnification showing in darkfield (A) the distribution of degenerating thalamocortical axons following destruction of the thalamic ventrobasal complex and their distribution only to the granular aggregations of the rat somatic sensory cortex, as seen in a Nissl stained section (B). Arrows indicate same landmarks. From Wise and Jones (1978).

the principal thalamic relay nucleus for the somatic sensory cortex, the ventrobasal complex, were distributed (as a series of columnlike bundles) only to the granular aggregations that form the basis of the body representation (Wise and Jones, 1978) (Fig. 15-3). Double-labeling experiments (Fig. 15-4) indicate little or no overlap between the two systems and any additional thalamic input to the agranular zones, if it exists, has not yet been identified.

The fairly strict alternation of the terminations of two afferent fiber systems raises the possibility that the pattern of distribution may be set up during the course of development by competitive interactions between the two systems. In our attempts to demonstrate such an interaction, we have been able to analyze the patterns of growth of the callosal and thalamocortical fiber systems. The time scale over which each system develops is unique, but the basic pattern is the same for both and seems to conform to a general pattern found in several axon systems and in other species. Our experiments involved the injection of axoplasmically transported markers in the cortex or thalamus of late fetal and newborn rats (Wise and Jones, 1976, 1978; Jones, unpublished). These materials accumulate in the growing tips of the callosal or thalamocortical axons

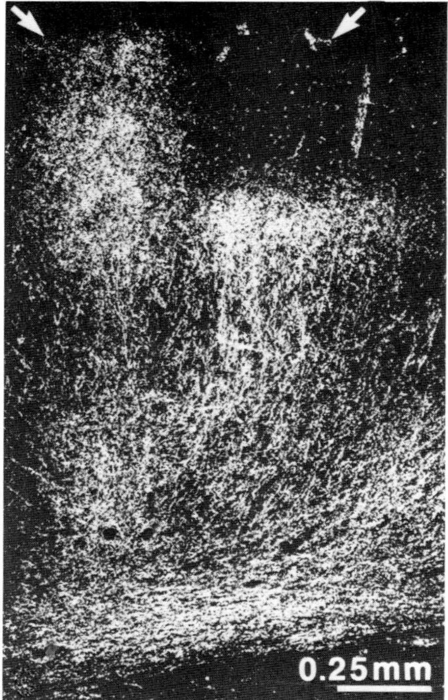

Figure 15.4 Segregation of a column of degenerating commissural axon terminal ramifications (left) from a column of degenerating thalamocortical axon ramifications (right). Note also the different layers of termination. Arrows indicate cortical surface. Rat somatic sensory cortex, lesions of thalamus and corpus callosum. From Wise and Jones (1978)

(Figs. 15-5 and 15-6) and serve to delimit the point to which the axons have grown at the time of killing the animal. Our results indicate that growth of the two sets of axons towards the cortex is an early phenomenon and that they have accumulated together in the presumptive white matter (the remains of the intermediate zone) beneath the cortical plate between the 17th and 20th days of gestation. Callosal fibers lag behind the thalamic, first entering the intermediate zone late on the eighteenth day. Then at about the day of birth (21st day of gestation), the thalamic fibers appear to receive some kind of selective signal that causes them alone to grow into the overlying cortex (Figs. 15-5 and 15-6). At the time of invasion, Layer V and Layer VI of the cortex can be identified, but the more superficial layers are represented by the homogeneous mass of relatively undifferentiated cells that constitute the remains of the cortical plate. Gradually, over the ensuing days, the thalamic fibers penetrate further into the cortex, piercing the remains of the cortical plate to reach Layer I by the third postnatal day. Their initial ingrowth is diffuse and there is little evidence of laminar, columnar, or zonal concentrations of terminal ramifications. The laminar concentrations in Layers I, III–IV, and VI start to appear on the fourth postnatal day as Layer III and Layer IV become distinguishable from the remains of the cortical plate. By

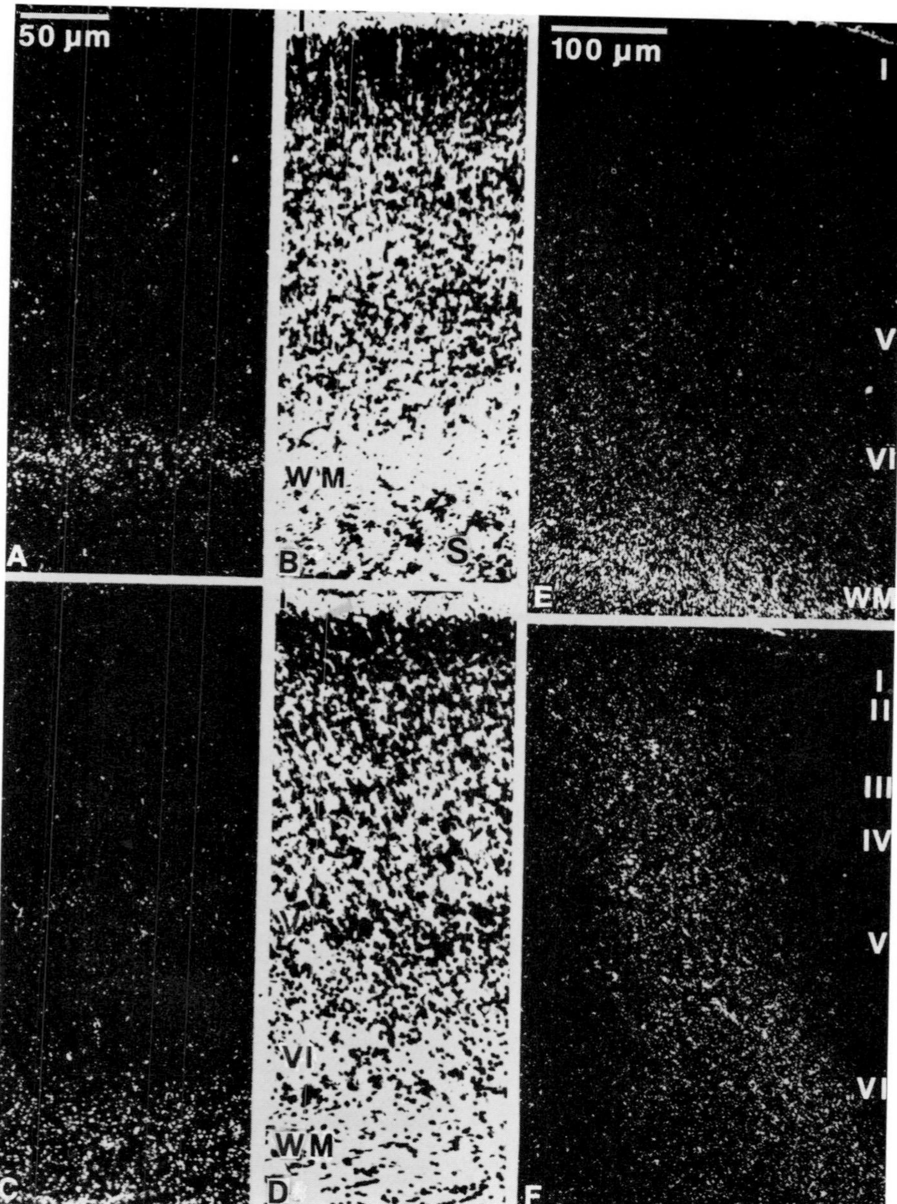

Figure 15.5 Figures from Wise and Jones (1976) showing labeled callosal fibers "waiting" in the white matter beneath the maturing cortex in one-day-old (A,B) and three-day-old (C,D) rats, invading the deeper layers of the cortex diffusely in a five-day-old rat (E) and forming columns as they reach Layer I in a seven-day-old rat (F).

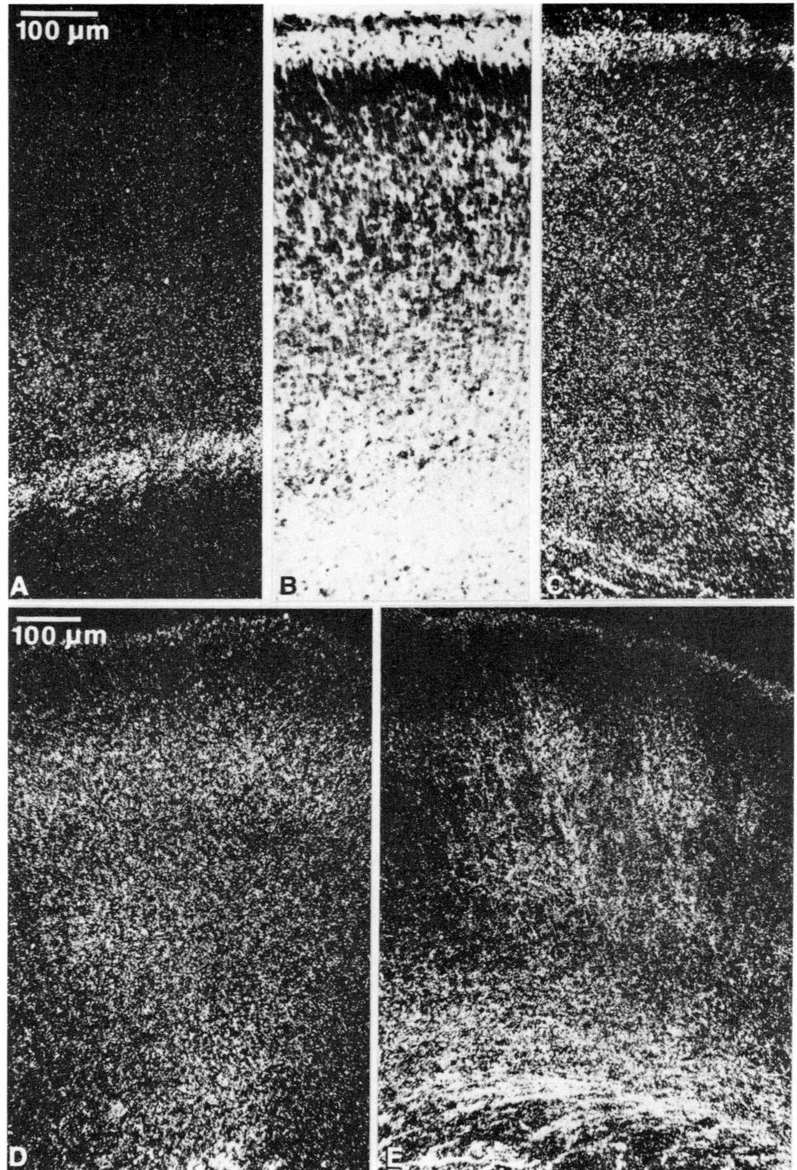

Figure 15.6 A series of figures similar to those in Figure 15.5, showing labeled thalamic fibers entering Layer VI of the presumptive somatic sensory cortex in a one-day-old rat (A,B), spreading diffusely through the cortex, except for the remains of the cortical plate, in a two-day-old rat (C), beginning to form laminar and columnar terminal aggregations in a four-day-old rat (D) and finally forming the definitive laminar and columnar pattern of Figure 15.3 in a six-day-old rat (E). From material illustrated in Wise and Jones (1978).

the fifth postnatal day, the terminations become confined to the newly appeared granule cell aggregates of Layer IV, and subsidiary columns are now visible within their zones of termination.

The pattern of early accumulation, of delayed growth into the cortex, of initial diffuse ingrowth, and of subsequent laminar, zonal, and columnar concentration is followed also by callosal fibers in the rat somatic sensory cortex (Wise and Jones, 1976). Accumulating in the intermediate zone over approximately the same period as the thalamocortical fibers, these fibers receive their signal to enter the cortex, nevertheless, some two to three days later than the thalamic fibers. The renewed spurt of growth that carries the commissural fibers into the cortex occurs on the fourth to fifth postnatal day, and the fibers only reach Layer I by the seventh to eighth postnatal day. The ingrowth is initially diffuse, and the fibers only resolve into columns in the agranular zones and into laminar concentrations in Layers II–III and V by the seventh day.

The nature of the signals that promote the two growth phases of the commissural and thalamocortical axons is unknown at present. Presumably the initial growth towards the cortical target is determined in large part by factors intrinsic to the parent cells, accompanied by those other as yet ill-defined factors, mechanical and chemical, that are thought to assist in pathway selection (Jacobson, 1979). It is possibly comparable to the early growth of spinal fibers towards the vicinity of a limb bud some time before the bud is evident (Hamburger, 1928). Whether the signal to make a second growth spurt into the target to form synapses is also genetic or conditioned by hormonal, nutritional, or by other environmental factors, some of which might be local, remains to be determined. What we do know is that the two-stage pattern seems to be followed by fibers directed towards other cortical areas of the rat (Lund and Mustari, 1977), by fibers directed towards cortical areas of other species (Rakic, 1976, 1977; Wise, Hendry, and Jones, 1977), and by fibers in other developing central nervous pathways. In the spinal cord of the neonatal rat, for example (Schreyer and Jones, 1980), fibers accumulate in the growing corticospinal tract two or more days before entering the gray matter of a particular segment.

Where the gestation period of an animal is longer than that of the rat, the "waiting period" prior to the second growth spurt of fibers into their cortical target is correspondingly prolonged. In the fetal cat, for example (Fig. 15-7), thalamocortical fibers have accumulated in the intermediate zone beneath the sensorimotor cortex some 22 days before the expected date of birth (at 65 days), yet they commence their second growth spurt into the cortex only a few days before birth (Wise, et al., 1977). In the visual cortex of the rhesus monkey, with a gestation period of approximately 165 days, Rakic (1976, 1977) has shown that geniculocortical fibers, labeled by the transneuronal transport of labeled precursors

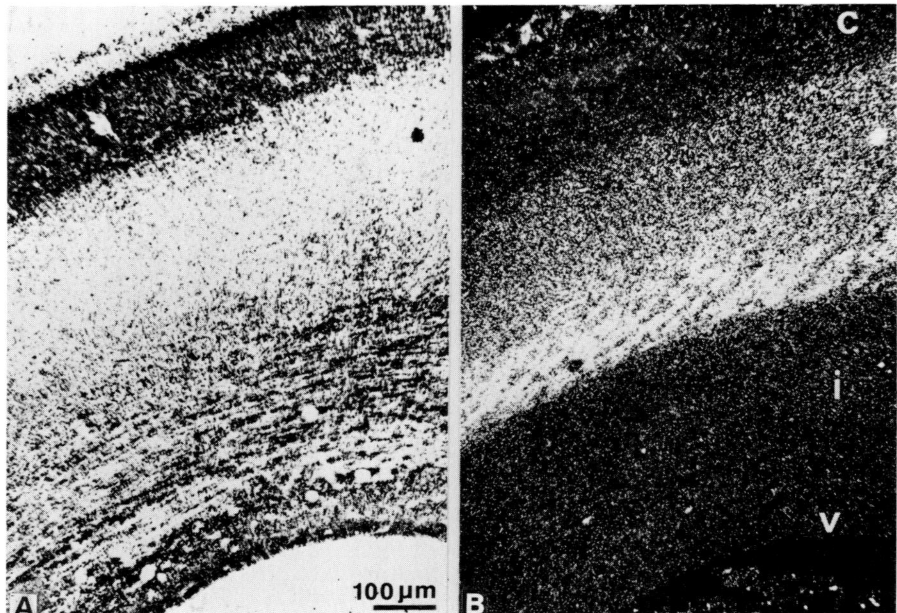

Figure 15.7 Autoradiographically labeled thalamocortical fibers accumulated in the intermediate zone of a cat fetus at 43 days of gestation. Note immature cortical plate (c) and large numbers of migrating cells in intermediate zone (i). From Wise et al. (1977).

from an injection into an eye, accumulate in the presumptive white matter many weeks before birth and commence their second growth spurt some 60 or so days before birth.

The early accumulation of thalamocortical fibers beneath the cortical plate occurs before all the cells destined to form the cerebral cortex have completed their migration from the underlying ventricular and subventricular zones. In the rhesus monkey, for example, cells destined to form Layer IV of the visual cortex commence their migration through the intermediate zone between gestational Day 70 and Day 80 (Rakic, 1974), some days after geniculocortical fibers have first been identified in the intermediate zone. It seems apparent, therefore, that in the monkey and probably in other species as well, many of the cells which are destined to receive the terminations of thalamocortical axons actually migrate through the growing axons. Because the cells reach the cortex before the axons commence their second growth spurt, no permanent connections can be made in the intermediate zone. However, we cannot rule out the formation of transitory connections which may permit the exchange of materials that subsequently enable cells and axons to recognize one another when they meet again in the cortex. There is no firm justification for this suggestion other than the morphological observation that junctional complexes, including synapselike formations, are seen transiently in a number of developing fiber tracts (Skoff and Hamburger, 1974; Knyihar, Csilik, and Rakic, 1978; Valentino and Jones, 1980).

As in the rat and cat, the ingrowth of thalamocortical fibers into the monkey visual cortex is diffuse, but just before birth the local aggregations of fibers that form the "ocular dominance" columns of the adult monkey begin to be discerned (Rakic, 1976, 1977). This correlates quite well with the pattern of early ingrowth and reorganization seen in the sensorimotor cortex of the cat and rat. It is clear, however, that the dimensions of the monkey ocular dominance columns can be subsequently modified by early neonatal visual experience (Hubel, Wiesel, and Le Vay, 1977). To what extent experience plays a role in establishing the zonal and columnar patterns of thalamic and callosal fiber distribution in the sensorimotor cortex of an animal such as the rat is not clear.

In all sensory systems, it is commonly believed that connection formation occurs from the periphery sequentially upwards to the cortex. The basis for this idea is, to a large extent, traditional rather than evidential, though newer data on synapse development in the visual system seem to support it (Cragg, 1975; Lund and Bunt, 1976). One wonders whether the formation of synapses by afferent fibers on thalamic relay cells (or of thalamocortical fibers on callosal cells) is the trigger that signals the axons of the cells to make their second growth spurt into the cortex?

Though the time course of synapse formation from periphery to center needs to be investigated in other systems, there is some reason to believe that the formation of those anatomical markers of topographic organization, such as axon bundling and "barrel" aggregations of cells, that characterize the somatic sensory system of the rat do appear sequentially from lower centers up to the cortex. Clustering of thalamocortical relay cells that already receive topographically ordered lemniscal inputs can be detected by retrograde labeling (Fig. 15-8) in the thalamic ventrobasal complex two or three days prior to the columnar clustering of their axons in the cortex (Wise and Jones, 1978). It seems likely that comparable clustering may occur in the brainstem relay nuclei at an even earlier stage, though the only study that might have provided this information unfortunately did not examine animals at a sufficiently early age (Belford and Killackey, 1979; Killackey and Belford, 1979). The bundling of axons and the development of the columns and "barrels" in which they terminate in the cortex are thus indications of a common organizing principle in operation from the periphery up to the cortex.

Other factors that may shape the distribution pattern of input fibers in the developing neocortex include competitive interactions between fibers belonging to the same or to different afferent systems. Competition

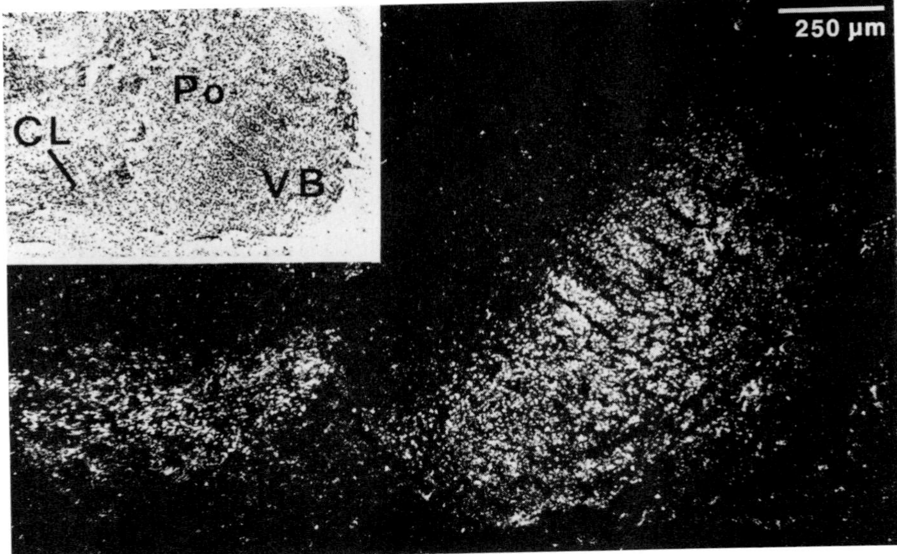

Figure 15.8 Retrogradely labeled thalamocortical relay cells at two days of age in the part of the rat thalamus indicated in the inset. Note that relay cells in the ventrobasal complex (VB) have aggregated into clusters prior to the formation of columns by their axons in the somatic sensory cortex (CL: central lateral nucleus; Po: posteromedial nucleus). From Wise and Jones (1978).

for developing synaptic sites has been postulated in order to account for the differences in density of termination of commissural and associational fibers along the rostrocaudal axis of the hippocampal formation (Gottlieb and Cowan, 1972). Successful competition also seemingly accounts for the failure of ocular dominance columns in the monkey visual cortex innervated by a normal eye to retract from territory that would normally be occupied by columns receiving input from the opposite eye, if this had not been deprived of visual experience in early neonatal life (Hubel, et al., 1977).

The differential distribution of thalamic and callosal fibers in the rat somatic sensory cortex, particularly when coupled with the earlier growth of thalamic fibers into the cortex, suggested to us that these patterns might be set up because the earlier growing thalamic fibers had gained some advantage over the callosal fibers in the competition for synaptic sites. Unfortunately, our attempts to induce aberrant distribution of callosal fibers in rats subjected to thalamotomy at birth (Wise and Jones, 1978) were negative. When examined in adulthood, these rats displayed a normal distribution of callosal fibers. There are a number of reasons that could account for this phenomenon. In the first place, synaptic sites normally destined to receive thalamic fibers may have failed to appear because they had not been exposed to some influence from the developing thalamic fibers that had only reached Layer VI at the time of thalamotomy. Second, such potential thalamic synaptic sites may have appeared and disappeared before the ingrowth of callosal fibers. Third, they may have been colonized by an additional competing fiber system, for example, by the ipsilateral associational system (Ryugo and Killackey, 1975), such as occurs in the hippocampal formation (O'Leary, et al., 1979; Stanfield and Cowan, 1979). Fourth, thalamic and callosal axons may be so rigidly specified that each is unable to terminate upon cells (or upon synaptic sites on the same cell) that are specified to receive the terminations of the other system (or, if they do terminate, they may be displaced by other, more appropriate, connections). If the fourth explanation were to be the case, one should expect that it would be possible to demonstrate competitive interactions between axon groups within a system. For example, to show that, following the destruction of one part of the somatic sensory cortex, callosal axons normally destined to terminate in that part would terminate in an atypical part of the cortex. Already there is some preliminary evidence that this may be so (Cusick and Lund, 1977).

A further influence that has been postulated in the shaping of connection patterns in the cerebral cortex is the noradrenergic innervation from the locus coeruleus. Fibers belonging to this system appear in large numbers in the cortex of the rat before the arrival of the other fiber systems mentioned above (Molliver and Kristt, 1975; Coyle and Molliver, 1977). Subsequently, they become concentrated in Layer IV, forming about 30 percent of the synaptic terminals in the cortex by the sixth postnatal day. However, apart from the fact of their early appearance, there have been no mechanisms suggested as to how these fibers could influence subsequent cortical maturation. The observation of Maeda and colleagues (1974) that after destruction of the locus coeruleus in rats, Layer VI cells extend dendrites abnormally towards Layer I seems without foundation, since as stressed by Ramón y Cajal (1911, page 574), many of these cells normally extend dendrites to that layer. The inability of Wendlandt and coworkers (1977) to duplicate the results of Maeda and others (1974) probably stems from failure to impregnate the (normal) ascending dendrite or from selection of a population of neurons whose dendrites do not ascend to Layer I.

STATUS OF OUTPUT CELLS DURING AXONAL GROWTH INTO THE CORTEX

"Begin at the beginning," the King said, very gravely, "and go on till you come to the end: then stop."

It is an old dictum, only rarely contested, that the axon of a nerve cell always establishes synaptic connections before the cell grows a substantial dendritic tree. Possibly the growth of the dendrites is modulated by signals from the periphery (Weiss, 1936). The evidence for dendritic growth being dependent on axonal connections seems to derive in particular from the work of Ramón y Cajal (1909), Barron (1943, 1946), and Hamburger and Keefe (1944) who showed that motoneurons in the chick embryo spinal cord grow their dendrites only after their axons have entered the limb musculature. The drawings by Ramón y Cajal of cerebellar Purkinje cells and neocortical pyramidal cells, which have seemingly long axons but only rudimentary dendritic arborizations, have suggested a comparable developmental sequence for cells that make their connections entirely within the central nervous system. In recent years, despite the appearance of more satisfactory methods for demonstrating axonal connections, there have been surprisingly few studies that have sought to correlate synapse formation by the axon with the onset of dendritic growth. There would seem to be some value in renewing such studies; for the sometimes lengthy "waiting period," which is observed by thalamocortical and callosal fibers and described above, shows that the presence of a set of axons near the target is no guarantee that synapse formation has occurred.

Our own studies on this topic have been concerned with the dendritic maturation of pyramidal cells in the rat sensorimotor cortex. These are the only cell class sending axons outside the cortex; and we have concentrated on those in Layer III, which give rise to a large proportion of the callosal axons, and on those in Layer V, which give rise to corticobulbar, corticospinal, and a smaller proportion of callosal axons (Wise and Jones, 1976, 1977). Our results confirm the predictions that can be made from observation of the drawings of Ramón y Cajal: the growth of a long axon is independent of the formation of a substantial dendritic tree. It is also independent of the development of synaptic inputs. As pointed out above, we have been able to show that callosal fibers have already crossed the midline and accumulated in the contralateral intermediate zone in the newborn rat (Wise and Jones, 1976). Similarly, corticobulbar and corticospinal axons, arising from cells in the immature Layer V, have already reached the medulla oblongata in the case of the former and the upper cervical segments in the case of the latter (Wise, Fleshman, and Jones, 1979a; Schreyer and Jones, 1980) (Figs. 15-9 and 15-10). In neither case, however, have the growing axons entered the gray matter of their respective targets. In the fetal cat, callosal, corticobulbar, corticospinal, and several other subcortical sets of axons have reached the white matter near their definitive targets as much as 20 days before birth (Wise, et al., 1977).

When the initial outgrowth of callosal axons occurs in the rat (between

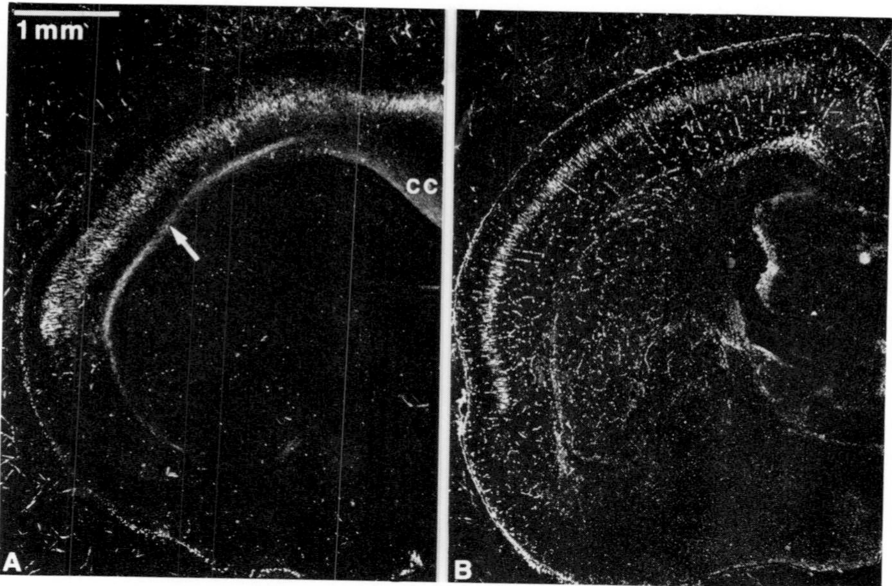

Figure 15.9 Darkfield photomicrographs showing uninterrupted distribution of retrogradely labeled callosal (A) and corticobulbar and corticospinal (B) neurons in two-day-old rats. In A the arrow indicates anterogradely labeled fibers of the corpus callosum (cc) that have not yet entered the cortex. (Cf. Figs. 15.16, 15.17.) From Wise et al. (1979).

the 18th and 19th day of gestation, Valentino and Jones, 1980), Layer III cells are still migrating into the cortical plate (Berry and Rogers, 1965). From our retrograde-labeling studies, it is our impression that the immature Layer III pyramids may not grow axons across the midline until their somata reach the part of the cortical plate destined to form Layer III, for in some of the less mature rats injected on one side with horseradish peroxidase at birth, retrogradely labeled cells appear on the other side predominately in the immature Layer V. In most newborns, however, two concentrations of retrogradely labeled somata are seen: one in the deeper part of the remaining cortical plate; the other in Layer V (Figs. 15-9A, 15-16A). At birth, Layer V also contains the only somata of corticospinal and corticobulbar cells (Figs. 15-9B, 15-16B). The parent cell in both Layer V and the deep cortical plate at the time its axon has reached the other side or the spinal cord is little more than a bipolar neuroblast (Fig. 15-11), though in both cases the apical dendritic process reaches the marginal layer and there is some evidence of beginning basal and secondary dendrite formation on the Layer V cells. The presence of a substantial apical dendrite and some rudimentary secondary dendrites

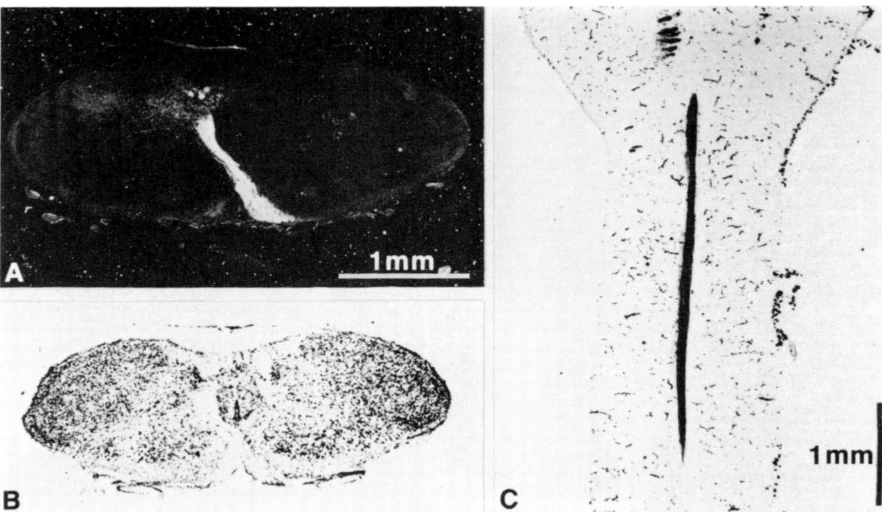

Figure 15.10 Extent of the corticospinal tract in newborn rats. A and B are darkfield (A) and brightfield (B) photomicrographs of adjacent transverse sections showing the autoradiographically labeled corticospinal fibers crossing at the pyramidal decussation. C shows in a horizontal section the decussation, and the tract reaching into the upper cervical segments, as labeled by anterograde transport of horseradish peroxidase. A and B from Wise et al. (1979); C from unpublished work of David L. Schreyer.

while the axons still lie waiting outside their gray matter targets, would tend to argue against synapse formation being a prerequisite for the initiation of dendritic growth. Although it is possible to argue that collateral branches of the axon close to the parent soma may form intracortical synapses that give the signal for the commencement of dendritic proliferation, in our material, such collaterals do not become obvious until the apical dendrite is well established.

The mode of formation of the axonal process of the immature cortical pyramidal cell is still unclear, though all workers describe such a process before the cell enters the cortical plate. Is it simply a trailing process of the migrating young pyramidal neuron that breaks its continuity with the neuroepithelium and proceeds to grow in a new direction? Or is it a new outgrowth altogether from the trailing edge of the migrating or recently migrated cell? Recent work on the rat (Derer, 1974) suggests that, at least in the case of the earlier migrating neurons, there may be a stage of rather profuse outgrowth of processes from the neuron before its migration from the neuroepithelium. After entry into the intermediate zone, however, all except one (the new axon) disappear and an apical dendrite commences as a new outgrowth. A preliminary proliferation

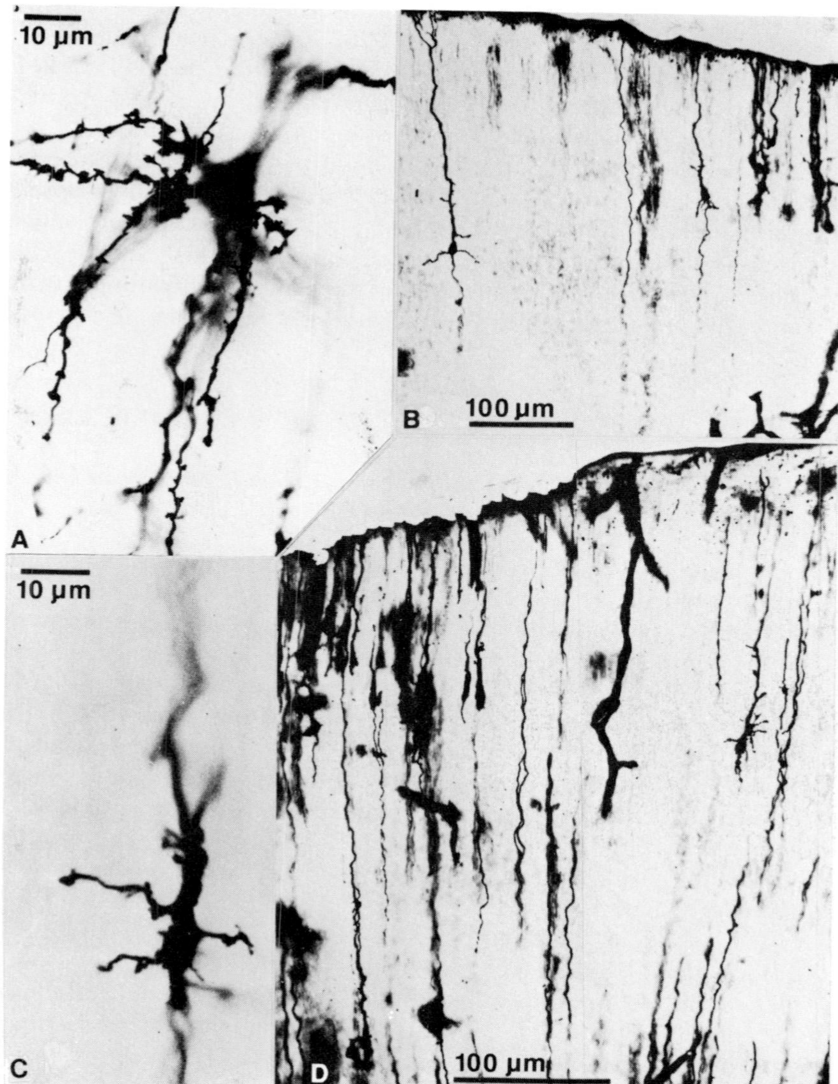

Figure 15.11 Golgi preparations of immature neurons in the presumptive somatic sensory cortex of the rat. A shows filopodia and other appendages on a pyramidal neuron at seven days of age, before the formation of substantial numbers of dendritic spines. B shows two immature Layer V pyramids at one day of age; the apical dendrite has reached Layer I and branched but the basal and other dendrites remain rudimentary; note the axon. C shows dendritic growth cones on a Layer V pyramid at one day of age. D shows several immature cortical plate neurons at the bipolar stage (left) and a slightly more advanced Layer V pyramid (right) at one day of age. From Wise et al. (1979).

of processes, most of which are lost during migration, has not been described by other workers, (e.g., Rakic, 1973; Stensaas, 1967) but it is clear from all studies that the outgrowth of the axon is determined long before a neuron reaches its final laminar destination in the cortex. Arrival at the target may be more delayed, as evidenced by our failure to retrogradely label presumptive Layer III callosal neurons in immature newborn animals. But the absence of retrogradely labeled callosal or corticospinal neurons in atypical laminae (or prelaminar regions) suggests that errors in the direction of early axonal outgrowth must be rare. Therefore, initial pathway selection by the axon may also be determined intrinsically and certainly long before the parent cell completes its migration.

DENDRITIC MATURATION OF OUTPUT (PYRAMIDAL) CELLS

> *. . . The other Messenger's called Hatta. I must have* two *you know–to come and go. One to come and one to go. . . .*

The continued survival of a neuron depends not only on the ability of its axon to establish synapses but also on the neuron receiving its normal complement of synaptic inputs (Cowan, 1978). Just as the establishment of connections by the axon has been thought to be in some way concerned with the initial outgrowth of the dendrites, so the subsequent growth and maturation of secondary and tertiary dendrites and dendritic spines has been thought to be inextricably bound up with the receipt of afferent synapses. Probably the most extensive studies on this issue have been carried out in the cerebellar cortex where the finer modeling of the dendritic trees of Purkinje cells and the orientations of the dendrites of stellate cells at different depths have been attributed to the successful formation of synapses by parallel fibers (Altman and Anderson, 1972; Rakic, 1972; Rakic and Sidman, 1973; Sotelo, 1975; Altman, 1976; Berry and Bradley, 1976). In the cerebral cortex, though there are now available several descriptive accounts of pyramidal and nonpyramidal cell maturation (e.g., Lund, Boothe, and Lund, 1977; Parnavelas, et al., 1978), the influence of afferent fibers upon this maturation has not been extensively investigated.

In his study, which was the first in modern times to thoroughly elucidate the light microscopic morphology of the growth cone and other growth protrusions of dendrites, Morest (1969a,b) was of the opinion that dendritic differentiation might be instigated by the arrival of afferent axons and guided by contact with axon branches. A similar viewpoint was adopted by Marin-Padilla (1970) in his study of the human fetal motor cortex. Marin-Padilla reported that extrinsic afferent fibers from the internal capsule and corpus callosum entered the cortex in the fifth month of gestation, before the development of any lamination in the cortical plate, and established contacts with Cajal-Retzius cells in Layer I. He was then able to describe dendritic maturation proceeding over the next few months from the deepest to the superficial layers as they differentiated from the cortical plate. The order of maturation he attributed to the prior arrival of thalamic afferents in Layer V and Layer IV and of association afferents in Layer III and Layer II. Later (Marin-Padilla, 1971) he reported comparable observations in fetal cats, in which extrinsic afferents appear in the cortex as early as 22 days of gestation, though no synapses appear in the region for another 22 days (Cragg, 1972).

In view of the evidence for the late ingrowth of thalamic and commissural axons reviewed in an earlier section, it seems unlikely that these sets of axons can have the fundamental determining effect upon dendrites that has been attributed to them, unless the effect is promoted when the immature neurons grow through the fibers as the latter lie in the intermediate zone. Also against the idea of thalamic or commissural afferents playing an initial role in dendrite formation in the cortex are the observations of DeLong, (1970) and of Seil, Kelly, and Leiman (1974). These workers noted that in cultures of reaggregating mouse cortical neurons or in explants of the same tissue, the gross features of lamination and of pyramidal cell morphology were preserved, though the cells or explants were removed from the brain before the ingrowth of substantial numbers of thalamic or callosal afferents could have occurred.

The origin of the extrinsic afferents observed by Marin-Padilla, is something of a mystery. It is possible that they are noradrenergic fibers from the locus coeruleus; these seem to be the earliest extrinsic afferents to enter the cortex, at least in the rat (Molliver and Kristt, 1975; Coyle and Molliver, 1977). However, whether they have the postulated inductive effect is unknown.

Despite the evidence against afferent fibers actually inducing dendrite formation after the neurons have arrived in the cortex, there is still reason to believe that extrinsic afferents may play some role in the finer modeling of the dendritic tree, at least of pyramidal cells. The maturation of pyramidal cells, starting with those in Layer VI and proceeding towards those in Layer II, was thoroughly documented by Ramón y Cajal (1906) and has been confirmed many times. In the last three years there have been a number of studies that, when taken in conjunction with the new connectional data, imply that the greatest growth of secondary and tertiary dendrites and of dendritic spines occurs in association with the accumulation of extrinsic afferent terminals in the cortex (Kristt, 1978; Juraska and Fifková, 1979a; Wise, et al., 1979a). It is still difficult, however to relate the one to the other in terms of cause and effect.

In our own studies (Figs. 15-11–15-14), we examined quantitatively, the state of the dendritic tree of Layer V and Layer III pyramidal cells

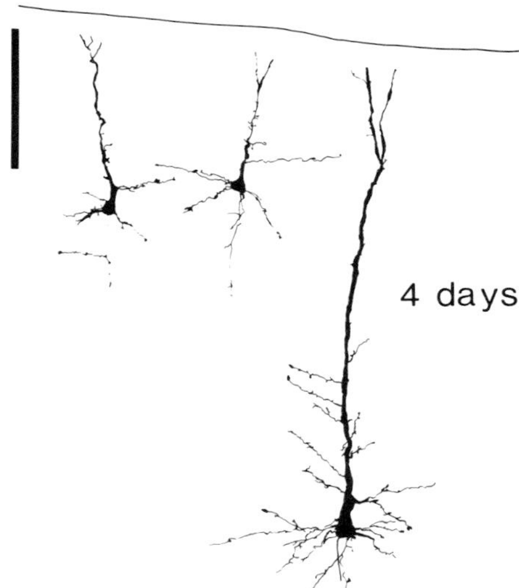

Figure 15.12 The more advanced state of dendritic maturation of a Layer V pyramid (right), in comparison with two in the presumptive Layer III (left) of a rat at four days of age. Camera lucida drawing from Wise et al. (1979). Bar represents 100 μm.

in the presumptive somatic sensory cortex of the neonatal rat over the period during which we knew thalamic and callosal axons were growing into the cortex (Wise, et al., 1979a). We observed that all pyramidal cells, irrespective of the depth of their somata in the cortex, undergo the following sequence of dendritic maturation: (1) the earliest outgrowth to occur is the apical dendrite, which ascends to Layer I and may even have a few rudimentary terminal branches there before the elaboration of basal dendrites; (2) basal dendrites next appear; (3) oblique branches of the apical dendrite then appear, and the basal and apical dendritic tufts undergo further branching; dendritic outgrowth and branching is preceded by the formation of terminal or preterminal growth cones; (4) filopodia and other irregular protrusions appear on the dendrites, followed somewhat later by the development of dendritic spines as dendritic length and branching complexity proceeds; (5) dendritic spine formation continues for many weeks.

The maturation of Layer V pyramids precedes that of Layer III pyramids by about three days: at birth or even before (Peters and Feldman, 1973) both have apical dendrites in Layer I but only Layer V pyramids have a commencing tuft of terminal branches there; rudimentary basal dendrites appear on Layer V pyramids at birth (PO) and on Layer III pyramids three days later (P3); oblique branches of the apical dendrites appear on Layer V pyramids on the third postnatal day (P2) and on Layer III pyramids on the fifth postnatal day (P4). By P7–P8, when cortical lamination has attained the adult form, the differences between the two

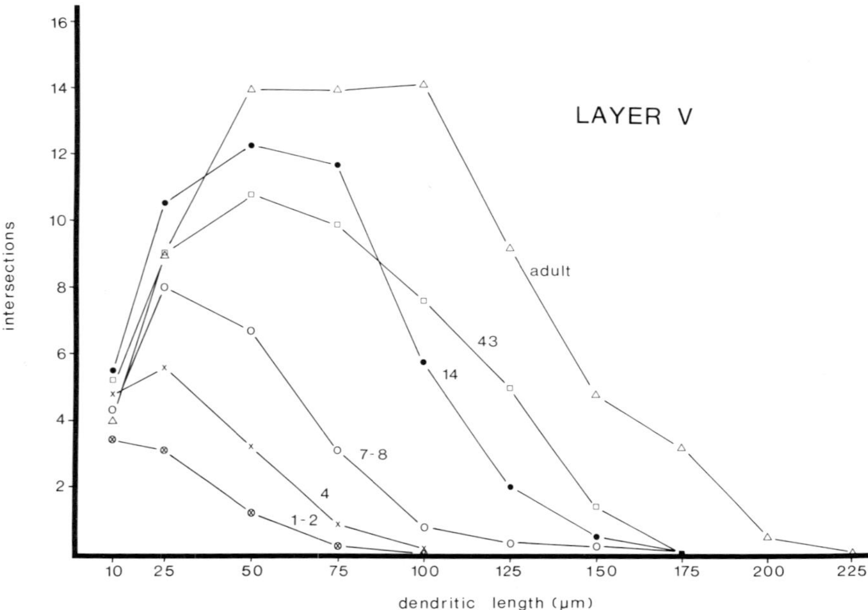

Figure 15.13 Graph showing the increasing length and complexity of branching of basal and oblique branches of the apical dendrite on Layer V pyramids at increasing postnatal ages (in days). Calculated by plotting number of dendrites intersecting rings of increasing diameter centered on the soma. From Wise et al. (1979).

are no longer clearly discernible, though each undergoes substantial further growth and dendritic branching and most of their spine formation occurs.

Our results, along with those of Kristt (1978), indicate that superficial and deep pyramidal cells mature in a temporal order that recapitulates their different times of gestation and migration from the neuroepithelium. (In the rat, the precursors of deep pyramids undergo their last mitosis between 16 and 18 days of gestation, those of superficial pyramids between 19 and 20 days of gestation; each reaches the cortical plate from 2 to 3 days later [Berry and Rogers, 1965].) We were unable to confirm the statement of Molliver and Van der Loos (1970) that deep pyramids merely appear to mature earlier because they elaborate a substantial basal dendritic tree before their apical dendrite reaches Layer I. Superficial pyramids, on the other hand, were said to grow their apical dendrites to Layer I first, before their basal dendrites develop. It seems to us that the interpretation of Molliver and Van der Loos, which is largely taken

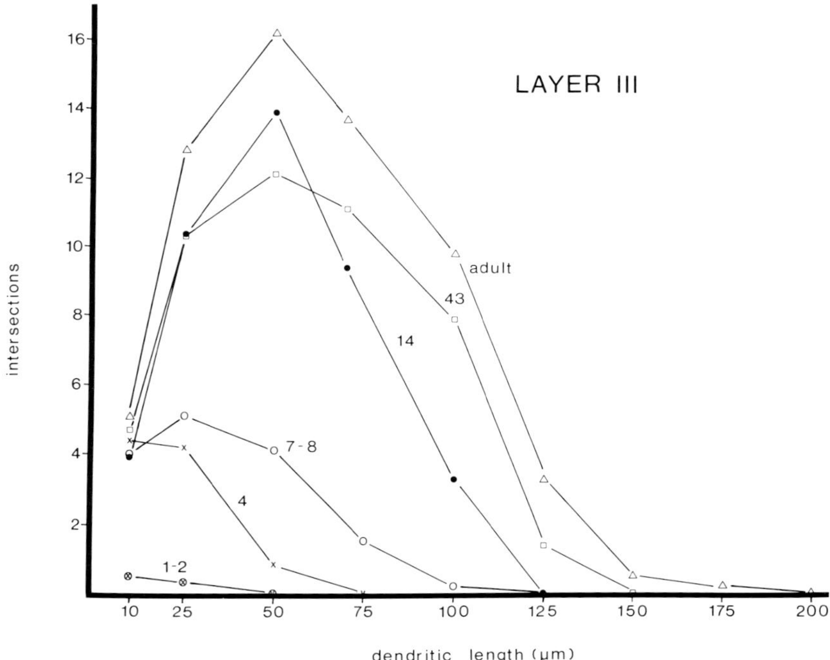

Figure 15.14 Figure comparable to Figure 15.13 but for Layer III pyramids. Note the lag in dendritic maturation in comparison with Layer V pyramids (Figure 15.13), a lag which coincides with the slightly later arrival of extrinsic afferent axons in Layer III. From Wise et al. (1979).

from an evaluation of the older literature, depends upon studies in which the total extents of the dendritic trees were not impregnated. Similarly, failure of complete Golgi impregnations almost certainly accounts for Eayrs and Goodhead's observations (1959), which suggest that pyramidal cell maturation in the rat parietal cortex occurs over a time scale approximately a week later than the more recent studies on sensorimotor and visual cortex indicate (Kristt, 1978: Juraska and Fifková, 1979a; Wise, et al., 1979a).

It is interesting to relate the pattern of pyramidal cell maturation described above to the pattern of ingrowth of afferent fibers. Although it was once fashionable to disclaim that cortical afferents terminated directly on pyramidal cells, there is now ample evidence in favor of this view (e.g., Jones and Powell, 1970; Peters and Feldman, 1978; White, 1979; Hendry and Jones, 1980). Hence, the relationship of these afferents to the maturing pyramidal cell becomes of some significance. The pyramidal cells of Layer V and Layer III are all relatively immature when thalamic and callosal fibers first enter the cortex of the rat (respectively, at 1 to 2 days and at 3 to 5 days of age), but they have elaborated their initial apical and basal dendritic ramifications. Hence, it is unlikely that the two sets of afferent fibers can play a primary inductive role, unless, as stated above, this has already occurred in the intermediate zone. On the other hand, the subsequent time scale of dendritic maturation and of afferent fiber development is such that the afferent fibers could play a major role in the finer modeling of the dendritic tree. The afferent terminal ramifications clearly start to concentrate at the junction of Layer V and Layer VI and in Layer III and Layer IV at just those times when the basal and the oblique branches of the apical dendrites are being elaborated in the same regions. Hence, by promoting the elongation of dendrites or their terminal branching (or both) and by maintaining certain populations of dendritic spines, the afferent fibers could determine some of the finer aspects of dendritic morphology, just as they seem to do in the cerebellum.

Our studies of the development of the efferent projections from the rat sensorimotor cortex show that most of the axons of Layer III and Layer V pyramids have reached their targets long before the parent neurons have developed a substantial dendritic tree. At birth in the rat, when the pyramidal cells have developed little more than an apical dendrite, efferent axons have crossed the corpus callosum, grown down to and across the corticospinal decussation, and have already entered some terminal sites such as the striatum, thalamus, and pontine nuclei (Wise and Jones, 1976; Wise, et al., 1979a; Donatelle, 1977; DeMeyer, 1967; Schreyer and Jones, 1980). Other terminal sites, such as the cortex, tectum, dorsal column nuclei, and spinal gray matter remain to be invaded (Wise and Jones, 1976: Schreyer and Jones, 1980). In the case of the striatum, thalamus, and pons, invasion of the target and possibly also synaptogenesis (for which there is, as yet, no evidence), have occurred before the maturation of the Layer V or Layer VI pyramids that give rise to these projections (Wise and Jones, 1977). In the case of the other sites, the invasion of the distal target seems to coincide with the peak of dendritic maturation. For example, corticospinal fibers commence invading the spinal central gray matter from postnatal Day 4 through Day 5, when the dendritic maturation of the parent Layer V pyramids is well established. One wonders, again, whether the growth into the gray matter may be triggered by synapse formation on the parent cell. Hence, if dendritic maturation has anything at all to do with connection formation, it is likely that it is a complicated process involving interactions between the cell, its target, its afferents, and their parent cells and involving possibilities for exchange of axoplasmically transported trophic factors and other signals on all sides.

REDISTRIBUTION OF OUTPUT CELLS DURING DEVELOPMENT

The other guests had taken advantage of the Queen's absence, and were resting in the shade: however, the moment they saw her, they hurried back to the game, the Queen merely remarking that a moment's delay would cost them their lives.

The vertical (laminar) distribution of pyramidal cells whose axons project to a particular site seems to be fixed by the time of their initial axonal outgrowth in the rat sensorimotor cortex. As mentioned above, retrograde-labeling experiments indicate that, from birth, callosally and subcortically projecting cell somata are found only in those layers (or the precursors of those layers) that contain them in the adult (Wise and Jones, 1976; Wise, et al., 1979a). For example, corticospinal cell somata are found from birth in Layer V only, never in Layer VI or in Layer II through Layer IV or their precursors. Therefore, there is no diffuseness in the vertical dimension of the cortex, again implying that the initial choice of a pathway by a particular axon is not a random event and that the direction of initial axonal ourgrowth may be predetermined.

By contrast with the vertical distribution, the horizontal distribution of output cells is far more diffuse in the neonatal rat than in the adult. Figures 15-15A and 15-15B show two "flattened" preparations of the cortex of adult rats in which callosal (Fig. 15-15A), and corticospinal and corticotrigeminal (Fig. 15-15B) neurons have been retrogradely labeled by appropriate injections of horseradish peroxidase. At this mature age, callosally projecting cells are distributed around and between the granular aggregations that form the heart of the body representation (Fig. 15-15A; Wise and Jones, 1976) whereas the subcortically projecting cells in Layer V lie mainly beneath the granular aggregations (Fig. 15-15B; Wise, 1975; Wise, Murray and Coulter, 1979b). In the newborn animal, however, we have shown that both callosally and subcortically projecting cells are distributed homogeneously across the presumptive sensorimotor cortex with no clusters and intervening gaps (Figs. 15-9, 15-16) as would have been predicted from the adult pattern of distribution (Wise and Jones, 1976; Wise, et al., 1979b). This has since been confirmed for callosal cells by Ivy, Akers, and Killackey (1979).

The change from the homogeneous neonatal to the mature dysjunctive pattern of distribution occurs rather suddenly. Retrograde labeling of callosally projecting cells remains homogeneous up until the eighth day of postnatal life, at which stage gaps commence appearing in the labeled pattern (Figs. 15-17, 15-18). This change in the labeled pattern is concomitant with the change in the pattern of distribution of the callosal axons in the opposite cortex from a diffuse to a columnar pattern, as described earlier. The change in the horizontal distribution pattern of subcortically

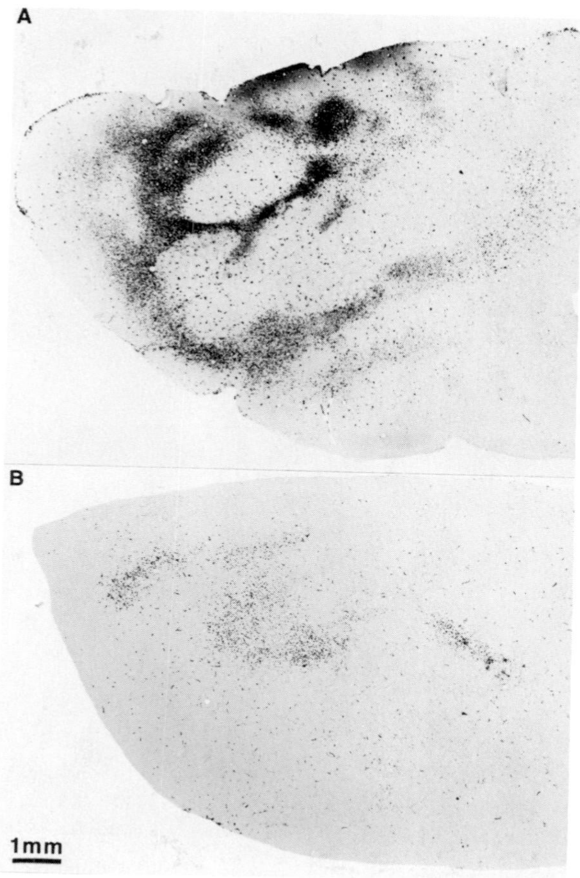

Figure 15.15 Sections from "flattened" preparations of adult rat brains showing dysjunctive pattern of distribution of retrogradely labeled callosal (A) and of corticospinal and corticobulbar (B) cells. In comparison with Figures 15.1 and 15.2, callosal cells are distributed to the agranular zones of the somatic sensory cortex and the corticospinal and corticobulbar cells beneath the intervening granular aggregations. A from unpublished material of the author; B kindly provided by Dr. S. P. Wise.

projecting cells seems to occur before this. On the fourth postnatal day the dysjunctive pattern becomes identifiable, that is, at about the time corticospinal axons in the upper cervical segments commence growing into the gray matter of the spinal cord (Schreyer and Jones, 1980) and at about the time the granular aggregates first become identifiable in the cortex (Wise and Jones, 1978).

Quite dramatic changes in the distribution patterns of callosally proj-

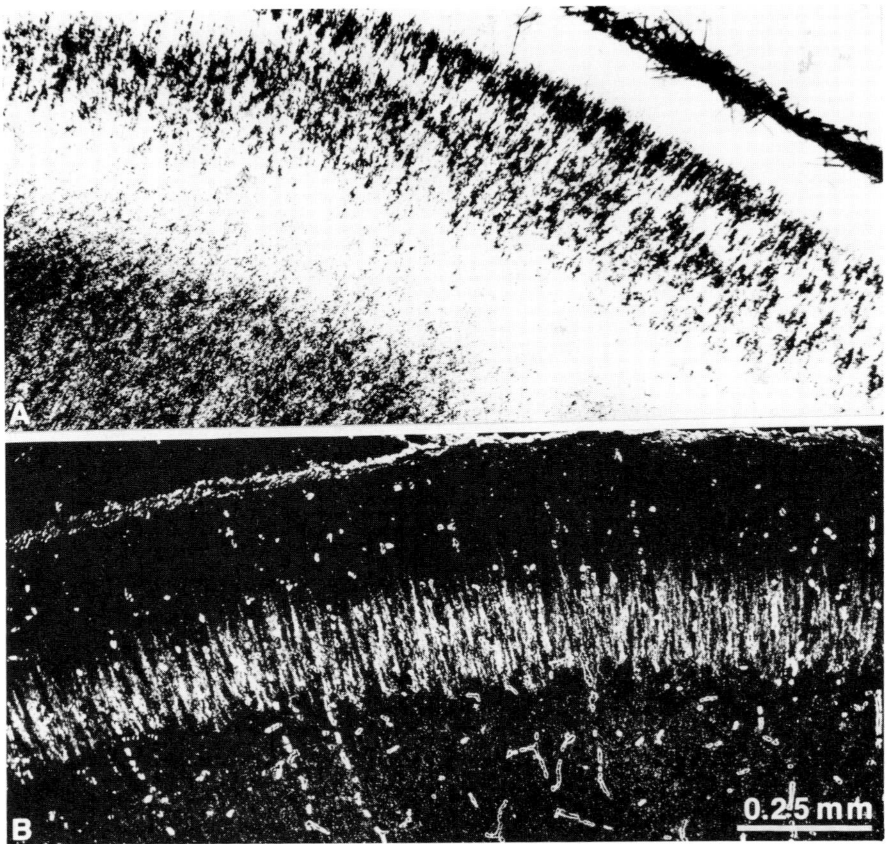

Figure 15.16 (A) homogeneous, bilaminar pattern of distribution of retrogradely labeled callosal cells in Layer V and the deeper half of the remains of the cortical plate in a two-day-old rat, before the entry of callosal axons, anterogradely labeled at bottom left. (B) homogeneous, unilaminar pattern of distribution of corticobulbar and corticospinal axons in Layer V of a two-day-old rat. (Cf. Figs. 15.9, 15.17, 15.18.)

ecting neurons are found in the somatic sensory and visual areas of the cortex of the neonatal cat. In our own work on newborn kittens (Fig. 15-19) we have found that callosal cells are evenly distributed throughout the representation of the forepaw, a region that is not normally connected in the adult (Jones and Powell, 1968). In the work of Innocenti and his colleagues (Innocenti, Fiore, and Caminiti, 1977) on the visual cortex of the kitten, they have found that callosally projecting cells extends into parts of the visual-field representation far beyond their normal adult distribution, which is close to the representation of the vertical meridian

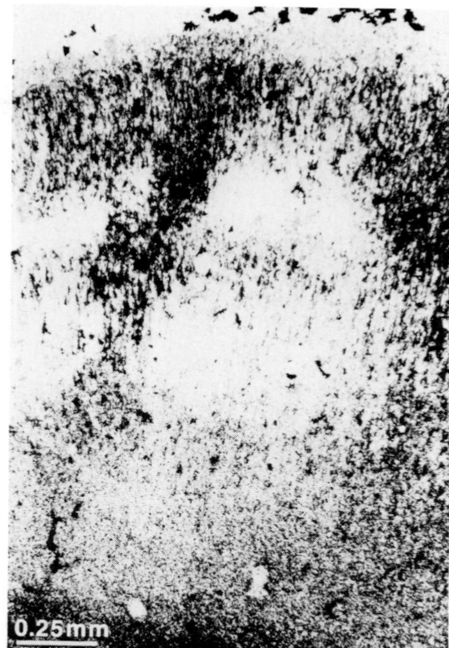

Figure 15.17 Vertical section through the somatic sensory cortex of a seven-day-old rat showing the commencing reorganization of retrogradely labeled callosal neurons to form columns with intervening gaps corresponding to the granular aggregations. (Cf. Figs. 15.9, 15.16, 15.18.) From material prepared by David L. Schreyer.

of the visual field. Their subsequent concentration around the representation of the vertical meridian would seem to be dependent upon peripheral cues of some kind, for the distribution of callosal axons (which can be expected to duplicate that of the callosally projecting cells) in the visual cortex can be modified by an artificially induced strabismus (Lund, Mitchell, and Henry, 1978) and in association with the inappropriately ordered retinogeniculate connections of the Siamese cat (Shatz, 1977).

In the cat, the disappearance of substantial numbers of callosal cells

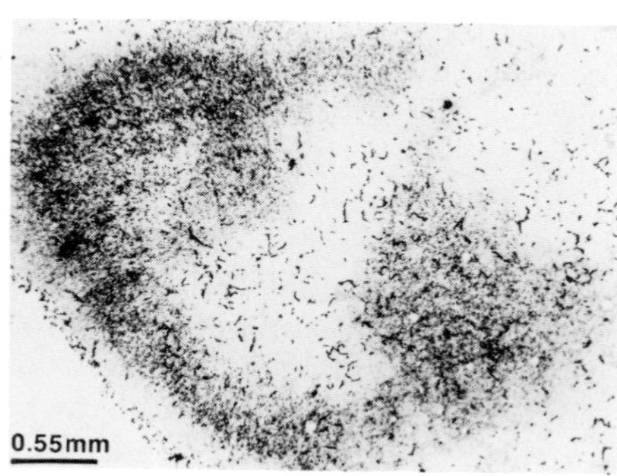

Figure 15.18 Section from a "flattened" preparation of a seven-day-old rat brain showing the first appearance of the dysjunctive pattern of retrogradely labeled callosal cells in a region corresponding to the lower left part of Figure 15.15A. From unpublished material of the author.

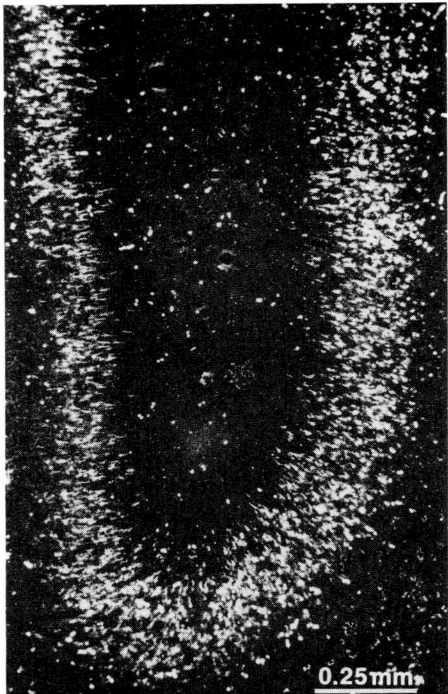

Figure 15.19 Frontal section through the coronal sulcus of a two-day-old kitten showing retrograde labeling of callosal cells in both the medial (left) and lateral (right) banks. In adult cats the medial bank (paw representation) contains no such cells. From unpublished material of the author.

from a large part of the sensory representation is perhaps less easy to comprehend than in the rat in which it is associated with a morphological reorganization into granular (thalamic-connected) and nongranular (callosal-connected) zones. It seems clear that the introduction of a substantial, "callosal-free" gap in the forepaw and peripheral visual field representations is not simply occasioned by expansive growth of the cortex nor by changes in the topography of the peripheral representation, for in both somatic sensory and visual areas, the relationship of the representation pattern to gross landmarks such as sulci and gyri seems fixed from the earliest times (Hubel and Wiesel, 1963; Rubel, 1971). It is possible that callosal cells in the forepaw and peripheral visual field representations simply die because they compete unsuccessfully for synaptic space with other axonal systems on the contralateral side. Death of substantial numbers of cells that are unable to form synaptic connections is a well-known phenomenon in developing systems, being first elucidated by Hamburger and Levi-Montalcini (1949) following Hamburger's (1934) original observation of motoneuronal hypoplasia after limb-bud ablation. It has since been confirmed many times (Hughes, 1961; Cowan and Wenger, 1967; Rogers and Cowan, 1973; Hamburger, 1975; Landmesser and

Pilar, 1974; Clarke, Rogers, and Cowan, 1976). An explanation of the reorganization of the callosal system in terms of death of cells that are unable to establish synaptic connections, or whose early connections are selectively eliminated, seems readily applicable to the rat somatic sensory cortex, in which there is a clear dissociation of callosal and thalamic inputs and a corresponding cytoarchitectonic specialization of the cortex; however, the reason for subcortically projecting cells being distributed mainly in association with the thalamic connected zones is not explicable in these terms. In the cat, on the other hand (and, for that matter, in other mammals such as monkeys; Jones, Coulter, and Wise, 1979), there seems to be little morphologically detectable difference between the thalamic connectivity of callosally and noncallosally connected zones and no obvious cytoarchitectonic variation. Moreover, so far as we can tell at the moment, thalamic fibers do not enter zones destined to lose callosal cells and callosal axon terminations before they enter zones that retain the callosal cells and axons. It is, therefore, necessary to suggest that there is something about thalamic or other unidentified connections that make them more successful at competing with callosal fibers for synaptic space in the distal limb and peripheral visual-field representations than in other parts of the representations.

Although some form of unsuccessful competition seems the best hypothesis to account for the disappearance of such large numbers of cells, it is still not clear that the cells actually die. Their apparent disappearance, particularly in the absence of any light microscopic evidence of massive cell death, conceivably could be due to no more than the loss of a callosally projecting collateral, the cell being sustained by the presence of further collaterals to other, presumably ipsilateral, sites. This potential explanation, which might also account for the selective loss of subcortically projecting cells, needs to be explored further, though our studies in monkeys do not lead us to favor extensive collateral projections of cells in one cortical layer to more than a single site (Jones and Wise, 1977).

DEVELOPMENT OF NONPYRAMIDAL CELLS

"Make a remark," said the red Queen: "it's ridiculous to leave all the conversation to the pudding."

The differentiation of the neurons whose axons do not leave the cortex and which form the several classes of nonpyramidal cells or cortical interneurons (Jones, 1975) has received very little attention. It is a much-quoted belief that these cells are formed later than the cortical projection (pyramidal) neurons, but there is really very little justification for it.

The idea derives mainly from autoradiographic studies (Angevine,

1965) of the genesis of the different layers of the cortex and seems to be based upon an analogy with the retina, cerebellar cortex, and certain other sites. It rests on the observed fact that in most regions of the brain, large neurons are formed before small. The theory becomes invalid, in the cortex when it assumes that all large and all projection neurons are found only in Layer V and Layer VI and that all interneurons are small and found in the later formed Layer II through Layer IV. Contrary to this: pyramidal cells in Layer II and Layer III can project axons to quite long distances, including to the opposite cortex; one class of interneuron, the basket cell, is often larger than the largest pyramidal cell; small and large interneurons are found in all cortical layers. Thus, while it may be true that large cortical neurons are generated before small, the idea that this reflects a difference in the temporal appearance of projection neurons and interneurons is not soundly based.

Just as it is often stated that cortical interneurons are generated last, so it is implied that their dendritic and axonal maturation occurs later than that of the pyramidal cells. While this may be true for the smaller classes of interneurons in many sites (Morest, 1969b), it need not be universally true. In order to prove or disprove the idea of a temporal segregation of differentiation of neuron types in the cortex, it is necessary to examine neuronal development by means of techniques, such as that of Golgi, which permit the identification of neurons in terms of their total morphology rather than simply in terms of somal size and laminar position. Regrettably, there have been few studies as comprehensive as those that have been carried out on the earliest stages of pyramidal cell development. In one of the few studies that have been made, Marin-Padilla (1970) considers that, in the human, a particular class of interneuron, the basket cell, matures *pari passu* with the pyramidal cells of the layer in which it lies because of the intimate synaptic connections between the two. Other interneuronal classes are said to develop later. A seemingly opposing view is taken by Morest (1969b), who stresses the late development of the Golgi II cell. It is unlikely, however, that he means to apply this to all classes of interneuron.

A number of workers have noted the presence of obvious nonpyramidal cells at very early phases in neocortical development, particularly in the marginal and intermediate zones on either side of the cortical plate. In the rat, for example, Rickman, Chronwall, and Wolff (1977) have described horizontally disposed neurons in the marginal zone two days before the first appearance of a cortical plate. With the development of the cortical plate, these neurons become separated into two strata: one in Layer I superficial to the cortical plate, and the second in the intermediate zone deep to it. Autoradiographic experiments suggest that neurons are added to each of these strata throughout cortical development, in a manner that is said to be independent of the now classical inside-out sequence of laminar development in the cortical plate. During this time, the neurons of the two strata develop into Cajal-Retzius cells in Layer I and into spiny, multipolar neurons in both strata. Similar observations have been made in fetal cats and humans by Marin-Padilla (1970, 1971) who describes axonal connections joining the two layers that later disappear as development of the cortical plate proceeds.

There has been a tendency to dismiss the Cajal-Retzius cells of Layer I as also of only transitory significance and certainly they are difficult to demonstrate in the adult brain. But there is no reason to believe that the large number of other nonpyramidal cells described by Rickman and colleagues also disappear. It may be necessary, therefore, to revise our ideas about the development of cortical interneurons, although it seems unlikely that all differentiate in Layer I and the intermediate zone. Unfortunately, the relative proportions and types of interneurons developing in and outside the cortical plate are at present unknown.

SYNAPSE FORMATION IN THE CEREBRAL CORTEX

> *"Well, now that we* have *seen each other," said the Unicorn, "if you'll believe in me, I'll believe in you. Is that a bargain?"*

The formation of synapses in the neocortex of many animals seems to be largely a postnatal event. Even in animals such as the cat, in which the cortical lamination is far more advanced at birth than in the rat, 98 percent of the synapses in the visual cortex appear more than eight days after birth (Cragg, 1972, 1975). Though synapses have been reported in several species, including man, days or months before birth, the numbers are extremely small and there is reason to believe that some of the contacts reported may be desmosomal, rather than classical synapses (Cragg, 1972; Molliver, Kostovic, and Van der Loos, 1973; Peters and Feldman, 1973).

At birth in rats, cats, rabbits, and dogs and at comparable prenatal stages of cortical lamination in the human, synapses are almost exclusively confined to the marginal layer and to the deep, presumptive Layer VI lying below the cortical plate. Virtually none are found in the plate itself (Cragg, 1972, 1975; Molliver and van der Loos, 1970; Molliver et al., 1973; Peters and Feldman, 1973; König, Roch, and Marty, 1975; Kristt and Molliver, 1976; Rickman et al., 1977; Vrensen, De Groot, and Nunes-Cardozo, 1977; Kristt, 1978; Juraska and Fifková, 1979b). The number of synapses seems to increase first in the deeper layers as they develop from the cortical plate and there may be a later burst of increase in the more superficial layers (Cragg, 1972). In the cat visual cortex the steepest increase in synapse density occurs in the four weeks succeeding eye opening, which has been regarded as significant in terms of the critical period during which visual experience can modify cortical organization (Cragg, 1972). In rats the steepest rise in synapse density occurs in the motor cortex between 4 days and 14 days postnatal (Armsbrong-James and Johnson, 1970) and in the parietal cortex after the 12th day (Aghajanian and Bloom, 1967). Adult levels are not established until much later.

The nature of the earliest synapses in the cortex is still uncertain, especially now that it can be fairly satisfactorily ruled out that they arise from the major classes of extrinsic afferents. One possibility is that they are formed by noradrenergic fibers, for synapses can be labeled with 5-hydroxydopamine in the rat cortex at birth (Molliver and Kristt, 1975; Coyle and Molliver, 1977). Another possibility is that they are intrinsic and derived from locally ramifying axons. For example, in Layer I of 19-day-old and 21-day-old rat fetuses, Peters and Feldman (1973) observed many axonal profiles but could not relate these to any substantial number of axons ascending through the cortical plate. It is likely that a substantial proportion of the early synapses in the deep synaptic stratum also arise from intrinsic axons, for Lund and Mustari (1977) report that only a very small percentage of the synapses there degenerate in the visual cortex of newborn rats following lesions of the lateral geniculate nucleus. One source of these remaining synapses is suggested by our findings in fetal cats (Fig. 15-20) in which labeled axons can be seen as two strata superficial and deep to the cortical plate, as they extend tangentially outwards from an injection of tritiated amino acids in an adjacent part of the cortex. These association fibers can be detected at times before the ingrowth of thalamic or callosal fibers into the cortex (Wise et al., 1977). The labeled axons might well be the early forming collaterals of pyramidal cells, the axons of basket cells or, in Layer I, the axons of the transient Cajal-Retzius cells.

The early increase in synaptic density in the deeper part of the maturing cortex is undoubtedly associated with the ingrowth and branching of extrinsic afferent fibers. These appear to commence forming synapses as soon as they enter (Lund and Mustari, 1977). Most of the earliest synapses occur on dendrites, and axodendritic and axospinous contacts seem to account for the greater part of the subsequent increase in synapse density (Cragg, 1975).

In an attempt to quantify the synaptic maturation of pyramidal cells in the rat somatic sensory cortex, we counted the numbers of dendritic spines on basal and on oblique branches of the apical dendrites of Layer III and Layer V pyramidal cells during the time at which afferent innervation was becoming established (Wise et al., 1979b) (Fig. 15-21). It seems clear that the earliest synapses do not occur on dendritic spines, for few or no spines are present until several days after the establishment of afferent innervation and the appearance of peripherally elicited evoked

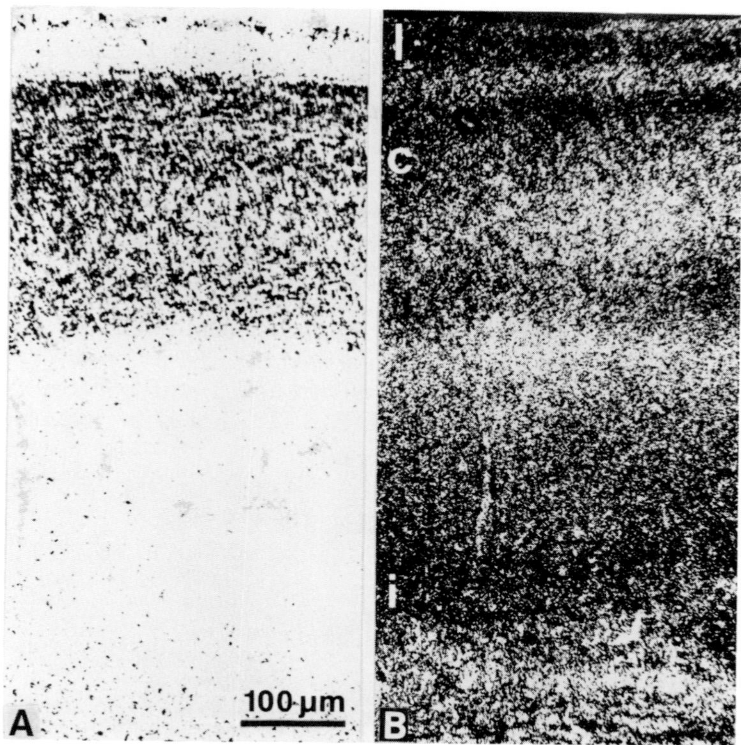

Figure 15.20 Anterograde labeling of axons leaving an injection site in the cerebral cortex out of the micrograph to the right. From a cat fetus at 55 days of gestation. Upper two strata show intracortical axons spreading out superficial and deep to remains of cortical plate (c); middle stratum in intermediate zone (i) is composed of corticocortical axons that will enter a distant cortical area; deepest stratum is composed of commissural, corticothalamic, and other efferent axons. From Wise et al. (1977).

potentials and unit activity in the somatic cortex (Verley and Axelrad, 1975; Armstrong-James, 1975; Verley, 1977). The late development of spines in the cortex was originally recognized by Ramón y Cajal (1906). The earliest synapses seem to occur predominantly on the filopodial and other protrusions that are far more common on the pyramidal cell dendrites during the first seven to ten days of postnatal development than are true spines (Fig. 15-11A). The filopodia can be recognized electron microscopically (Fig. 15-22) by their relatively dense matrix and lack of microtubules. In these features, they resemble dendritic spines though they do not have a narrow stalk and no spine apparatus is present. It is, therefore, uncertain whether the filopodia are necessarily spine precursors, as some have suggested.

After the first week, filopodia have largely disappeared, though spine density is still relatively low (Fig. 15-21) and only reaches something approaching adult levels by the end of the fourth week. If we may correlate the increase in spine numbers with an increase in synapse formation, then it is clear that synaptogenesis continues on pyramidal cells for a rather protracted period after the time when the extrinsic afferents set up their definitive laminar and zonal patterns of distribution. The time scale correlates quite well with that obtained from direct counts of synapses (Armstrong-James and Johnson, 1970) or of synaptic membrane thickenings (Aghajanian and Bloom, 1967), though it is appreciated that a large proportion of the new synapses being formed may be on nonpyramidal cells as well. The small decline in spine density on the pyramidal cells between the end of the first month and maturity correlates with the observations of Feldman and Dowd (1975) and perhaps reflects a

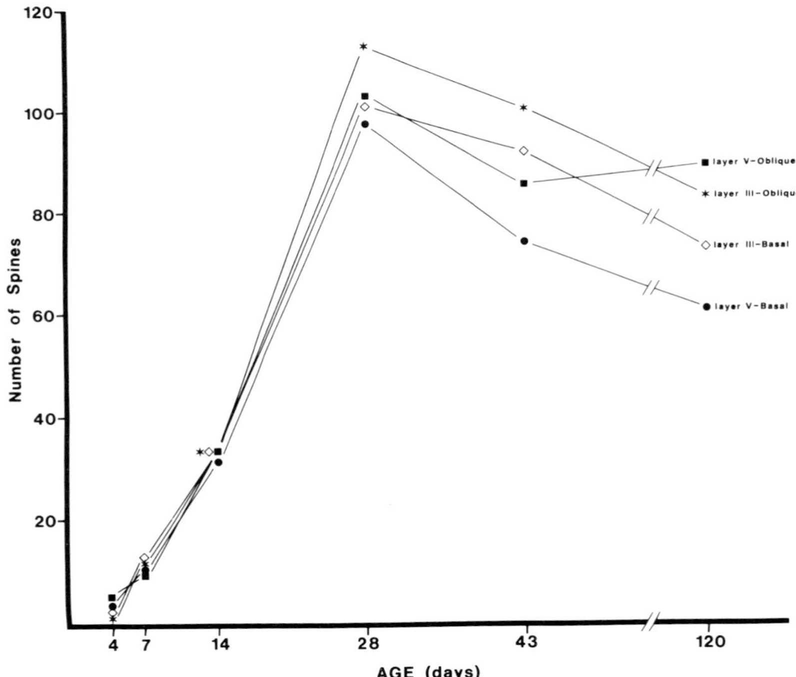

Figure 15.21 Graph showing number of dendritic spines per 50 μm length of basal or oblique branches of apical dendrites of Layer III and Layer V pyramidal cells in the somatic sensory cortex of a series of rats. Note that a substantial increase occurs only after the invasion and establishment of the definitive zonal and laminar distribution of extrinsic afferents. Note also the decline in numbers with advancing age. From data reported in Wise et al. (1979).

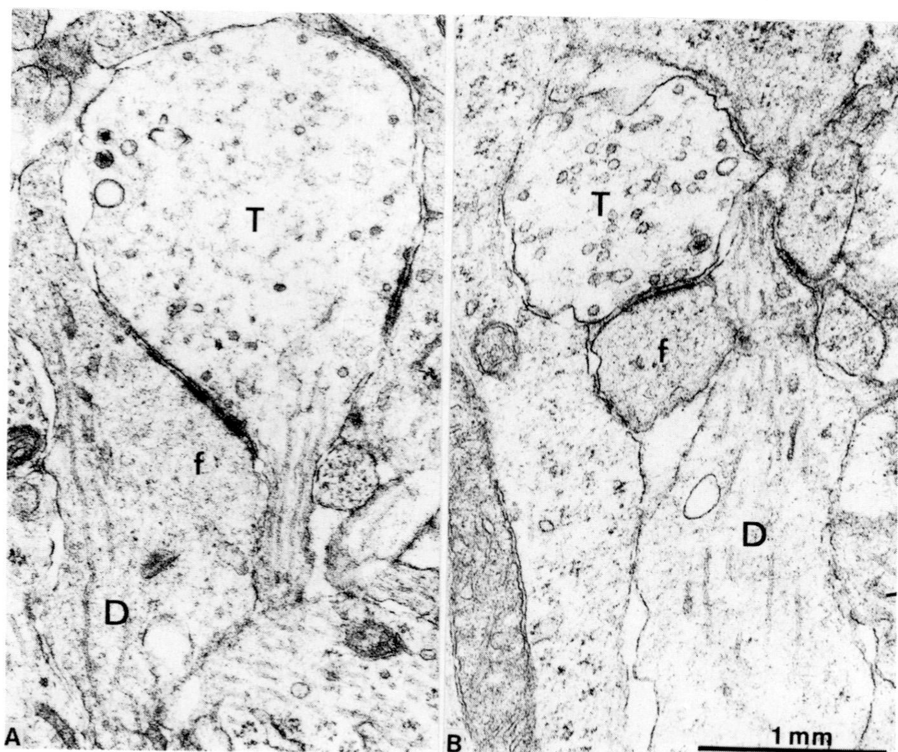

Figure 15.22 Electron micrographs from Layer IV of the somatic sensory cortex of a six-day-old rat showing that before spine formation, the earlier synapses are found on irregular dendritic protrusions that lack microtubules. From unpublished material of the author.

decline in synaptic density similar to that occurring after 36 days of life in cats (Cragg, 1972, 1975).

We are aware that spine numbers may not give a perfectly accurate estimate of synapse numbers, for some may not receive synaptic contacts. Transient spines are often seen in immature animals on populations of neurons, such as cortical nonpyramidal cells, that do not normally possess them in the adult (Scheibel, Davies, and Scheibel, 1973; Jones, 1975; Lund, Boothe, and Lund, 1977). Similarly, Purkinje cells that fail to receive their normal complement of parallel fiber synapses may acquire excessive numbers of spines (Rakic and Sidman, 1973; Sotelo, 1975). Hence, the decline in spine number on the mature pyramidal cells might be accounted for by their failure to receive synapses. The general increase in spine numbers must, however, reflect a valid increase in synaptic density occurring after the arrival of extrinsic afferents, for no spine has been found to be devoid of a synapse in the adult cortex (Jones and Powell, 1969; Peters and Kaiserman-Abramof, 1969). But the reservations expressed above do not permit us to conclude automatically that the extrinsic afferents are concerned in the induction of spine protrusion.

SUMMARY

1. In describing the results of experiments on rats and cats carried out over the last five years in the author's laboratory, an attempt is made to review the current state of knowledge regarding the development of input and output connections in the mammalian neocortex.

2. The two principal extrinsic afferent fiber systems, the thalamocortical and the callosal, grow very early towards their cortical targets but accumulate over a protracted period in the intermediate zone beneath the cortical plate before receiving a signal to resume growth and enter the cortex. In the early stages of accumulation, cells upon which the axons are ultimately destined to terminate may grow through the axons and into the cortex without permanent contacts being made.

3. A similar waiting period followed by a second growth spurt into the target is observed by growing fibers of the corticospinal tract.

4. The signal to make a second growth spurt into the cortex or spinal gray matter is a differential one, for thalamic fibers grow into the cortex before commissural fibers. The nature of the signal is unclear; it may be related to synapse formation on the parent cells of the growing fibers.

5. In both cortical afferent systems, the initial ingrowth is diffuse; zonal, laminar, and columnar patterns of distribution develop only later. Experiments designed to test for competitive interactions between thalamic and callosal fibers during synapse formation in the cortex have, to date, proven negative.

6. In the early phases of axon outgrowth, callosal and corticospinal neurons, though found only in cortical layers comparable to the adult, are homogeneously distributed in the horizontal dimension. The change from a diffuse to the columnar, dysjunctive pattern characteristic of the adult is rapid. It is uncertain whether it is due to massive cell death or to the breaking of an inappropriate collateral axon owing to failure to compete successfully for available synaptic sites.

7. The ability to grow a long axon is independent of the growth of dendrites, for callosal and corticospinal cells have sent axons long distances towards their targets before the growth of more than an apical dendrite. Synapse formation by the axon, however, does not appear to induce dendritic outgrowth by the parent cell.

8. Similarly, dendritic differentiation does not appear to be initiated by extrinsic afferent fibers, though subsequent dendritic maturation pro-

ceeds in conjunction with the ingrowth of afferents. There appears to be substantial synapse formation long after the afferents have set up their adult patterns of distribution.

ACKNOWLEDGMENTS

The work reported here could not have been completed without the continuing efforts of a series of associates, all of whom have been at some time students of Viktor Hamburger: James W. Fleshman, Jr., Stewart H.C. Hendry, David L. Schreyer, Karen L. Valentino, and Steven P. Wise. I am also grateful for the continuing technical support of Bertha McClure and Ronald L. Steiner. Overall, the work was supported by grant numbers NS10526, NS15070, and T32NS0757 from the National Institutes of Health, United States Public Health Service.

REFERENCES

Aghajanian, G. K., Bloom F. E. 1967. The formation of synaptic junctions in developing rat brain: A quantitative electron microscopic study. *Brain Res.* 6:716–27

Akers, R. M., Killackey, H. P. 1978. Organization of corticocortical connections in the parietal cortex of the rat. *J. Comp. Neur.* 181:513–38

Altman, J. 1975. Experimental reorganization of the cerebellar cortex. VII. Effects of late x-irradiation schedules that interfere with cell acquisition after stellate cells are formed. *J. Comp. Neurol.* 165: 65–75

Altman, J., Anderson, W. J. 1972. Experimental reorganization of the cerebellar cortex. I. Morphological effects of elimination of a microneuron with prolonged x-irradiation at birth. *J. Comp. Neurol.* 146:355–406

Angevine, J. B., Jr. 1965. Time of neuron origin in the hippocampal region. An autoradiographic study in the mouse. *Exp. Neurol. Suppl.* 2:1–70

Armstrong-James, M. 1975. The functional status and columnar organization of single cells responding to cutaneous stimulation in neonatal rat somatosensory cortex S1. *J. Physiol.* (London) 246:501–38

Armstrong-James, J., Johnson, R. 1970. Quantitative studies of postnatal changes in synapses in rat superficial motor cerebral cortex. *Z. Zellforsch Mikrosk. Anat.* 40: 559–68

Barron, D. H. 1943. The early development of the motor cells and columns in the spinal cord of the sheep. *J. Comp. Neurol.* 18:1–26

Barron, D. H. 1946. Observations on the early differentiation of the motor neuroblasts in the spinal cord of the chick. *J. Comp. Neurol.* 85:149–69

Belford, G. R., Killackey, H. P. 1979. Vibrissae representation in subcortical trigeminal centers of the neonatal rat. *J. Comp. Neurol.* 183:305–64

Berry, M., Bradley, P. 1976. The growth of the dendritic trees of Purkinje cells in irradiated agranular cerebellar cortex. *Brain Res.* 116: 361–87

Berry, M., Rogers, H. W. 1965. The migration of neuroblasts in developing cerebral cortex. *J. Anat.* 99:691–709

Clarke, P. G. H., Rogers, L. A., Cowan, W.M. 1976. The time of origin and the pattern of survival of neurons in the isthmo-optic nucleus of the chick. *J. Comp. Neurol.* 167:125–41

Cowan, W. M. 1978. Aspects of neural development. In *International Review of Physiology*, vol. 17, Neurophysiology III, ed. R. Porter, pp. 149–91. Baltimore, Md.: Univ. Park Press

Coyle, J. T., Molliver, M. E. 1977. Major innervation of newborn rat cortex by monoaminergic neurons. *Science* 196:444–46

Cragg, B. G. 1972. The development of synapses in cat visual cortex. *Vision Res.* 11:377–85

Cragg, B. G. 1975. The development of synapses in the visual system of the cat. *J. Comp. Neurol.* 160:147–66

Cusick, C., Lund, R.[D]. 1977. Plasticity and specificity of connections of the corpus callosum in the albino rat following neonatal lesions. *Neurosci. Abstr.* 3:424

DeLong, R. G. 1970. Histogenesis of fetal mouse isocortex and hippocampus in reaggregating cell culture. *Dev. Biol.*, 22:563–83

DeMeyer, W. 1967. Ontogenesis of the rat corticospinal tract. *Arch. Neurol.* Chicago 16:203–11

Derer, P. 1974. Histogenèse du néocortex du rat albinos durant la période foetale et néonatale. *J. Hirnforsch.* 15:49–74

Donatelle, J. M. 1977. Growth of the corticospinal tract and the development of placing reactions in the postnatal rat. *J. Comp. Neurol.* 175:207–32

Eayers, J. T., Goodhead, B. 1959. Postnatal development of the cerebral cortex in the rat. *J. Anat.* 93:385–402

Feldman, J. L., Dowd, C. 1975. Loss of dendritic spines in aging cerebral cortex. *Anat. Embryol.* 148:279–301

Gottlieb, D. I., Cowan, W. M. 1972. Evidence for a temporal factor in the occupation of available synaptic sites during the development of the dentate gyrus. *Brain Res.* 41:452–56

Hamburger, V. 1928. Die Entwicklung experimentell erzeugter nervloser und schwach innervierter Extremitäten von Anuren. *Arch. Entwicklungs Mech. Org.* 114: 272–363

Hamburger, V. 1934. The effects of wing bud extirpation on the development of the central nervous system in chick embroys. *J. Exp. Zool.* 68:449–94

Hamburger, V. 1975. Cell death in the development of the lateral motor column of the chick embryo. *J. Comp. Neurol.* 160:535–46

Hamburger, V., Keefe, E. L. 1944. The effects of peripheral factors on the proliferation and differentiation in the spinal cord of the chick embryo. *J. Exp. Zool.* 96: 223–42

Hamburger, V., Levi-Montalcini, R. 1949. Proliferation, differentiation and degeneration in the spinal ganglia of the chick embryo under normal and experimental conditions. *J. Exp. Zool.*, 111: 457–502

Hendry, S. H. C., Jones, E. G. 1980. Electron microscopic demonstration of thalamic axon terminations on identified commissural neurons in monkey somatic sensory cortex. *Brain Res.* 196:253–7

Hubel, D. H., Wiesel, T. N. 1963. Receptive fields of cells in striate cortex of very young, visually inexperienced kittens. *J. Neurophysiol.* 26:994–1002

Hubel, D. H., Wiesel, T. N., LeVay, S. 1977. Plasticity of ocular dominance columns in monkey striate cortex. *Philos. Trans. R. Soc. London Ser. B.* 278:371–409

Hughes, A. F. 1961. Cell degeneration in the larval ventral horn of *Xenopus laevis* (Daudin). *J. Embryol. Exp. Morphol.* 9:269–84

Innocenti, G. M., Fiore, L., Caminiti, R. 1977. Exuberant projection into the corpus callosum from the visual cortex of newborn cats. *Neurosci. Lett.* 4:237–42

Ivy, G. O., Akers, R. M., Killackey, H.P. 1979. Differential distribution of callosal projection neurons in the neonatal and adult rat. *Brain Res.* 173:532–37

Jacobson, M. 1978. *Developmental Neurobiology,* 2nd ed. New York: Plenum 562 pp.

Jones, E. G. 1975. Varieties and distribution of non-pyramidal cells in the somatic sensory cortex of the squirrel monkey. *J. Comp. Neurol.* 160:205–68

Jones, E. G. 1981. Anatomy of cerebral cortex: Columnar input-output relations. In *The Organization of the Cerebral Cortex,* ed. F.O. Schmitt, F.G. Worden, G. Adelman, S.G. Dennis, pp. 199–235. Cambridge, Mass.: MIT

Jones, E. G., Coulter, J. D., Wise, S. P. 1979. Commissural columns in the sensory-motor cortex of monkeys. *J. Comp. Neurol.* 188:113–36

Jones, E. G., Powell, T.P.S. 1968. The commissural connexions of the somatic sensory cortex in the cat. *J. Anat.* 103:433–55

Jones, E. G., Powell, T. P. S. 1969. Morphological variations in the dendritic spines of the neocortex. *J. Cell Sci.* 5:509–29

Jones, E. G., Powell, T. P. S. 1970. An electron microscopic study of the laminar pattern and mode of termination of the afferent fibre pathways to the somatic sensory cortex. *Philos. Trans. R. Soc. London Ser. B.* 257:45–62

Jones, E. G., Wise, S. P. 1977. Size, laminar and columnar distribution of efferent cells in the sensory-motor cortex of primates. *J. Comp. Neurol.* 175:391–438

Juraska, J. M., Fifková, E. 1979a. A Golgi study of the early postnatal development of the visual cortex of the hooded rat. *J. Comp. Neurol.* 183:247–56

Juraska, J .M., Fifková, E. 1979b. An electron microscope study of the early postnatal development of the visual cortex of the hooded rat. *J. Comp. Neurol.* 183:257–68

Killackey, H. P., Belford, G. R. 1979. The formation of afferent patterns in the somatosensory cortex of the neonatal rat. *J. Comp. Neurol.* 183:285–304

Knyihar, E., Csilik, B., Rakic, P. 1978. Transient synapses in the embryonic primate spinal cord. *Science* 202:1206–8

König, N., Roch, G., Marty, R. 1975. The onset of synaptogenesis in rat temporal cortex. *Anat. Embryol.* 148:73–87

Kristt, D. A. 1978. Neuronal differentiation in somatosensory cortex of the rat. I. Relationship to synaptogenesis in the first postnatal week. *Brain Res.* 150:467–86

Kristt, D. A., Molliver, M. E. 1976. Synapses in newborn rat cerebral cortex: A quantitative ultrastructural study. *Brain Res.* 108:180–86

Landmesser, L., Pilar, G. 1974. Synaptic transmission and cell death during normal ganglionic development. *J. Physiol.* (London) 241:737–49

Lund, J. S., Boothe, R. G., Lund, R. D. 1977. Development of neurons in the visual cortex (area 17) of the monkey *(Macaca nemestrina):* A Golgi study from fetal day 127 to postnatal maturity. J. Comp. Neurol. 176:149–81

Lund, R. D., Bunt, A. H. 1976. Prenatal development of central optic pathways in albino rats. *J. Comp. Neurol.* 165:247–64

Lund, R. D., Mitchell, D. E., Henry, G. H. 1978. Squint-induced modification of callosal connections in cats. *Brain Res.* 144:169–72

Lund, R. D., Mustari, M. J. 1977. Development of the geniculocortical pathway in rats. *J. Comp. Neurol.* 173:289–305

Maeda, T., Tohyama, M., Shimizu, N. 1974. Modification of postnatal development of neocortex in rat brain with experimental deprivation of locus coeruleus. *Brain Res.* 70:575–20

Marin-Padilla, M. 1970. Prenatal and early postnatal ontogenesis of the human motor cortex: A Golgi study. I. The sequential development of the cortical layers. *Brain Res.* 23:167–83

Marin-Padilla, M. 1971. Early prenatal ontogenesis of the cerebral cortex (neocortex) of the cat *(Felis domestica).* A Golgi study. *Z. Anat. Entwicklungs Gesch.* 134: 117–45

Molliver, M. E., Kostovic, I., Van der Loos, H. 1973. The development of synapses in cerebral cortex of the human fetus. *Brain Res.* 50: 403–7

Molliver, M. E., Kristt, D. A. 1975. The fine structural demonstration of monoaminergic synapses in immature rat neocortex. *Neurosci. Lett.* 1:305–10

Molliver, M. E., Van der Loos, H. 1970. The ontogenesis of cortical circuitry: The spatial distribution of synapses in somesthetic cortex of newborn dog. *Ergeb. Anat. Entwicklungs Gesch.* 42:1–53

Morest, D. 1969a. The differentiation of cerebral dendrites: A study of the post-migratory neuroblast in the medial nucleus of the trapezoid body. *Z. Anat. Entwicklungs Gesch.* 128:271–89

Morest, D. 1969b. The growth of dendrites in the mammalian brain. *Z. Anat. Entwicklungs Gesch.* 128:290–317

O'Leary, D. D. M., Fricke, R. A., Stanfield, B. B., Cowan, W. M. 1979. Changes in the associational afferents to the dentate gyrus in the absence of its commissural input. *Anat. Embryol.* 156:283–300

Parnavelas, J. G., Bradford, R., Mounty, E. J., Lieberman, A. R. 1978. The development of non-pyramidal neurons in the visual cortex of the rat. *Anat. Embryol.* 155:1–14

Peters, A., Feldman, M. 1973. The cortical plate and molecular layer of the late rat fetus. *Z. Anat. Entwicklungs Gesch.* 141:3–38

Peters, A., Feldman, M. 1977. The projection of the lateral geniculate nucleus to area 17 of the rat cerebral cortex. IV. Terminations upon spiny dendrites. *J. Neurocytol.* 6:669–89

Peters, A., Kaiserman-Abramof, I. R. 1969. The small pyramidal neuron of the rat cerebral cortex. The synapses upon dendritic spines. *Z. Zellforsch. mikrosk. Anat.* 100:487–506

Rakic, P. 1972a. Mode of cell migration to the superficial layers of fetal monkey cortex. *J. Comp. Neurol.* 145:61–84

Rakic, P. 1972b. Extrinsic cytological determinants of basket and stellate cell dendritic pattern in the cerebellar molecular layer. *J. Comp. Neurol.* 146:335–54

Rakic, P. 1974. Neurons in rhesus monkey visual cortex: Systematic relation between time of origin and eventual disposition. *Science* 183:425–26

Rakic, P. 1976. Prenatal genesis of connections subserving ocular dominance in the rhesus monkey. *Nature* (London) 261:467–71

Rakic, P. 1977. Prenatal development of the visual system in rhesus monkey. *Philos. Trans. R. Soc. London Ser. B.* 278:245–60

Rakic, P., Sidman, R. L. 1973. Organization of cerebellar cortex secondary to deficit of granule cells in weaver mutant mice. *J. Comp. Neurol.* 152:133–62

Ramón y Cajal, S. 1906. *Studien über die Hirnrinde des Menschen. 5. Vergleichende Strukturbeschreiburg und Histogenesis der Hirnrinde.* Leipzig: J. A. Barth

Ramón y Cajal, S. 1909–1911. *Histologie du Systéme Nerveux de l'Homme et des Vertébrés,* vols. 1 and 2, tr. S. Azoulay. Paris: Maloine

Rickmann, M., Chronwall, B. M. Wolff, J. R. 1977. On the development of non-pyramidal neurons and axons outside the cortical plate: The early marginal zone as a pallial anlage. *Anat. Embryol.* 151:285–307

Rogers, L. A., Cowan, W. M. 1973. The development of the mesencephalic nucleus of the trigeminal nerve in the chick. *J. Comp. Neurol.* 147:291–319

Rubel, E. W. 1971. A comparison of somatotopic organization in sensory neocortex of newborn kittens and adult cats. *J. Comp. Neurol.* 143:447–80

Scheibel, M. E., Davies, T. L., Scheibel, A. B. 1973. Maturation of reticular dendrites: Loss of spines and development of bundles. *Exp. Neurol.* 38:301–10

Schreyer, D. S., Jones, E. G. 1980. Growth of the corticospinal tract in neonatal rats. *Neurosci. Abstr.* 6:649

Seil, F. J., Kelly, J. M., Leiman, A. L. 1974. Anatomical organization of cerebral neocortex in tissue culture. *Exp. Neurol.* 45:435–50

Shatz, C. 1977. A comparison of visual pathways in Boston and midwestern Siamese cats. *J. Comp. Neurol.* 171:205–28

Skoff, R. P., Hamburger, V. 1974. Fine structure of dendritic and axonal growth cones in embryonic chick spinal cord. *J. Comp. Neurol.* 153:107–48

Sotelo, C. (1975) Formation and maintenance of Purkinje spines in the cerebellum of "mutants" and experimental animals. *Exp. Brain Res. Suppl.* 1:133–37

Stanfield, B., Cowan, W. M. 1979. Evidence for sprouting of entorhinal afferents into the "hippocampal zone" of the molecular layer of the dentate gyrus. *Anat. Embryol.* 156:37–52

Stensaas, L. J. 1967. The development of hippocampal and dorsolateral pallial regions of the cerebral hemisphere in fetal rabbits. II. 20 mm stage; neuroblast morphology. *J. Comp. Neurol.* 129:71–84

Valentino, K. L., Jones, E. G. 1980. An electron microscopic study of the developing corpus callosum in fetal and neonatal rats. *Neurosci. Abstr.* 6:487

Verley, R. 1977. The post-natal development of the functional relationships between the thalamus and the cerebral cortex in rats and rabbits. *Electroencephalog. Clin. Neurophysiol.* 43:679–90

Verley, R., Axelrad, H. 1975. Postnatal ontogenesis of potentials elicited in the cerebral cortex by afferent stimulation. *Neurosci. Lett.* 1:99–104

Vrensen, G., De Groot, D., Nunes-Cardozo, J. 1977. Postnatal development of neurons and synapses in the visual and motor cortex of rabbits: A quantitative light and electron microscopic study. *Brain Res. Bull.* 2:405–16

Weiss, P. 1936. Selectivity controlling the central-peripheral relations in the nervous system. *Biol. Rev.* 11:494–531

Welker, C. 1976. Receptive fields of barrels in the somatosensory neocortex of the rat. *J. Comp. Neurol.* 166:173–90

Welker, C., Woolsey, T. A. 1974. Structure of layer IV in the somatosensory neocortex of the rat: Description and comparison with the mouse. *J. Comp. Neurol.* 158:437–54

Wendlandt, S., Cron, T. J., Stirling, R. V. 1977. The involvement of the noradrenergic system arising from the locus coeruleus in the postnatal development of the cortex in rat brain. *Brain Res.* 125:1–9

White, E. L. 1978. Identified neurons in mouse SmI cortex which are postsynaptic to thalamocortical axon terminals: A combined Golgi-electron microscopic and degeneration study. *J. Comp. Neurol.* 181:627–62

Wise, S. P. 1975. The laminar organization of certain afferent and efferent fiber systems in the rat somatosensory cortex. *Brain Res.* 90:139–42

Wise, S. P., Fleshman, Jr., J. W., Jones, E. G. 1979a. Maturation of pyramidal cell form in relation to developing afferent and efferent connections of rat somatic sensory cortex. *Neuroscience* 4: 1275–97

Wise, S. P., Hendry, S. H. C., Jones, E. G. 1977. Prenatal development of sensorimotor cortical projections in cats. *Brain Res.* 138:538–44

Wise, S. P., Jones, E. G. 1976. The organization and postnatal development of the commissural projection of the rat somatic sensory cortex. *J. Comp. Neurol.* 163: 313–43

Wise, S. P., Jones, E. G. 1977. Cells of origin and terminal distribution of descending projections of the rat somatic sensory cortex. *J. Comp. Neurol.* 175:129–58

Wise, S. P., Jones, E. G. 1978. Developmental studies of thalamocortical and commissural connections in the rat somatic sensory cortex. *J. Comp. Neurol.* 178:187–208

Wise, S. P., Murray, E. A., Coulter, J. D. 1979b. Somatotopic organization of corticospinal and corticotrigeminal neurons in the rat. *Neuroscience* 4:65–78

16 / *Further Observations on the Development of the Dentate Gyrus*

W.M. COWAN
B.B. STANFIELD
D.G. AMARAL

INTRODUCTION

The terminal segment of the hippocampal formation, which is variously known as the *dentate gyrus*, the *fascia dentata*, or more generally, as the *area dentata*, has proved to be one of the most useful model systems for the analysis of cortical development in the mammalian brain. There are several reasons for this, but two are particularly pertinent. First, and perhaps most importantly, the structure of the dentate gyrus is considerably simpler, and much better understood, than that of any other part of the cerebral cortex. And second, much of its development occurs at a relatively late stage; indeed, in most of the common laboratory mammals many of the critical events in its development occur after birth; thus they can often be manipulated experimentally without recourse to intrauterine surgery or to costly labeling procedures involving both the mother and fetus.

Although we are still far from being able to give a complete account of the development of the dentate gyrus in any species, we can now at least indicate (as a result of the work of the past decade) the major features in its development and some of the consequences of the late formation

* It is a pleasure to dedicate this review to Professor Viktor Hamburger in appreciation of his many contributions to the neurosciences in general, and more particularly, to our own education in developmental neurobiology: his influence is to be found, either directly or indirectly, in almost every aspect of our developmental studies.

of its principal cell type—the granule cells. Many of these features have been reviewed recently elsewhere (Cowan, Stanfield, and Kishi, 1980), but as we have had an opportunity to reexamine several points since completing that review, we shall give in what follows a relatively brief account of the overall development of the dentate gyrus, and describe, in rather more detail, our recent observations. These concern the site and mode of origin of the cells of the dentate gyrus; the time of arrival of certain of its extrinsic afferents; some consequences of the late development of the granule cells; and the changes that may occur in the development of one class of afferents when the ingrowth of a second class that is normally coextensive with it is prevented. However, before dealing with these aspects of the development of the dentate gyrus, a brief account of its normal morphology (based mainly on its appearance in adult rodents, which have been used in most of our studies) will provide the necessary background for this review.

THE MORPHOLOGY OF THE DENTATE GYRUS

The general appearance of the rodent dentate gyrus is illustrated in Figure 16-1, which is a low-power photomicrograph of a horizontal section through the hippocampal region in the rat. It is evident from this that the dentate gyrus is essentially a trilaminar structure consisting of (1) a centrally placed *granule cell layer* (or *stratum granulosum*) formed by the somata of the dentate granule cells, which are by far the most numerous cell type in the area, (2) a superficial, relatively cell-free, *molecular layer,* in which the dendrites of the granule cells ramify and in which most of the extrinsic afferents to the dentate gyrus terminate, and (3) an inner *polymorphic layer,* which occupies the deepest part of the so-called hilar region.

By far the commonest cell type is the granule cell: in the albino rat it has been estimated that there are between 6×10^5 and 1×10^6 such cells (Schlessinger, Cowan, and Gottlieb, 1975; Gaarskjaer, 1978), and in different strains of mice the number is said to vary from 2.7×10^5 to 4.5×10^5 cells (Wimer et al., 1978). The cells have relatively small, ovoid somata ($\sim$ 8–10 μm in diameter) and typically have only ascending dendrites that extend superficially towards either the hippocampal fissure or the pia mater. The appearance of the dendrites varies considerably from cell to cell. Two fairly typical dendritic patterns are shown in Figure 16-2. As a rule, the deeper-lying granule cells (like that on the right in the figure) have a single stem dendrite that only begins to branch on entering the molecular layer, whereas the more superficially located cells may give off several stem dendrites from the superficial aspect of their somata. Quantitative estimates indicate that the total dendritic length of different cells may vary by as much as a factor of two or more. The

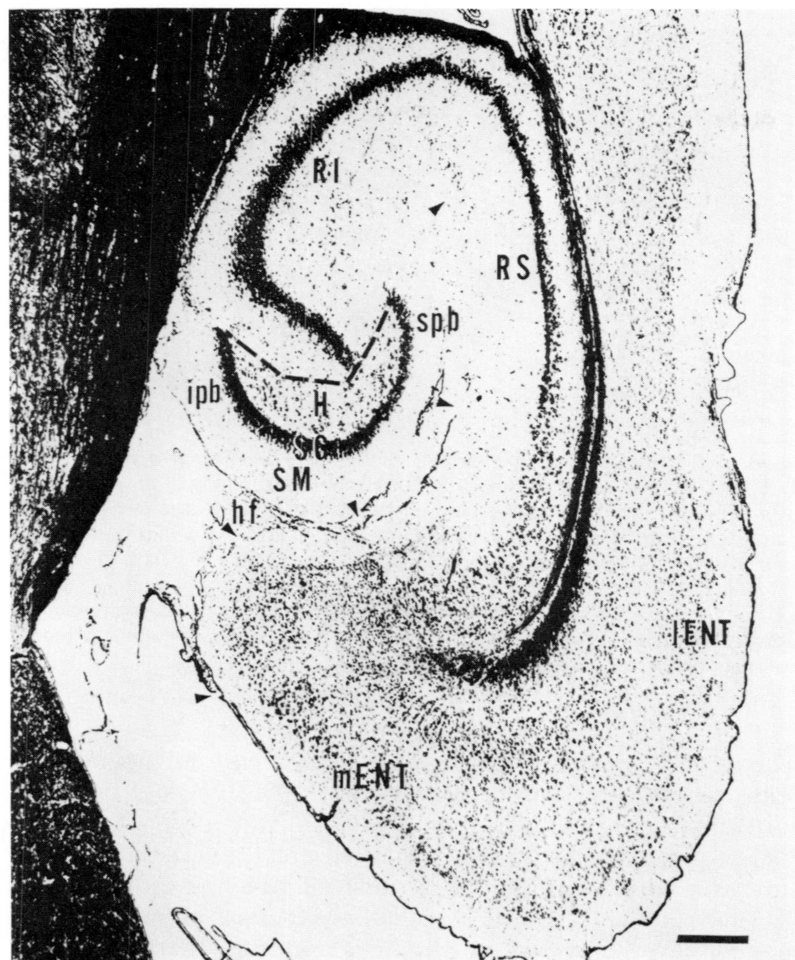

Figure 16.1 A Klüver–Barrera-stained horizontal section through the temporal portion of the hippocampal formation in an adult rat to show the general topographic relationships of the dentate gyrus. The broken line marks the limits of the *area dentata*. lENT, mENT: lateral and medial entorhinal areas; hf: hippocampal fissure; RS, RI: *regio superior* and *regio inferior* of the hippocampus; H: hilar region of the dentate gyrus; SG: *stratum granulosum;* SM: *stratum moleculare;* ipb, spb: infrapyramidal and suprapyramidal blades of the dentate gyrus. Scale: 250 μm. From Cowan *et al.* (1980).

proximal parts of the dendrites bear short, sessile spines. Further out, the spines become appreciably longer and more numerous; however, towards the ends of the longer dendrites the length and density of the spines again declines. Most of the synapses that the cells receive are axospinous and of the asymmetric variety, but a smaller number of sym-

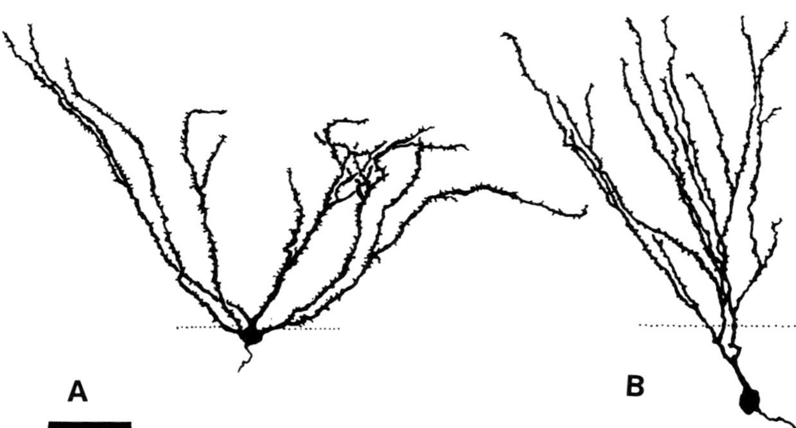

Figure 16.2 Camera lucida drawings of the two Golgi-impregnated neurons from the dentate gyrus of a young adult rat to show something of the morphological variation found among granule cells. Cells like that on the left are usually found close to the border between the granule cell and molecular layers (which is marked by the dotted line); these cells are among the first to be generated. Cells like that on the right are usually generated much later and, in addition to having their somata deep within the granule cell layer, they commonly have a single stem dendrite, which breaks up into branches only when it reaches the molecular layer. Scale: 50 μm. From Cowan *et al.* (1980).

metric synapses are found on the cell somata and on the shafts of the proximal dendrites (Laatsch and Cowan, 1966; Gottlieb and Cowan, 1972a; Tömböl et al., 1978).

For the present purpose only two other cell types need be described. The first is a variety of interneuron (one of which is shown in Fig. 16-3). The somata of most of these cells lie either in, or just deep to, the granule cell layer, and they usually have dendrites that reach both into the molecular layer and into the subjacent polymorphic layer. Their axons, which commonly arise from the apical dendrite, ascend into or through the stratum granulosum and there synapse upon the somata and proximal dendrites of the granule cells; electrophysiological studies indicate that these "basket cells" are inhibitory (Andersen, Holmqvist, and Voorhoeve, 1966). The second type of cell is quite different: their somata lie deep to the granule cell layer, and their dendrites ramify extensively across much of the infragranular zone. These neurons (which have been termed *mossy cells*—Amaral, 1978) have been identified recently as the source of the commissural and associational afferents to the dentate gyrus.

One of the most interesting features of the morphology of the dentate gyrus is the laminar arrangement of most of its major extrinsic afferents (Fig. 16-3). These afferents are derived from five sources: the medial and lateral parts of the entorhinal cortex; the polymorphic zone of the dentate gyrus of both sides; and the supramammillary region of the hypothalamus. In addition, there are three classes of afferents that lack this type of laminar pattern; these are derived from the medial septal/diagonal band complex, the locus coeruleus, and the raphe nuclei of the brainstem.

The afferents from the entorhinal cortex arise from stellate cells in Layer II of both the lateral and medial entorhinal areas (Area 28b and Area 28a of Krieg, 1946 a,b; Steward and Scoville, 1976) those from the lateral area terminate within the outer third of the molecular layer, while those from the medial entorhinal cortex end in its middle third (Raisman, Cowan, and Powell, 1965; Hjorth-Simonsen, 1972; Hjorth-Simonsen and Jeune, 1972; Steward, 1976). The afferents from the polymorphic zone of the area dentata of the two sides comprise the commissural and associational inputs whose zones of termination are coextensive within the inner third of the molecular layer (Blackstad, 1956; Raisman et al., 1965; Zimmer, 1971; Gottlieb and Cowan, 1972b, 1973; Hjorth-Simonsen and Laurberg, 1977; Swanson, Wyss, and Cowan, 1978; West et al., 1979; Laurberg, 1979). The hypothalamic input from the supramammillary region terminates in a narrow zone immediately deep to this, partly within the deepest part of the molecular layer and partly in the outer third of the granule cell layer (Segal, 1979; Wyss, Swanson, and Cowan, 1979).

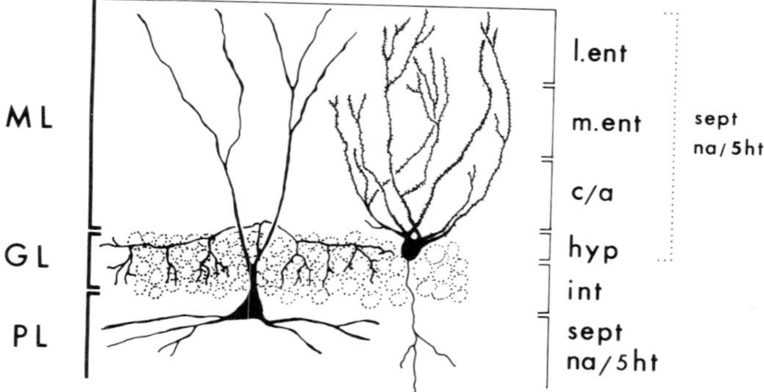

Figure 16.3 A schematic drawing to illustrate the laminar distribution of the major extrinsic inputs to the granule cells of the dentate gyrus. The square brackets on the right mark the limits of each of the inputs while those on the left indicate the boundaries of the molecular layer (ML), granule cell layer (GL), and the polymorphic layer (PL). The cell drawn on the right is a typical dentate *granule cell,* while that on the left is characteristic of the so-called *pyramidal-basket cells* whose axons form a dense plexus around the somata of adjacent granule cells. l.ent, m.ent: afferents from the lateral and medial entorhinal areas, respectively; c/a: location of the commissural and associational afferents; hyp: hypothalamic afferents from the supramammillary region; int: interneuronal inputs; sept, na/5ht: septal, noradrenergic and serotonergic inputs, respectively. From Amaral *et al.* (1980), with permission.

In preparations stained by the Timm method (Fig. 16-4), the zone of termination of the afferents from the lateral entorhinal cortex in the outer part of the molecular layer is stained a moderately deep brown color; internal to this there is an essentially unstained region in which the medial entorhinal afferents terminate; and this in turn is bounded on its deep aspect by a second deeply stained zone in which the commissural and associational fibers end (Fig. 16-4; Haug, 1974). No correlate for the recently described hypothalamic input has as yet been identified in Timm's preparations, although one candidate is the brown precipitate that has been observed in association with the granule cell somata in the outer part of the stratum granulosum (Haug, 1974; Geneser-Jensen, Haug, and Danscher, 1974).

The afferents from the basal forebrain and brainstem are less distinctly organized (see Figure 16-3). There is still some uncertainty about the

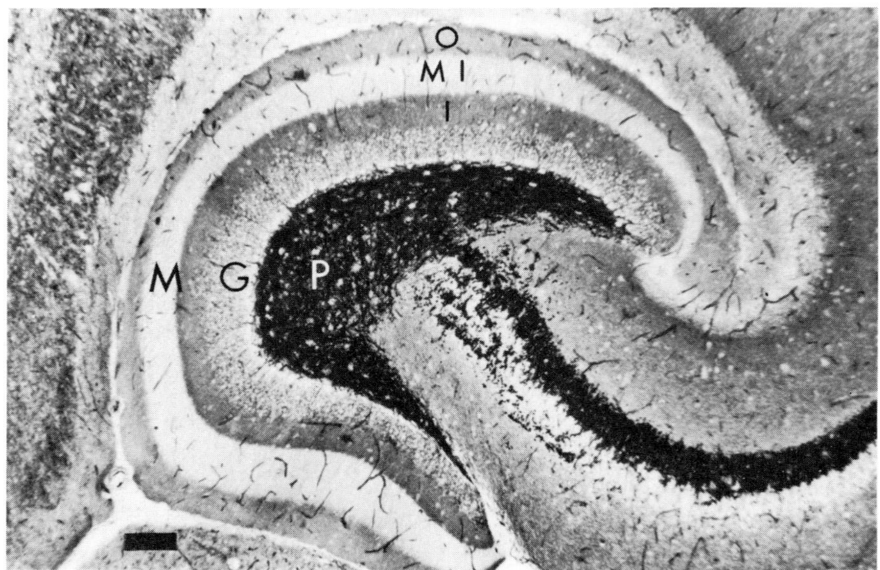

Figure 16.4 A Timm-stained preparation of a horizontal section through the dentate gyrus in an adult rat to show the characteristic laminar staining of the molecular layer (M). Note the outer darkly staining zone (0), which corresponds to the zone of termination of the afferents from the lateral entorhinal cortex; the very pale intermediate zone (MI), in which the fibers from the medial entorhinal cortex end; and the inner more deeply stained zone (I), in which the associational and commissural afferents terminate. The granule cell layer is largely unstained, but the axons of the granule cells (the so-called mossy fibers) are stained very intensely as they travel through the polymorphic layer (P) and on into the hippocampus. M, G and P: molecular, granule cell, and polymorphic layers. Scale: 100 μm.

distribution of the septal afferents although the best available evidence suggests that most of the fibers terminate in a narrow zone just deep to the granule cell layer, while a smaller number of fibers seem to be distributed diffusely throughout the molecular layer (Swanson and Cowan, 1979). The fibers from the locus coeruleus provide the noradrenergic input to the dentate gyrus; while they mainly innervate the polymorphic layer, some fibers also distribute to the stratum granulosum and to the molecular layer (Swanson and Hartman, 1975; Jones and Moore, 1977; Koda and Bloom, 1977; Loy et al., 1980). The serotonergic input from the raphe nuclei has a more homogeneous distribution to all parts of the dentate gyrus (Conrad, Leonard, and Pfaff, 1974; Moore and Halaris, 1975).

THE SITE OF ORIGIN OF THE CELLS IN THE DENTATE GYRUS

Because of the complex infolding of the hippocampal formation, the site of origin of the cells that comprise the dentate gyrus has been debated for some years. It is now clear, however, that like the rest of the hippocampus, both the granule cells and the cells in the hilar region are ultimately derived from the neuroepithelium of the ventricular zone, which is close to the telachoroidea of the lateral ventricle (Fig. 16-5). ^{3}H-thymidine autoradiography shows that at an early stage (until about the 13th day of gestation—E13—in the rat) the proliferation in this region displays the characteristic interkinetic nuclear migration found in other parts of the ventricular zone. Subsequently, an increasing number of cells, or at least their nuclei, appear to be displaced from the ventricular zone, which can now be seen to overlie a distinctive, cell-free region that will later form the fimbria of the hippocampus (Fig. 16-5).

The progressive displacement of labeled nuclei from this part of the ventricular zone and the concomitant appearance of labeled nuclei deep to it, near the developing pia, have generally been interpreted as evidence that a sizable proportion of the neuroepithelial cells lose their attachment at the ventricular surface and migrate across the region of the future fimbria (Altman and Das, 1965, 1966; Altman, 1966; Schlessinger et al., 1975; Cowan et al., 1980) in a manner comparable to the migration of postmitotic neurons in other regions of the brain (Sidman and Rakic, 1973; Cowan, 1978). However, another possibility is suggested by the appearance of the cells in this region in the scanning electron microscope, which provides an especially striking view of the entire hippocampal complex during mid-to-late fetal development. As Figure 16-6 (which is taken from an E16 rat fetus) shows, some of the cells in this region appear to have "trailing" processes that extend superficially towards the ventricular zone, although the perikarya are already close to the pial

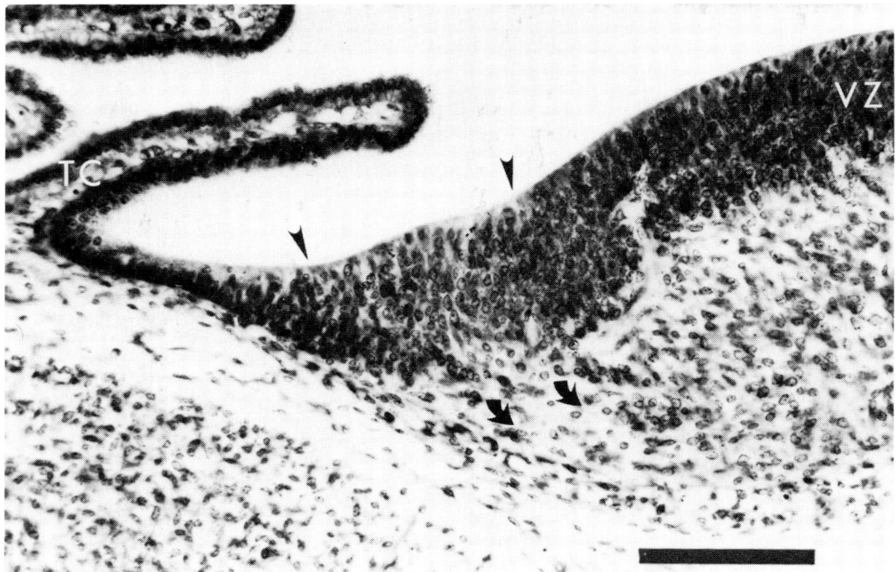

Figure 16.5 A low-power photomicrograph from a 16-day-old rat fetus through the region of the ventricular zone (VZ) from which the dentate gyrus arises. The arrowheads mark the approximate limits of the area that gives rise to the dentate gyrus; this is bounded to the left by the developing tela choroidea (TC) of the lateral ventricle and to the right by the region from which the *regio inferior* of the hippocampus is derived. The arrows mark the direction of migration of the cells that will form the anlage of the dentate gyrus. Scale: 100 μm.

surface of what will later become the *proliferative zone* of the dentate gyrus. This appearance suggests that the initial reorganization in this region may involve not a true migration of the precursor cells of the dentate gyrus but rather a translocation of the nuclei from the ventricular to the basal aspects of the cells. If such a nuclear translocation occurs and if this were to be followed by the detachment of the ventricular (or apical) processes of the cells, the net effect would be the establishment of a new proliferative region, comparable in most respects to the subventricular zone found in other parts of the forebrain, in which many of the late-formed, smaller neurons and glial cells are generated (Sidman and Rakic, 1973). With the material available at present, it is difficult to resolve this issue, and until serial transmission electron micrographs are available, the precise manner in which the secondary proliferative zone of the dentate gyrus is set up will remain uncertain. Although this may seem to be a rather trivial matter, if it could be established that the initial step in the setting up of the dentate proliferative zone is a

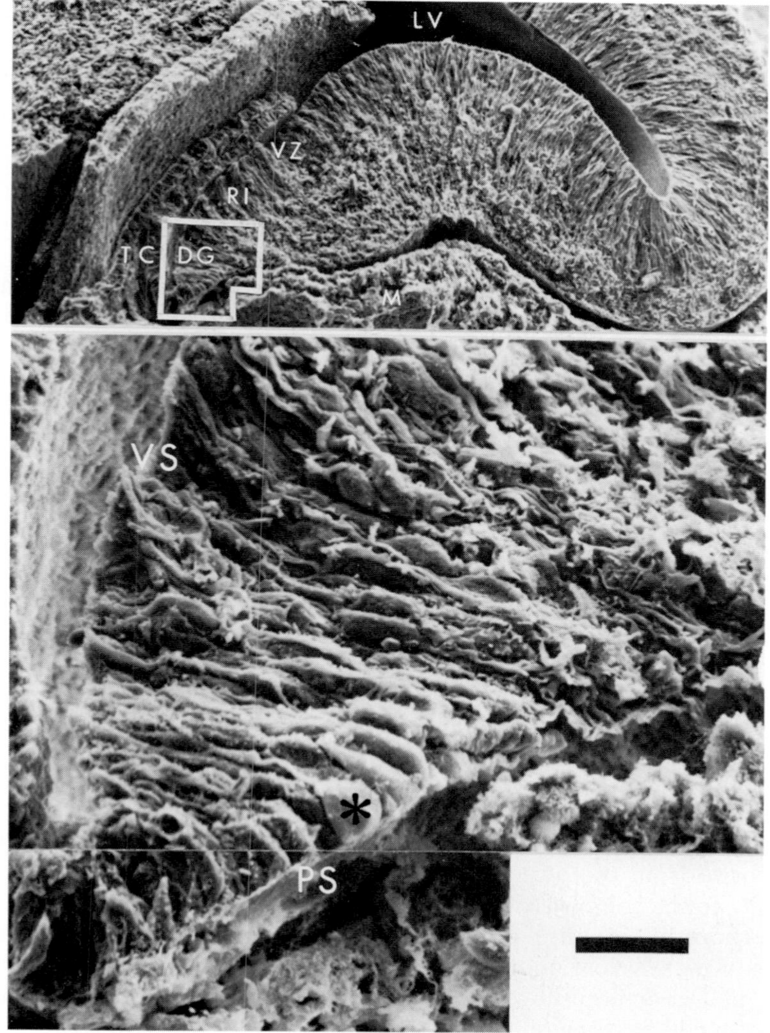

Figure 16.6 (UPPER) A low-power scanning electron micrograph (SEM) from a 16-day-old rat fetus of the region from which the dentate gyrus arises. Note: the ventricular zone (VZ); the tela choroidea of the lateral ventricle (TC) and the approximate areas from which the *regio inferior* of the hippocampus (RI) and the dentate gyrus (DG) are derived. M: menintes; LV: lateral ventricle. The area outlined is shown at higher magnification in Figure 16.6 (lower).
(LOWER) This SEM montage shows the area marked in Figure 16.6 (upper) at higher magnification. Note the presence of cells whose somata appear to be near the region in which the anlage of the dentate gyrus will appear; these cells appear to have trailing processes that extend superficially towards the ventricular zone. VS: ventricular surface; PS: pial surfaces. (The asterisk marks one such cell.) The possible significance of this appearance is discussed in the text. Scale, 100 μm for upper micrograph; 15 μm for lower.

nuclear translocation, rather than a true cellular migration, the composition of the proliferative zone would be clarified considerably, and proliferation in the dentate gyrus could be considered essentially the same as in other cortical regions (the only significant difference being the topographic displacement of the dentate subventricular region) instead of appearing to be completely different in character from proliferation elsewhere in the central nervous system (save perhaps for the external granule zone of the developing cerebellar cortex).

The situation is complicated by the fact that the two major populations of neurons in the area dentata—the granule cells and the cells in the polymorphic zone—appear to develop quite differently. As we shall point out in the following section, the cells in the polymorphic zone are among the earliest neurons in the hippocampal formation to become postmitotic, whereas the majority of the granule cells are not formed until several days later (some granule cells, in fact, are still being formed 2 or 3 months later). In view of this, it seems not unlikely that the cells in the polymorphic zone lose their capacity for DNA synthesis while still being located within the ventricular zone. If this is so, then their movement away from the ventricular zone to the anlage of the dentate gyrus would be analogous, in all respects but one, to the migration of the cells that give rise to the adjoining *regio inferior* of the hippocampus: the only significant difference is that the future hippocampal pyramids aggregate to form a distinct *cortical plate* (as in other parts of the cerebral cortex), but there appears to be no comparable structure in the development of the dentate gyrus.

In summary, the cells of the polymorphic zone and the precursor population of neuroblasts from which the dentate granule cells are derived, arise from the portion of ventricular zone of the lateral ventricle, which is contiguous with the region from which the hippocampus itself arises; there are reasons for thinking that the cells of the polymorphic zone completely lose their capacity for division while lying within the ventricular zone and subsequently migrate as young postmitotic neurons across the cell-free region from which the fimbria is derived. The precursor cells that in time give rise to the dentate granule cells may follow a similar migratory course, but it is possible that they undergo an initial nuclear translocation and that following the separation of their ventricular processes they set up a second proliferative locus analogous in most respects to the subventricular region, which is seen elsewhere in the forebrain.

THE TIME OF ORIGIN OF THE CELLS IN THE DENTATE GYRUS

The time of origin of the neurons in the dentate gyrus, or, more correctly, the times at which these cells lose their capacity for DNA synthesis and withdraw from the mitotic cycle, has now been determined by using the technique of ^{3}H-thymidine autoradiography in mice (Angevine, 1965; Stanfield and Cowan, 1979b), in rats (Altman and Das, 1965; Altman, 1966; Schlessinger et al., 1975; Schlessinger, Cowan, and Swanson, 1978), and in monkeys (Rakic and Nowakowski, 1981; Nowakowski and Rakic, 1981). The evidence from all these species is consonant in showing that the cells in the hilar region of the dentate gyrus (including the polymorphic zone) are among the first neurons in the hippocampal formation to become postmitotic and that the genesis of the dentate granule cells continues for substantially longer than that of almost any other class of neuron in the mammalian forebrain, with the possible exception of the granule cells in the olfactory bulb. In addition, by examining the final disposition of the cells generated at different stages in development, it is now known that the cells are generated along at least two, and possibly three, morphogenetic gradients.

The Time of Origin of the Cells in the Hilar Region of the Dentate Gyrus

The hilar region of the dentate gyrus is a complex zone containing a variety of different cell types (Amaral, 1978), some of which are more closely related to the regio inferior of the hippocampus than to the dentate gyrus but some of which—and especially those in the so-called deep hilar region—are intimately related to the dentate granule cells. Many of the latter are interneurons that provide for various forms of feedforward or feedback inhibition of the granule cells, but at least one class, the so-called mossy cells, appears to give rise to the associational and commissural afferents to the dentate gyrus whose development is considered below (Laurberg and Sørensen, 1981).

In the rat, for which our data are most complete, the cells in the hilar region are generated over a relatively brief period, between day E14 and day E19 (Schlessinger et al., 1978). In fact, although some labeled cells can be identified in the hilar region after ^{3}H-thymidine injections on day E18 and day E19, about 90 percent of the cells in the hilar zone (including those in the deep hilar region) cease DNA synthesis on, or before, day E17, when about 10 percent or less of the granule cells have been formed. One feature of the development of this region which came out of this study (Schlessinger et al., 1978) is that there is a tendency for the earliest-formed hilar neurons (including those generated on day E14 and day E15) to be located immediately beneath the suprapyramidal blade of the dentate gyrus, while those that are formed towards the end of the proliferative period for hilar cells (on day E16 and day E17) tend to be distributed deep to the infrapyramidal blade (see Figure 16-7). If this trend involves the cells of origin of the associational and commissural afferents, as seems likely, this may well prove to be a significant

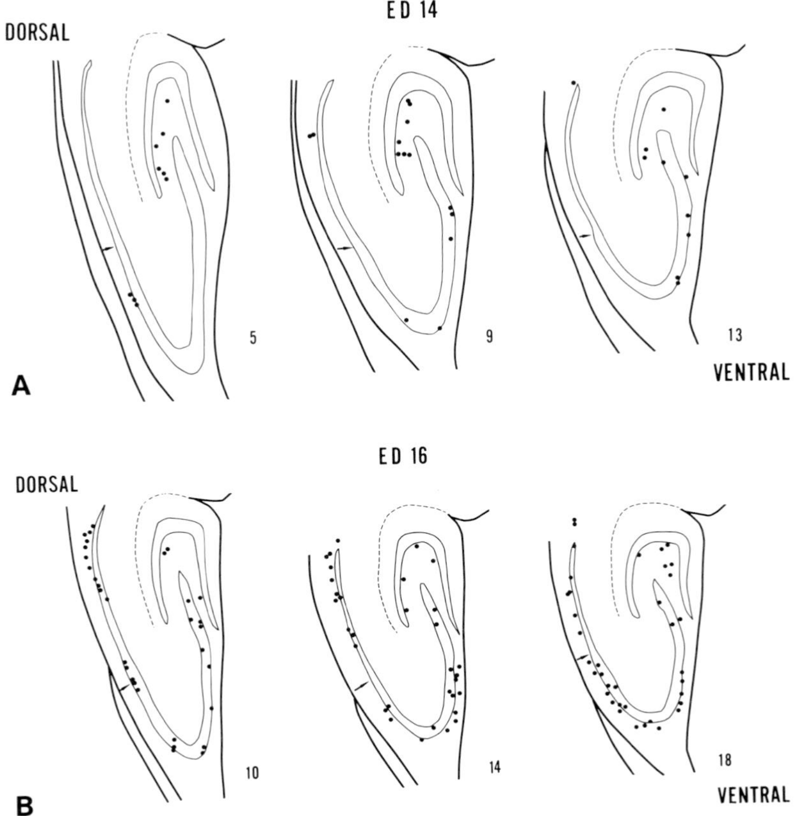

Figure 16.7 These tracings are from autoradiographs of horizontal sections from the dorsal, mid, and ventral portions of the hippocampal formation of two four-week-old rats whose mothers were injected with ^{3}H-thymidine on the 14th and 16th days of gestation, respectively. They illustrate a general gradient in the time of origin of the hilar region cells by which those below the suprapyramidal limb of the granule cell layer (in these diagrams this limb is below the dashed line, which defines the hippocampal fissure) pass their final mitosis earlier than those cells subjacent to the infrapyramidal limb of the granule cell layer. In this, and the subsequent two figures, each dot marks the location of a heavily labeled cell; though the dots are not to scale, their position is accurately plotted. From Schlessinger *et al.* (1978), with permission.

factor in determining the distribution of these two pathways to the suprapyramidal and infrapyramidal blades of the dentate gyrus itself.

An interesting feature of the development of the neurons of the polymorphic layer derives from the fact that their final mitosis occurs a week or so before the assembly of the definitive stratum granulosum beneath which they are normally located. What role, if any, these cells play in determining the structure of the overlying stratum granulosum is unknown. It is of interest in this regard that when the stratum granulosum that comprises the infrapyramidal blade of the dentate gyrus is prevented from forming (by early postnatal x-irradiation), certain of the afferent fibers that project to the molecular layer still grow into this region (Laurberg and Hjorth-Simonsen, 1977) and presumably synapse with the dendrites of the subjacent hilar cells. It will be of interest to determine the later fate of these neurons in the absence of the granule cells that comprise the major (and perhaps the sole) target for their axons.

The Time of Origin of the Dentate Granule Cells

Unlike the neurons in the hilar region, the dentate granule cells are generated over a prolonged period of almost three weeks in the mouse (Angevine, 1965; Stanfield and Cowan, 1979b) of up to several weeks in the rat (Schlessinger et al., 1975; Kaplan and Hinds, 1977) and of up to five to six months in the monkey (Rakic and Nowakowski, 1981). In the rat the earliest granule cells arise on, or just before, day E14. At this stage a small number of cells, destined mainly for the suprapyramidal blade, can be shown to have withdrawn from the mitotic cycle (Fig. 16-8). On subsequent days increasing numbers of granule cell precursors become postmitotic, but even by the end of the 21st day of gestation less than 20 percent of the total number of granule cells have been generated (Bayer and Altman, 1974; Schlessinger et al., 1975). In the rat the peak period of granule cell proliferation occurs during the first postnatal week, and at its height something in excess of 50,000 cells are generated each day. During the second postnatal week, the rate of neuronogenesis drops sharply and has fallen to a very low level by the beginning of the third week. However, Kaplan and Hinds (1977) have recently reported that a small number of granule cells continue to be generated as late as the third month postnatally.

The Location of Granule Cell Proliferation

From experiments in which ^{3}H-thymidine was administered to neonatal rats and mice, within an hour or two of death, it is evident that although a majority of the dividing cells are located in the secondary proliferative zone we have described above, some cells that have already entered the developing stratum granulosum are capable of DNA synthesis *in situ.* It is difficult to determine what proportion of the total number of granule cells arises within the stratum granulosum, but since it is in this region that proliferation continues well into postnatal life, it seems likely that a fairly significant number of the cells are generated locally. In particular, it seems probable that local proliferation of this kind accounts for most

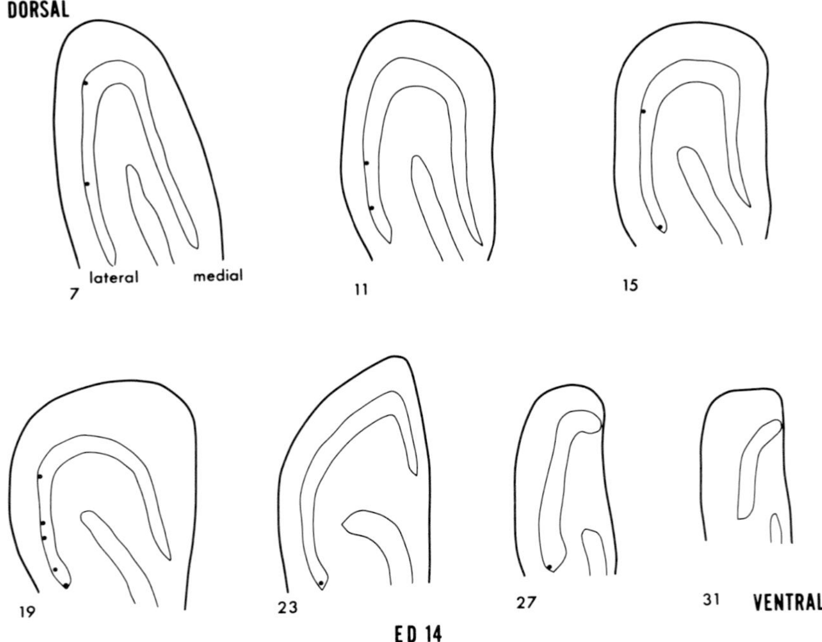

Figure 16.8 These tracings show the location of the first postmitotic granule cells in the rat dentate gyrus, which were labeled with ^{3}H-thymidine on the 14th day of gestation. From Schlessinger *et al.* (1975), with permission.

of the deeply placed granule cells (i.e., those in the inner part of the stratum granulosum) that are generated later in the proliferation period.

Temporospatial Patterns in Granule Cell Proliferation

In every species that has been examined, one of the most striking features in the development of the dentate gyrus is that it occurs along at least two, and possibly three, distinct gradients. The most obvious of these gradients is that which occurs along the transverse axis of the dentate gyrus, from the lateral margin of the suprapyramidal blade to the infrapyramidal blade. Although some cells destined for both blades are generated throughout the proliferative period, in general the cells in the suprapyramidal blade become postmitotic earlier than those that constitute the infrapyramidal blade. This is illustrated for the rat in Figure 16-9, and comparable figures for the mouse have been published by Angevine (1965) and by Stanfield and Cowan (1979b). The second consistent gradient is that found within the radial dimension of the stratum granulosum.

Throughout the dentate gyrus, the more superficially placed granule cells (i.e., those near the interface between the stratum granulosum and the stratum moleculare) arise earlier than the more deeply placed cells that adjoin the polymorphic zone. This "outside-in" sequence is the reverse of that seen elsewhere in the cerebral cortex (including the adjoining fields of the hippocampus) and appears to be a direct consequence of the pattern of outward migration of the granule cells from the proliferative zone in the hilar region (Schlessinger et al., 1975). In the rat and the mouse we have found evidence of a third gradient along the temporal to septal (or caudal to rostral) dimension of the dentate gyrus (Schlessinger et al., 1975; Stanfield and Cowan, 1979b). This gradient is less obvious than the other two, but on average the cells in the temporal part of the dentate gyrus appear to be generated earlier than those near its

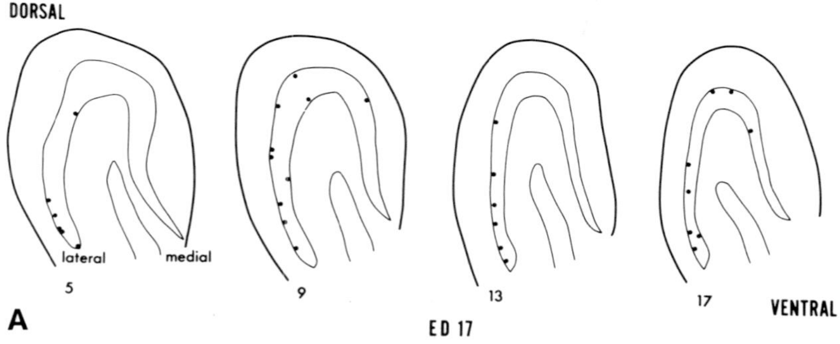

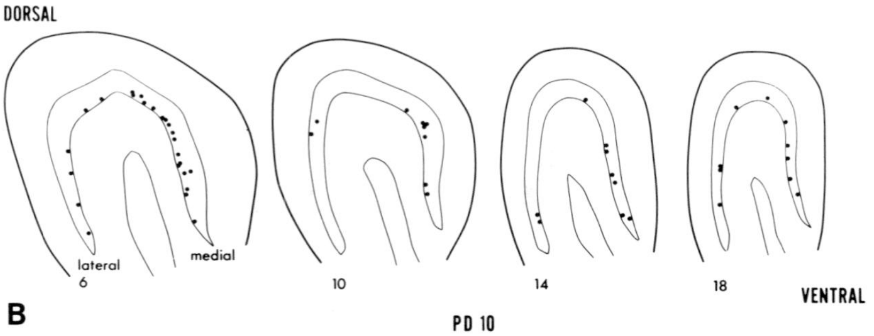

Figure 16.9 These tracings are from the brains of young adult rats that had been labeled with ^{3}H-thymidine on the 17th day of gestation (ED17) and the 10th postnatal day (PD10), respectively, and show the general suprapyramidal to infrapyramidal gradient in the genesis of granule cells. From Schlessinger *et al.* (1975).

septal pole. Although a comparable gradient has not been observed in the monkey (Rakic and Nowakowski, 1981) there is evidence that in the rat it may be a significant factor in determining the longitudinal distribution of the commissural and associational afferents (Fricke and Cowan, 1978).

The morphogenetic mechanisms responsible for these complex patterns of cell proliferation are not known, but, as we shall point out below, it is clear that since they determine the final disposition of the granule cells they are critical in determining certain features of the connections of the granule cells and probably also the arrangement and distribution of their dendrites (Stanfield and Cowan, 1979a).

THE ASSEMBLY OF THE STRATUM GRANULOSUM

Because of the prolonged period of cell proliferation, the appearance of the granule cell layer as a distinct cytoarchitectonic entity is delayed until long after the establishment of most cortical laminae, including the *stratum pyramidale* in the adjoining hippocampal fields. Furthermore, a direct consequence of the sequence of cell proliferation that we have described is that the two blades of the dentate gyrus appear at different times. In the rat, many of the granule cells destined for the *suprapyramidal blade* have migrated out from the proliferative zone and formed, by birth, a recognizable cellular layer. At this stage there is little indication of the *infrapyramidal blade* and, indeed, it is not until near the end of the first postnatal week that this blade becomes clearly recognizable. The actual time at which corresponding regions of the dentate gyrus are assembled in the other species that have been examined differs from case to case, but the sequential appearance of the two blades is essentially the same.

This finding is of some interest since ^{3}H-thymidine autoradiographic studies indicate that some of the granule cells destined for the infrapyramidal blade are generated at an early stage (as early as day E14 and day E15 in the rat). So the question arises as to what happens to these early-formed cells between their withdrawal from the mitotic cycle and their appearance in the newly assembled stratum granulosum. In a more general way, one might ask what determines the initiation of cell migration following the cessation of DNA synthesis. Most neurons appear to migrate away from the germinal zone in which they arise, almost immediately after their final division; but evidently some—like the dentate granule cells—may remain in, or at least close to, the germinal zone for a substantial period of time and apparently move away from this zone only in response to some (as yet unknown) external stimulus. In the case of the dentate gyrus, one consequence of this late migration is that many of the granule cells do not appear to form mature dendrites until several days (up to a week or more) after their withdrawal from the mitotic cycle. This delay in forming mature dendrites is evident from the late appearance of the molecular layer overlying the infrapyramidal blade, which is generally not recognizable until some days after the appearance of the stratum granulosum; it is also evident from the immature appearance of the granule cells in this region in Golgi preparations.

THE DEVELOPMENT OF THE AFFERENT CONNECTIONS OF THE DENTATE GYRUS

The development of the afferent connections to the dentate gyrus has been studied mainly in the rat with a variety of direct and indirect methods. Of the *indirect methods,* the most generally useful has been the analysis of material prepared by the Timm method. As we and others have reviewed this material elsewhere (Zimmer and Haug, 1978; Zimmer, 1978; Stanfield and Cowan, 1979b; Cowan et al., 1980), we need not comment further upon it here, except to say that it is now clear that the distinctive laminar patterns that can be seen with this histochemical procedure do not appear until some time after the arrival of the major classes of afferents which occupy each of the various laminae. Of the *direct methods,* the analysis of axonal and terminal degeneration (in preparations stained by one or other variant of the Nauta method—e.g., after lesions of the entorhinal area, or the contralateral hippocampus) has been perhaps the least useful, since the earliest signs of axonal degeneration can usually be seen only some days after the relevant afferents have grown into the dentate gyrus (Singh, 1977; Loy, Lynch, and Cotman, 1977). The other two direct methods involve either the *retrograde* transport of horseradish peroxidase (HRP) or the *anterograde* transport of radioactively labeled proteins; these have given essentially consonant results, and since these appear to be the most reliable methods available at present, in what follows we shall concentrate on the findings that are based on their use.

In an earlier study (Fricke and Cowan, 1977) we described the appearance of autoradiographs of the dentate gyrus following injections of ^{3}H-proline into the entorhinal cortex, or the hippocampus, on the 3rd, 6th, and 12th postnatal days.[1] The principal findings in that study were as follows: (1) At least some of the entorhinal afferents appeared to have reached the dentate gyrus by postnatal Day 3, but no indication of the

[1] In that study, the day of birth was regarded as the first postnatal day (Day 1). Here, as in our later papers, we shall term the day of birth, Day 0, so that Day 3 in Fricke and Cowan's study (1977), corresponds to Day 2 in the present account.

commissural input was obtained at that stage. (2) From the earliest stage at which they could both be identified, the commissural and entorhinal afferents bore the same topographic relation to each other as they have in mature animals. (3) The associational afferents to the inner part of the molecular layer were present before the commissural fibers reached it. Loy and her colleagues (1977), using the same approach, identified commissural fibers in the molecular layer on Day 4, and they concluded from their material that there was relatively little expansion of the zone occupied by these fibers during the following three or four weeks, although the molecular layer as a whole expanded in thickness by more than 100 percent during this period.

Because both of these studies were preliminary and involved background levels of radioactivity that were rather high (owing in large part to the proximity of the injection sites to the dentate gyrus and to the amounts of label injected), we have recently reexamined this issue in a larger series of rats that were injected with smaller amounts of ^{3}H-proline on each successive day from the day of birth (Day 0) until the end of the first postnatal week. Among these experiments is a series of brains in which isotope injections had been made into the hypothalamus; as most of these had involved the supramamillary region we have been able to use them to define the time of arrival of the hypothalamodentate afferents. In addition to this material we have also prepared a series of animals of various ages in which HRP was injected into the hippocampus several hours before sacrifice. Collectively, this material provides a fairly complete picture of the sequence of arrival and of the initial disposition of all the major extrinsic afferents to the dentate gyrus, excluding the monoaminergic inputs.

Injections of ^{3}H-proline into the entorhinal cortex on the day of birth (Day 0) clearly result in the anterograde transport of labeled proteins in the perforant path to the dentate gyrus. In the autoradiographs of these brains, labeling is most evident in the molecular layer of the suprapyramidal blade, at temporal levels; it becomes progressively less distinct at more septal levels, and even in the temporal region it is less sharply focused towards the "crest" region of the gyrus.

Following injections into the hippocampus itself at this stage, we can see, on the one hand, some indication of labeling of the ipsilateral associational input to the dentate gyrus, at some distance from the injection site (although the precise septotemporal extent of this projection is unclear because of the inevitably high background levels of radioactivity so close to the injection site). On the other hand, it is clear from the autoradiographs of the contralateral side of the brain, that few, if any, commissural fibers to the dentate gyrus are labeled, even though the commissural projections to the adjoining fields of the hippocampus and to the hilar region are distinctly recognizable. That commissural afferents have reached the contralateral hippocampus is borne out by the finding that after fairly large injections of HRP into the hippocampus on one side on Day 0, substantial numbers of retrogradely labeled cells can be seen in the regio inferior of the contralateral hippocampus. It is worth adding in this context, that in none of our experiments with HRP injections have we observed retrogradely labeled cells in other fields of the hippocampus, and in particular, we have never seen labeled cells in the *regio superior*, which in adult animals receives a commissural input but does not contribute to the crossed projection (Gottlieb and Cowan, 1973; Swanson et al., 1978). This contrasts with the findings of Innocenti and colleagues (1977) who have reported that all parts of Area 17 contribute to the projection to the opposite side in neonatal kittens, whereas at later stages the callosal zone becomes confined to a narrow region along the border between Area 17 and Area 18.

Injections of ^{3}H-proline into the hippocampus on the day after birth (Day 1) clearly result in axonal transport to the inner part of the molecular layer of the ipsilateral dentate gyrus (owing to labeling of the associational afferents), and there is some indication of label near the tip of the suprapyramidal blade on the contralateral side. The labeling on the contralateral side is not confined to the innermost part of the molecular layer, as it is in mature animals, but occupies about two-thirds of its cross-sectional width. In addition, it does not extend throughout the mediolateral or rostrocaudal extent of the molecular layer; it becomes increasingly diffuse and ill defined towards the crest region, and is only present in the septal one-fourth to one-third of the dentate gyrus. However, it is clear from these experiments that the first commissural fibers reach the dentate gyrus no later than the day after birth, which is some days earlier than has been reported previously (Fricke and Cowan, 1977; Loy et al., 1977).

In a series of rats in which isotope injections were made into the caudal hypothalamus on Day 1, we have been unable to detect labeling in the fibers of the supramamillary projection to the dentate gyrus, although there is clearly labeling of certain of the other diencephalic inputs to the hippocampal formation.

The commissural projection is more distinctly labeled after isotope injections on Day 2, but in these cases it is still confined to the suprapyramidal blade in the septal half of the dentate gyrus (Fig. 16-10). In this region the "above-background" labeling extends across the greater part of the thickness of the stratum moleculare and is certainly not limited to its inner third.

The first indication of the arrival of the supramamillary input to the dentate gyrus is found after injections on Day 3. At this stage there is clear labeling over the stratum granulosum, and, in the autoradiographs, silver grains can be seen to extend outwards for some distance into the molecular layer—certainly well beyond the immediate supragranular

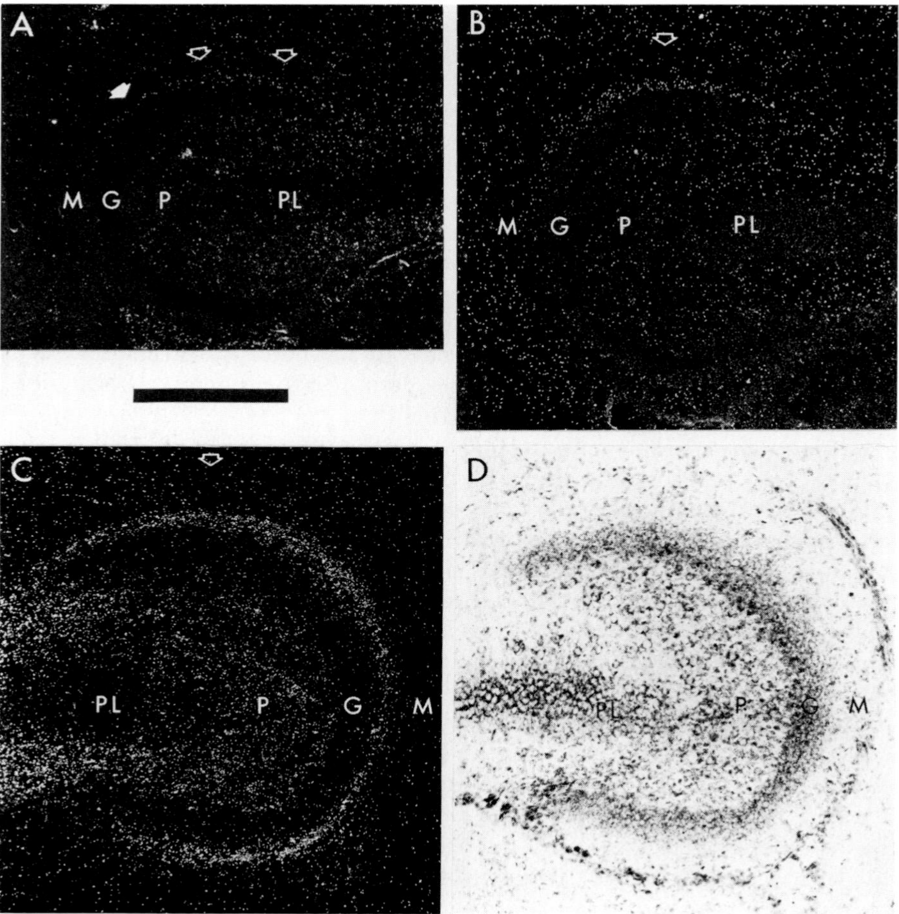

Figure 16.10 These low-power photomicrographs are from autoradiographs of the dentate gyrus of young rats in which the commissural projection was labeled by the injection of ³H-proline on postnatal Days 2, 5, and 7, respectively, into the contralateral hilar region. In the darkfield photomicrographs (A, B, and C) the silver grains over the molecular layer appear as bright dots; D is a brightfield photomicrograph of the same section shown in C. M, G, and P: molecular, granule cell, and polymorphic layers, respectively; PL: end of the pyramidal cell layer of the *regio inferior* of the hippocampus. The open arrows indicate the position of the hippocampal fissure, while the filled-in arrow in A, marks the limit of the commissural projection to the suprapyramidal blade at postnatal Day 2. Scale: 250 μm.

zone that this projection normally occupies in adult animals (Segal, 1979; Wyss et al., 1979).

By Day 5 all of the major extrinsic afferents can be clearly labeled by appropriately placed isotope injections and their topographic relationships to each other have become clearly recognizable. By this stage, and certainly by Day 7, the entorhinal afferents have become restricted to the outer three-fourths of the molecular layer, the major focus of the commissural afferents is now clearly within the inner part of the layer, and the supramamillary projection is largely (but not yet exclusively) limited to the deepest part of the molecular layer and the subjacent stratum granulosum. Furthermore, although additional fibers are probably added to these projections for some days, by Day 7 each pathway can be labeled along the entire transverse and longitudinal axes of the gyrus (Fig. 16-11).

The later development of these projections has already been described by Fricke and Cowan (1977), who have shown that by the end of the second postnatal week all the afferent pathways examined have assumed their adult form and have become restricted to their appropriate zones of termination. What our more recent evidence has provided is: (1) The clear demonstration that certain of the pathways (notably the commissural input) reach the dentate gyrus some days earlier than had previously been recognized. (2) Evidence that although the major pathways to the molecular layer always bear the same general relationship to each other as they do in the adult, at the time of their initial ingrowth they overlap each other extensively and that the sharply laminated arrangement normally seen within the adult molecular layer results from the progressive restriction of the various afferents to their definitive zones of termination. And (3) that the supramamillary projection to the dentate gyrus arrives on, or about, Day 3 and that from the time of its arrival it is located in the deepest part of the molecular layer and the underlying part of the stratum granulosum.

The finding that the different classes of afferent fibers to the molecular layer overlap each other extensively when they first arrive has several implications both for the development of the dentate gyrus and for its sustained capacity for morphological plasticity. The initial impetus for our developmental studies was to account for the unusually precise lamination of the afferents to the pyramidal cells in the hippocampus and to the dentate granule cells. When this precise lamination was first observed, three possibilities were suggested to account for it. Initially, it was thought that the overall pattern is established from the very beginning through the precise matching of the incoming afferent fibers with the appropriate receptor sites on the dendrites of their target cells. Alternatively, it seemed possible that the afferent fibers enter the hippocampal formation in a quite random fashion and that only subsequently do they

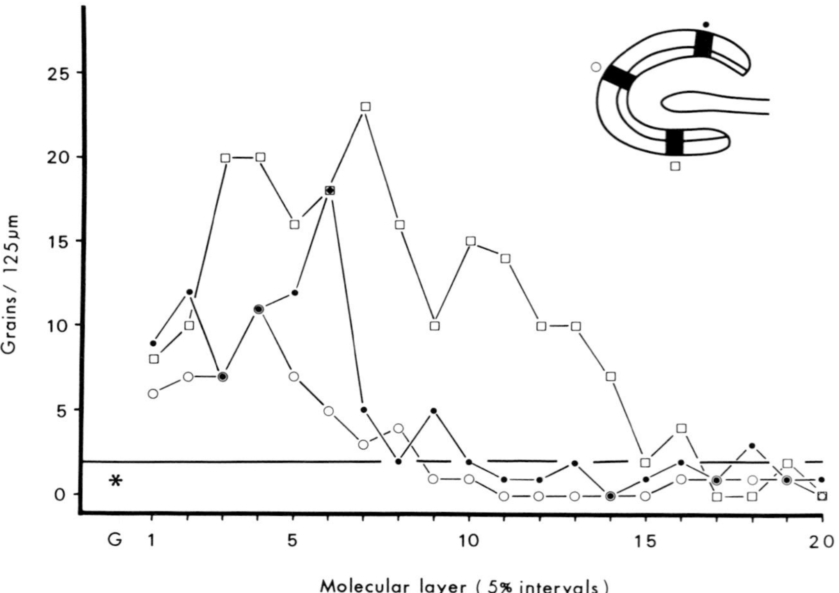

Figure 16.11 These three graphs illustrate grain density traverses across the granule cell and molecular layers of the dentate gyrus at the sites indicated in the insert drawings. These data are taken from a seven-day-old rat in which the cells of origin of the commissural projection to the dentate gyrus had been labeled with ^{3}H-proline six hours before sacrifice. The asterisk indicates the background level of radioactivity in this preparation and the horizontal line the twice background level above which the grain counts were considered significant. Whereas by this age the transported radioactivity in the commissural fibers to the suprapyramidal region (filled circles) and to the crest of the gyrus (open circles) is more or less confined to the inner one-third of the molecular layer, note that in the infrapyramidal blade (open squares) a significant number of grains are seen as far out as the junction of the middle and outer thirds of the molecular layer; in this region it would overlap extensively with the zone of termination of the entorhinal afferents.

become rearranged under the influence of some undefined morphogenetic or functional mechanism. A third possibility was that the fibers enter the hippocampal formation in a reasonably ordered manner, but their final distribution within their various zones of termination only becomes established secondarily, in a manner comparable to the refinement that is now known to occur in other contiguous and essentially nonoverlapping fiber systems such as the eye-dominance columns in the primate visual cortex (Hubel, Wiesel, and LeVay, 1977), in the various layers of the dorsal lateral geniculate nucleus (Rakic, 1977), and in the primary olfactory complex (Price, Moxley, and Schwob, 1976).

It is now almost certain that the third alternative is correct. When the afferents enter the molecular layer of the dentate gyrus, they have more or less the same topographic arrangement as they bear to each other in the mature brain. That is, the major part of the supramamillary input always occupies the deepest zone within the molecular layer (and the outer part of the stratum granulosum) while the commissural and associational fibers occupy the region between the supramamillary projection and the entorhinal afferents. But at this stage they overlap extensively, and it is not until well into the second postnatal week that the boundaries between the adjoining fiber systems become sharply defined.

These results have important implications for what has come to be known as "neuromorphological plasticity" in the dentate gyrus. Several studies have shown that after the chronic removal of one class of afferent, the afferents within neighboring laminar zones may "sprout" and, in time, come to occupy a significant proportion of the deafferented zone (see Lynch, Gall, and Dunwiddie, 1978; Cotman and Nadler, 1978; Stanfield and Cowan, 1979c, Amaral, Avendaño, and Cowan, 1980). It is generally agreed that although this capacity for reactive synaptogenesis persists into adult life, it is most vigorous in immature animals and is especially marked when the deafferentation is performed in the immediate postnatal period (Lynch, Stanfield, and Cotman, 1973; Zimmer and Hjorth-Simonsen, 1975). Our developmental findings suggest that in part, at least, the apparent sprouting after neonatal deafferentation is, in fact, due to the persistence of fibers that would normally be eliminated or relocated. In this sense the apparent expansion of, say, the commissural input to the dentate gyrus after lesions of the entorhinal cortex in neonates is analogous to the apparent expansion of the eye-dominance columns seen in the monkey visual cortex after early unilateral visual deprivation (Hubel et al., 1977).

SYNAPTOGENESIS IN THE DENTATE GYRUS

It is important to emphasize that what we have actually described in the previous section is an overlap in the distribution of the *fibers* in the different projections; at present we have no evidence that the overlapping fibers form synapses upon the granule cell dendrites in an overlapping pattern. In fact, what evidence we have concerning synaptogenesis in the molecular layer of the dentate gyrus suggests that all the various afferents reach it several days before significant numbers of morphologically recognizable synapses are detectable (Crain et al., 1973; Cowan et al., 1980). This point is clearly indicated in Table 16-1, which is taken from the work of our colleague, Dr. Kiyoshi Kishi. On the day of birth (when the entorhinal and the ipsilateral associational fibers are already

TABLE 16-1. Synaptic density (synapses/100 μm²) in the molecular layer over the suprapyramidal (SPB) and infrapyramidal (IPB) blades of the rat dentate gyrus at 1, 5, 10, 21, and 41 days of age

Age in Days		Rostral	Middle	Caudal	Total
1	SPB	0.27	0.0	1.0	0.4
5	SPB	3.2	4.1	2.5	3.2
	IPB	0.82	0.88	0.6	0.8
10	SPB	12.2	8.8	12.1	11.0
	IPB	5.4	8.0	5.0	6.1
21	SPB	32.5	37.7	36.6	35.6
	IPB	36.4	28.4	36.9	33.9
41	SPB	36.6	35.6	38.2	36.8
	IPB	37.0	32.6	39.3	36.3

From Cowan et al. (1980), with permission.

present in the molecular layer) very few synapses can be seen, and, of course, those that are present are confined to the suprapyramidal blade. By the fifth postnatal day, when all the major afferents to the dentate gyrus have invaded the molecular layer, the overall density of synapses in this layer is very low; indeed, in the suprapyramidal blade it is less than 10 percent of the final density, and only about 2 percent to 3 percent in the infrapyramidal region. The major phase of synaptogenesis in the molecular layer occurs towards the end of the second postnatal week and continues through the end of the third week, by which time it is essentially complete. Between the fifth and tenth days there is a fivefold increase in synaptic density and this pace is maintained throughout the third postnatal week. During this whole period the volume of the molecular layer is also changing rapidly (from 0.246 mm³ to 1.053 mm³ between Day 5 and Day 10; from 1.053 mm³ to 1.875 mm³ between Day 10 and Day 21; and from 1.875 mm³ to 2.671 mm³ at Day 41). Thus the total number of synapses increases dramatically—by about 16-fold between Day 5 and Day 10; about 8-fold between Day 10 and Day 21; and approximately 1.5-fold between Day 21 and Day 41.

The largest spurt in synapse formation occurs at the time that most of the granule cells are forming dendritic spines. As Fricke (1975) has shown, there are relatively few spines on the dendrites of the granule cells before the end of the first postnatal week; thereafter they appear in large numbers and many of the granule cells in the suprapyramidal blade acquire an essentially complete dendritic tree (as estimated by their total dendritic length and their spine density) before the end of the second week. The appearance of dendritic spines manifests itself in a marked change in the relative numbers of "spine" and "shaft" synapses during the first three weeks of postnatal life. These changing patterns are indicated in Table 16-2, which also demonstrates very clearly the delay in synaptogenesis in the infrapyramidal blade compared to that in the suprapyramidal blade.

It follows from this that the laminar segregation of the different classes of afferents occurs more-or-less concurrently with the peak period of synaptogenesis. However, it is not clear what the basis for the segregation may be. It is tempting to suggest that it is the result of a competitive interaction between the neighboring classes of afferents and that what they are competing for is the limited number of synaptic sites on the dendrites of the granule cells. As we have pointed out elsewhere (Stanfield, Caviness, and Cowan, 1979), such competitive interactions alone would more likely result in the overlap rather than the mutually exclusive lamination of the zones of afferent termination and, at present, the evidence that competition is the basis for segregation is wholly circumstantial as, indeed, it is in most of the systems in which process withdrawal has been described. (See the review by Purves in this volume.)

It has commonly been assumed that different segments of the granule cell dendrites are in some way "specified" for the receipt of specific classes of inputs. Again, there is no direct evidence for this view, but

Table 16-2. Percentages of shaft (sh) and spine (sp) synapses in the developing molecular layer of the rat dentate gyrus

Age in Days			Rostral	Middle	Caudal	Means
1	SPB	sh	100.0	0.0	50.0	
		sp	0.0	0.0	16.7	
5	SPB	sh	29.2	46.7	74.1	50.0
		sp	41.7	46.7	22.2	36.9
	IPB	sh	62.5	76.9	71.4	70.3
		sp	12.5	15.4	0.0	9.3
10	SPB	sh	31.6	42.9	47.9	40.8
		sp	56.8	36.0	48.5	47.1
	IPB	sh	49.9	56.5	45.5	50.6
		sp	39.3	37.3	47.5	41.4
21	SPB	sh	17.5	12.1	10.7	13.4
		sp	80.0	86.3	86.7	84.3
	IPB	sh	10.4	16.5	9.9	12.3
		sp	88.3	81.8	88.3	86.1
41	SPB	sh	8.4	10.5	11.8	10.2
		sp	89.7	88.8	86.8	88.4
	IPB	sh	14.2	7.7	9.2	10.4
		sp	84.7	89.6	88.0	87.4

From Cowan et al. (1980), with permission.

two observations may be cited in support of it. The first is the finding that the commissural and associational afferents which are now known to be collaterals derived from the same neurons in the hilar region are coextensive in their distribution within the inner third of the molecular layer (Gottlieb and Cowan, 1972a). As these fibers arise from the same cells, it would seem reasonable to assume that they share the same cytochemical specificity (if such specificities exist). The second observation is that in the reeler mouse, in which many of the granule cells are displaced into the hilar zone, the commissural and associational afferents appear not only to retain their topographic relationship to the entorhinal afferents but also to display, to some extent, a normal differential distribution upon the granule somata and dendrites relative to the afferents from the supramamillary region (Stanfield et al., 1980). To account for this it has been suggested that, as in the neocortex, the different classes of afferents to the reeler dentate gyrus specifically "seek out" the appropriate dendritic segments upon which to synapse even though to do so they may pursue quite aberrant routes.

THE DEVELOPMENT OF THE MOSSY FIBERS

The axons of the dentate granule cells are collectively referred to as the mossy fibers (Ramón y Cajal, 1911). Their development has been studied previously in Golgi preparations by Minkwitz (1976) and in Timm sulphide silver preparations (in which they stain intensely) by Zimmer (1978), Zimmer and Haug (1978) and Stanfield and Cowan (1979b) and electron microscopically by Stirling and Bliss (1978). Because we have previously reviewed this earlier work (Cowan et al., 1980), we shall confine ourselves in this report to summarizing the results of a recent, detailed analysis of mossy fiber development from our own laboratory (Amaral and Dent, 1981).

The earliest mossy fibers are probably formed prenatally, and certainly by the day of birth. Timm preparations through the temporal portion of the hippocampal formation reveal a densely stained zone extending from the developing suprapyramidal part of the stratum granulosum, through the hilar region, towards the regio inferior. Here they form a reasonably distinct band in the zone immediately superficial to the stratum pyramidale, which, in time, will become the *stratum lucidum.* Golgi preparations from newborn rats similarly reveal the presence of scattered mossy fibers in the hilar region, but at this stage most of them lack the characteristic swellings and protrusions from which the fibers derive their name. Electron micrographs from the incipient stratum lucidum at this stage show many, small expansions along the length of the mossy fibers; the largest of these swellings are only about one-tenth the size of the fully developed mossy fiber expansions, but they contain appreciable numbers of spherical synaptic vesicles and appear to form both symmetric and asymmetric contacts with the primary dendritic shafts of hilar cells and the hippocampal pyramids.

Over the course of the next two weeks the entire mossy fiber system develops rapidly. In Timm preparations this is marked by (1) a progressive increase in the density of the sulphide silver deposits and an increase in the size of the individual stained particles, (2) an expansion in the thickness of the suprapyramidal bundle within the stratum lucidum, (3) a progressive rostral extension towards the septal pole of the hippocampus, and (4), with the appearance of the infrapyramidal blade of the dentate gyrus, the formation of a distinct infrapyramidal bundle of mossy fibers that in time extends across much of the extent of the regio inferior, especially at septal levels. As Zimmer and Haug (1978) have emphasized, the sequential changes in the pattern of Timm staining parallel the suprapyramidal-to-infrapyramidal and the temporal-to-septal gradients in the time of origin of the granule cells. In fact, the temporal-to-septal progression in the appearance of the mossy fiber bundles in these preparations provides one of the strongest lines of evidence for the validity of the neurogenetic gradient along the longitudinal dimension of the dentate gyrus, which we have previously reported (Schlessinger, et al., 1975) but which others have failed to observe. By the end of the third postnatal week, the appearance of the mossy-fiber system closely resembles that seen in fully mature animals, but when one compares the brains of three-week-old to four-week-old animals with those from very old rats (up to 420 days), it is evident that the system as a whole continues to grow at a slow, but steady, pace associated with the progressive addition of granule cells to the stratum granulosum and with the growth and elaboration of individual mossy fibers.

Amaral and Dent's Golgi preparations (1981) indicate that the first distinct fascicles of mossy fibers appear in the stratum lucidum around Day 3. At this stage, individual fibers still lack distinct swellings along their length, but most have recognizable growth cones at their tips. Characteristic expansions along the mossy fibers are first seen towards the end of the first postnatal week, but after this they grow rapidly and reach their mature size by about Day 14. Interestingly, the large, thornlike excrescences upon the proximal parts of the apical dendritic shafts of the hippocampal pyramidal cells (with which the mature mossy fibers synapse) do not appear until about the middle of the second postnatal week, which is several days after synaptic contacts first appear between the mossy fibers and the pyramidal cells and is also some days after the smaller, more typical dendritic spines are first found on the secondary and tertiary dendrites of the pyramidal cells.

During this period there is also a marked increase in the mean cross-sectional area of the *en passant* presynaptic mossy fiber profiles seen in the electron microscope. Thus, between Day 1 and Day 9 the mean cross-sectional area increases about fivefold, whereas there is only a doubling in area between Day 9 and the end of the fourth postnatal week. Electronmicrographs from a three-day-old rat and a mature rat are shown in Figure 16-12. Beginning around Day 9, dendritic protrusions can be seen forming fingerlike invaginations into the mossy fiber expansions. The protruding processes are generally free of distinctive organelles, but they mark the first indication of the thorny excrescences that represent the major locus of synaptic articulation between the mossy fibers and the pyramidal cells in the regio inferior. Over the course of the next week, these invaginating protrusions become progressively larger and increasingly complex in form and develop a variety of organelles including multivescicular bodies, coated vesicles, and finally a distinctive "spine apparatus." By the end of the second postnatal week, the definitive arrangement of these synapses is established with clear asymmetric membrane specializations on the apposing surfaces of the thorns and the mossy fiber expansions, together with irregular, symmetrically disposed *puncta adhaerentia* between the expansions and the dendritic shafts. At later stages, the mossy fiber synapses become increasingly irregular in shape (owing to the formation of further invaginations by dendritic thorns) and show an appreciable increase in the number of synaptic vesicles (Fig. 16-12).

The principal difference between these observations (which are schematically summarized in Fig. 16-13) and those previously reported in the literature is the finding that the axons of the granule cells form morphologically distinct synapses at an appreciably earlier stage than has generally been recognized. However, it remains to be determined when these synapses become functional and also when the transmitter involved is first synthesized and becomes available for release.

ON THE SIGNIFICANCE OF TEMPORAL FACTORS IN THE DEVELOPMENT OF THE DENTATE GYRUS

The prolonged period of neurogenesis in the dentate gyrus, taken together with the finding that there are distinct gradients in the sequence of granule cell generation and the assembly of the stratum granulosum, appears to have important consequences for the distribution of certain of the afferent connections to the granule cells, just as it has for the distribution of the mossy fibers (see the previous section above). This first came to light in Gottlieb and Cowan's autoradiographic study (1972a) of the commissural and associational connections, where it was observed that after injections of tritiated amino acids into the hilar region at one

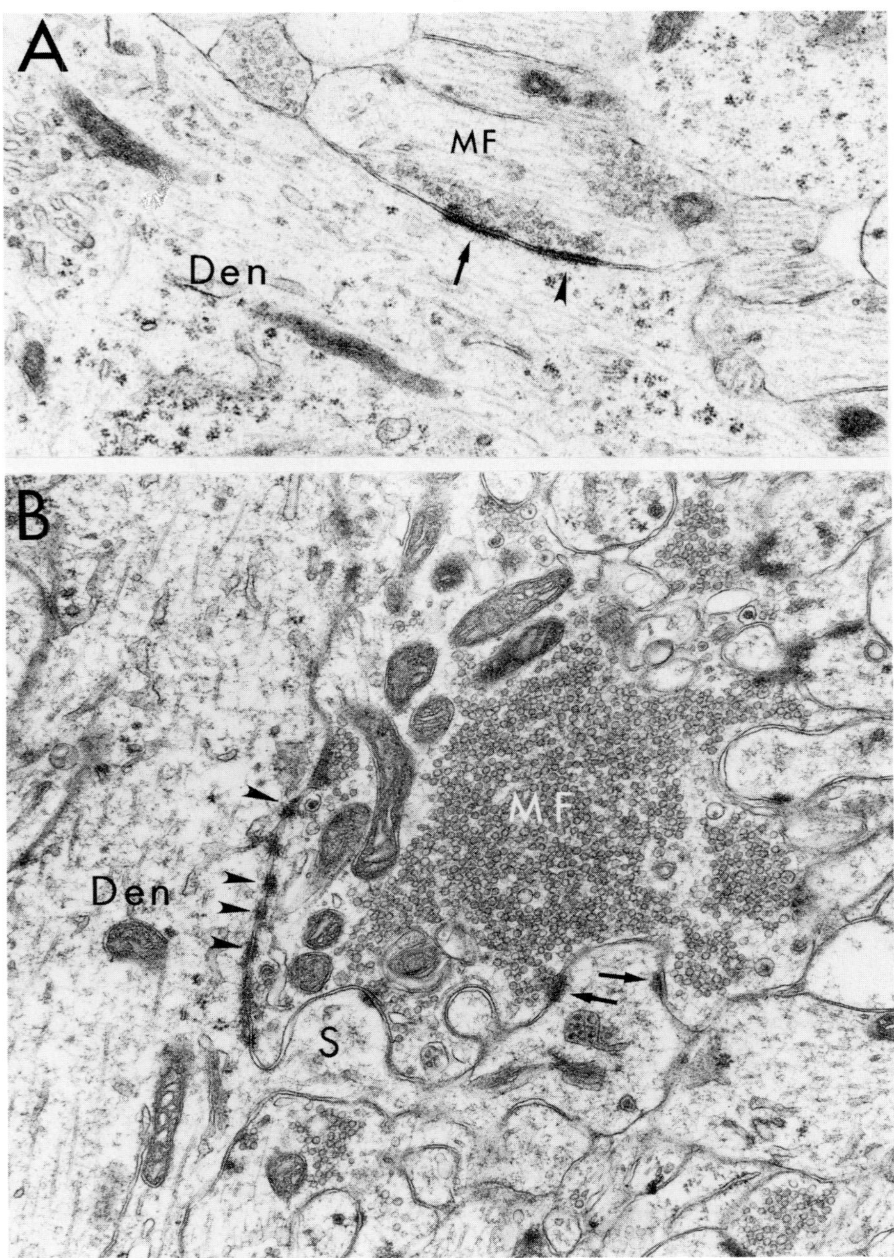

Figure 16.12 These two electron micrographs show the characteristic appearance of mossy fiber (MF) synapses at postnatal Day 3 (A) and Day 35 (B). The arrows point to sites where typical *asymmetric* synapses are formed, while the arrowheads mark *symmetric* contacts. (See text for further details.) S: dendritic spine; Den: dendrite. Magnification: ×29,400. From Amaral and Dent (1980), with permission.

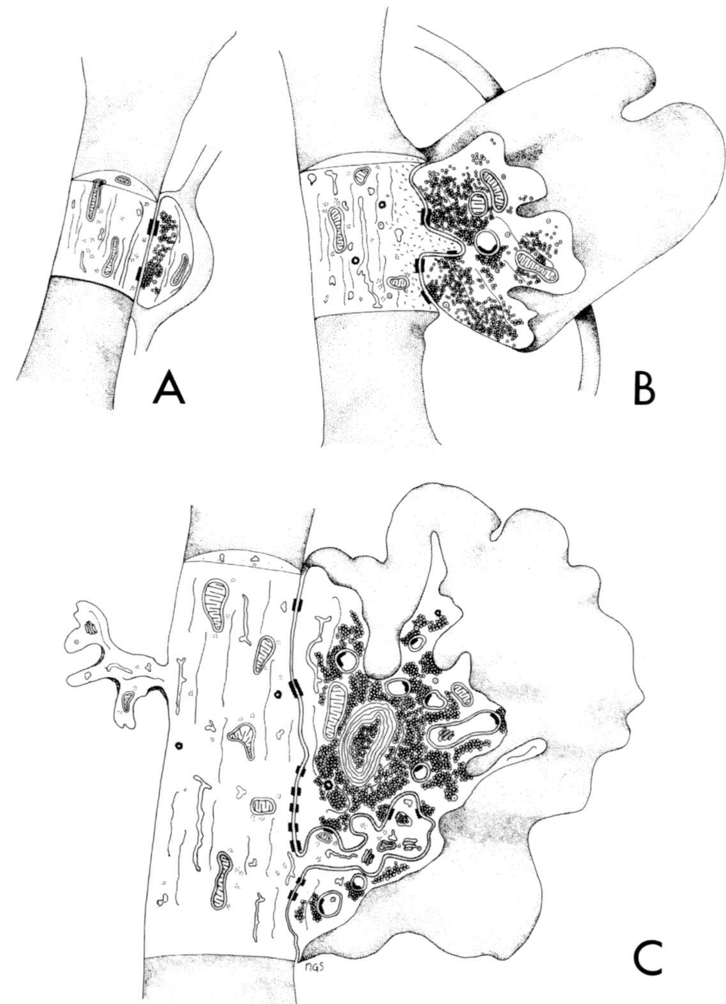

Figure 16.13 Diagrammatic representation of three stages in the maturation of the mossy fiber synapse. In the earliest period (A), small, mossy fiber expansions make both symmetric and asymmetric contacts directly with dendritic shafts. After a period of pronounced growth and vesicle accumulation (B), ''fingers'' of dendritic outgrowth begin to indent the presynaptic profile. These precursor dendritic spines are devoid of organelles except for a dense background of microfilamentous material. During this period, the asymmetric synapses come to be associated with the indenting dendritic processes. In the final stage of development (C), the spines (thorny excrescences) enlarge and become lobulated and their cytoplasm accumulates a variety of organelles including the spine apparatus, ribosomes, smooth endoplasmic reticulum, mitochondria, and dense core vesicles. By this stage, all asymmetric synapses are made with dendritic spines while the symmetric junctions are maintained on the dendritic shaft.

level there are systematic differences in the distribution of the axonally transported proteins within the inner part of the molecular layer of the dentate gyrus. In a typical experiment, after an injection is made near the middle of the septotemporal extent of the hilar region, grain density measurements that are done on autoradiographs from about the middle of the septal third of the gyrus indicate that there is usually about three times as much transported radioactivity near the tip of the suprapyramidal blade on the side of the injection (owing to transport in the associational afferents) as there is over the equivalent region on the opposite side (owing to transport in the commissural fibers). Near the crest region the corresponding grain density ratio is close to 2:1, while over the infrapyramidal blade there is roughly the same amount of transported material on the two sides. Although a number of explanations could be put forward to account for such systematic grain density patterns, as Gottlieb and Cowan (1972a) first pointed out, these data are most plausibly interpreted by assuming: (1) that the patterns reflect differences in the relative numbers of synaptic connections made by the two groups of fibers; and (2) that these differences are the result of a competitive interaction between the two afferent pathways, the outcome of which is determined, on the one hand, by the time at which each pathway arrives and, on the other hand, by the time at which synaptic sites upon the granule cell dendrites become available for innervation. The associational afferents, which have the shortest distance to grow were assumed to arrive earlier than the commissural fibers, which have to grow across the midline from the opposite side of the brain; and so the former command a majority of the synaptic sites in the earlier formed suprapyramidal blade. Conversely, by the time the infrapyramidal blade is ready for innervation (late in the first postnatal week), the commissural fibers were thought to be present in the molecular layer and hence in a position to compete (presumably on an equal footing) with the associational fibers.

This "temporal hypothesis," as it has come to be known, rested upon two assumptions: first, that the associational afferents reach the dentate gyrus earlier than the commissural fibers; and second, that grain density ratios, estimated from autoradiographs of the type used in these experiments, bear a close relationship to the relative numbers of axon terminals derived from the two pathways. As we have seen, there is now direct evidence that the first assumption is correct (see Fricke and Cowan, 1977). More recently, we have been able to show that in this system, at least, the relative numbers of silver grains seen in light microscopic autoradiographs are closely paralleled by the relative numbers of labeled terminal axons and presynaptic processes seen in adjoining ultrathin sections viewed in the electron microscope (Kishi, Stanfield, and Cowan, 1980).

The temporal hypothesis has the merit that it can be tested experimentally. If the pattern of labeling seen after injections of tritiated amino acids is, indeed, due to a competitive interaction between the commissural and associational afferents, the elimination of the competition (by excluding one or the other class of afferent) should produce a significant, and predictable, alteration in the observed grain density pattern. O'Leary et al. (1979) have succeeded in demonstrating that this, in fact, occurs. As Figure 16-14, which is taken from their work, shows, following isotope injections at the level indicated in normal young adult rats, the ratio of the number of silver grains seen near the tip of the suprapyramidal blade to that seen near the tip of the infrapyramidal blade, on the side of the injection, is close to 2:1. The constancy of this ratio from experiment to experiment (provided that in each case the injections and the grain density measurements are made at the corresponding levels) and the fact that the ratio is quite independent of the absolute grain densities (which often vary considerably depending on the amounts of radioactivity injected and the duration of the exposure of the autoradiographs) strongly suggest that it is attributable to some underlying feature in the organization of the associational afferents. Eliminating the commissural input to the dentate gyrus, which is done by destroying the hippocampus and fimbria on one side on the day of birth, results in a significantly different organization of the associational afferents. Thus, when the isotope is injected, at the usual level, into the hilar region of the unoperated side, the number of silver grains seen in autoradiographs at the level used for the analyses is found to be uniform around the entire suprapyramidal to infrapyramidal length of the gyrus (Fig. 16-15); and the normally observed ratio of grain densities between the suprapyramidal and infrapyramidal blades of about 1.88:1 is changed (in five separate experiments) to a mean value of 1.05:1.

Since, as we have seen, the commissural fibers do not reach the septal pole of the dentate until the second postnatal day, this reorganization of the associational projection is not due to a secondary sprouting of fibers in response to a denervation of the previously occupied synaptic site (although this does occur—see Lynch et al., 1976; O'Leary, Stanfield, and Cowan, 1980) but is apparently attributable to the innate capacity of the associational fibers to form considerably more synapses than they normally do, and, at least in the case of the infrapyramidal blade, they have the capacity to form almost twice as many synapses as they do when confronted with competition from the commissural fibers.

A second, less direct, test of the temporal hypothesis has been reported by Fricke and Cowan (1978). As we have discussed above, in the rat there is a fairly distinct gradient in the time of origin of the granule cells along the longitudinal axis of the dentate gyrus from its septal to its temporal pole. If the temporal hypothesis is generally correct, then one would expect to find: (1) that there are relatively fewer commissural fibers towards the temporal pole of the gyrus than at its septal end (since

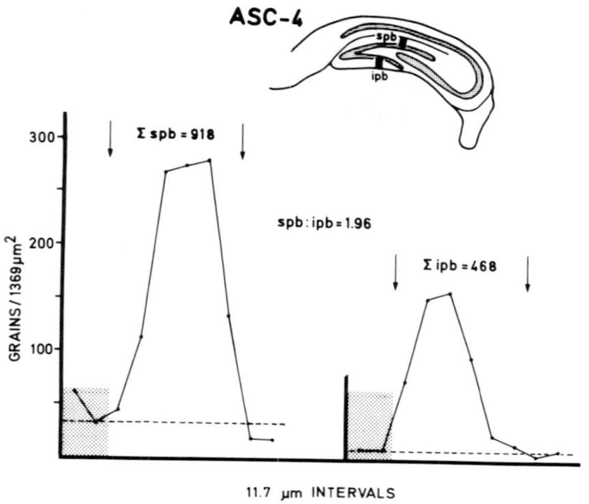

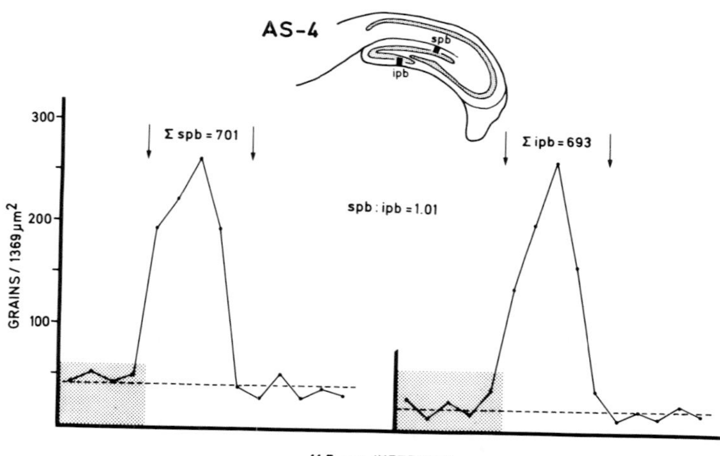

Figures 16.14 (top) and 16.15 These four graphs illustrate the relative grain densities seen over the molecular layer after injections of ^{3}H-proline that labeled the cells of origin of the associational afferents to the dentate gyrus some distance caudal to that shown in the drawings of the sections through the gyrus. In a normal, control animal (ASC-4 in Fig. 16.14) there are usually twice as many grains over the suprapyramidal as there are over the infrapyramidal blade at the sites marked in the drawings (in this case 918 grains compared to 468). Some weeks after destroying the hippocampus on one side (so as to eliminate the commissural fibers to the dentate gyrus), there is a clear reorganization of the associational projection; this is shown by the grain density traverses in Fig. 16.15, in which the ratio of grains over the suprapyramidal and infrapyramidal blades is 1.01 : 1. (See text for discussion.) From O'Leary *et al.* (1979), with permission.

the commissural fibers approach the gyrus from its rostral pole); (2) that the associational fibers extend further rostrally than caudally, since the rostrally directed axons would encounter more and more recently formed granule cells, while those that were directed caudally would always be competing with other, earlier formed afferents; and (3) that the relative numbers of associational fibers in the molecular layer of the infrapyramidal blade would be greater at temporal than at septal levels. All three of these predictions have, in fact, been demonstrated in experiments designed to label the cells of origin or the two pathways at different temporal-to-septal levels (Fricke and Cowan, 1978).

SOME REMAINING PROBLEMS IN THE DEVELOPMENT OF THE DENTATE GYRUS

It should be evident from the foregoing that we can now give a reasonably adequate account of the development of the dentate gyrus, at least at a descriptive level. But there are a number of issues that remain to be addressed. Chief among these are the following.

The Time of Arrival of the Septal and Aminergic Afferents

At present we have no direct information on either of these points, although from studies of the time of the appearance of acetylcholinesterase staining in the dentate gyrus (Mellgren, 1973; Matthews et al., 1974; Vijayan, 1979), of the appearance of cholinergic receptors (Hunt and Schmidt, 1971), and of the changes in the levels of choline acetyltransferase in the hippocampal region (Nadler et al., 1974) and choline uptake (Shelton, Nadler, and Cotman, 1979), it is evident that the cholinergic system (which is either derived from or at least regulated by the septal nuclei) is present at a very early stage (around the second postnatal day) but does not become fully established until near the end of the third postnatal week. An autoradiographic study of the time of arrival of the septal input, comparable to those we have reported here for the entorhinal and commissural/associational systems, should not only settle this point but also further clarify the relationship between the septal input and the cholinergic mechanisms that have been the subject of so much interest to those studying neuronal plasticity. Similarly, it should be relatively easy to determine the time of arrival of the noradrenergic and serotonergic inputs to the dentate gyrus by examining the sequential appearance of the appropriate transmitters (Loy and Moore, 1979) or the enzymes involved in their synthesis (Swanson and Hartman, 1975).

The Nature of the Neurotransmitters Utilized by the Granule Cells and Their Various Afferent Inputs

One of the major limitations to further progress in our understanding of the assembly of the dentate gyrus is our ignorance of the transmitters used by most of the relevant neurons, either extrinsic or intrinsic. We have already considered the cholinergic input to the dentate and have pointed out that we still have no direct evidence that the cholinergic fibers have their origin in the medial septal/diagonal band complex. The origin and distribution of the aminergic inputs to the dentate gyrus are well established, but their possible role in the development of the other connections of the gyrus has not yet been specifically addressed. The entorhinal, and possibly the commissural and associational, afferents are thought to utilize either glutamate or aspartate as their transmitter, but conclusive evidence on this point is still lacking. Recently, Gall and colleagues (1981) have reported that at least some of the afferents from the lateral entorhinal area (and the adjoining perirhinal region) display enkephalinlike immunoreactivity, as do the granule cells and the mossy fibers. The latter finding is of especial importance since it provides not only the first indication of the nature of the transmitter used by the granule cells but also a potential new approach to the study of their development. We may confidently anticipate a number of significant advances in this direction in the near future, and with them the possibility of new insights into the functional role of the dentate gyrus within the hippocampal formation as a whole.

The Development of Interneurons and of Local Circuitry within the Dentate Gyrus

Apart from the rather sketchy information we now have about the time of origin of some of the interneurons, we know virtually nothing about the growth and maturation of these cells—when (and from whence) they receive their afferents and when they establish connections with the granule cells or with the cells of origin of the associational and commissural inputs. Since such interneurons can at present only be clearly recognized in Golgi preparations (and so far their appearance in Golgi studies of the immature dentate has been too infrequent to permit any sequential analysis of their development), this problem is likely to prove fairly intractable unless some new technical approach is forthcoming. One promising line again involves the identification of the neurotransmitter(s) that these cells utilize and the perfection of techniques for their identification at various stages of development. Unfortunately, until the connections of these cells in the mature dentate gyrus have been established, progress in this direction is likely to be limited.

The Onset of Function in, and the Physiological Maturation of, the Dentate Gyrus

Because we know so little about the function of the hippocampal formation as a whole, and in particular because there is no way to "physiologically activate" the granule cells, very little is known about this important aspect of the development of the dentate gyrus. In two pioneering studies, Bliss, Chung, and Stirling (1974) and Stirling and Bliss (1978) have attempted to determine when electrical stimulation of the dentate gyrus can first activate the pyramidal cells in the regio inferior. Their evidence suggests that in the rat the earliest synaptic activation may occur around postnatal Day 5. However, it was not until Day 7 that electrical stimulation of the gyrus reliably resulted in the depolarization of the pyramidal cells, and even at this stage the cells did not fire consistently in response to a mossy fiber volley. By the end of the second postnatal week, the system appeared to be functionally mature, at least in the sense that population spikes could be regularly generated in the regio inferior when the mossy fibers were activated.

There appears to have been no comparable study of the activation of the granule cells from any of their known afferent inputs, and considering the small size of the granule cells, this may be difficult. Nevertheless, although such electrophysiological studies may present a number of technical difficulties, they are critical for the evaluation of the existing morphological data. For while it is important to determine when a group of afferents reaches its target zone, and to establish when the relevant afferents develop morphologically recognizable synapses, it is a quite different and ultimately much more significant matter to determine when the pathway in question can be functionally activated and to know when it begins to play its normal role in the functional economy of the neural system of which it forms a part.

ACKNOWLEDGMENTS

The work reported in this chapter was supported in part by grants EY-01255 and NS-10943, from the National Institutes of Health, and DA-00259 from ADAMAH and was carried out while B.B.S. was supported on training grant NSO-7071 and while D.G.A. was in receipt of the NIH postdoctoral fellowship F32-NSO-5765.

REFERENCES

Altman, J. 1966. Autoradiographic and histological studies of postnatal neurogenesis II. A longitudinal investigation of the kinetics, migration and transformation of cells incorporating tritiated thymidine in infant rats, with special reference to postnatal neurogenesis in some brain regions. *J. Comp. Neurol.* 128:431–74

Altman, J., Das, G. D. 1965. Autoradiographic and histological evidence of postnatal hippocampal neurogenesis in rats. *J. Comp. Neurol.* 124:319–36

Altman, J., Das, G. D. 1966. Autoradiographic and histological studies of postnatal neurogenesis. I. A longitudinal investigation of the kinetics, migration and transformation of cells incorporating tritiated thymidine in neonate rats, with special reference to postnatal neurogenesis in some brain regions. *J. Comp. Neurol.* 126:337–90

Amaral, D. G. 1978. A Golgi study of cell types in the hilar region of the hippocampus in the rat. *J. Comp. Neurol.* 182:851–914

Amaral, D. G., Avendaño, C., Cowan, W. M. 1980. The effects of neonatal 6-hydroxydopamine treatment on reactive synaptogenesis in the dentate gyrus of the rat following entorhinal lesions. *J. Comp. Neurol.* 194:171–81

Amaral, D. G., Dent, J. A. 1981. Development of the mossy fibers of the dentate gyrus: I. A light and electron microscopic study of the mossy fibers and their expansions. *J. Comp. Neurol.* 195:51–86

Anderson, P., Holmqvist B., Voorhoeve, P. D. 1966. Entorhinal activation of dentate granule cells. *Acta Physiol. Scand.* 66:448–60

Angevine, J. B., Jr. 1965. Time of neuron origin in the hippocampal region. An autoradiographic study in the mouse. *Exp. Neurol. 13, Supp.* 2:1–70

Bayer, S. A., Altman, J. 1974. Hippocampal development in the rat: Cytogenesis and morphogenesis examined with autoradiography and low-level x-irradiation. *J. Comp. Neurol.* 158:55–80

Blackstad, T. W. 1956. Commissural connections of the hippocampal region in the rat, with special reference to their mode of termination. *J. Comp. Neurol.* 105:417–537

Bliss, T. V. P., Chung, S. H., Stirling, R. V. 1974. Structural and functional development of the mossy fibre system in the hippocampus of the post-natal rat. *J. Physiol. London* 239:92–94

Conrad, L. C. A., Leonard, C. M., Pfaff, D. W. 1974. Connections of the median and dorsal raphe nuclei in the rat: An autoradiographic and degeneration study. *J. Comp. Neurol.* 156:179–206

Cotman, C. W., Nadler, J. V. 1978. Reactive synaptogenesis in the hippocampus. In *Neuronal Plasticity,* ed. C. W. Cotman, pp. 227–71. New York: Raven

Cowan, W. M. 1978. Aspects of neural development. *Int. Rev. Physiol.* 17:149–91

Cowan, W. M., Stanfield, B. B., Kishi, K. 1980. The development of the dentate gyrus. In *Current Topics in Developmental Biology,* ed. R. K. Hunt, pp. 103–57. New York: Academic

Crain, B., Cotman, C., Taylor, D., Lynch G. 1973. A quantitative electron microscopic study of synaptogenesis in the dentate gyrus of the rat. *Brain Res.* 63:195–204

Fricke, R. A. 1975. Studies of the morphology and development of the dentate gyrus. Doctoral Dissertation, Washington Univ., St. Louis, Mo

Fricke, R., Cowan, W. M. 1977. An autoradiographic study of the development of the entorhinal and commissural afferents to the dentate gyrus of the rat. *J. Comp. Neurol.* 173:231–50

Fricke R., Cowan, W. M. 1978. An autoradiographic study of the commissural and ipsilateral hippocampo-dentate projections in the adult rat. *J. Comp. Neurol.* 181: 253–70

Gaarskjaer, F. B. 1978. Organization of the mossy fiber system of the rat studied in extended hippocampi. I. Terminal area related to number of granule and pyramidal cells. *J. Comp. Neurol.* 178:49–72

Gall, C., Brecha, N., Karten, H. J., Chang K.-J. 1981. Localization of enkephalin-like immunoreactivity to identified axonal and neuronal populations of the rat hippocampus. *J. Comp. Neurol.* 198:335–350.

Geneser-Jensen, F. A., Haug, F.-M. Š., Danscher, G. 1974. Distribution of heavy metals in the hippocampal region of the guinea pig. A light microscope study with Timm's sulphide silver method. *Z. Zellforsch. Mikrosk. Anat.* 147:441–78

Gottlieb, D. I., Cowan, W. M. 1972a. Evidence for a temporal factor in the occupation of available synaptic sites during the development of the dentate gyrus. *Brain Res.* 41:452–56

Gottlieb, D. I., Cowan, W. M. 1972b. On the distribution of axonal terminals containing spheroidal and flattened synaptic vesicles in the hippocampus and dentate gyrus of the rat and cat. *Z. Zellforsch. Mikrosk. Anat.* 129:413–29

Gottlieb, D. I., Cowan, W. M. 1973. Autoradiographic studies of the commissural and ipsilateral association connections of the hippocampus and dentate gyrus of the rat. I. The commissural connections. *J. Comp. Neurol.* 149:393–422

Haug, F.-M. Š. 1974. Light microscopical mapping of the hippocampal region, the pyriform cortex and the corticomedial amygdaloid nuclei of the rat with Timm's sulphide silver method. I. Area dentata, hippocampus and subiculum. *Z. Anat. Entwicklungsgesch.* 145:1–27

Hjorth-Simonsen, A. 1972. Projection of the lateral part of the entorhinal area to the hippocampus and fascia dentata. *J. Comp. Neurol.* 146:219–32

Hjorth-Simonsen, A., Jeune, B. 1972. Origin and termination of the hippocampal perforant path in the rat studied by silver impregnation. *J. Comp. Neurol.* 144:215–32

Hjorth-Simonsen, A., Laurberg, S. 1977. Commissural connection of the dentate area in the rat. *J. Comp. Neurol.* 174:591–606

Hubel, D. H., Wiesel, T. H., LeVay, S. 1977. Plasticity of ocular dominance columns in monkey striate cortex. *Philos. Trans. R. Soc. London. Ser. B.* 278:377–409

Hunt, S., Schmidt, J. 1979. The relationship of α-Bungarotoxin binding activity and cholinergic termination within the rat hippocampus. *Neurosci.* 4:585–92

Innocenti, G. M., Fiore, L., Caminiti, R. 1977. Exuberant projection into the corpus callosum from the visual cortex of newborn cats. *Neurosci. Lett.* 4:237–42

Jones, B. E., Moore, R. Y. 1977. Ascending projections of the locus coeruleus in the rat. II. Autoradiographic study. *Brain Res.* 127:23–53

Kaplan, M. S., Hinds, J. W. 1977. Neurogenesis in the adult rat: Electron microscopic analysis of light radioautographs. *Science* 197:1092–94

Kishi, K., Stanfield, B. B., Cowan, W. M. 1980. A quantitative EM autoradiographic study of the commissural and associational connections of the dentate gyrus in the rat. *Anat. Embryol.* 160:173–186

Koda, L. Y., Bloom, F. E. 1977. A light and electron microscopic study of noradrenergic terminals in the rat dentate gyrus. *Brain Res.* 120:327–50

Krieg, W. J. S. 1946a. Connections of the cerebral cortex. I. The albino rat. A. Topography of the cortical areas. *J. Comp. Neurol.* 84:221–76

Krieg, W. J. S. 1946b. Connections of the cerebral cortex. I. The albino rat. B. Structure of the cortical areas. *J. Comp. Neurol.* 84:277–323

Laatsch, R. H., Cowan, W. M. 1966. Electron microscopic studies of the dentate gyrus of the rat. I. Normal structure with special reference to synaptic organization. *J. Comp. Neurol.* 128:359–96

Laurberg, S. 1979. Commissural and intrinsic connections of the rat hippocampus. *J. Comp. Neurol.* 184:685–708

Laurberg, S., Hjorth-Simonsen, A. 1977. Growing central axons deprived of normal target neurones by x-ray irradiation still terminate in a precisely laminated fashion. *Nature* (London) 269:158–60

Laurberg, S., Sørensen, K. E. 1981. Associational and commissural collaterals of neurons in the hippocampal formation (hilus fasciae dentate and subfield CA3) *Brain Res.* 212:287–300.

Loy, R., Koziell, D. A., Lindsey, J. D., Moore, R. Y. 1980. Noradrenergic innervation of the adult rat hippocampal formation. *J. Comp. Neurol.* 189:699–710

Loy, R., Lynch, G., Cotman, C. W. 1977. Development of afferent lamination in the fascia dentata of the rat. *Brain Res.* 121:229–43

Loy, R., Moore, R. Y. 1979. Ontogeny of the noradrenergic innervation of the rat hippocampal formation. *Anat. Embryol.* 157:243–53

Lynch, G., Gall, C., Dunwiddie, T. V. 1978. Neuroplasticity in the hippocampal formation. *Prog. Brain Res.* 48:113–28

Lynch, G., Gall, C., Rose, G., Cotman, C. W. 1976. Changes in the distribution of the dentate gyrus associational system following unilateral or bilateral entorhinal lesions in the adult rat. *Brain Res.* 110:57–71

Lynch, G., Stanfield, B., Cotman, C. W. 1973. Developmental differences in post-lesion axonal growth in the hippocampus. *Brain Res.* 59:155–68

Matthews, D. A., Nadler, J. V., Lynch, G. S., Cotman, C. W. 1974. Development of cholinergic innervation in the hippocampal formation of the rat—I. Histochemical demonstration of acetylcholinesterase activity. *Dev. Biol.* 36:130–41

Mellgren, S. I. 1973. Distribution of acetylcholinesterase in the hippocampal region of the rat during postnatal development. *Z. Zellforsch. Mikrosk. Anat.* 141:375–400

Minkwitz, H-G. 1976. Zur Entwicklung der Neuronenstruktur des Hippocampus während die prä- und postnatalen Ontogenese der Albinoratte. I. Mitteilung: Neurohistologische Darstellung der Entwicklung langaxoniger Neurone aus den Regionen CA_3 und CA_4. *J. Hirnforsch.* 17:213–31

Moore, R. Y., Halaris, A. E. 1975. Hippocampal innervation by serotonin neurons in the midbrain raphe in the rat. *J. Comp. Neurol.* 164:171–84

Nadler, J. V., Matthews, D. A., Cotman, C. W., Lynch, G. S. 1974. Development of cholinergic innervation in the hippocampal formation of the rat—II. Quantitative changes in choline acetyltransferase and acetylcholinesterase activities. *Dev. Biol.* 36:142–54

Nowakowski, R. S., Rakic, P. 1981. The

site of origin and route and rate of migration of neurons to the hippocampal region of the rhesus monkey. *J. Comp. Neurol.* 196:129–54

O'Leary, D. D. M., Fricke, R. A., Stanfield, B. B., Cowan, W. M. 1979. Changes in the associational afferents to the dentate gyrus in the absence of its commissural input. *Anat. Embryol.* 156:283–99

O'Leary, D. D. M., Stanfield, B. B., Cowan, W. M. 1980. Evidence for the sprouting of the associational fibers to the dentate gyrus following removal of the commissural afferents in adult rats. *Anat. Embryol.* 159:151–61

Price, J. L., Moxley, G. F., Schwob, J. E. 1976. Development and plasticity of complementary afferent fiber systems to the olfactory cortex. *Exp. Brain Res. Suppl.* 1:148–54

Raisman, G., Cowan, W. M., Powell, T. P. S. 1965. The extrinsic afferent, commissural and association fibres of the hippocampus. *Brain* 88:963–96

Rakic, P. 1977. Prenatal development of the visual system in the rhesus monkey. *Philos. Trans. R. Soc. London Ser. B.* 278:245–60

Rakic, P., Nowakowski, R. S. 1981. The time of origin of neurons in the hippocampal region of the rhesus monkey. *J. Comp. Neurol.* 196:99–128

Ramón y Cajal, S. 1911. Histologie du système nerveux de l'homme et des vertebrés, vol. 2, Paris: A. Maloine 992 pp.

Schlessinger, A. R., Cowan, W. M., Gottlieb, D. I. 1975. An autoradiographic study of the time of origin and the pattern of granule cell migration in the dentate gyrus of the rat. *J. Comp. Neurol.* 159:149–76

Schlessinger, A. R., Cowan, W. M., Swanson, L. W. 1978. The time of origin of neurons in Ammon's horn and the associated retrohippocampal fields. *Anat. Embryol.* 154:153–73

Segal, M. 1979. A potent inhibitory monosynaptic hypothalamo-hippocampal connection. *Brain Res.* 162:137–41

Shelton, D. L., Nadler, J. V., Cotman, C. W. 1979. Development of high affinity choline uptake and associated acetylcholine synthesis in the rat fascia dentata. *Brain Res.* 163:263–75

Sidman, R. L., Rakic, P. 1973. Neuronal migration, with special reference to developing human brain: A review. *Brain Res.* 62:1–35

Singh, S. C. 1977. The development of olfactory and hippocampal pathways in the brain of the rat. *Anat. Embryol.* 151:183–99

Stanfield, B. B., Caviness, V. S., Jr., Cowan, W. M. 1979. The organization of certain afferents to the hippocampus and dentate gyrus in normal and reeler mice. *J. Comp. Neurol.* 185:461–84

Stanfield, B. B., Cowan, W. M. 1979a. The morphology of the hippocampus and dentate gyrus in normal and reeler mice. *J. Comp. Neurol.* 185:393–422

Stanfield, B. B., Cowan, W. M. 1979b. The development of the hippocampus and dentate gyrus in normal and reeler mice. *J. Comp. Neurol.* 185:423–60

Stanfield, B., Cowan, W. M. 1979c. Evidence for the sprouting of entorhinal afferents into the "hippocampal zone" of the molecular layer of the dentate gyrus. *Anat. Embryol.* 156:37–52

Stanfield, B. B., Wyss, J. M., Cowan, W. M. 1980. The projection of the supramammillary region upon the dentate gyrus in normal and reeler mice. *Brain Res.* 198:196–203

Steward, O. 1976. Topographic organization of the projections from the entorhinal area to the hippocampal formation of the rat. *J. Comp. Neurol.* 167:285–314

Steward, O., Scoville, S. A. 1976. Cells

of origin of entorhinal cortical afferents to the hippocampus and fascia dentata of the rat. *J. Comp. Neurol.* 169:347–70

Stirling, R. V., Bliss, T. V. P. 1978. Hippocampal mossy fiber development at the ultrastructural level. *Prog. Brain Res.* 48:191–98

Swanson, L. W., Cowan, W. M. 1979. The connections of the septal region in the rat. *J. Comp. Neurol.* 186:621–56

Swanson, L. W., Hartman, B. K. 1975. The central adrenergic system. An immunofluorescence study of the location of cell bodies and their efferent connections in the rat utilizing dopamine-β-hydroxylase as a marker. *J. Comp. Neurol.* 163:467–506

Swanson, L. W., Wyss, J. M., Cowan, W. M. 1978. An autoradiographic study of the organization of intrahippocampal association pathways in the rat. *J. Comp. Neurol.* 181:681–716

Tömböl, T., Somogyi, G., Hajdu, F., Madarász, M. 1978. Granule cells, mossy fibers and pyramidal neurons: an electron microscopic study of the cat's hippocampal formation. *Acta Morphol. Acad. Sci. Hung.* 26:291–310

Vijayan, V. K. 1979. Distribution of cholinergic neurotransmitter enzymes in the hippocampus and the dentate gyrus of the adult and the developing mouse. *Neurosci.* 4:121–37

West, J. R., Nornes, H. O., Barnes, C. L., Bronfenbrenner, M. 1979. The cells of origin of the commissural afferents to the area dentata in the mouse. *Brain Res.* 160:203–16

Wimer, R. E., Wimer, C. C., Vaughn, J. E., Barber, R. P., Balvanz, B. A., Chernow, C. R. 1978. The genetic organization of neuron number in the granule cell layer of the area dentata in house mouse. *Brain Res.* 157:105–22

Wyss, J. M., Swanson, L. W., Cowan, W. M. 1979. Evidence for an input to the molecular layer and the *stratum granulosum* of the dentate gyrus from the supramammillary region of the hypothalamus. *Anat. Embryol.* 156:165–76

Zimmer, J. 1971. Ipsilateral afferents to the commissural zone of the fascia dentata, demonstrated in decommissurated rats by silver impregnation. *J. Comp. Neurol.* 142:393–416

Zimmer, J. 1978. Development of the hippocampus and fascia dentata: morphological and histochemical aspects. *Prog. Brain Res.* 48:171–89

Zimmer, J., Haug, F.-M. Š.: 1978. Laminar differentiation of the hippocampus, fascia dentata and subiculum in developing rats, observed with the Timm sulphide silver method. *J. Comp. Neurol.* 179:581–618

Zimmer, J., Hjorth-Simonsen, A. 1975. Crossed pathways from the entorhinal area to the fascia dentata. II. Provokable in rats. *J. Comp. Neurol.* 161:71–102

17 / Critical Notes on the Neurology and Neuroembryology of René Descartes: 1596–1650

THOMAS S. HALL

STUDENTS OF THE PHYSIOLOGY OF DESCARTES have not agreed on its merits, some having found it imaginative and forward looking; others, too ingenious—and even bizarre.[1] But I think that some of Descartes's depreciators have displayed a misguided tendency to judge him in the light of later discoveries—to which he pointed the way but understandably could not attain. In neurology, in any case, his chief contributions were two. First, he asked many "right questions" about the nervous system, questions still central to investigative neurology today. Second, he was the first to try to answer such questions in a wholly mechanical way, that is, in terms of matter in motion.

DESCARTES'S NEUROLOGICAL PROGRAM

Cartesian Questions

The questions that Descartes asked had to do with the material movements he regarded as responsible for sensory reception, afferent and efferent conduction, and the processing of nervous information by the brain. On the sensory side, he sought a physical basis in peripheral nerve activity for both intermodal and intramodal discrimination and for such distinctions (common to all sense modalities) as weak versus strong and painful versus pleasant, the latter distinction involving him in mathematical analysis of visual and acoustical esthetics. He looked for material causes of the phenomena we speak of today as reciprocal innervation, neuromyal excitation, the "phantom limb" phenomenon, the simulation of normal nerve function by stimuli applied at any point along the length of a nerve, and dozens of others. The brain appeared to him to be a highly organized ensemble of subvisible material arrangements for the circuiting of automatic responses and for interaction between man's material body and his supposedly nonmaterial mind (Descartes was, of course, the major Renaissance proponent of the "dualist-interactionist" hypothesis, as Popper and Eccles call it today). He thought it the business of science to "discover" the subvisible brain mechanisms and to use them in explaining such major thought modalities as ideation, imagination, association, memory, and waking consciousness (as distinct from dreaming).[2]

Descartes's Neurology and Ours

Descartes's neurological program as outlined was more than superficially similar to that of investigative neurology today. In the immediately preceding paragraph, for example, one could, without depriving it of meaning, substitute "the modern neurologist" for "Descartes" and change the time frame from past to present. But the posing by 17th-century thinkers of what seem to be 20th-century questions could elicit only 17th-century answers. The physics to which Descartes sought to reduce nervous action was not modern—but pre-Newtonian and specifically Cartesian—physics. Moreover, although the mechanisms he described were largely "microscopic" they preceded actual microscopy by several decades. They also antedated by nearly 200 years the general cell theory of Schwann (1839) and by an even longer interval the neuron theory of Ramón y Cajal (ca. 1900).

In retrospect, Descartes's conclusions have a markedly conjectural cast that has made them seem, to some critics, rather contrived. Yet it was precisely through them that he introduced into physiology the practice of mechanical model building, that is, of interpreting sensible functions in terms of hypothetical, insensible mechanical arrangements—and it would be difficult to think of a development in Renaissance physiology more fateful for its later development than that. The rest of this paper will illustrate Descartes's procedures by synoptically presenting and criticizing his "micromechanical" interpretations (1) of the routing of nervous activity, and (2) of the embryonic origins of the brain, sense organs, and nerves.

Subvisible Organization

A nerve, for Descartes, was a tube whose outer membrane was ultimately continuous with the outer membrane of the brain, the "dura" (he got this idea from Galen). The tube contained several smaller tubules whose membranes were continuous with the brain's inner membrane, the "pia." Each tubule contained a central longitudinal bundle or "marrow composed of very fine threads" which entered and composed the solid substance of the brain. Intracerebrally all the fibers terminated in a single plane (like cut-off broom whiskers), and this plane represented the intracerebral ventricular wall. The whole neural and cerebral fibrillar system was thought to be bathed in a volatile "animal spirit," a material substance whose constituent particles were intermediate in size, Descartes believed, between those of fire and of air. This spirit likewise fully filled the ventricular cavities themselves. Its source was the pineal gland, mistakenly believed by nearly all pre-Cartesian thinkers—and by Descartes himself—to be situated within the ventricular system between the third ventricle and the fourth. The gland filtered off material spirit particles from the blood and discharged them into the ventricles and thence, through interfibrillar conduits or channels in the brain substance, into the lumina of individual peripheral nerve tubules.[3]

Descartes's physical model of the brain as thus sketched may seem gratuitous, at first, but it was a step forward from the "custard"-like consistency that, as Boyle was later to note, had characterized most earlier models.[4] Descartes realized that the brain's soft and optically rather homogeneous texture offered no explanatory basis for understanding the patterned connections that the normal brain must subserve. The question he was endeavoring to answer was whether the brain's marvelous ability to make functional connections between its varied and numerous inputs and likewise numerous outputs could be explained mechanically. To this question he ventured an affirmative answer, admitting its conjectural status and acknowledging God as the divine artificer of whatever arrangements were actually there.

The Physical Basis of Conscious Perception

Descartes thought that all sensory reception came down to mechanical displacements of the peripheral termini of fibrils present in an appropriate organ of sense. Such peripheral dislocations caused simultaneous movements of the same fibrils where they ended centrally at the ventricular boundaries within the brain. These patterned central movements had the effect of enlarging certain interfibrillar channels whilst diminishing others. Spirit emanating from the pineal gland moved most abundantly toward channels whose openness facilitated the flow. The mind, notoriously placed by Descartes in the gland, was aware of the differential pattern of the efflux of spirit and this awareness was conscious perception.[5]

Voluntary Motion

By the same token, the mind could initiate a differential efflux of spirit and direct its flow through the ventricle to particular interfibrillar channels. And since these channels were continuous with the lumina of nerve tubules, an increased flow could be sent to a particular nerve or nerves. The slight increase in outflow of spirit through a nerve was sufficient to cause the contraction of one of a pair of muscle antagonists and the simultaneous relaxation of the other, thus extending or flexing a moving part of the body. Descartes showed how events of this sort could cause proper movements of the eyes, the eyelids, the thorax (in breathing), and the limbs. The inflation and consequent contraction of one antagonist at the expense of the other was brought about by a transfer of contained animal spirit from the latter to the former through subvisible intermuscular channels whose valves operated in response to the nerves.[6]

Cerebral Projection of the Visual Image

On the just-outlined pattern of supposed physical events, Descartes rang numerous changes. In particular, he regarded the surface of the gland as a kind of map or stencil used by the mind in sensation and voluntary motion. Thus a retinal image produced a rearrangement of fibers at the ventricular wall and a correspondingly patterned efflux of spirits from the gland which—because the surface of the gland was a map—permitted the mind to perceive the arrangement of objects in the field. The "field map" (our term, not Descartes's) on the surface of the gland (Fig. 17-1) likewise permitted the mind, by flexing the gland in this direction or that, to accelerate the efflux of spirit toward particular muscles and to orient just those muscles toward particular objects.[7] The problem that Descartes deals with here—the problem of the brain's physical intermediation of images for use in perception and volition—is still urgent and is at present being pursued promisingly by neurologists.

The Mind and the Pineal Gland

If Descartes's solutions seem, in retrospect, "wrong," they were not wrong in relation to Cartesian physics or in relation to what was then known about the organization of the brain. Nor, from this perspective, does his location of the mind in the pineal gland seem farfetched. The hypothesis that animal spirit intermediated between the immaterial soul and the material body had a long history dating from Alexandria (300 B.C.) or perhaps even from Athens (350 B.C., since Aristotle had intimated something of the sort). What better locus for soul-qua-mind than the

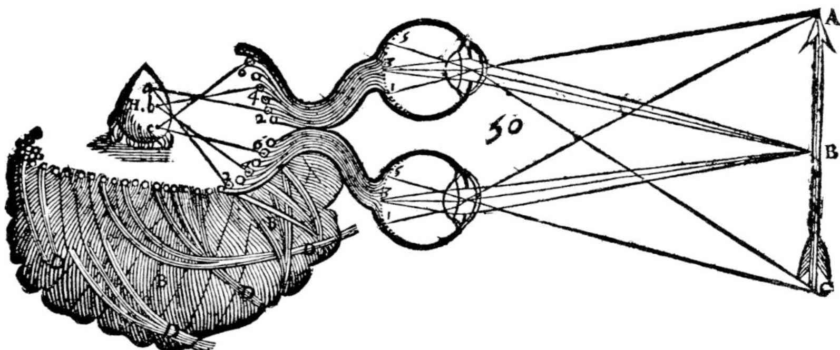

Figure 17.1 The physical object (A, B, C) and three projections thereof, namely, on the retina (1, 3, 5), the ventricular surface (2, 4, 6) and the surface of the pineal gland (a, b, c). The lines within the ventricle show currents flowing out toward interfibrillar channels (2, 4, 6) widened by sense-related fibrillar displacements. The figures in Descartes's *L'Homme* (1664), of which this one is representative, were frankly intended to represent dynamic rather than precise spatial relations; hence the awkward position of the left eye above the right one.

gland in which the intermediating animal spirit was separated off from the blood and in which the movements of this spirit could be monitored and controlled?[8]

Reflex Theory

The details imagined by Descartes had the added advantage of being intricate enough to permit a vast array of responses not requiring the intervention of mind—responses occurring, as we should say, reflexly. Descartes did not use the term "reflex" to designate such autonomous responses but he did suppose that innately open interfibrillar channels permitted many useful flow patterns of spirit, as exemplified by the pattern involved, for instance, in the automatic withdrawal of a hand or a foot from contact with fire. Descartes also suggested that, with use, channels other than innate ones could be more or less permanently opened so that the repertory of automatic actions was augmented as experience kept imprinting *(imprimer)* the brain.[9]

Concluding Note on Descartes's Neurological Program

The chief novelty in Descartes's theory was its substitution of subvisible material arrangements for presumed "other-than-material" causes of nervous action and conduction. According to the conventional wisdom—which was largely Galenic at the time—both conscious and nonconscious functions required the active presence in the tissues of a physiologically interventive anima or "life-soul." The soul had three "powers" (animal, vital, and natural), each of which, when further examined, was a constellation of constituent subpowers—one each for almost every identifiable function of the body, motive, visual, imaginative, mnemonic, digestive, assimilative, hemopoetic, secretory, and so on.[10] It was in lieu of such powers and subpowers that Descartes suggested a hypothetical subvisible material organization of the body, a suggestion that succeeded in the sense that later physiologists less and less frequently invoked "powers" of the soul for functions other than cogitant ones, which continued to be regarded as operations of mind.

Where Descartes's physics chiefly misled him was in the picture he drew of an everchanging network of currents and crosscurrents traversing the spirit-filled ventricles of the brain. He was doubtless influenced to think in terms of them by his preoccupation as a physicist with hydrodynamics, which he interpreted elsewhere in terms of his particulate theory of matter.[11] So fluid a medium was not really suited, even by Cartesian standards, to serve for the patterned routing of nervous activity between the brain's sensory inputs and motor outputs. Another shortcoming, the failure to distinguish motor from sensory nerves, was not wholly illegitimate for a period in which a number of authors were departing from Galen's classification of nerves into "hard" (motor), "soft" (sensory), and intermediate (mixed). These authors were arguing that sensory versus motor function was determined not by compositional differences in the nerves themselves but by the peripheral organ of attachment.[12] Incidentally, Descartes's identification of motor conduction with a momentarily accelerated efflux of spirit through the nerve was soon to give way to various mostly percussive theories of conduction.[13] Meanwhile his hydrodynamic ideas formed the basis for his account of fetal development, to which we turn in the section that follows.

DESCARTES'S NEUROEMBRYOLOGICAL PROGRAM

The preformation-epigenesis issue had been aired by Aristotle in the fourth century B.C. Today this issue no longer applies because during development, as we now understand it, complexity increases at the visible level under instructions from subvisible (informational) complexity already present in the genome of the germ cells. Descartes acknowledged the possibility of particle patterns existing preformed in the solid seeds of plants. With respect to animals and man, he was an epigenesist in the old-fashioned Aristotelian sense, and he wrote the first book-length, step-by-step, committedly materialistic account of what it was then conventional to call "the formation of the fetus."[14] This was a courageous endeavor. A moment's reflection will remind us that real progress in the causal analysis of development still lay, at the time of Descartes, more than two hundred years in the future. Aristotle had called it "no easy puzzle," and had "solved" it by invoking "final causes," "entelechies," "substantial forms" and other extramaterial causal agents that Descartes was determined to avoid. The post-Cartesian refractoriness of the problem is reflected in the 18th- and the early 19th-century invocations of various "vital" or "formative" principles, properties, and powers, which include the "vis essentialis" (of C. F. Wolff), the "nisus formativus" (of J. F. Blumenbach), and the paraphernalia of vitalism properly so called.[15] It will not surprise us therefore if Descartes's attempted "mechanization" of neuroembryology turns out to be less seductive than his neurology as outlined above. Nor is it irrelevant to compare his embryology with the more or less contemporary epigenetic embryology of William Harvey whose *Generation of Animals* (1651) represented a retreat—in an animistic and teleological direction—from the mechanical approach he had used in his triumphant earlier studies on the flow of the blood (1628).

In any case, Descartes tried to do what the 20th-century embryologist is still attempting, namely, to explain differentiative development in terms of the particle physics that was at hand. Ideally we should have liked him to explain both the mapping out of the principal organs and, if possible, the filling in of the map with functional detail. Actually, pleading inadequate acquaintance with the subject, he put off treating development at all, from 1630 when his interest in physiology began, until 1648, just two years before he died. And when he finally got around to discussing it, he was willing to take only the first of the indicated steps, namely, that of accounting for the formation of a general plan. Only in a few particulars did he suggest first steps toward functional organization.

Descartes on Differentiation

Descartes thought the fetus took form in a blended mixture *(mélange confus)* of male and female seminal fluids. Unlike Democritus, 400 B.C. (the presumed founder of the "similar semina" theory of conception), Descartes did not derive the two fluids from all the varied body parts of the father and mother; he was not a "pangenesist." The source of the fluids was not a particular concern for Descartes because he thought that their developmental potentialities resided wholly in the hydrodynamics—the particle mechanics—of the fluids themselves. Once the developmental process began, everything else followed according to God-given laws of physical motion.

The first event was a heat-productive fermentive action of each seminal fluid on the other. The result was a local, heat-caused expansion of seminal matter, resisted by counterpressure from the surrounding residual semen, and the consequent formation of a vesicle (the rudimentary heart) filled with warm incipient blood. The rest of organ formation proceeded, with variations, as follows. Heat-expanded blood surged upward and out from the primary vesicle (the left ventricle) but soon circled back and reentered the heart. There it was repeatedly heated and reheated in such a way as to cause the cardiac pulse. Presently, the principal (aortic) stream gave rise to branches and morphogenesis followed: first, with the acquisition by these currents of solidified vascular walls; and, second, by the spinning out, through pores in these walls, of complexly interlocking fibers.[16] These fibers formed, according to Descartes, the *tissu* of the organs.[17]

Note that, in depicting all these events, Descartes remained wholly loyal to his own theory of the physics of particle movement. The fundamental process involved was streaming accompanied by morphogenetic solidification. A modified application of the same processes was central to the formation of the brain, sense organs, and nerves.

The Mapping Out of the Nervous System

As the aortic current flexed downward, its most volatile particles—those destined to form animal spirit—pushed farther upward to become the "Anlage," as we should say, of the brain *(la place où doit estre après le cerveau).* But some of this upwardly surging spirit was, like aortic blood, turned back downward, and, flowing along the spine, it came to constitute the spinal cord. Its downflow was facilitated by a loose and agitated motion of particles in the local seminal matrix, a condition caused by heat coming from the nearby aortic blood. The cordal current met occasional resistances that caused lateral streams to branch out through the semen and become the ramifying spinal nerves. Meanwhile the ascending current of spirit in the brain turned back in two descending streams that pushed out at various brain levels to form the cranial nerves and paired sense organs related thereto.[18]

From the perspective of modern studies of the outgrowth of nerves, it is interesting that Descartes answered—almost without asking—the question whether their course was guided by intrinsic or extrinsic factors. The only extrinsic factor he identified was that of physical resistance—with consequent furcation and ramification. Directionality was, for him, otherwise a matter of intrinsic obedience of the particles that constituted the current to the laws of motion that held for matter in general. The guidance we see as provided by genes he assigned, indirectly, to God who with infinite wisdom had created matter, had arranged it, and had instigated in it the ordered motions that underlie all phenomena with the single exception of human awareness, or thought. Descartes liked to point to manmade machines (clocks, automated toys, and the like) that, once set going, could run on their own for hours or days. Why, he asked, could not almighty God initiate in machines made by Him—

namely, newborn living machines—movements lasting through each individual's life? Embryologists would later inquire, God and metaphysics aside, exactly what the initiating movements were and in what kinds of matter, how arranged, such movements occurred.

Not seldom in the history of science a crucial insight, or constellation of insights, has occurred to someone before the means were at hand to confirm it. In this sense, Descartes's belief in a possible physiological subscience of developmental mechanics was prescient but premature. What matters historically is that he publicly raised the possibility for realization later.

NOTES

Abbreviations used. H: *René Descartes, Treatise of Man, French Text with Translation and Commentary,* by Thomas Steele Hall, Cambridge (Harvard University Press), 1972; cited by page.
AT: Charles E. Adam and Paul Tannery, eds., *Oeuvres de Descartes,* Paris (Cerf), 1897–1910, republ. Paris (Vrin), 1956–1957 and 1964–1967; cited by volume and page.

1. Thus Claude Bernard, *Leçons de pathologie expérimentale . . . ,* Paris (Baillière), 1872, p. 481. For leads to criticism of Descartes's physiology, see Thomas S. Hall, "Descartes's physiological method . . . ," *J. Hist. Biol.,* 3 (1), 1970, p. 55.
2. Descartes treated these questions most fully in H and raised some of them in his *Principles of Philosophy* (1644), pt. 4, principles 188–97, and in his *Passions of the Soul* (1649), pt. 1, articles 10–17 and 31–37.
3. H: pp. xxxvi–xxxix containing references to Descartes's text.
4. Robert Boyle, "The Christian Virtuoso," *Works of the Honourable R. B.,* London (Rivington), 1772, v. 6, p. 746.
5. H: 77–78.
6. H: 24–35 and 106–8.
7. H: 93–100.
8. H: 36–37 and 86 n. 135.
9. H: 33, 88 n. 137, and AT: xi, 360. For Descartes on the nervous reflex, see T. S. Hall in *Founders of Experimental Physiology,* J. W. Boylan, ed., Munich (Lehmann), 1971, pp. 9–14.
10. Galen scattered his treatment of soul powers through many of his extant works for which see the entries under *facultas* in the indices of A. Brassavola (1609) and K. G. Kuhn (1821–1833) to their editions of the works of Galen. The edition of 1609 is that most likely to have been used by Descartes. For a critical note on Galen's power or faculty theory, see M. T. May, ed., *Galen On the Usefulness of the Parts of the Body,* Ithaca (Cornell University Press), 1968, vol. 1, pp. 49–50 and for the later development of this theory, O. W.

Temkin, *Galenism,* Ithaca, (Cornell University Press), 1973, index entries under "Soul" and "Faculty(ies)."

11. See esp. the *Principles of Philosophy,* pt. 2, principles 25–64.
12. See, e.g., A. Piccolomini, *Anatomicae praelectiones . . . ,* Rome (Bonfadini), 1586, pp. 261–262; A. du Laurens, *Historia anatomica humani corporis . . . ,* Frankfurt (Becker), 1600, p. 161; and C. Bartholinus, *Anatomicae institutiones . . . ,* Strasbourg (Scher), 1626, p. 395.
13. On this subject, see K. E. Rothschuh, "Vom Spiritus animalis zum Nervenreaktionsstrom," *Ciba-Zeitschrift,* 8, 1958, pp. 2950–80.
14. René Descartes, *La description du corps humain,* first published posthumously (together with Descartes's *l'Homme*), Paris (Angot; LeGras; Girard), 1664. The original edition shows the just-indicated title on the title page but all succeeding pages bear the heading *"La formation du foetus."*
15. For vitalistic interpretations of epigenesis, see T. S. Hall, *Ideas of Life and Matter,* Chicago (University of Chicago Press), 1969, vol. 2, pp. 89–90, 93, 99–103, and for later developments 219–84.
16. *Description du corps* (n. 14), sections xxviii–xxx (early stages), xxxiii–xxxv (early angiology), xlviii–lviii (later angiology), and esp. lx–lxxiv *("formation des solides").*
17. H: 77 and n. 123.
18. *Description du corps,* sections xxviii–xlvii.

Index